NUMERICAL MATHEMATICS AND SCIENTIFIC COMPUTATION

Series Editors

G. H. GOLUB A. GREENBAUM

A. M. STUART E. SÜLI

NUMERICAL MATHEMATICS AND SCIENTIFIC COMPUTATION

Books in the series

Monographs marked with an asterix (∗) appeared in the series 'Monographs in Numerical Analysis' which has been folded into, and is continued by, the current series.

*P. Dierckx: *Curve and surface fittings with splines*
*J. H. Wilkinson: *The algebraic eigenvalue problem*
*I. Duff, A. Erisman, and J. Reid: *Direct methods for sparse matrices*
*M. J. Baines: *Moving finite elements*
*J. D. Pryce: *Numerical solution of Sturm–Liouville problems*

K. Burrage: *Parallel and sequential methods for ordinary differential equations*
Y. Censor and S. A. Zenios: *Parallel optimization: theory, algorithms and applications*
M. Ainsworth, J. Levesley, W. Light and M. Marletta: *Wavelets, multilevel methods and elliptic PDEs*
W. Freeden, T. Gervens, and M. Schreiner: *Constructive approximation on the sphere: theory and applications to geomathematics*
Ch. Schwab: *p- and hp- finite element methods: theory and applications to solid and fluid mechanics*
J. W. Jerome: *Modelling and computation for applications in mathematics, science, and engineering*
Alfio Quarteroni and Alberto Valli: *Domain decomposition methods for partial differential equations*
G. E. Karniadakis and S. J. Sherwin: *Spectral/hp element methods for CFD*
I. Babuška and T. Strouboulis: *The finite element method and its reliability*
B. Mohammadi and O. Pironneau: *Applied shape optimization for fluids*
S. Succi: *The lattice Boltzmann equation for fluid dynamics and beyond*
P. Monk: *Finite element methods for Maxwell's equations*
A. Bellen and M. Zennaro: *Numerical methods for delay differential equations*
J. Modersitzki: *Numerical methods for image registration*
M. Feistauer, J. Felcman and I. Straškraba: *Mathematical and computational methods for compressible flow*
W. Gautschi: *Orthogonal polynomials: computation and approximation*
M. K. Ng: *Iterative methods for Toeplitz systems*
Michael Metcalf, John Reid, and Malcolm Cohen: *Fortran 95/2003 explained*
George Em Karniadakis and Spencer Sherwin: *Spectral/hp element methods for CFD, second edition*
Dario A. Bini, Guy Latouche, and Beatrice Meini: *Numerical methods for structured Markov chains*
Howard Elman, David Silvester, and Andy Wathen: *Finite Elements and Fast Iterative Solvers: with applications in incompressible fluid dynamics*
Moody Chu and Gene Golub: *Inverse Eigenvalue Problems: Theory, Algorithms, and Applications*

Inverse Eigenvalue Problems: Theory, Algorithms, and Applications

Moody T. Chu
North Carolina State University

Gene H. Golub
Stanford University

OXFORD
UNIVERSITY PRESS

This book has been printed digitally and produced in a standard specification in order to ensure its continuing availability

OXFORD
UNIVERSITY PRESS

Great Clarendon Street, Oxford OX2 6DP

Oxford University Press is a department of the University of Oxford.
It furthers the University's objective of excellence in research, scholarship,
and education by publishing worldwide in

Oxford New York

Auckland Cape Town Dar es Salaam Hong Kong Karachi
Kuala Lumpur Madrid Melbourne Mexico City Nairobi
New Delhi Shanghai Taipei Toronto
With offices in
Argentina Austria Brazil Chile Czech Republic France Greece
Guatemala Hungary Italy Japan South Korea Poland Portugal
Singapore Switzerland Thailand Turkey Ukraine Vietnam

Published in the United States
by Oxford University Press Inc., New York

Reprinted 2012

ISBN 978-0-19-856664-9

Printed and bound by CPI Group (UK) Ltd, Croydon, CR0 4YY

To Joyce

for her love, patience, care, and prayer

For Neal

my beloved nephew, a gentle giant

PREFACE

In one of the classic books for student of linear algebra, *Finite Dimensional Vector Spaces*, Halmos (1974) wrote,

Almost every combination of the adjectives proper, latent, characteristic, eigen and secular, with the nouns root, number and value, has been used in the literature for what we call a proper value.

This interesting comment on the nomenclature of eigenvalue echoes the enigmatic yet important role that eigenvalues play in nature. This entity of eigenvalues has been recognized under so many different names because its existence has been found in settings of widely varied disciplines. One instance, as Parlett (1998) put it, is that,

Vibrations are everywhere, and so too are the eigenvalues associated with them.

In our fervent pursuit of the Knowledge of Nature, it often becomes necessary to first understand the spectral properties of the underlying physical system. It thus follows that considerable research effort has been expended on eigenvalue computation. The applications of this research furnish critical insight into the understanding of many vital physical systems.

An inverse eigenvalue problem, in contrast, concerns the reconstruction of a physical system from prescribed spectral data. The spectral data involved may consist of the complete or only partial information of eigenvalues or eigenvectors. It is obvious that the construction must be subject to some corporeal constraints due to, for instance, the structure or feasibility of the system. The objective of an inverse eigenvalue problem is to construct a physical system that maintains a certain specific structure as well as that given spectral property.

Inverse eigenvalue problems arise in a remarkable variety of applications, including system and control theory, geophysics, molecular spectroscopy, particle physics, structure analysis, and so on. Generally speaking, the basic goal of an inverse eigenvalue problem is to reconstruct the physical parameters of a certain system from the knowledge or desire of its dynamical behavior. Since the dynamical behavior often is governed by the underlying natural frequencies and/or normal modes, the spectral constraints are thus imposed. On the other hand, in order that the resulting model is physically realizable, additional structural constraints must also be imposed upon the construction. Depending on the application, inverse eigenvalue problems appear in many different forms. Our basic assumption in this presentation is that the underlying physical system is somehow represented in terms of matrices. The subsequent discussion therefore centers around eigenvalue problems, and particularly the inverse eigenvalue problems, for matrices.

Associated with any inverse eigenvalue problem are two fundamental questions – the theoretic issue on solvability and the practical issue on computability. Solvability concerns obtaining a necessary or a sufficient condition under which an inverse eigenvalue problem has a solution and whether a solution is unique. Computability concerns developing a procedure by which, knowing *a priori* that the given spectral data are feasible, a matrix can be constructed numerically. Both questions are difficult and challenging, and we still do not have complete answers. Additionally, except for a few cases most inverse eigenvalue problems have multiple solutions. The very hard yet important consideration of its sensitivity analysis should not be overlooked.

In this note our emphasis is to provide an overview of the vast scope of this fascinating problem. The fundamental questions, some known results, many applications, mathematical properties, a variety of numerical techniques, as well as several open problems will be discussed. We have to acknowledge that merely getting the current materials organized has been a formidable task since the beginning of this project. Each cross-section of this immense subject is in fact a major research effort itself with many variations. Theories and methods vary accordingly but sometimes share surprising and subtle similarities. We feel that it might be helpful to at least categorize the problems by some kinds of characteristics. As such, we divide the inverse eigenvalue problems into those that are attributed by parameters in Chapter 3, those that carry specific structures in Chapter 4, those that are characterized by partial information of eigenvalues and eigenvectors in Chapter 5, those that are of least squares nature in Chapter 6, those that are spectrally constrained in Chapter 7, those that are of low ranks in Chapter 8, and those that are specified by orbits of group actions in Chapter 9. No doubt such a classification will never be perfect. It does appear that, instead of setting forth a systematic theory and practical algorithms, we are proffering a problem book to readers. Though some of these chapters imbricate and refer to each other, readers might find some relief in knowing that each chapter can be rendered independently of each other.

We wish to have accomplished three goals in this treatise: First, we desire to demonstrate the breadth of areas where inverse eigenvalue problems can arise. The discipline ranges from practical engineering applications to abstract algebraic theorization. Secondly, we want to corroborate the depth of intricacy of inverse eigenvalue problems. While the setup of an inverse eigenvalue problem seems relatively easy, the solution is not straightforward. The instruments employed to solve such a problem are quite sophisticated, including techniques from orthogonal polynomials, degree theory, optimization, to differential geometry, and so on. Finally and most importantly, we want to arouse interest and encourage further research into this topic. Throughout the text, we wish to convey the message that there is much room for further study of the numerical development and theoretical understanding of this fascinating inverse problem.

This book is an accumulation of many years' research supported in part by the National Science Foundation under the grants DMS-9803759 and DMS-0073056. The book is based on a series of lectures first presented at the Istituto per Ricerche di Matematica Applicata (IRMA), Bari, Italy, in the summer of 2001 under the encouragement of Fasma Diele. The success of presenting those lectures was made possible by Roberto Peluso at the IRMA and Dario Bini at the Universitá di Pisa with the support from Il Consiglio Nazionale delle Ricerche (CNR) and the Gruppo Nazionale per il Calcolo Scientifico (GNCS) under the project "Matrici con struttura, analisi, algoritmi e applicazioni". Later in the fall of 2001 the same series was presented at the National Center of Theoretical Sciences (NCTS), Hsinchu, Taiwan, upon the invitation by Wen-Wei Lin at the Tsinhua University. At about the same time, we received a summons from Arieh Iserles to write a treatise for *Acta Numerica.* This sequence of events inadvertently kindled the fire within us to further extend and complete this project. Many other people have made various contributions to this project. We are especially indebted to Hua Dai at the Nanjing University of Aeronautics and Astronautics, Graham Gladwell at the University of Waterloo, Robert Plemmons at the Wake Forest University, Yitshak Ram at the Louisiana State University, and Shufang Xu at the Peking University, for their comments, suggestions and generous assistance. The heartfelt kindness and encouragement received from these many dear colleagues are greatly appreciated.

Moody T. Chu and Gene H. Golub
Raleigh, North Carolina and Stanford, California
October, 2004

CONTENTS

LIST OF ACRONYMS

IEP	Inverse Eigenvalue Problem
ISVP	Inverse Singular Value Problem

AIEP	Additive IEP
ECIEP	Equality Constrained IEP
ISEP	Inverse Singular/Eigenvalue Problem
JIEP	Jacobi IEP
LiPIEP	Linear Parameterized IEP
LSIEP	Least Square IEP
LSPDIEP	Least Square Partially Described IEP
MIEP	Multiplicative IEP
MQIEP	Monic Quadratic IEP
MVIEP	Multi-Variate IEP
NIEP	Nonnegative IEP
NNMF	Nonnegative Matrix Factorization Problem
NNMP	Nearest Normal Matrix Problem
PAP	Pole Assignment Problem
PDIEP	Partially Described IEP
PEIEP	IEP with Prescribed Entries
PEISVP	ISVP with Prescribed Entries
PIEP	Parameterized IEP
PISVP	Parameterized ISVP
QIEP	Quadratic IEP
RNIEP	Real-valued Nonnegative IEP
SCAP	Spectrally Constrained Approximation Problem
SHIEP	Schur-Horn IEP
SIEP	Structured IEP
SLRAP	Structured Low Rank Approximation Problem
SNIEP	Symmetric Nonnegative IEP
SQIEP	Standard Quadratic IEP
StIEP	Stochastic IEP
STISVP	Sing-Thompson ISVP
SVCAP	Singular Value Constrained Approximation Problem
ToIEP	Toeplitz IEP
UHIEP	Unitary Hessenberg IEP

LIST OF FIGURES

LIST OF TABLES

1

INTRODUCTION

In a mathematical model, we generally assume that there is a correspondence between the endogenous variables, that is, the internal parameters, and the exogenous variables, that is, the external behavior. The process of analyzing and deriving the spectral information and, hence, inducing the dynamical behavior of a system from *a priori* known physical parameters such as mass, length, elasticity, inductance, capacitance, and so on, is referred to as a *direct* problem. The *inverse* problem, in contrast, is to validate, determine, or estimate the parameters of the system according to its observed or expected behavior. The concern in the direct problem is to express the behavior in terms of parameters whereas in the inverse problem the concern is to express the parameters in term of behavior. The inverse problem is just as important as the direct problem in applications. In the former the behavior usually is a deterministic consequence of the parameters. In the latter the inverse problem often turns out to be ill-posed in that it has multiple solutions.

There are many kinds of inverse problems across a wide variety of disciplines. A picturesque illustration by Bolt (1980) might shed some light on the general difference between a direct problem and an inverse problem: "Suppose that we ski down a mountain trail. As long as we neglect friction and know the undulations in the trail, we can calculate exactly the time it will take to travel from a point on the mountain side to the valley below. This is a direct problem. It was difficult for Galileo's contemporaries. It is, however, now old hat and what is of current deeper interest is the following problem. If we start skiing from different places on the slope and on each occasion we time our arrival at a fixed place on the valley floor, how can we calculate the topographic profile of undulations of the trail? This is the inverse problem. It is certainly a practical problem. It is also challenging and difficult and, indeed, in the general sense, it has no unique solution."

Among the inverse problems of various nature, this book focuses mainly on the particular class of eigenvalue problems associated with matrices. In this context, an inverse eigenvalue problem (**IEP**) concerns the reconstruction of a matrix from prescribed spectral data.

We must point out immediately that there is a well developed and critically important counterpart of eigenvalue problem associated with differential systems. The Sturm–Liouville problem,

$$-\frac{d}{dx}\left(p(x)\frac{du}{dx}\right) + q(x)u = \lambda\omega(x)u, \tag{1.1}$$

for u on an interval $x \in (a, b)$ and the Helmholtz equation,

$$\nabla \cdot (\tau \nabla \Phi) + \rho \omega^2 \Phi = 0, \tag{1.2}$$

for Φ over a bounded domain V under various boundary conditions, for example, are classical eigenvalue problems in the setting of ordinary and partial differential equations, respectively. Both direct and inverse eigenvalue problems of these types traditionally are cast and studied in functional space. In order to solve these problems numerically (Ames, 1992; Pryce, 1993), the discretization procedure usually leads to a matrix eigenvalue problem, although in return one should be very cautious in using the discrete information to interpret the continuous problem (Osborne, 1971).

There has been vast interest in the inverse eigenvalue problem, including the pole assignment problem. Earlier work in the context of mechanics application can be found in the book by Gladwell (1984) with a follow-up in Gladwell (1996). An approach to inverse problem in geophysics, including uniqueness and existence results for the inverse eigenvalue problem arising from the vibrating string and the vibrating beam equations is surveyed in Barcilon (1986). A comprehensive review of the inverse spectral and inverse scattering problems, including the inverse Sturm–Liouville problems can be found in Chadan et al. (1997). A related inverse nodal problem for an elliptic equation in potential form with Dirichlet boundary conditions over a rectangular domain has been carefully discussed in Hald and McLaughlin (1996). A review of classical theories of vibration or buckling of structures, including rods, beams, plates, or shells with specified cross-sections, along with the associated inverse problems and a huge collection of references from the engineering perspective are given in Elishakoff (2000). Some general reviews and extensive bibliographies of inverse eigenvalue problems for matrices can be found, for example, in the books by Zhou and Dai (1991) and Xu (1998) and the more recent papers by Chu (1998) and Chu and Golub (2002).

1.1 Direct problem

It is perhaps valid to state that eigenvalues and singular values are two of the most distinguished characteristics in any given general square matrix. We shall call collectively either of these two characteristics (and the associated eigenvectors and singular vectors) the *spectral information* of the matrix, although the word "spectrum" usually refers to the eigenvalues only and a rectangular matrix has singular eigenvalues only.

A direct problem amounts to computing the eigenvalues, eigenvectors, singular values, or singular vectors of a given matrix. Spectral analysis has a wide range of applications. It simplifies the representation of complicated systems, sheds light on the asymptotic behavior of differential equations, and facilitates the performance evaluation of important numerical algorithms, to mention a few. Information on singular values, on the other hand, assumes a critical role whenever there is the presence of roundoff error or inexact data. Many fundamental topics in linear algebra and important applications in practice are best understood when formulated in terms of the singular value decomposition. The profound impact of spectral information in the sciences or engineering is well documented throughout the literature.

Over the years, one of the most fruitful developments in numerical linear algebra, which serves as a computational platform for a variety of application problems, is that of efficient and stable algorithms for the computation of eigenvalues and singular values of a given matrix. We shall make no attempt here to discuss the development of numerical methods for retrieving spectral information of a matrix for two reasons: one is that the thrust of the book is limited to inverse problems; the other is that studies of direct problems have been extensive and spread over the literature. It would take volumes to describe the many advances made in this regard. We suggest only three books as general background references: the book by Wilkinson (1965) is the classic in eigenvalue computation, the book by Golub and Van Loan (1996) contains extensive discussion of numerical algorithms and an extended collection of bibliographies related to numerical linear algebra, and the book by Horn and Johnson (1991) contains a detailed layout of theoretical matrix theory. Be it sufficient to mention that state-of-the-art software packages, for instance, LAPACK (Anderson et al., 1999), and the associated theories, have already been developed for the direct eigenvalue problem.

We do want to make a note that the spectral information corresponding to matrices with certain special structures sometimes are also structured (Bini et al., 2003). The content of "structure" should be taken relative to the ambient space in which the matrix resides. Real-valued symmetric matrices, for example, have a structure of symmetry in $\mathbb{R}^{n\times n}$. Generally the spectra of matrices in $\mathbb{R}^{n\times n}$ are distributed over the complex field, but eigenvalues of symmetric matrices are always real. We consider this real spectrum structured in comparison with the generic complex-valued ones. Symmetric Toeplitz matrices, on the other hand, are further structured in the space of symmetric matrices. The spectra of symmetric Toeplitz matrices enjoy a special parity structure (Cantoni and Bulter, 1976) which we shall discuss later. Other examples of structured matrices with structured spectra include the fact that eigenvalues of a real Hamiltonian matrix appear in pairs (Faßbender, 2000), the spectra of stochastic matrices must be bounded within a specific region in the complex plane (Karpelevič, 1951), and that all circulant matrices have the same set of eigenvectors, forming columns of the so-called Fourier matrix (Van Loan, 1992). Direct eigenvalue problems for structured matrices form an interesting research subject in their own right. The efforts in that regard have been centered mainly around taking advantage of the structure to induce effective computation, but it is not always clear how the structure of the corresponding spectra should be categorized. We shall see in the sequel that the structure of a matrix plays a significant role in the theory of IEPs. Without a structure, an IEP simply cannot be formulated properly. Since we generally do not know how to characterize the spectra of matrices of a given structure, to prove the existence of a matrix with a certain structure and with a given set of spectra generally is a very challenging task.

1.2 Inverse problem

It is clear that an IEP could trivially be solved if the matrices were subject to no restriction on structure. For the problem to be more meaningful, either physically or mathematically, it is often necessary to confine the construction to certain special classes of matrices. Matrices with a specified structure, for example, constitute a special class. The confinement of construction to a special class of matrices implies that, generally speaking, an IEP is always a *structured* inverse eigenvalue problem. So that we can better classify IEPs according to the different classes, be aware that we reserve the term structured inverse eigenvalue problem (SIEP) specifically for a more narrowly defined class of structures in Chapter 4 while other structures are named differently. It is critical to keep this notion in mind that the solution to an IEP must satisfy two constraints – the *spectral constraint* referring to the prescribed spectral data, and the *structural constraint* referring to the desirable structure. The variation of these constraints defines the variety of IEPs, some of which will be surveyed in this book.

Corresponding to almost all types of IEPs discussed henceforth where a matrix is to be constructed with prescribed eigenvalues, there is a similar formulation of inverse singular value problem (**ISVP**) where a matrix is to be constructed with prescribed singular values. We want to quickly point out at this very beginning stage of this book that very little is known about the ISVP in the literature. This area is widely open for further research.

1.2.1 *Constraints*

More should be said about the two constraints that are involved in the formulation of an IEP. First, we may assume in a loose sense that the structural constraint and the spectral constraint define, respectively, two smooth manifolds in the space of matrices of a fixed size. If the sum of the dimensions of these two manifolds exceeds the dimension of the ambient space, then under some mild conditions one can argue that the two manifolds must intersect and the IEP must have a solution. A more challenging situation is when the sum of dimensions emerging from both structural and spectral constraints does not add up to the transversal property. In that case, it is much harder to tell whether or not an IEP is solvable.

Second, we note that in a complicated physical system it is not always possible to attain knowledge of the entire spectrum. On the other hand, especially in structural design, it is often demanded that certain eigenvectors should also satisfy some specific conditions. The spectral constraints involved in an IEP, therefore, may consist of complete or only partial information on eigenvalues or eigenvectors.

We further observe that, in practice, it may occur that one of the two constraints in an IEP should be enforced more critically than the other, due to physical realizability, say. Without this, the physical system simply cannot be built. There are also situations when one constraint could be more relaxed than the other, due to the physical uncertainty, say. The uncertainty arises when there is simply no accurate way to measure the spectrum, or no reasonable means to obtain the information. When the two constraints cannot be satisfied simultaneously, the IEP could be formulated in a least squares setting, in which a decision is made as to which constraint could be compromised.

Finally, inasmuch as every IEP should be regarded as a structured inverse eigenvalue problem, the meaning of "being structured" can be taken very liberally. Some of the structures, such as Jacobi or Toeplitz, result in matrices forming linear subspaces; some structures, such as nonnegative or stochastic, limit entries of matrices in a certain range; while others, such as matrices with prescribed entries or with prescribed singular values or a specific rank, lead to some implicitly defined structural constraints. In this book, we shall reserve the acronym SIEP to the very narrowly defined class of structured IEPs delineated in Chapter 4. Other types of IEPs, though also structured, will be given different names as we shall see in Section 1.3.

1.2.2 *Fundamental issues*

Associated with any IEP are four fundamental questions. These are issues concerning:

- the theory of *solvability*,
- the practice of *computability*,
- the analysis of *sensitivity*, and
- the reality of *feasibility*.

A major effort in solvability has been to determine a necessary or a sufficient condition under which an inverse eigenvalue problem has a solution. Related to the solvability is the issue of uniqueness of a solution. The main concern associated with computability, on the other hand, has been to develop procedures by which, knowing *a priori* that the given spectral data are feasible, a matrix can be constructed in a numerically stable fashion. The discussion on sensitivity concerns perturbation analysis when an IEP is modified by changes in the spectral data. The feasibility is a matter of differentiation between whether the given data are exact or approximate, complete or incomplete, and whether an exact value or only an estimate of the parameters of the physical system is needed. Each of these four questions is essential but challenging to the understanding of a given IEP. We are not aware of many IEPs that are comprehensively understood in all these four aspects. Rather, considerably more work remains to be done. For the very same reason, we cannot possibly treat each IEP evenhandedly in this book.

With different emphases and different formulations, studies of IEPs have been intensive and scattered, ranging from acquiring a pragmatic solution to a real-world application to discussing the mathematical theory of an abstract formulation. A timely review of progress made that better defines the regimen of IEPs as a whole is critical for further research and understanding. Earlier endeavors in this regard include the book by Gladwell (1986b), where the emphasis was on applied mechanics, the survey by Boley and Golub (1987), where the emphasis was on numerical computation, the book by Zhou and Dai (1991), which pointed to many publications in Chinese that were perhaps unknown to the West, and the article by Gladwell (1996), which reviewed activities and literature between 1985 and 1995 as a ten-year update of his previous book.

In a recent review article, Chu (1998) briefly described a collection of thirty-nine IEPs. These problems were categorized roughly according to their characteristics into three types of IEPs, that is, parameterized, structured, and partially described. Since then, many more old results have been unearthed, while new articles have continued to appear, notably the treatise by Ikramov and Chugunov (2000), translated from Russian with the emphasis on finitely solvable IEPs and rational algorithms, and the book by Xu (1998), where many results on the sensitivity issue by Chinese mathematicians are made known in English for the first time. It becomes clear that there is a need to update the history and recent developments in both theory and application. The recent article by Chu and Golub (2002) concerns only the one paradigm of SIEP. This book is an attempt to cover a much larger ground of IEPs and to prepare readers to launch into a much larger field for further research.

We shall touch upon a variety of IEPs by describing their formulations, highlighting some theories or numerical procedures, and suggesting some pertinent references. Additionally, we shall outline some applications of IEPs from selected areas of disciplines. From time to time, we shall point out some open questions. While we sometimes seem to be concentrating on one particular numerical method applied to one particular problem, often the method has enough generality that with some suitable modifications it can also be applied to other types of problems. We choose not to encumber readers with the details.

We hope that this monograph, along with previous treatments mentioned above, will help to inspire some additional interest and to stimulate further research.

1.3 Nomenclature

For ease of identifying the characteristics of various IEPs, we suggest using a unified name scheme ***IEP#** to categorize an IEP (Chu, 1998). When singular values are involved in the spectral constraint, we distinguish ISVPs from IEPs. A letter or letters replacing the symbol "$*$" in front of the word IEP registers the type of problem. The numeral "#" following IEP, if any, indicates the sequence

Table 1.1. *Summary of acronyms used in the book*

Acronym	Meaning	Reference
AIEP	Additive IEP	Section 3.3
ECIEP	Equality Constrained IEP	Page 21
ISEP	Inverse Singular/Eigenvalue Problem	Section 4.9
ISVP	Inverse Singular Value Problem	Section 4.8
JIEP	Jacobi IEP	Section 4.2
LSIEP	Least Square IEP	Chapter 6
MIEP	Multiplicative IEP	Section 3.4
MVIEP	Multi-Variate IEP	Not discussed
NIEP	Nonnegative IEP	Section 4.4
PAP	Pole Assignment Problem	Page 11
PEIEP	IEP with Prescribed Entries	Section 4.7
PIEP	Parameterized IEP	Chapter 3
PDIEP	Partially Described IEP	Chapter 5
RNIEP	Real-valued Nonnegative IEP	Page 94
SHIEP	Schur–Horn IEP	Page 113
SIEP	Structured IEP	Page 4
SNIEP	Symmetric Nonnegative IEP	Page 94
StIEP	Stochastic IEP	Section 4.5
STISVP	Sing–Thompson ISVP	Page 116
ToIEP	Toeplitz IEP	Section 4.3
UHIEP	Unitary Hessenberg IEP	Section 4.6

of variation within type "*IEP". For the convenience of later reference, we summarize the acronyms appearing in this book in Table 1.1. Also indicated are the page numbers or the section numbers where the problems are first described or where more detailed discussion can be found.

The diagram in Figure 1.1 depicts a possible inclusion relationship between different problems. In particular, the diagram intends to imply the following points:

- Multivariate IEPs include univariate IEPs. This book only discusses univariate IEPs.
- All problems have a natural generalization to a least squares formulation.
- The structural constraint involved in the IEPs can appear in various forms and hence defines different IEPs.
- Replacing the eigenvalue constraint by the singular value constraint, there is a counterpart ISVP corresponding to any structured IEP. Very little is known about these structured ISVPs (and hence no diagrams).

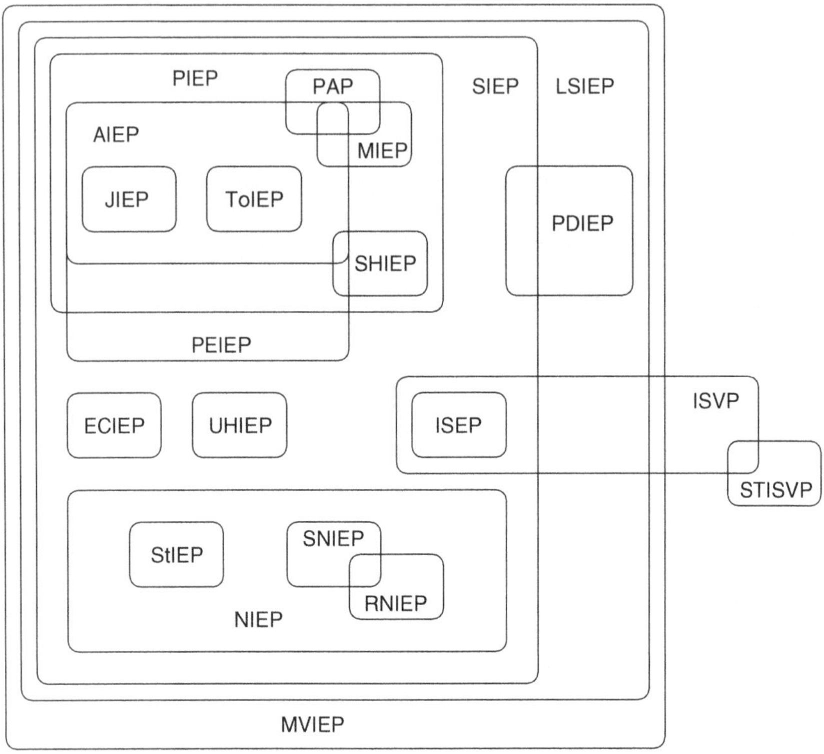

Figure 1.1. *Classification of inverse eigenvalue problems*

- The class of PIEPs is considered to be a subset of general SIEPs while the PIEP itself contains many classical IEPs as special cases.
- The relationship depicted in Figure 1.1 is not necessarily definitive because many characteristics may overlap.

In his review article, Gladwell (1986c) advocated that various IEPs should be differentiated according to the type of the mechanical system they represent, that is, continuous or discrete, damped or undamped, and the type of the prescribed data, i.e., spectral, modal, or nodal, complete or incomplete. References in Gladwell (1996) are divided according to this concept. This interesting notion and our way of classification should complement each other to offer a fairly broad view of research activities in this area.

The problems discussed in this book definitely are not all the possible IEPs. Our criteria of selection are simply those problems that are representative of a variety of structural constraints and are slightly better studied in the literature. We shall consider these problems slightly more in breadth and depth with regard to the four issues on solvability, computability, sensitivity, and applicability.

Some main results, applications, and algorithmic issues will also be presented. It is inevitable that the formations and algorithms will differ noticeably from problem to problem, depending upon the degree we understand them. It should be pointed out that we choose not to include the pole assignment problems in this discussion because that classical topic alone has been well discussed in many other places (Byers and Nash, 1989; Byrnes, 1989; Kautsky et al., 1985; Sun, 1996).

1.4 Summary

We believe that inverse eigenvalue problems should always be structured problems. In this book, we try to explain, motivate, and review only a small segment of the full scope of inverse eigenvalue problems. The structures we selected for study in this presentation are by no means emblematic, but rather reflect our personal preferences. The ideas introduced in this monograph are by no means inclusive, but rather divulge our limited understanding of this subject. We have collected an extensive bibliography of more than 400 papers on this topic. We will keep a constantly updated list at the site *http://www4.ncsu.edu/~mtchu*. Even so, the list is far from being comprehensive as we have already overlooked much of the engineering literature. See, for example, the vast references listed in (Elishakoff, 200; Gladwell, 1986b,c, 1996; Ram, 2003). We invite reader to inform us of any work that we have overlooked.

Note the distinctions between problems where the entire spectrum is known and where only partial knowledge of eigenvalues and eigenvectors is given. Note also the distinctions between problems where the structural constraint must be satisfied and where only approximate structure is sufficient. There are many open areas for further research. The discussion in this book is but a demonstration of the wide scope of techniques and applications that inverse eigenvalue problems can involve. The theory employed to answer an inverse eigenvalue problem is certainly not trivial and far from being complete. There is much room for further study of either the numerical development or the theoretical understanding of these fascinating inverse problems.

2

APPLICATIONS

2.1 Overview

Inverse eigenvalue problems arise from a remarkable variety of applications. The list includes, but is not limited to, control design, system identification, seismic tomography, principal component analysis, exploration and remote sensing, antenna array processing, geophysics, molecular spectroscopy, particle physics, structural analysis, circuit theory, and mechanical system simulation.

A common phenomenon that stands out in most of these applications is that the physical parameters of the underlying system are to be reconstructed from knowledge of its dynamical behavior. The meaning of "dynamical behavior" can be qualified in a number of ways. For example, vibrations depend on natural frequencies and normal modes, stability controls depend on the location of eigenvalues, and so on. As such, the spectral information used to affect the dynamical behavior varies in various ways. If the physical parameters can be, as they often are, described mathematically in the form of a matrix, then we have an IEP. The structure of the matrix is usually inherited from the physical properties of the underlying system.

In this chapter we briefly highlight a few applications that, in our judgment, should be of general interest to readers. Again, we stress that the reason for our selection is solely based on the simplicity of these examples. We have chosen to ignore many real-world applications that are large, complicated, and require more analysis even in setting them up. Still, so as not to lose sight of the notion of an IEP, it is clear that we have to sacrifice technical details even in the description of these simpler applications. The purpose of this chapter is to illustrate how some practical applications lead to an IEP.

For the present, our emphasis is on the formulation of the underlying IEP. We shall put off any theoretical discussion or numerical computation to the later part of this book. We shall divide the discussions into six categories: pole assignment problem, applied mechanics, inverse Sturm–Liouville problem, applied physics, numerical analysis, and signal and data processing. Each category covers some additional problems.

For convenience, we shall adopt henceforth the notation $\sigma(M)$ to represent the entire spectrum of a given square matrix M. For a general rectangular matrix N, $\varpi(N)$ denotes the collection of all nonzero singular values.

2.2 Pole assignment problem

Consider first the following dynamic state equation:

$$\dot{\mathbf{x}}(t) = A\mathbf{x}(t) + B\mathbf{u}(t), \tag{2.1}$$

where $\mathbf{x}(t) \in \mathbb{R}^n$ denotes the state of a certain physical system to be controlled by the input $\mathbf{u}(t) \in \mathbb{R}^m$. The two given matrices $A \in \mathbb{R}^{n\times n}$ and $B \in \mathbb{R}^{n\times m}$ are invariant in time. One classical problem in control theory is to select the input $\mathbf{u}(t)$ so that the dynamics of the resulting $\mathbf{x}(t)$ is driven into a certain desired state. Depending on how the input $\mathbf{u}(t)$ is calculated, there are generally two types of controls, both of which have been extensively studied and documented in the literature.

2.2.1 *State feedback control*

In state feedback control, the input $\mathbf{u}(t)$ is selected as a linear function of current state $\mathbf{x}(t)$, that is,

$$\mathbf{u}(t) = F\mathbf{x}(t). \tag{2.2}$$

In this way, the system (2.1) is changed to a closed-loop dynamical system:

$$\dot{\mathbf{x}}(t) = (A + BF)\mathbf{x}(t). \tag{2.3}$$

A general goal in such a control scheme is to choose the *gain matrix* $F \in \mathbb{R}^{m\times n}$ so as to achieve stability or to speed up response. This control problem has been a major research topic for a long time, due to its critical role in a wide range of important applications. There are many ways to achieve this goal. One way is to minimize a certain cost function in the so-called linear–quadratic regulator. Another way is to directly translate the task to the selection of F so that the spectrum $\sigma(A+BF)$ is bounded in a certain region of the complex plan. Obviously, in the latter the choice of the region affects the degree of difficulty of control. We can further restrict the location of the spectrum by reassigning eigenvalues of the matrix $A+BF$ to a prescribed set. This leads to a special type of inverse eigenvalue problem usually referred to in the literature as the state feedback pole assignment problem (**PAP**).

> **Problem 2.1** *(State feedback PAP)*
> Given $A \in \mathbb{R}^{n\times n}$ and $B \in \mathbb{R}^{n\times m}$ and a set of complex numbers $\{\lambda_k\}_{k=1}^n$, closed under complex conjugation, find $F \in \mathbb{R}^{m\times n}$ such that
>
> $$\sigma(A + BF) = \{\lambda_k\}_{k=1}^n.$$

2.2.2 *Output feedback control*

It is often the case in practice that the state $\mathbf{x}(t)$ is not directly observable. Instead, only the output $\mathbf{y}(t)$ that is related to $\mathbf{x}(t)$ via

$$\mathbf{y}(t) = C\mathbf{x}(t) \tag{2.4}$$

is available. In the above, $C \in \mathbb{R}^{p\times n}$ is a known matrix. The input $\mathbf{u}(t)$ must now be chosen as a linear function of the current output $\mathbf{y}(t)$, that is,

$$\mathbf{u}(t) = K\mathbf{y}(t). \tag{2.5}$$

The closed-loop dynamical system thus becomes

$$\dot{\mathbf{x}}(t) = (A + BKC)\mathbf{x}(t). \tag{2.6}$$

The goal now is to select the *output matrix* $K \in \mathbb{R}^{m\times p}$ so as to reassign the eigenvalues of $A + BKC$. This output feedback PAP once again gives rise to a special type of IEP.

Problem 2.2 *(Output feedback PAP)*
Given $A \in \mathbb{R}^{n\times n}$, $B \in \mathbb{R}^{n\times m}$, and $C \in \mathbb{R}^{p\times n}$, and a set of complex numbers $\{\lambda_k\}_{k=1}^{n}$, closed under complex conjugation, find $K \in \mathbb{R}^{m\times p}$ such that

$$\sigma(A + BKC) = \{\lambda_k\}_{k=1}^{n}.$$

There is great interest in the subject of PAP alone. This is still a very active and well-studied research area even to this date. We would suggest the papers by Byrnes (1989) and Rosenthal and Willems (1999) for theories, and Kautsky et al. (1985) and Syrmos et al. (1997) for methods, which gave an excellent account of activities in this area, as a starting point for further exploration. Thanks to the widely available literature of this subject, we shall not give any further details on PAPs except to mention that PAPs are a special case of what we call the PIEP in the later part of this book.

One important remark should be made at this point. The PAPs, as well as many other IEPs, usually have multiple solutions. Among these multiple solutions, the one that is *least* sensitive to perturbations of problem data is perhaps most critical from a practical point of view. Such a solution, termed the *robust solution* in the literature, is usually found by minimizing the condition number associated with the solution. In other words, there are two levels of work when solving an IEP for a robust solution. The first is to develop a means to find a solution, if there is any at all; the second is to use optimization techniques to minimize the condition number associated with the solution. Most of the numerical methods discussed in this book are for the first task only. Except for the PAPs (Kautsky et al., 1985), the second task for general IEPs has not been fully explored as yet.

We should also point out that, in contrast to what has been described earlier as structural constraint for IEPs, the matrices involved in the above context of PAPs usually do not carry any further structure at all. Specifically, there is no specific restrictions on either F or K. The PAP will become a much harder IEP if either F or K needs to satisfy some additional structural constraints. For the state feedback problem, there has been some interest in the case where K is structured. One such application is the so-called *decentralized dynamic* PAP where K is a diagonal matrix. Some background information can be found in a recent paper by Ravi et al. (1995). Numerical algorithms for this type of structured PAP are needed.

2.3 Applied mechanics

Interpreting the word "vibration" in a broad sense, we see applied mechanics everywhere. The transverse motion of masses on a string, the buckling of structures, the transient current of electric circuits, and the acoustic sound in a tube are just a few instances of vibration. One of the basic problems in classical vibration analysis is to determine the natural frequencies and normal modes of a vibrating body. But inverse problems are concerned with the construction of a model of a given type, for example, a mass–spring system, a string, an IC circuit, and so on, with prescribed spectral data. Such a reconstruction, if possible, would have practical value to applied mechanics and structure design.

2.3.1 *A string with beads*

Consider the vibration of beads on a taut string illustrated in Figure 2.1. Assume that the beads, each with mass m_i, are placed along the string with equal horizontal spacing h and are subject to a constant horizontal tension F. Then the equations of motion (for four beads) are given by:

$$
\begin{aligned}
m_1\frac{d^2x_1}{dt^2} &= -F\frac{x_1}{h} + F\frac{x_2 - x_1}{h}, \\
m_2\frac{d^2x_2}{dt^2} &= -F\frac{x_2 - x_1}{h} + F\frac{x_3 - x_2}{h}, \\
m_3\frac{d^2x_3}{dt^2} &= -F\frac{x_3 - x_2}{h} + F\frac{x_4 - x_3}{h}, \\
m_4\frac{d^2x_4}{dt^2} &= -F\frac{x_4 - x_3}{h} - F\frac{x_4}{h}.
\end{aligned}
$$

The equations of motion can easily be generalized to the case of n beads which can conveniently be described in matrix form:

$$
\frac{d^2\mathbf{x}}{dt^2} = -DJ_0\mathbf{x}, \tag{2.7}
$$

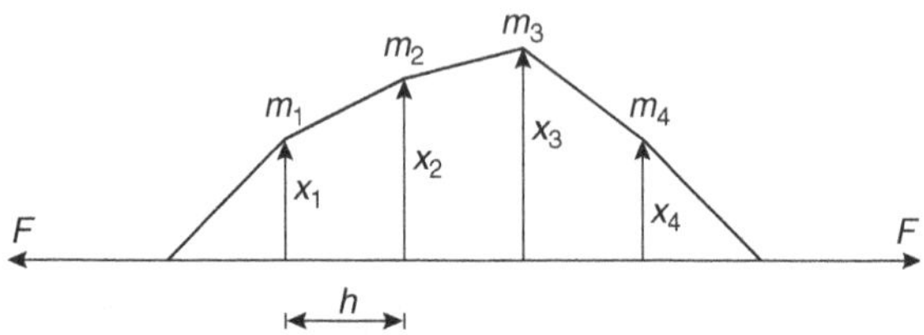

Figure 2.1. *Vibration of beads on a string*

where $\mathbf{x} = [x_1, x_2, \ldots, x_n]^{\mathsf{T}}$, J_0 is the Jacobi matrix

$$J_0 = \begin{bmatrix} 2 & -1 & 0 & & & \\ -1 & 2 & -1 & & & \\ 0 & -1 & 2 & \ldots & & 0 \\ \vdots & & & \ddots & & \\ 0 & & & & 2 & -1 \\ 0 & & & & -1 & 2 \end{bmatrix}, \tag{2.8}$$

and $D = \operatorname{diag}(d_1, d_2, \ldots, d_n)$ is a diagonal matrix with $d_i = F/m_i h$. We remark that the system (2.7) may also be thought of as the method of lines applied to the one-dimensional wave equation. Eigenvalues of the matrix product DJ_0 are precisely the squares of the so-called *natural frequencies* of the system. An interesting inverse problem, originally considered by Gantmacher and Krein (2002) and is a special case of the so-called multiplicative IEP (MIEP), concerns placing the weights m_i, $i = 1, \ldots, n$, appropriately so that the resulting system has a prescribed set of natural frequencies.

Problem 2.3 *(String with weighted beads)*
Given J_0 as in (2.8) and a set of positive numbers $\{\lambda_k\}_{k=1}^n$, find a diagonal matrix D with positive entries so that

$$\sigma(DJ_0) = \{\lambda_k\}_{k=1}^n.$$

Note that the underlying matrix $-DJ_0$ has a unique structure in that D is a diagonal matrix and J_0 is a fixed Jacobi matrix. An even more fundamental question related to the solvability is whether such a string can have arbitrarily prescribed natural frequencies by adjusting the diagonal matrix D and, if not, what are the reachable frequencies. Recently, Gladwell (2004) derived a closed form procedure to construct a string with minimal mass from the given spectrum.

2.3.2 *Quadratic eigenvalue problem*

More generally, the equation of motion arising in many mechanics applications appears as a linear second-order differential system:

$$M\ddot{\mathbf{x}} + C\dot{\mathbf{x}} + K\mathbf{x} = f(t), \tag{2.9}$$

where $\mathbf{x} \in \mathbb{R}^n$ and M, C, $K \in \mathbb{R}^{n\times n}$. Usually, the mass matrix M is diagonal, both the damping matrix C and the stiffness matrix K are symmetric and tridiagonal, M is positive definite and K is positive semidefinite. It is known that the general solution to the homogeneous equation is of vital importance to the stability of the subsequent dynamical behavior. Toward that end, the fundamental solution can be derived by proposing a solution of the form

$$\mathbf{x}(t) = \mathbf{v}e^{\mu t}.$$

Upon substitution, it turns out that $\mathbf{v}$ and μ are solutions to the quadratic eigenvalue problem

$$(\mu^2 M + \mu C + K)\mathbf{v} = 0. \tag{2.10}$$

Assuming the case that all eigenvalues are distinct, then a general solution to the homogeneous system is given by the superposition principle as

$$\mathbf{x}(t) = \sum_{k=1}^{2n} \alpha_k \mathbf{v}_k e^{\mu_k t},$$

where $(\mu_k, \mathbf{v}_k)$, $k = 1, \ldots, 2n$, are the eigenpair solutions to (2.10).

In the undamped system where $C = 0$, the quadratic eigenvalue problem is reduced to the generalized eigenvalue problem,

$$(K - \omega^2 M)\mathbf{v} = 0, \tag{2.11}$$

if we write $\lambda = i\omega$. In this case, ω is precisely the natural frequency of the system and $\mathbf{v}$ is the corresponding natural mode. Let $\lambda = \omega^2$, $J := M^{-1/2}KM^{-1/2}$, and $\mathbf{z} = M^{1/2}\mathbf{x}$. The generalized eigenvalue problem can be further reduced to the Jacobi eigenvalue problem

$$J\mathbf{z} = \lambda\mathbf{z}. \tag{2.12}$$

At this point, there are two ways to formulate IEPs in the above context. First, note that the stiffness matrix K normally is more complicated than the mass matrix M. The requirement of maintaining physical feasibility also imposes constraints on the stiffness matrix, making it less flexible and more difficult to construct. Thus, one usual way of forming an IEP is to have the stiffness matrix K determined and fixed from the existing structure, that is, the static constraints, and we want to find the mass matrix M in (2.11) so that some desired natural frequencies are achieved. This inverse problem is equivalent to the MIEP discussed earlier.

Problem 2.4 *(Inverse generalized eigenvalue problem)*
Given a set of complex numbers $\{\lambda_k\}_{k=1}^n$, closed under complex conjugation, a fixed symmetric matrix K that has positive diagonal entries, negative off-diagonal entries, and is weakly diagonally dominant, find a symmetric and positive definite matrix M such that

$$\sigma(K + \lambda M) = \{\lambda_k\}_{k=1}^n.$$

An alternative way of formulation is to construct an unreduced, symmetric, and tridiagonal matrix J from its eigenvalues and the eigenvalues of its first leading principal submatrix. This is a special case of the so-called Jacobi IEP (JIEP). We shall illustrate in Section 4.2.2 that such an inverse problem can be identified as configuring a mass–spring system from its spectrum and the spectrum of the same system but with the last mass fixed to have no motion.

The inverse problem for a damped system is considerably more complicated (Datta and Sarkissian, 2001; Nylen, 1999; Ram, 2003). Assuming that M is normalized to the identity, an analogue of the JIEP for the damped system concerns the reconstruction of matrices C and K from the given spectral information of the damped system. A particular formulation is given as SIEP6b in Section 4.2.1. Quadratic inverse eigenvalue problems with partially described spectral information will be discussed in Sections 5.3 and 5.4.

Sometimes it becomes desirable to modify the dynamics of (2.9) by applying a control force of the form $B\mathbf{u}(t)$, where $B \in \mathbb{R}^{n\times m}$ is called a *control matrix* and $\mathbf{u}(t) \in \mathbb{R}^m$ is a state feedback control vector, in addition to the external force $f(t)$. One choice of the feedback control appears in the form:

$$\mathbf{u}(t) = F^\top \dot{\mathbf{x}}(t) + G^\top \mathbf{x}(t), \tag{2.13}$$

where $F, G \in \mathbb{R}^{n\times m}$ are constant matrices. The associated quadratic pencil becomes

$$Q_c(\mu) = \mu^2 M + \mu(C - BF^\top) + (K - BG^\top). \tag{2.14}$$

In the scenario of *model updating*, only a few spectral properties of an existing model (2.9) need to be modified while the rest is kept invariant. This leads to a broad range of eigenstructure assignment problems generalizing the conventional PAPs. We only mention two cases (Datta et al., 2000; Datta and Sarkissian, 2001).

Problem 2.5 *(Partial eigenvalue assignment problem)*
Let $\{\mu_1, \ldots, \mu_{2n}\}$ denote the set of eigenvalues for the system (2.10) with fixed symmetric matrices $M, C, K \in \mathbb{R}^{n\times n}$ and M being positive definite. Given $B \in \mathbb{R}^{n\times m}$ and a subset $\{\lambda_1, \ldots, \lambda_p\} \subset \mathbb{C}$ that is closed under complex conjugation, find real feedback matrices F and G so that

$$\sigma(Q_c(\mu)) = \{\mu_1, \ldots, \mu_p; \lambda_{p+1}, \ldots, \lambda_{2n}\}.$$

Problem 2.6 *(Partial eigenstructure assignment problem)*
Let $\{(\mu_i, \mathbf{v}_i)\}_{i=1}^{2n}$ denote the set of eigenpairs for the system (2.10) with fixed symmetric matrices $M, C, K \in \mathbb{R}^{n\times n}$ and M being positive definite. Suppose the subsets $\{(\mu_i, \mathbf{v}_i)\}_{i=1}^{p}$ and $\{(\lambda_i, \mathbf{u}_i)\}_{i=1}^{p}$ are closed under complex conjugation. Find $B \in \mathbb{R}^{n\times m}$ real feedback matrices F and G so that

$$\begin{cases} Q_c(\lambda_i)\mathbf{u}_i = 0, i = 1, \ldots, p, \\ Q_c(\mu_i)\mathbf{v}_i = 0, i = p+1, \ldots, 2n. \end{cases}$$

We do have to caution about one unpleasant feature in the model (2.14). That is, the matrices $C - BF^\top$ and $K - BG^\top$ generally are no longer symmetric, which often is critical in application. A related partially described IEP that maintains the symmetry will be discussed in Section 5.3.

2.3.3 *Engineering applications*

There are many other types of engineering applications into which an IEP formulation and its solution could offer useful insight that, in turn, could lead to better control of performance, safety, or effects of the system. The area is very broad. We mention only a few cases.

A recent paper by Tisseur and Meerbergen (2001) offers an excellent survey of quadratic eigenvalue problems and related applications. The review by Elishakoff (2000) discusses classic theories for direct, semi-inverse and inverse eigenvalue problems of structures.

Specific applications of IEPs to structure design problems can be found in Joseph (1992) as well as the conference collection (Motershead and Friswell, 2001). By measuring the changes in the natural frequencies, the IEP idea can be employed to detect the size and location of a blockage in a duct or a crack in a beam. See Wu (1990), Gladwell and Morassi (1999), and Gladwell (1996) for additional references.

Studies on IEPs with applications to mechanics are especially flourishing. The research began in the former Soviet Union with the work of Kreĭn (1933). It first became known in the West through the 1960 (German) translation of the book by Gantmacher and Krein (2002). The book by Gladwell (1986b) and his follow-up review (Gladwell, 1996) cover a broad scope of practices and references of IEPs for small mechanical systems. Other individual articles such as Barcilon (1979), Dai (1995), Gladwell (1984), Gladwell and Gbadeyan (1985), Gladwell (1986a, 1997, 1999), Ram and Caldwell (1992), and Ram and Gladwell (1994) and the web site by (Ram, 2003) represent a short list of typical applications of IEPs to vibrating rods and beams.

Applications of IEPs to model updating problems and fault detection problems for machine and structure diagnostics are discussed in Datta (2002), Starek and Inman (2001). Discussion for higher dimensional problems can be found in Barcilon (1990), Gladwell and Zhu (1992), Knobel and McLaughlin (1994),

McLaughlin et al. (1994), McLaughlin and Hald (1995), Zayed (1993). Important extensions of the Jacobi-type analysis to a tree-like system can be found in Duarte (1989), Nylen and Uhlig (1997a, b) Nylen (1999). A more comprehensive bibliography can be found at our web site *http://www4.ncsu.edu/~mtchu*, which will be updated from time to time.

2.4 Inverse Sturm–Liouville problem

Much of the discussion for IEPs in the literature has been due to an interest in the inverse Sturm–Liouville problem. A classical regular Sturm–Liouville problem concerns a differential equation of the form:

$$-\frac{d}{dx}\left(p(x)\frac{du(x)}{dx}\right) + q(x)u(x) = \lambda u(x), \quad a < x < b, \tag{2.15}$$

where $p(x)$ and $q(x)$ are piecewise continuous on $[a, b]$ and appropriate boundary conditions are imposed. As a direct problem, it is known that eigenvalues of the system (2.15) are real, simple, countable, and tend to infinity. As an inverse problem, the question is to determine the potential function $q(x)$ from eigenvalues. This inverse problem has generated much interest in the field, for example, the work contained in (Andrew, 1994; McLaughlin and Handelman, 1980; Paine, 1984; Zhornitskaya and Serov, 1994), notably the celebrated work by Gel′fand and Levitan (1955) in which the fundamental fact that *two* sets of data sequences are required to uniquely determine a potential is settled. A quick introduction to the inverse Sturm–Liouville problem can be found in the book by Chadan et al. (1997) Chapter 3. A more thoroughgoing discussion was done in the translated book by Levitan (1987).

When a numerical solution is sought, the Sturm–Liouville problem is discretized (Pryce, 1993). Likewise, the inverse problem leads to a matrix analogue IEP. Assuming that $p(x) \equiv 1$, $[a, b] = [0, 1]$, and mesh size $h = 1/(n + 1)$, the differential equation (2.15) is reduced by the central difference scheme, for instance, to the matrix eigenvalue problem

$$\left(-\frac{1}{h^2}J_0 + X\right)\mathbf{u} = \lambda\mathbf{u}, \tag{2.16}$$

where J_0 is given by (2.8) and X is the diagonal matrix representing the discretization of $q(x)$. The inverse problem is to determine a diagonal matrix X so that the matrix on the left side of (2.16) possesses a prescribed spectrum. This is a special case of the so-called additive IEP (AIEP).

Problem 2.7 *(Discrete inverse Sturm–Liouville problem)*
Given J_0 as in (2.8), $h > 0$, and a set of negative numbers $\{\lambda_k\}_{k=1}^n$, determine a positive diagonal matrix X so that

$$\sigma\left(-\frac{1}{h}J_0 + X\right) = \{\lambda_k\}_{k=1}^n.$$

It should be cautioned that there is a fundamental difference between the behavior of the discrete problem and that of the continuous case. See the discussion in Hald (1972), Osborne (1971). The number of eigenvalues involved in a discrete problem is at most finite while the Sturm–Liouville theory asserts that the continuous problem has always infinitely many eigenvalues. Eigenvalues from the discrete model mimic only the first few *smallest* eigenvalues of the continuous model. The matrix analogue IEP such as (2.16), however, is of interest in its own right.

We mention one application to geophysics. Assuming that the Earth has spherical symmetry, geophysicists want to infer its internal structure from the frequencies of spheroidal and torsional modes of oscillations. One such model leads to the generalized Sturm–Liouville problem, that is,

$$u^{(2k)} - (p_1 u^{(k-1)})^{(k-1)} + \cdots + (-1)^k p_k u = \lambda u.$$

Following the work of Gel′fand and Levitan (1955), Barcilon (1974a) suggested that $k+1$ spectra, associated with $k+1$ distinct sets of boundary conditions, must be present to construct the unknown coefficients $p_1, \ldots, p_k$. See also (Barcilon, 1974b). At this moment, it is not clear how the matrix analogue for this high-order problem should be formulated.

2.5 Applied physics

The IEP formulation can sometimes be used to explore and alleviate some difficult computational problems in applied physics. We demonstrate two applications in this section.

2.5.1 *Quantum mechanics*

We first describe an application to quantum mechanics. The material is taken from the paper by Deakin and Luke (1992). In computing the electronic structure of an atom, one usually expands the atom's state vector over a convenient basis. The expansion coefficients are determined by solving the eigenvalue problem for a Hamiltonian matrix H. It is known that these expansion coefficients are sensitive to the diagonal elements of H. Yet, in many cases of interest, the diagonal elements of H cannot be determined to sufficient accuracy. On the other hand, eigenvalues of H correspond to energy levels of an atom that can usually be measured to a high degree of accuracy. The idea now is to use these measured energy levels to correct diagonal elements. Furthermore, for practical purposes, all matrices involved are required to be real. Under such a constraint, it is almost always impossible to match the eigenvalues exactly. We therefore formulate a least squares IEP (LSIEP) as follows.

Problem 2.8 *(LSIEP arising from quantum mechanics)*
Given a real symmetric matrix H and a set of real values $\boldsymbol{\lambda} = [\lambda_1, \ldots, \lambda_n]^\top$, find a real diagonal matrix D such that

$$\|\sigma(H + D) - \boldsymbol{\lambda}\|_2 \tag{2.17}$$

is minimized.

We need to give an explanation of the above notation. Without causing ambiguity, we have used the spectrum (set) notation $\sigma(M)$ to indicate also the column (vector) formed by the eigenvalues of M. Be aware that included in the minimization process of (2.17) is also a rearrangement of eigenvalues in the spectrum for each fixed ordering in $\boldsymbol{\lambda}$. More details for LSIEPs will be discussed in Chapter 6.

2.5.2 *Neuron transport theory*

We next describe an application to neuron transport theory. This application is considered in the work by Li (1997b). One model for the dynamics in an additive neural network is the differential equation

$$\frac{d\mathbf{u}}{dt} = -A\mathbf{u} + \Omega\mathbf{g}(\mathbf{u}) + \mathbf{p}, \tag{2.18}$$

where $A = \operatorname{diag}(a_1, \ldots, a_n)$ denotes the decaying factor, $\Omega = [\omega_{ij}]$ denotes connection coefficients between the neurons, $\mathbf{g}(\mathbf{u}) = [g_1(u_1), \ldots, g_n(u_n)]^\top$ denotes the squashing function in which each g_i is strictly increasing but bounded in u_i, and $\mathbf{p}$ is a constant input. One of the design problems in transport theory is that, given quantities A, $\mathbf{g}$, and $\mathbf{p}$, choose the connection matrix Ω so that a predestined point $\mathbf{u}^* \in \mathbb{R}^n$ is a stable equilibrium. This requirement translates into two conditions that must be satisfied simultaneously. First, the linear equation

$$-A\mathbf{u}^* + \Omega\mathbf{g}(\mathbf{u}^*) + \mathbf{p} = 0 \tag{2.19}$$

must hold for Ω. Second, all eigenvalues of the Jacobian matrix,

$$\Upsilon = -A + \Omega G(\mathbf{u}^*), \tag{2.20}$$

where $G(\mathbf{u}^*) = \mathrm{diag}(g_1'(\mathbf{u}^*), \ldots, g_n'(\mathbf{u}^*))$, must lie in the left half-plane. Upon rearranging the terms, it is easy to see that (2.19) can be rewritten as

$$\Upsilon \mathbf{x} = \mathbf{y}, \tag{2.21}$$

where $\mathbf{x} = G^{-1}(\mathbf{u}^*)\mathbf{g}(\mathbf{u}^*)$ and $\mathbf{y} = A\mathbf{u}^* - AG^{-1}(\mathbf{u}^*)\mathbf{g}(\mathbf{u}^*) - \mathbf{p}$ are known vectors. This is a special case of the equality constrained IEP (**ECIEP**) which will be shown in Section 4.10 to be equivalent to a state feedback PAP.

> **Problem 2.9** *(ECIEP)*
> Given two sets of real vectors $\{\mathbf{x}_i\}_{i=1}^p$ and $\{\mathbf{y}_i\}_{i=1}^p$ with $p \leq n$, and a set of complex numbers $\{\lambda_1, \ldots, \lambda_n\}$, closed in conjugation, find a real matrix Υ such that
>
> $$\begin{cases} \Upsilon \mathbf{x}_i = \mathbf{y}_i, & i = 1, \ldots, p, \\ \sigma(\Upsilon) = \{\lambda_k\}_{k=1}^n. \end{cases}$$

A similar matrix approximation problem with linearly constrained singular values is discussed in Nievergelt (1997).

2.6 Numerical analysis

We point out that even within the field of numerical analysis the notion of IEP helps to shed additional insight into a numerical method and stabilizes some numerical algorithms. We comment on three applications: preconditioning, derivation of high-order stable Runge–Kutta schemes, and Gaussian quadrature.

2.6.1 *Preconditioning*

One of the main ideas in preconditioning a linear equation $Ax = b$ is to transform the original system into an equivalent system that is easier (quicker) to solve with an iterative scheme. The preconditioning of a matrix A can be thought of as implicitly multiplying A by M^{-1} where M is a matrix for which hopefully $Mz = y$ can easily be solved, $M^{-1}A$ is not too far from normal, and $\sigma(M^{-1}A)$ is clustered. This final hope, that the eigenvalues of a preconditioned system should be clustered, is a loose MIEP criterion. Although, in the context of preconditioning, the locations of eigenvalues need not be exactly specified, the notion of MIEP can certainly help us to see what is to be expected of the ideal preconditioner. Many types of unstructured preconditioners have been proposed, including the low-order (coarse-grid) approximation, SOR, incomplete LU factorization, polynomial, and so on. It would be interesting to develop another category of preconditioners where the M is required to possess a certain structure (Forsythe and Straus, 1955; Greenbaum and Rodrigue, 1989).

Problem 2.10 *(Structured preconditioner)*
Given a square matrix A, find a matrix M so that M^{-1} has a certain structure and that $\sigma(M^{-1}A)$ is clustered, or that the condition number of $M^{-1}A$ is minimized among all matrices of the same structure.

Another related problem that has potential application to optimization is to regulate the singular value by a rank-one modification. This is a special case of the inverse eigenvalue problem that we shall discuss in Section 4.8. We characterize the problem as follows.

Problem 2.11 *(Rank-one ISVP)*
Given a matrix $C \in \mathbb{R}^{m \times n}$ and a constant vector $\mathbf{b} \in \mathbb{R}^m$, find a vector $\mathbf{x} \in \mathbb{R}^n$ such that the rank-one updated matrix $\mathbf{b}\mathbf{x}^\top + C$ has a prescribed set of singular values.

2.6.2 *Numerical ODEs*

Recall that the Butcher array

$$\begin{array}{c|cccc} c_1 & a_{11} & a_{12} & \dots & a_{1s} \\ c_2 & a_{21} & a_{22} & \dots & a_{2s} \\ \vdots & \vdots & & & \vdots \\ c_s & a_{s1} & a_{s2} & \dots & a_{ss} \\ \hline & b_1 & b_2 & \dots & b_s \end{array}$$

uniquely determines an s-stage Runge–Kutta method for solving ordinary differential systems. Let $A = [a_{ij}]$, $\mathbf{b} = [b_1, \dots, b_s]^\top$ and $\mathbf{1} = [1, \dots, 1]^\top$. It is well established that the stability function for an s-state Runge–Kutta method is given by

$$R(z) = 1 + z\mathbf{b}^\top (I - zA)^{-1}\mathbf{1}. \tag{2.22}$$

See, for example, the book by Lambert (1991). To attain numerical stability, implicit methods are preferred. However, fully implicit methods are too expensive. Diagonally implicit methods (DIRK) where A is low triangular with *identical* diagonal entries are computationally more efficient, but difficult to construct. As an alternative, it is desirable to develop singly implicit methods (SIRK) in which the matrix A need not be lower triangular but must have an s-fold eigenvalue. Such a consideration can be approached by an IEP formulation with prescribed entries, as is done by Müller (1992).

Problem 2.12 *(PEIEP arising from Runge–Kutta method)*
Given the number s of stages and the desired order p of the method, define $k = \lfloor (p-1)/2 \rfloor$ and constants $\xi_j = \frac{1}{2}(4j^2-1)^{-1/2}$, $j = 1, \ldots, k$. Find a real number λ and $Q \in \mathbb{R}^{(s-k)\times(s-k)}$ such that $Q + Q^\top$ is positive semidefinite and $\sigma(X) = \{\lambda\}$, where $X \in \mathbb{R}^{s\times s}$ is of the form:

$$X = \left[\begin{array}{cccc|c} 1/2 & -\xi_1 & & & \\ \xi_1 & 0 & & & \\ 0 & & & & \\ \vdots & & \ddots & & \\ & & & 0 & -\xi_k \\ \hline 0 & & & \xi_k & Q \end{array}\right],$$

and $q_{11} = 0$ if p is even.

Note that only the submatrix Q at the lower-right corner needs to be completed for X. This is a special case of the IEP with prescribed entries (PEIEP) to be discussed in Section 4.7, except that in the above formulation the value of the s-fold eigenvalue λ is part of the unknowns to be determined.

2.6.3 *Quadrature rules*

Given a weight function $\omega(x) \geq 0$ on $[a, b]$, an n-point Gaussian quadrature rule for the integral

$$\mathcal{I}(f) = \int_a^b \omega(x) f(x)\, dx \tag{2.23}$$

is a formula of the form

$$\mathcal{G}_n(f) = \sum_{i=1}^{n} w_i f(\lambda_i) \tag{2.24}$$

with selected abscissas $\{\lambda_k\}_{k=1}^n$ and weights $\{w_1, \ldots, w_n\}$ so that

$$\mathcal{G}_n(f) = \mathcal{I}(f) \tag{2.25}$$

for all polynomials $f(x)$ of degree no higher than $2n-1$. With respect to the given $\omega(x)$, a sequence of orthonormal polynomials $\{p_k(x)\}_{k=0}^\infty$ satisfying

$$\int_a^b \omega(x) p_i(x) p_j(x)\, dx = \delta_{ij} \tag{2.26}$$

can be defined. Recall that orthogonal polynomials play a crucial role in the development of Gaussian quadrature rules. It is an established fact that roots of each $p_k(x)$ are simple, distinct, and lie in the interval $[a, b]$. Indeed, in order that the resulting quadrature should achieve the highest degree of precision $2n-1$, these roots $\{\lambda_i\}_{i=1}^n$ of a fixed $p_n(x)$ should be precisely those Gaussian abscissas.

On the other hand, it is also known that, with $p_0(x) \equiv 1$ and $p_{-1}(x) \equiv 0$, orthogonal polynomials always satisfy a three-term recurrence relationship:

$$p_n(x) = (a_n x + b_n)p_{n-1}(x) - c_n p_{n-2}(x). \tag{2.27}$$

Let $\mathbf{p}(x) = [p_0(x), p_1(x), \ldots, p_{n-1}(x)]^{\mathsf{T}}$. This relationship can be written in matrix form as

$$x\mathbf{p}(x) = \underbrace{\begin{bmatrix} \frac{-b_1}{a_1} & \frac{1}{a_1} & 0 & & 0 \\ \frac{c_2}{a_2} & \frac{-b_2}{a_2} & \frac{1}{a_2} & & \\ 0 & & & & \\ \vdots & & & \ddots & \vdots \\ 0 & & & & \frac{1}{a_{n-1}} \\ 0 & & \cdots & \frac{c_n}{a_n} & \frac{-b_n}{a_n} \end{bmatrix}}_{\mathsf{T}} \mathbf{p}(x) + \begin{bmatrix} 0 \\ 0 \\ \\ \vdots \\ 0 \\ \frac{1}{a_n} p_n(x), \end{bmatrix}, \tag{2.28}$$

where T is a tridiagonal matrix. Observe that $p_n(\lambda_j) = 0$ if and only if

$$\lambda_i \mathbf{p}(\lambda_i) = \mathsf{T}\mathbf{p}(\lambda_i),$$

that is, roots of $p_n(x)$ are precisely eigenvalues of T. It is known that the matrix T can be symmetrized by diagonal similarity transformation into a Jacobi matrix J and that the weight w_j in the quadrature is given by

$$w_i = q_{1i}^2, \quad i = 1, \ldots, n,$$

where $\mathbf{q}_i$ is the i-th normalized eigenvector of J. This gives rise to an interesting inverse problem: Given a quadrature with abscissas $\{\lambda_k\}_{k=1}^n$ and weights $\{w_1, \ldots, w_n\}$ satisfying $\sum_{i=1}^n w_i = 1$, determine the corresponding orthogonal polynomials and the corresponding weight function $\omega(x)$. This has been a classical problem, as has been discussed in Ferguson (1980), Kautsky and Elhay (1984).

We illustrate one interesting application to the derivation of the Gauss–Kronrod quadrature rule. Given a Gaussian quadrature (2.24), the associated Gauss–Kronrod quadrature is a $(2n+1)$-point integral rule

$$\mathcal{K}_{2n+1}(f) = \sum_{i=1}^{n} \tilde{w}_i f(\lambda_i) + \sum_{j=1}^{n+1} \hat{w}_j f(\hat{\lambda}_j) \tag{2.29}$$

that is exact for all polynomials of degree at most $3n+1$. Note that the original abscissas $\{\lambda_k\}_{k=1}^n$ form a subset of the new abscissas in $\mathcal{K}_{2n+1}$. It has been argued by Laurie (1997) that the existence of a Gauss–Kronrod quadrature rule with real distinct abscissas and positive weights is equivalent to the existence of a real solution to the following special IEP with prescribed entries.

Problem 2.13 *(PEIEP arising from Gauss–Kronrod quadrature rule)* Determine an $n \times n$ symmetric tridiagonal matrix with prescribed first $n-1$ entries (counting row-wise in the upper tridiagonal portion) and prescribed eigenvalues $\{\lambda_k\}_{k=1}^n$.

Computational details of this special PEIEP can be found in the paper by Calvetti et al. (2000).

2.7 Signal and data processing

Finally, we note that the problem of low rank approximation also belongs to the realm of IEP. The reason is that in a low rank matrix approximation a section of the spectrum has been preset to zero. This is a kind of IEP where only a portion of the eigenvalue information is known. The desirable rank depends on the underlying physical specifics which we now exemplify below. More details will be discussed in Chapter 8.

2.7.1 *Signal processing*

Low rank approximation is a common tool used for noise removal in signal processing or image enhancement processing. Research activities in this area have been vigorous. Many results can be found, particularly in the signal processing literature. We mention (de Beer, 1995; Burg et al., 1982; Cadzow, 1988; Cybenko, 1982; Li et al., 1999b; Shaw and Kumaresan, 1988; Schulz, 1997; Williams and Johnson, 1993) as a few starting points for further study. These are not necessarily classical references, but should provide sufficient background information. Because research on this topic is extensive and literature is abundantly available, we shall not discuss any specific application to save space. Generally speaking, the underling covariance matrix often has Toeplitz or block Toeplitz structure (Li et al., 1999a; Williams and Johnson, 1993). The rank to be removed corresponds to the noise level where the signal to noise ratio (SNR) is low (Tufts and Shah, 1993).

Likewise, low rank approximation can be used for model reduction problems in speech encoding and filter design (De Moor, 1994; Tufts and Shah, 1993; Scharf and Tufts, 1987). The underlying matrix often is of Hankel structure (Dendrinos et al., 1991; Li, 1997a; Park et al., 1999; Shaw and Kumaresan, 1988). The rank to be restored corresponds to the number of sinusoidal components contained in the original signal.

2.7.2 *Computer algebra*

Structured low rank approximation also finds applications to computer algebra. One fundamental question in algebra is to compute the greatest common divisor

(GCD) of polynomials. To compute the exact GCD for a system of polynomials, it suffices to know how to compute the GCD for two given polynomials. Suppose

$$p(x) = a_n \prod_{i=1}^{n} (x - \alpha_i),$$

$$q(x) = b_m \prod_{j=1}^{m} (x - \beta_j).$$

It is known that the resultant of $p(x)$ and $q(x)$ with respect to x,

$$a_n^m b_m^n \prod_{i=1}^{n} \prod_{m=1}^{m} (\alpha_i - \beta_j),$$

is zero if and only if $p(x)$ and $q(x)$ have common divisors. Furthermore, since the resultant is the determinant of the Sylvester matrix $S(p(x), q(x))$ where

$$S(p(x), q(x)) = \begin{bmatrix} a_n & a_{n-1} & \dots & a_0 & 0 & \dots & 0 \\ 0 & a_n & \dots & & a_0 & & \\ \vdots & & \ddots & & & & \\ 0 & 0 & 0 & a_n & a_{n-1} & \dots & a_0 \\ b_m & b_{m-1} & \dots & & b_0 & \dots & 0 \\ \vdots & & \ddots & & & \ddots & \\ 0 & 0 & 0 & b_m & & \dots & b_0 \end{bmatrix}, \tag{2.30}$$

the rank deficiency of $S(p(x), q(x))$ is precisely the degree of the GCD Corless et al. (1995), Karmarkar and Lakshman (1998).

Suppose now the coefficients of the given polynomials are inexact. An interesting question is to compute a pair of polynomials with a non-trivial common divisor close to the given polynomials. In this way, the problem of approximating the GCD for polynomials can be formulated as a low rank approximation problem with Sylvester structure. A similar formulation can be set up for multivariate polynomials.

Problem 2.14 *(Low rank approximation of Sylvester structure)*
Given a matrix $S_0 \in \mathbb{R}^{(n+m)\times(n+m)}$ and a positive integer $k < \min\{n, m\}$, minimize the Frobenius norm $\|X - S_0\|_F$ among all Sylvester matrices $X \in \mathbb{R}^{(n+m)\times(n+m)}$ of rank k.

We stress again that in the above problem the spectral constraint imposed on X is the rank condition and the structural constraint is the prescribed zeros and the repeated rows as are indicated in (2.30), while the approximation problem itself is a least squares problem.

2.7.3 *Molecular structure modelling*

Low rank approximation can also be applied to the molecular structure modelling for the protein folding problem in $\mathbb{R}^3$. Assuming that the sequence of amino acid molecules are located at points $\mathbf{x}_1, \ldots, \mathbf{x}_n$ in $\mathbb{R}^3$, their relative positions then give rise to the Euclidean distance matrix $D = [d_{ij}] \in \mathbb{R}^{n\times n}$ where $d_{ij} = \|\mathbf{x}_i - \mathbf{x}_j\|_2^2$ for $i, j = 1, \ldots, n$. It can be proved that D is of rank no more than 5 (Gower, 1982). On the other hand, it is known that the three dimensional shape of a protein largely determines how the protein functions or acts. It is an incredibly important problem to determine the three dimensional structure of a protein.

Obviously, if we knew all the interatomic distances in the protein, the three dimensional structure would be relatively easy to generate. The trouble is that with current technologies, such as x-ray crystallography, we cannot "see" well the entries in the matrix D. Thus we have an observed matrix D_0 that is not quite a distance matrix. A symmetric and nonnegative matrix of rank 5 is a necessary condition for the approximation (Glunt et al., 1990).

> **Problem 2.15** *(Nearest Euclidean distance matrix approximation)*
> Given a matrix $D_0 \in \mathbb{R}^{n\times n}$, minimize the Frobenious norm $\|D - D_0\|_F$ among all Euclidean distance matrices.

2.7.4 *Principal component analysis, data mining and others*

Principal component analysis has long been used as a mathematical procedure that transforms a larger number of possibly correlated variables into a smaller number of uncorrelated principal components. The objectives are to detect structure in the relationships between variables, to reduce the dimensionality of the data set and to identify new meaningful underlying variables. This technique is ubiquitous for data analysis. Examples of its many applications include data compression, image processing, visualization, exploratory data analysis, pattern recognition and time series prediction.

Often it is the case that the components are ranked or discovered according to their significance measured by some spectral information. The first principal component accounts for as much of the variability in the data as possible, and each succeeding component accounts for as much of the remaining variability as possible. The number of components determines the rank of the approximation.

In a similar way, data mining analyzes relationships and patterns in stored transaction data based on open-ended user queries. Generally, data mining seeks relationships such as classes, clusters, associations, or sequential patterns. Latent semantic indexing (LSI), for example, looks at *similarity* values between queries and a pool of keywords. The indexing matrix records which keywords a document contains. The method examines the document collection as a whole to see which other documents contain some of those same words. LSI considers documents that have many words in common to be semantically close, and ones with few words in common to be semantically distant. Its task of extracting, transforming, and loading from a multidimensional database system would require an enormous amount of computing power. Additionally, not all words are created equal. Consequently, it is often desirable to further refine the indexing matrix by a low rank approximation whose principal terms are supposed to capture the inherent nature of the original indexing matrix (Berry et al., 1999; Zha and Simon, 1999).

Also, there are some discussions on using structure preserving rank reduction computation as a regularization tool in the solution of certain ill-posed inverse problems (Hansen, 1987, 1990). All of these can be considered as structured IEPs with partial spectrum identically zero.

2.8 Summary

We have illustrated that IEPs can arise from a variety of important applications. We have demonstrated also that different applications lead to different IEP formulations. Each of these IEP formulations carries some unique traits of characteristics. We will see more applications in the sequel as we move across different chapters of this book. It would be a challenge to address the four fundamental issues defined in Chapter 1 to each of these inverse problems. We shall begin to cast these problems in a slightly more general framework in the next few chapters, treating some of their mathematical properties and a variety of numerical techniques.

We conclude this introduction with one additional remark by Gladwell (1996) who suggested that, for application purposes, there should also be a distinction between *determination* and *estimation* in the nature of an inverse problem. When the given data is exact and complete so that the system can be precisely determined, the IEP is said to be *essentially mathematical.* In contrast, we say that we have an *essentially engineering* IEP when the data is only approximate and often incomplete, in which only an estimate of the parameters of the system is sought and the resulting behavior is expected to agree only approximately with the prescribed data.

3

PARAMETERIZED INVERSE EIGENVALUE PROBLEMS

3.1 Overview

Starting with this chapter, we shall begin to examine the various IEPs in more detail. Because of the vast diversity of problems, theories, algorithms, and open questions, it is very difficult to maintain any uniformity in the presentation. Following the diagram in Figure 1.1, it seems appropriate to begin the discussion with the class of parameterized inverse eigenvalue problems (**PIEP**), that includes many other types of IEPs as special cases. Indeed, by considering unknown entries in the matrix to be constructed as "parameters", one can argue that an inverse eigenvalue problem is generally a parameter estimation problem. This view of not taking the matrix structure into account but casting the IEP as a nonlinear algebraic system is perhaps too general to be effective. For the very same reason that we normally do not approach a direct eigenvalue problem as a problem solving nonlinear algebraic equations, we want to consider PIEPs in the nature of matrices. While a catholic theory or method, if any, for a general nonlinear algebraic system could be applied to the inverse problem, we expect that more could be done for a structured problem. Thus, by a "parameterized" IEP, we intend to accentuate the meticulous ways that these parameters regulate the problem. In other words, the structural constraint is regulated by a set of parameters.

The way the parameters give particulars of a structure can vary. Currently, the progress of study seems to mainly center around linear problems. That is, the problem is demarcated by its linear dependence on the parameters. This subclass of problems embraces many of the most extensively studied IEPs, including the additive inverse eigenvalue problem and the multiplicative inverse eigenvalue problem. If we can develop any results on the four fundamental issues for the PIEPs in general, those results would be readily applicable to the special cases. To set that stage, we shall address those four issues for PIEPs more carefully than for other problems and we still need to accentuate that not many theories are complete.

To facilitate the discussion, we shall adopt the following notation hereinafter:

- Boldfaced **F** represents the scalar field of either real $\mathbb{R}$ or complex $\mathbb{C}$.
- Roman uppercase letters $A, B, \ldots$ denote matrices.
- Boldfaced lowercase letters $\mathbf{b}^{(\nu)}, \mathbf{q}_i^{(\nu)}, \mathbf{v}_i, \ldots, \boldsymbol{\alpha}$, or $\boldsymbol{\lambda}$ denote vectors.

- Unless specified otherwise, $\|\cdot\|$ denotes either the 2-norm of a vector or the Frobenius norm of a matrix.
- Calligraphic uppercase letters $\mathcal{M}, \mathcal{N}, \ldots$ denote certain subsets of square matrices of which the size is clear from the context. In particular, we reserve the following notation for this chapter and more to be defined in the sequel:
  - $\mathcal{S}(n) := \{\text{All symmetric matrices in } \mathbb{R}^{n\times n}\}$,
  - $\mathcal{O}(n) := \{\text{All orthogonal matrices in } \mathbb{R}^{n\times n}\}$,
  - $\mathcal{D}_{\mathcal{R}}(n) := \{\text{All diagonal matrices in } \mathbb{R}^{n\times n}\}$,
  - $\mathcal{H}(n) := \{\text{All Hermitian matrices in } \mathbb{C}^{n\times n}\}$,
  - $\mathcal{D}_{\mathcal{C}}(n) := \{\text{All diagonal matrices in } \mathbb{C}^{n\times n}\}$.

3.1.1 *Generic form*

We first give the description of a generic PIEP from which many variations will germinate. In view of the various forms of IEPs arising from applications illustrated in Chapter 2, we have to point out immediately that the definition of PIEP below, though referred to as generic, is in fact quite restrictive. One trait of its predisposition is that only eigenvalues are used in the spectral constraint. It would be an entirely different class of consideration if eigenvectors are taken into account in the spectral constraint and we do not think this type of PIEP with partially described eigenvalues and eigenvectors (see the PDIEP in Chapter 5) has been studied yet in the literature.

Problem 3.1 *(Generic PIEP)*
Given a family of matrices $A(\mathbf{c}) \in \mathcal{M}$ with parameters $\mathbf{c} \in \mathbb{F}^m$ and a set of scalars $\Omega \subset \mathbb{F}$, find values of the parameter $\mathbf{c}$ such that

$$\sigma(A(\mathbf{c})) \subset \Omega.$$

In the above description, the subset $\mathcal{M}$ is used only to impose some basic structure, such as real or complex, symmetric or Hermitian, of the underlying matrices. Such a distinction is necessary because, in certain cases, we want to discern whether a problem is real solvable, symmetric solvable, or complex solvable. Within this basic structure $\mathcal{M}$, it is the characteristics of how the matrix A depends on the parameters $\mathbf{c}$ that delimits the structure constraint. We stress again that the desired structure is embodied in $A(\mathbf{c})$, although not necessarily that structure is immediately apparent and noticeable.

Note also in the generic form that the degree m of free parameters need not be the same as the number of unknown entries in the matrix, leaving the possibility that the problem could be over- or under-determined. For an over-determined PIEP, an alternative formulation is to seek least squares solutions. We shall discuss this approach in Chapter 6. For an under-determined PIEP, there will be generally multiple solutions. A desirable second stage of computation then would be to seek a robust solution which is least sensitive to perturbation. For most PIEPs we are about to examine below, we have not seen much discussion, either theoretically or numerically, on finding a robust solution to a PIEP. This might very well be an open area for further research.

The spectral constraint is specified by inclusion in the subset Ω. Obviously, the relationship of inclusion covers the relationship of equality. Some commonly used spectral constraint Ω include:

- $\Omega = \{\lambda_k\}_{k=1}^n$, where n is exactly the size of the matrix. This is the situation where the complete spectrum is specified.
- $\Omega = \{\lambda_k\}_{k=1}^n$, where n is larger than the size of the matrix. This amounts to the situation where the spectrum of $A(\mathbf{c})$ is limited to but is relatively free within a certain finite set of numbers.
- Ω is the left-half complex plane. This corresponds to the situation where only the stability (of a certain dynamical system) is of concern.

Be aware that the reversed inclusion relationship, that is,

$$\sigma(A(\mathbf{c})) \supset \Omega$$

constitutes another interesting class of IEPs in which only a partial spectrum of $A(\mathbf{c})$ is specified. We shall discuss this type of IEP in Chapter 5.

Depending upon how $A(\mathbf{c})$ is specifically defined, the PIEP can appear in very different forms. We exemplify several classical variations in the next section.

3.1.2 *Variations*

One of the most important types of PIEP is when the family of matrices $A(\mathbf{c})$ depends linearly (or, to be more precise, affinely) on the parameters $\mathbf{c}$. That is, given a sequence of square matrices $\{A_i\}_{i=0}^m \subset \mathcal{M}$,

$$A(\mathbf{c}) := A_0 + c_1 A_1 + \cdots + c_m A_m, \tag{3.1}$$

if $\mathbf{c} = [c_1, \ldots, c_m]^\mathsf{T} \in \mathbb{F}^m$. An IEP being parameterized in this way will be called a linear PIEP (**LiPIEP**).

Problem 3.2 *(LiPIEP)*
Given $A(\mathbf{c})$ as in (3.1) and scalars $\{\lambda_k\}_{k=1}^n$, find values of $\mathbf{c} := [c_1, \ldots, c_m]^\mathsf{T} \in \mathbb{F}^m$ such that

$$\sigma(A(\mathbf{c})) \subset \{\lambda_k\}_{k=1}^n.$$

Note that the basis matrices $\{A_i\}_{i=0}^m$ need not have any inherent structure in themselves. They are just the constituent elements that make up $A(\mathbf{c})$. The so-called structure in $A(\mathbf{c})$ is simply referring to the aggregated effect these basis matrices can bring forth when combined linearly in the way of (3.1). That is, assuming that the basis matrices $\{A_i\}_{i=0}^m$ are linearly independent, then all matrices $A(\mathbf{c})$ lie on a specific m-dimensional affine subspace. The structural constraint restricts matrices to that subspace.

This particular model of LiPIEP has found applications in a variety of areas, including the educational testing problem in social sciences, the communality problem in factor analysis, and pole assignment problem in control. A more detailed exposition was given by Friedland et al. (1986). Several special subcases of LiPIEP deserve further attention, which we outline below.

Problem 3.3 *(LiPIEP1)*
Solve the LiPIEP where $\mathcal{M} = \mathbb{R}^{n\times n}$, $\mathbf{F} = \mathbb{R}$, and $m = n$.

Problem 3.4 *(LiPIEP2)*
Solve the LiPIEP where $\mathcal{M} = \mathcal{S}(n)$, $\mathbb{F} = \mathbb{R}$, and $m = n$.

Both LiPIEP1 and LiPIEP2 concern real-valued matrices and expect an exact match between $\sigma(A(\mathbf{c}))$ and $\{\lambda_k\}_{k=1}^n$. Clearly, for a PIEP to make sense in this case, the prescribed set $\{\lambda_k\}_{k=1}^n$ is necessarily closed under complex conjugation. The main difference between LiPIEP1 and LiPIEP2 is that in the latter the basic structure is further limited to symmetric matrices and as such the prescribed eigenvalues $\{\lambda_k\}_{k=1}^n$ are necessarily real. Recall that direct eigenvalue problems for symmetric matrices are well conditioned problems (Parlett, 1998). It would be interesting to know how sensitive the inverse problems are to perturbations. We shall explore this problem in Section 3.2.2.

Rewriting the basis matrices in a more complicated way, the following generalized PAP arising from descriptor systems is still a LiPIEP (Wilhelmi, 1974) with matrices $\{K_i\}_{i=1}^q$ as its parameters.

Problem 3.5 *(LiPIEP3)*
Given matrices $A \in \mathbb{C}^{n\times n}$, $B_i \in \mathbb{C}^{n\times m_i}$, $C_i \in \mathbb{C}^{l_i\times n}$, $i = 1,\dots,q$, and scalars $\{\lambda_k\}_{k=1}^n \subset \mathbb{C}$, find matrices $K_i \in \mathbb{C}^{m_i\times l_i}$ such that

$$\sigma(A + \sum_{i=1}^{q} B_i K_i C_i) = \{\lambda_k\}_{k=1}^n.$$

The LiPIEP3 includes both state feedback and output feedback PAPs as special cases when $q = 1$. It should also be noted that the parameters $\{K_i\}_{i=1}^q$ in LiPIEP3 are arranged in matrix form and that the number of parameters $\sum_{i=1}^q m_i l_i$ could easily be more than the dimension n of the problem. The PAPs have been extensively studied via a number of means. Techniques employed range from linear system theory, combinatorics, and functional analysis to algebraic geometry. It is interesting to note that the theories and numerical methods developed for the PIEP seem to offer yet another perspective on the PAPs. We choose not to treat PAPs in this book because its studies are well documented in many other resources. See, for example, Byers (1988), Byrnes (1989), Kautsky et al. (1985).

One special PIEP that has attracted long-standing interest due to its ubiquitous and important applications in various field is the additive inverse eigenvalue problem (**AIEP**). The AIEP can be interpreted as modifying a given matrix A_0 by the addition of a certain type of matrix so as to maintain a specified spectrum. Likewise, an AIEP can also be considered as a matrix completion problem where some missing data are to be filled in. By being more specific on the type of candidates that can be added on, the AIEP traditionally has been studied under the context of its individual subclasses. We shall consider some of them in Sections 3.3 and 4.7. For the time being, a generic AIEP is formulated as follows:

Problem 3.6 *(Generic AIEP)*
Given a square matrix A_0 of size n, a special class of matrices $\mathcal{N}$ of size n, and a set of scalars $\{\lambda_k\}_{k=1}^n$, find $X \in \mathcal{N}$ such that

$$\sigma(A_0 + X) = \{\lambda_k\}_{k=1}^n.$$

In contrast to the AIEP, a multiplicative inverse eigenvalue problem (**MIEP**) stands out when the task is to *pre-multiply* a given matrix A_0 by a specially structured matrix X to reposition or to precondition the distribution of its eigenvalues.

Problem 3.7 *(Generic MIEP)*
Given a square matrix A_0 of size n, a special class of matrices $\mathcal{N}$ of size n, and a set of scalars $\{\lambda_k\}_{k=1}^n$, find $X \in \mathcal{N}$ such that

$$\sigma(XA_0) = \{\lambda_k\}_{k=1}^n.$$

We have already seen in Chapter 2 several situations from which an MIEP can arise. Applications of the MIEP include the preconditioning process, the inverse generalized eigenvalue problem, and the loading of beads on a string. Note that an MIEP is still an LiIEP because the product XA_0 can still be expressed as a linear combination of some basis matrices A_i, $i = 1, \ldots, m$. Suppose, for example, that $X = \text{diag}\{x_1, \ldots, x_n\}$ is a diagonal matrix. Then we may write

$$XA_0 = \sum_{k=1}^{n} x_k \underbrace{(\mathbf{e}_k \mathbf{a}_k^\top)}_{A_k}, \tag{3.2}$$

where $\mathbf{e}_k$ is the standard k-th unit column vector and $\mathbf{a}_k^\top$ is simply the k-th row of A_0. Any theory we can develop for a general PIEP therefore should be applicable to an MIEP. More about this will be discussed in Section 3.4.

3.2 General results for linear PIEP

We now begin to present some general results for PIEPs. We shall touch upon issues of existence, sensitivity, and computation in the next few sections. It would be nice to be able to address all fundamental issues concerning the PIEP by some major theorems. Such an attempt, however, probably will never succeed because the scope of a generic PIEP is simply too general. Searching through the literatures reveals that pieces of understanding of its individual variations are scattered around. The information is so diverse and massive that we find it extremely difficult to perceive any general conformity. It appears, though, that a good collection of problems exhibits linearity in parameters. Since the model LiPIEP contains a broad range of traditional IEPs studied separately in the literature, our concentration on the LiPIEP should serve as a reasonable framework that unifies various work that has been scattered across the field.

Obviously, whatever is known about the LiPIEP applies to both AIEP and MIEP as well, but the converse is not true. Some of the results known specifically for the AIEP or MIEP cannot be generalized to the larger class of LiPIEP. For that reason, we need to examine the two problems AIEP and MIEP separately.

Be warned that we choose not to provide a proof for every theorem mentioned in this book. Such a task often would require much additional background material to prepare for the proofs. To avoid distracting readers from the main course of this presentation, we have decided to sacrifice a great many details and to highlight only some essential concepts. Of course, pertaining references will be given whenever it is possible.

3.2.1 *Existence theory*

Before we present the existence theory, we hasten to point out that finding a solution over the real field $\mathbb{R}$ is more complicated and difficult than over the complex field $\mathbb{C}$. Part of the reason is that eigenvalues of a matrix are related to the coefficients of its corresponding characteristic polynomial. Since $A(\mathbf{c})$ is linear in $\mathbf{c}$, the parameters $\mathbf{c}$ are related to the prescribed eigenvalues of $A(\mathbf{c})$ via a system of polynomials. As such, the solvability of an LiPIEP is generally expected through solving a polynomial system over an algebraically closed field $\mathbb{C}$. In practice, we usually prefer to see real solvability, which is precisely the core of difficulty for a PIEP. We shall see subsequently that the presence of multiple eigenvalues in the real case can also make a difference.

Complex solvability Recall that the issue of solvability concerns developing necessary or sufficient conditions under which an LiIEP has a solution. If there are more undetermined parameters than the number of (independent) equations that the spectral constraint can impose, it is conceivable that the problem is solvable due to the extra degrees of freedom. On the other hand, an over-determined system requires some additional consistency conditions in order that a solution exists. To avoid the possible complication due to over- or under-determination, we shall consider the case where the number of parameters is exactly the same as the dimension of the underlying matrices, that is, we shall consider only the case where

$$A(\mathbf{c}) := A_0 + c_1 A_1 + \cdots + c_n A_n, \tag{3.3}$$

and each A_i is of size $n \times n$. Using the notion of "almost everywhere" to indicate that its converse forms a set of measure zero in the ambient space, we set forth our first major theorem as follows.

Theorem 3.1. (Xu, 1998) Given a set of n complex numbers $\{\lambda_k\}_{k=1}^n$, then for almost all $A_i \in \mathbb{C}^{n\times n}$, $i = 0, 1, \ldots, n$, there exists $\mathbf{c} \in \mathbb{C}^n$ such that $\sigma(A(\mathbf{c})) = \{\lambda_k\}_{k=1}^n$. Furthermore, there are at most $n!$ distinct solutions.

The basic idea behind Theorem 3.1 is the well-known Bézout theorem in classical algebraic geometry. The set of simultaneous solutions of a polynomial system is called a *algebraic variety*. The Bézout theorem offers a count of the intersection points with appropriate multiplicity. In our case, upon expanding the characteristic polynomial of $A(\mathbf{c})$ and comparing its coefficients with the basic symmetric functions formed from $\{\lambda_k\}_{k=1}^n$, we obtain a polynomial system in $\mathbf{c}$ where the Bézout theorem can be applied. Despite its generality, Theorem 3.1 is not practical in two aspects: that it involves only complex solvability and that it is not constructive.

It might be interesting to compare Theorem 3.1 with the following result by Helton et al. (1997, Theorem 2.4).

Theorem 3.2. (Helton et al., 1997) For almost all $A_0 \in \mathbb{C}^{n\times n}$ and for almost all $\{\lambda_k\}_{k=1}^n$, there is a $\mathbf{c} \in \mathbb{C}^n$ such that $\sigma(A(\mathbf{c})) = \{\lambda_k\}_{k=1}^n$ if and only if the following two conditions hold:

(i) The matrices $A_1, \ldots, A_n$ are linearly independent.
(ii) At least one of the matrices $A_1, \ldots, A_n$ has nonzero trace.

Real solvability The issue of real solvability is not as straightforward as that of complex solvability. To describe a sufficient condition, it is conventional to normalize the diagonals of the basis matrices first. This normalization can be done as follows. Let the entries of each matrix A_k, $k = 1, \dots, n$, be denoted by

$$A_k := \left[a_{ij}^{(k)}\right]_{i,j=1}^{n}, \quad \left(\text{or simply } A_k = [a_{ij}^{(k)}] \text{ for generic } i, j\right).$$

Record the diagonal of A_k as the k-th row of E, that is,

$$E := \left[a_{ii}^{(k)}\right]_{i,k=1}^{n}.$$

Assuming that the inverse matrix $E^{-1} = [\ell_{ij}]$ exists, define new parameters

$$\tilde{\mathbf{c}} := E\mathbf{c} = [\tilde{c}_1, \dots, \tilde{c}_n]^\top.$$

Then we may rewrite

$$\begin{aligned} A(\mathbf{c}) &= A_0 + \sum_{k=1}^{n} c_k A_k = A_0 + \sum_{k=1}^{n} \left(\sum_{j=1}^{n} \ell_{kj} \tilde{c}_j \right) A_k \\ &= A_0 + \sum_{j=1}^{n} \tilde{c}_j \underbrace{\left(\sum_{k=1}^{n} \ell_{kj} A_k \right)}_{\tilde{A}_j}. \end{aligned}$$

It is easy to check that the diagonals of the new basis matrices $\tilde{A}_k$ satisfy the normalized condition $\operatorname{diag}(\tilde{A}_k) = \mathbf{e}_k$ for $k = 1, \dots, n$. We shall use the notation "diag" in the following convenient way: when $\mathbf{c}$ is a vector, $\operatorname{diag}(\mathbf{c})$ indicates a square matrix with elements of $\mathbf{c}$ on the main diagonal; when M is a matrix, $\operatorname{diag}(M)$ indicates the column vector formed from the main diagonal of M. Therefore, $\operatorname{diag}(\operatorname{diag}(M))$ is the diagonal matrix of M. Recall also that, while keeping elements in their original order, we sometimes use the same set notation $\boldsymbol{\lambda} = \{\lambda_k\}_{k=1}^{n}$ to represent a column vector $\boldsymbol{\lambda} = [\lambda_1, \dots, \lambda_n]^\top$. To facilitate the description, we introduce some additional notation. Let

$$\begin{aligned} |M| &:= [|m_{ij}|]_{i,j=1}^{n}, \\ \pi(M) &:= \|M - \operatorname{diag}(\operatorname{diag}(M))\|_\infty, \qquad (3.4) \\ S &:= \sum_{k=1}^{n} |A_k|, \\ d(\boldsymbol{\lambda}) &:= \min_{i \neq j} |\lambda_i - \lambda_j|. \qquad (3.5) \end{aligned}$$

The following theorem gives rise to one of the sufficient conditions for real solvability.

Theorem 3.3. (Biegler-König, 1981b) Given n real numbers $\boldsymbol{\lambda} = \{\lambda_k\}_{k=1}^n$, and $n+1$ matrices $A_i \in \mathbb{R}^{n\times n}$, $i = 0, 1, \ldots, n$, assume that

(i) $\operatorname{diag}(A_k) = \mathbf{e}_k$, $k = 1, \ldots, n$,
(ii) $\pi(S) < 1$,
(iii) the gap $d(\boldsymbol{\lambda})$ is sufficiently large, that is,

$$d(\boldsymbol{\lambda}) \geq 4\frac{\pi(S)\|\boldsymbol{\lambda} - \operatorname{diag}(A_0)\|_\infty + \pi(A_0)}{1 - \pi(S)}.$$

Then the LiPIEP (with $m = n$) has a real solution $\mathbf{c} \in \mathbb{R}^n$.

Proof The original proof is quite insightful, so we outline the proof below. For convenience, denote

$$\epsilon := \frac{\pi(S)\|\boldsymbol{\lambda} - \operatorname{diag}(A_0)\|_\infty + \pi(A_0)}{1 - \pi(S)}.$$

The proof actually shows that the LiPIEP has a solution inside the ball

$$\mathcal{B}_\epsilon := \{\mathbf{c} \in \mathbb{R}^n \mid \|\mathbf{c} - (\boldsymbol{\lambda} - \operatorname{diag}(A_0))\|_\infty \leq \epsilon\}. \tag{3.6}$$

To achieve that goal, we divide the proof into two parts. We first argue that for every $\mathbf{c} \in \mathcal{B}_\epsilon$, the spectrum of $A(\mathbf{c})$ is real. We then show that $T(\mathcal{B}_\epsilon) \subset \mathcal{B}_\epsilon$ where T is the continuous mapping defined by

$$T(\mathbf{c}) = \boldsymbol{\lambda} + \mathbf{c} - \sigma(A(\mathbf{c})). \tag{3.7}$$

Observe that

$$\begin{aligned}
\epsilon &= \frac{1}{2}\left(\pi(S)\|\boldsymbol{\lambda} - \operatorname{diag}(A_0)\|_\infty + \pi(A_0)\right)\left[1 + \frac{1+\pi(S)}{1-\pi(S)}\right] \\
&= \frac{1}{2}\left(\left(\pi(S)\|\boldsymbol{\lambda} - \operatorname{diag}(A_0)\|_\infty + \pi(A_0)\right) + \epsilon(1+\pi(S))\right) \\
&= \pi(S)\left(\|\boldsymbol{\lambda} - \operatorname{diag}(A_0)\|_\infty + \epsilon\right) + \pi(A_0).
\end{aligned}$$

Thus for every $\mathbf{c} \in \mathcal{B}_\epsilon$, we have that

$$d(\boldsymbol{\lambda}) - 2\epsilon \geq 2\left(\pi(S)\|\mathbf{c}\|_\infty + \pi(A_0)\right) \geq 2\pi(A(\mathbf{c})). \tag{3.8}$$

The Gershgorin disks of $A(\mathbf{c})$ are given by

$$G_i := \left\{\mathbf{z} \in \mathbb{C} \mid |\mathbf{z} - (a_{ii}^{(0)} + c_i)| \leq r_i\right\},$$

with radii

$$r_i := \sum_{\substack{j=1\\ j\neq i}}^{n} |a_{ij}^{(0)} + \sum_{k=1}^{n} c_k a_{ij}^{(k)}| \leq \pi(A(\mathbf{c})), \tag{3.9}$$

for $i = 1, \ldots, n$, respectively. The minimal distance between the centers of these disks is measured by $d(\text{diag}(A_0) + \mathbf{c})$. By (3.6), the minimal distance satisfies

$$d(\text{diag}(A_0) + \mathbf{c}) \geq d(\boldsymbol{\lambda}) - 2\epsilon, \tag{3.10}$$

if $\mathbf{c} \in \mathcal{B}_\epsilon$. Combining facts (3.6), (3.9), and (3.10), we conclude that these n Gershgorin disks are mutually disjoint. Since each disk contains exactly one eigenvalue of $A(\mathbf{c})$, we have proved that for $\mathbf{c} \in \mathcal{B}_\epsilon$, the spectrum of $A(\mathbf{c})$ must be distinct and real. Indeed,

$$\|\sigma(A(\mathbf{c})) - (\text{diag}(A_0 + \mathbf{c}))\| \leq \pi(A(\mathbf{c})) \leq \pi(S)\|\mathbf{c}\|_\infty + \pi(A_0).$$

To see that $T(\mathcal{B}_\epsilon) \subset \mathcal{B}_\epsilon$, observe that

$$\begin{aligned}\|T(\mathbf{c}) - (\boldsymbol{\lambda} - \text{diag}(A_0))\|_\infty &= \|(\mathbf{c} + \text{diag}(A_0)) - \sigma(A(\mathbf{c}))\|_\infty \\ &\leq \pi(S)\|\mathbf{c}\|_\infty + \pi(A_0) \\ &\leq \pi(S)\left(\|\boldsymbol{\lambda} - \text{diag}(A_0)\|_\infty + \epsilon\right) + \pi(A_0) \\ &= \epsilon.\end{aligned}$$

It then follows from the Brouwer fixed point theorem that there exists a vector $\tilde{\mathbf{c}} \in \mathcal{B}_\epsilon$ such that $T(\tilde{\mathbf{c}}) = \tilde{\mathbf{c}}$. That is, $\sigma(A(\tilde{\mathbf{c}})) = \boldsymbol{\lambda}$. □

There are quite a few other types of sufficient conditions for the real solvability of an LiPIEP. See, for example, the books by Zhou and Dai (1991) and by Xu (1998). A common feature, however, appears to be that eigenvalues in the prescribed spectrum must be separated in some way. In the above theorem, for instance, we see a quantification on how much the separation of eigenvalues $d(\boldsymbol{\lambda})$ is sufficiently large to guarantee the existence of a solution. These sufficient conditions do not shed light on what will happen when the separation is not sufficiently large. At the extreme, when eigenvalues coalesce into multiple eigenvalues, we get into another interesting phenomenon which we now discuss.

Multiple eigenvalue Consider the LiPIEP associated with basis matrices $A_i \in \mathbb{R}^{n \times n}$, $i = 0, 1, \ldots, m$, and k real eigenvalues $\boldsymbol{\lambda} = \{\lambda_1, \ldots, \lambda_k\}$ where each λ_i has multiplicity $r_i \geq 0$ so that $r_1 + \cdots + r_k = n$. Let $r = \max\{r_1, \ldots, r_k\}$ denote the maximal multiplicity.

Theorem 3.4. (Shapiro, 1983; Sun and Ye, 1986) The LiPIEP is unsolvable almost everywhere if

$$n - m + r(r-1) > 1. \tag{3.11}$$

In addition, in the case $n = m$, the LiPIEP is unsolvable almost everywhere if and only if $r > 1$.

The main tool used in the proof is the Sard theorem from differential geometry. The argument should center around showing that the entirety of basis matrices and eigenvalues

$$(A_0, A_1, \ldots, A_m, \boldsymbol{\lambda}) \in \underbrace{\mathbb{R}^{n\times n} \times \cdots \times \mathbb{R}^{n\times n}}_{m+1} \times \mathbb{R}^k$$

at which the LiPIEP as described is solvable forms a set of measure zero in the ambient space. We shall refer readers to the original papers for details.

There has been some argument about how a PIEP should be posed when multiple eigenvalues are necessarily to present in applications. Note that in the above, the entire spectrum (with multiplicity) is given. In the presence of multiple eigenvalues, one way to avoid hitting the wall of insolvability is to free up some other eigenvalues as was demonstrated in Friedland et al. (1986). We shall discuss some of the formulations in Sections 3.2.4 and 4.8.

We conclude this section with an interesting open question: what can be said about the real solvability of Problem 3.2 if $m > n$?

3.2.2 *Sensitivity analysis*

The issue of sensitivity concerns how the solution to an IEP is modified by changes in the problem data. In the context of LiPIEP, we want to know how a solution is modified when either the basis matrices A_i, $i = 0, 1, \ldots, m$, or the prescribed spectrum $\{\lambda_k\}_{k=1}^n$ are changed. We remark without being specific that generally an IEP is ill-posed, because usually the solution to an IEP is not unique. Even if we can argue that the solution of an LiPIEP, being the roots of a polynomial system, should depend continuously upon the problem data, its numerical solutions, especially those obtained by iterative methods, could differ significantly from small perturbations due to the ill-posedness.

Forward problem for general $A(\mathbf{c})$ The sensitivity analysis for the forward problem is quite straightforward. We can do so even for a general $A(\mathbf{c})$. Earlier groundwork in this regard (on a single parameter) is due to Lancaster (1964) and Kato (1966). Sun (1985) generalized the analysis to matrices depending on several parameters. We present only one major result. Similar results can be found in Dieci and Eirola (1999) and Zhou and Dai (1991) and in other contexts of application such as the analytic singular value decomposition (Bunse-Gerstner et al., 1991) which will be used in Section 4.5.

Theorem 3.5. Assume that $A(\mathbf{c}) \in \mathbb{C}^{n\times n}$ is analytic in $\mathbf{c} \in \mathbb{C}^m$ over a neighborhood of $\mathbf{0} \in \mathbb{C}^m$ and that λ_0 is a *simple* eigenvalue of $A(\mathbf{0})$. Let $\mathbf{x}_0$ and $\mathbf{y}_0$ denote the right and left eigenvector of $A(\mathbf{0})$ corresponding to λ_0, respectively, and are normalized to satisfy $\mathbf{y}_0^\top \mathbf{x}_0 = 1$. Then

(i) There exists an analytic function $\boldsymbol{\lambda}(\mathbf{c})$ in a neighborhood $\mathcal{N}$ of $\mathbf{0} \in \mathbb{C}^m$ such that $\boldsymbol{\lambda}(\mathbf{c})$ is a simple eigenvalue of $A(\mathbf{c})$ with $\boldsymbol{\lambda}(\mathbf{0}) = \lambda_0$.

(ii) There exist analytic functions $\mathbf{x}(\mathbf{c})$ and $\mathbf{y}(\mathbf{c})$ in $\mathcal{N}$ such that $\mathbf{x}(\mathbf{c})$ is a right eigenvector corresponding to $\boldsymbol{\lambda}(\mathbf{c})$ and $\mathbf{y}(\mathbf{c})$ is a left eigenvector corresponding to $\boldsymbol{\lambda}(\mathbf{c})$ with $\mathbf{x}(\mathbf{0}) = \mathbf{x}_0$, $\mathbf{y}(\mathbf{0}) = \mathbf{y}_0$, respectively.

Furthermore, the sensitivity of $\boldsymbol{\lambda}(\mathbf{c})$ can be measured by

$$\left(\frac{\partial \boldsymbol{\lambda}(\mathbf{c})}{\partial c_i}\right)_{\mathbf{c}=0} = \mathbf{y}_0^\top \left(\frac{\partial A(\mathbf{c})}{\partial c_i}\right)_{\mathbf{c}=0} \mathbf{x}_0. \tag{3.12}$$

Proof We shall prove only the existence of an analytic eigenvalue and right eigenvector. A similar proof works for the left eigenvector. Once the differentiability is established, (3.12) follows from directly differentiating both sides of $A(\mathbf{c})\mathbf{x}(\mathbf{c}) = \boldsymbol{\lambda}(\mathbf{c})\mathbf{x}(\mathbf{c})$ and using the fact that $\mathbf{y}(\mathbf{c})^\top A(\mathbf{c}) = \boldsymbol{\lambda}(\mathbf{c})\mathbf{y}(\mathbf{c})^\top$.

By assumption, there exists a constant nonsingular matrix $V \in \mathbb{C}^{n\times n}$ in the form

$$V = [\mathbf{x}_0, M],$$

with $M \in \mathbb{C}^{n\times(n-1)}$, such that

$$V^{-1}A(\mathbf{0})V = \begin{bmatrix} \lambda_0 & \mathbf{0}^\top \\ \mathbf{0} & N \end{bmatrix},$$

where $N \in \mathbb{C}^{(n-1)\times(n-1)}$ and $\lambda_0 \notin \sigma(N)$. Define $\tilde{A}(\mathbf{c}) := V^{-1}A(\mathbf{c})V$ and express it as

$$\tilde{A}(\mathbf{c}) = \begin{bmatrix} \tilde{a}_{11}(\mathbf{c}) & \tilde{\mathbf{a}}_{12}^\top(\mathbf{c}) \\ \tilde{\mathbf{a}}_{21}^\top(\mathbf{c}) & \tilde{A}_{22}(\mathbf{c}) \end{bmatrix},$$

with $\tilde{a}_{11}(\mathbf{c}) \in \mathbb{C}$ and others are of appropriate sizes. Note that $\tilde{a}_{11}(\mathbf{0}) = \lambda_0$. Define the function $\mathbf{f} : \mathbb{C}^{n-1} \times \mathbb{C}^m \to \mathbb{C}^{n-1}$ by

$$\mathbf{f}(\mathbf{z}, \mathbf{c}) = \tilde{\mathbf{a}}_{21}(\mathbf{c}) + (\tilde{A}_{22}(\mathbf{c}) - \tilde{a}_{11}I)\mathbf{z} - \mathbf{z}\tilde{\mathbf{a}}_{12}^\top\mathbf{z}. \tag{3.13}$$

Note that $\mathbf{f}$ is analytic in both $\mathbf{z}$ and $\mathbf{c}$. Furthermore, observe that $\mathbf{f}(\mathbf{0}, \mathbf{0}) = \mathbf{0}$ and that

$$\left.\frac{\partial \mathbf{f}}{\partial \mathbf{z}}\right|_{(\mathbf{0},\mathbf{0})} = \tilde{A}_{22} - \lambda_0 I$$

is nonsingular. By the implicit function theorem, there exists a neighborhood $\mathcal{N}$ of $\mathbf{0} \in \mathbb{C}^m$ and a function $\mathbf{z}(\mathbf{c})$ analytic in $\mathcal{N}$ such that

$$\mathbf{f}(\mathbf{z}(\mathbf{c}), \mathbf{c}) = \mathbf{0}.$$

With this $\mathbf{z}(\mathbf{c})$ and (3.13), it can be checked that

$$\tilde{A}(\mathbf{c}) \begin{bmatrix} 1 \\ \mathbf{z}(\mathbf{c}) \end{bmatrix} = (\tilde{a}_{11}(\mathbf{c}) + \tilde{\mathbf{a}}_{12}^\top \mathbf{z}(\mathbf{c})) \begin{bmatrix} 1 \\ \mathbf{z}(\mathbf{c}) \end{bmatrix}.$$

Since $\tilde{A}(\mathbf{c})$ and $A(\mathbf{c})$ are similar, the analytic function

$$\boldsymbol{\lambda}(\mathbf{c}) := \tilde{a}_{11}(\mathbf{c}) + \tilde{\mathbf{a}}_{12}^{\top}\mathbf{z}(\mathbf{c}) \tag{3.14}$$

is the eigenvalue of $A(\mathbf{c})$. Upon substitution, the vector

$$\mathbf{x}(\mathbf{c}) := \mathbf{x}_0 + N\mathbf{z}(\mathbf{c}) \tag{3.15}$$

is a corresponding right eigenvector which obviously is also analytic. □

We remark that the technique used to prove the above theorem can be carried over to the real field $\mathbb{R}$, if $A(\mathbf{c}) \in \mathbb{R}^{n\times n}$ and λ_0 is a real-valued simple eigenvalue. In particular, a result similar to Theorem 3.5 remains true (Zhou and Dai, 1991) if $A(\mathbf{c}) \in \mathcal{S}(n)$.

Example 3.1. The assumption that λ_0 is simple is critical. Consider the family of real symmetric matrices

$$A(\mathbf{c}) = \begin{bmatrix} c_1 & c_2 \\ c_2 & -c_1 \end{bmatrix}$$

with $\mathbf{c} = [c_1, c_2]^{\top} \in \mathbb{R}^2$. It is easy to see that

$$\sigma(A(\mathbf{c})) = \left\{ \pm\sqrt{c_1^2 + c_2^2} \right\}.$$

Clearly, neither eigenvalue of $A(\mathbf{c})$ is differentiable at $\mathbf{c} = \mathbf{0}$. There is no way to define an analytic $\boldsymbol{\lambda}(\mathbf{c})$ in any neighborhood of $\mathbf{0}$.

An interesting application of Theorem 3.5 is that we can now measure how sensitively an eigenvalue of a given matrix A_0 responds to the perturbation of one particular section of a matrix. Let E denote the matrix with entries one at those particular positions of interest and zero elsewhere. Consider $A(\epsilon) = A_0 + \epsilon E$. Suppose λ_0 is a simple eigenvalue of A_0 with corresponding right and left eigenvector $\mathbf{x}_0$ and $\mathbf{y}_0$, respectively. Then the rate of change of eigenvalue λ_0 with respect to entries of A_0 at positions in E is given by the quantity $\mathbf{y}_0^{\top} E\mathbf{x}_0/\mathbf{y}_0^{\top}\mathbf{x}_0$, which is well known in the literature (Wilkinson, 1965).

Inverse problem for LiPIEP2 The issue of sensitivity for inverse problems is more subtle than for the forward problem, mainly because an inverse problem frequently has multiple solutions. Comparing far apart multiple solutions while another solution is nearby is simply not sensible. When talking about changes of a solution due to perturbations to the problem data, we need to carefully discern which original solution is being changed. An ordering of the eigenvalues is necessarily implied.

For the sake of practicality, we shall limit the discussion to the real case. So that we can maintain simplicity, we shall include an additional structure and consider the symmetric LiPIEP. That is, the problem that we are most comfortable in dealing with is the restricted LiPIEP2. In view of Theorem 3.4, it is reasonable to consider only the case that the prescribed eigenvalues $\{\lambda_k\}_{k=1}^n$ are all distinct. The following result, due to Xu (1998), is an excellent characterization of the perturbation effect.

Theorem 3.6. (Xu, 1998) Assume that all matrices are in $\mathcal{S}(n)$. Assume also that the LiPIEP with $A(\mathbf{c})$ defined in (3.3) is solvable for a prescribed set of distinct eigenvalues $\boldsymbol{\lambda} = \{\lambda_k\}_{k=1}^n$. Let

$$A(\mathbf{c}) = Q(\mathbf{c})\mathrm{diag}(\boldsymbol{\lambda})Q(\mathbf{c})^\top$$

be the spectral decomposition of $A(\mathbf{c})$ where $Q(\mathbf{c}) = [\mathbf{q}_1(\mathbf{c}), \ldots, \mathbf{q}_n(\mathbf{c})] \in \mathcal{O}(n)$ and $\mathbf{q}_i(\mathbf{c}) \in \mathbb{R}^n$, $i = 1, \ldots, n$, are orthonormal eigenvectors of $A(\mathbf{c})$. Define

$$J(\mathbf{c}) := \left[\mathbf{q}_i(\mathbf{c})^\top A_j \mathbf{q}_i(\mathbf{c})\right]_{i,j=1}^n,$$

$$\mathbf{b}(\mathbf{c}) := \left[\mathbf{q}_1(\mathbf{c})^\top A_0 \mathbf{q}_1(\mathbf{c}), \ldots, \mathbf{q}_n(\mathbf{c})^\top A_0 \mathbf{q}_n(\mathbf{c})\right]^\top.$$

Suppose that $J(\mathbf{c})$ is nonsingular and that the change of problem data

$$\delta = \|\boldsymbol{\lambda} - \tilde{\boldsymbol{\lambda}}\|_\infty + \sum_{i=0}^n \|A_i - \tilde{A}_i\|_2$$

is sufficiently small. Then:

(i) The LiPIEP2 associated with the perturbed basis matrices $\{\tilde{A}_i\}_{i=0}^n \subset \mathcal{S}(n)$ and eigenvalues $\tilde{\boldsymbol{\lambda}} = \{\tilde{\lambda}_i\}_{i=1}^n$ is solvable.
(ii) There is a solution $\tilde{\mathbf{c}}$ of the perturbed problem which is in a neighborhood of $\mathbf{c}$ and satisfies

$$\frac{\|\mathbf{c} - \tilde{\mathbf{c}}\|_\infty}{\|\mathbf{c}\|_\infty} \le \kappa_\infty(J(\mathbf{c})) \left(\frac{\|\boldsymbol{\lambda} - \tilde{\boldsymbol{\lambda}}\|_\infty + \|A_0 - \tilde{A}_0\|_2}{\|\boldsymbol{\lambda} - \mathbf{b}(\mathbf{c})\|_\infty} + \frac{\sum_{k=1}^n \|A_k - \tilde{A}_k\|_2}{\|J(\mathbf{c})\|_\infty} \right) + O(\delta^2),$$

where $\kappa_\infty(J(\mathbf{c})) = \|J(\mathbf{c})\|_\infty \|J(\mathbf{c})^{-1}\|_\infty$ denotes the condition number of the matrix $J(\mathbf{c})$.

Proof We shall assume for convenience that all spectra mentioned in the proof below are automatically arranged in ascending order.

Given an arbitrary but fixed vector $\boldsymbol{\epsilon} \in \mathbb{R}^n$, define for $t \in (-1, 1)$,

$$\boldsymbol{\mu}(t) = \boldsymbol{\lambda} + t\boldsymbol{\epsilon},$$

where recall that we use the same set notation $\boldsymbol{\lambda}$ for a column vector in $\mathbb{R}^n$. By selecting elements of ϵ small enough, we may assume that $\boldsymbol{\mu}(t)$ maintains it ordering, that is,

$$\mu_1(t) < \cdots < \mu_n(t),$$

for all $t \in (-1,1)$. Given arbitrary but fixed matrices $\{E_k\}_{k=0}^n \subset \mathcal{S}(n)$, define

$$B_k(t) := A_k + tE_k, \quad k = 0, 1, \dots, n.$$

Consider the family of matrices

$$B(t, \mathbf{x}) := B_0(t) + \sum_{k=1}^{n} x_k B_k(t) \tag{3.16}$$

and the function $\mathbf{f} : (-1,1) \times \mathbb{R}^n \to \mathbb{R}^n$ defined by

$$\mathbf{f}(t, \mathbf{x}) := \sigma(B(t, \mathbf{x})) - \boldsymbol{\mu}(\mathrm{t}). \tag{3.17}$$

Because $B(t,\mathbf{x})$ is analytic in $(t,\mathbf{x})$ and because all eigenvalues of $B(0,\mathbf{c}) = A(\mathbf{c})$ are distinct, we know from (the real version of) Theorem 3.5 that there exists a neighborhood $\mathcal{N} \subset (-1,1) \times \mathbb{R}^n$ of $(0,\mathbf{c})$ such that for all $(t,\mathbf{x}) \in \mathcal{N}$ the spectrum $\sigma(B(t,\mathbf{x}))$ is analytic. It follows that $\mathbf{f}(t,\mathbf{x})$ is analytic in $\mathcal{N}$.

Furthermore, observe that

$$\left.\frac{\partial \mathbf{f}}{\partial \mathbf{x}}\right|_{(0,\mathbf{c})} = J(\mathbf{c}) \tag{3.18}$$

is nonsingular by assumption. By the implicit function theorem, we know that there exists a neighborhood $\mathcal{I} \subset (-1,1)$ of 0 and an analytic function $\mathbf{x}(t)$ defined over $\mathcal{I}$ such that $\mathbf{x}(0) = \mathbf{c}$ and $\mathbf{f}(t, \mathbf{x}(t)) = 0$. That is,

$$\sigma(B(t, \mathbf{x}(t)) = \boldsymbol{\mu}(t).$$

We have thus proved that for $t \in \mathcal{I}$, the symmetric LiPIEP with basis matrices $\{B_k(t)\}_{k=0}^n$ and prescribed spectrum $\boldsymbol{\mu}(t)$ has a solution $\mathbf{x}(t)$. Note that $\{B_k(t)\}_{k=0}^n$ is in a neighborhood of $\{A_k\}_{k=0}^n$ and that $\boldsymbol{\mu}(t)$ is in a neighborhood of $\boldsymbol{\lambda}$ when t is small.

It only remains to estimate the distance of $\mathbf{x}(t)$ from $\mathbf{c}$. Toward that end, recall that by Theorem 3.5 it is also true that $B(t,\mathbf{x}(t))$ should have an analytic spectral decomposition

$$B(t, \mathbf{x}(t)) = U(t)\mathrm{diag}(\boldsymbol{\mu}(t))U(t)^\top \tag{3.19}$$

with some analytic $U(t) \in \mathcal{O}(n)$ satisfying $U(0) = Q(\mathbf{c})$. In particular, it is true that

$$\boldsymbol{\lambda} = J(\mathbf{c})\mathbf{c} + \mathbf{b}(\mathbf{c}). \tag{3.20}$$

It is easy to see from (3.19) that

$$\left.\frac{d\boldsymbol{\mu}}{dt}\right|_{t=0} = \mathrm{diag}\left(Q(\mathbf{c})^\top \left.\frac{dB(t,\mathbf{x}(t))}{dt}\right|_{t=0} Q(\mathbf{c}) \right),$$

whereas

$$\left.\frac{dB(t,\mathbf{x}(t))}{dt}\right|_{t=0} = E_0 + \sum_{k=1}^{n} x_k(0) E_k + \sum_{k=1}^{n} \left.\frac{dx_k}{dt}\right|_{t=0} A_k.$$

We thus obtain a relationship

$$\epsilon = \tilde{\mathbf{b}}(\mathbf{c}) + \tilde{J}(\mathbf{c})\mathbf{c} + J(\mathbf{c}) \left.\frac{d\mathbf{x}}{dt}\right|_{t=0},$$

where

$$\tilde{J}(\mathbf{c}) := [\mathbf{q}_i(\mathbf{c})^\top E_j \mathbf{q}_i(\mathbf{c})]_{i,j=1}^n,$$
$$\tilde{\mathbf{b}}(\mathbf{c}) := [\mathbf{q}_1(\mathbf{c})^\top E_0 \mathbf{q}_1(\mathbf{c}), \dots, \mathbf{q}_n(\mathbf{c})^\top E_0 \mathbf{q}_n(\mathbf{c})]^\top.$$

It now follows that

$$\begin{aligned}\|\mathbf{x}(t) - \mathbf{x}(0)\|_\infty &\leq \| \left.\frac{d\mathbf{x}}{dt}\right|_{t=0} + O(t^2)\|_\infty \\ &\leq \|J(\mathbf{c})^{-1}\|_\infty \left(\|t\epsilon\|_\infty + \|t\tilde{\mathbf{b}}(\mathbf{c})\|_\infty + \|t\tilde{J}(\mathbf{c})\|_\infty \|\mathbf{c}\|_\infty\right) + O(t^2).\end{aligned}$$

By definition, it is clear that

$$\|\tilde{\mathbf{b}}(\mathbf{c})\|_\infty \leq \|E_0\|_2,$$
$$\|\tilde{J}(\mathbf{c})\|_\infty \leq \sum_{k=1}^n \|E_k\|_2.$$

It is also clear from (3.20) that

$$\|\boldsymbol{\lambda} - \mathbf{b}(\mathbf{c})\|_\infty \leq \|J(\mathbf{c})\|_\infty \|\mathbf{c}\|_\infty.$$

We therefore have derived the estimate that

$$\frac{\|\mathbf{x}(t) - \mathbf{c}\|_\infty}{\|\mathbf{c}\|_\infty} \leq \kappa_\infty(J(\mathbf{c})) \left(\frac{\|t\epsilon\|_\infty + \|tE_0\|_2}{\|\boldsymbol{\lambda} - \mathbf{b}(\mathbf{c})\|_\infty} + \frac{\sum_{k=1}^n \|tE_k\|_2}{\|J(\mathbf{c})\|_\infty}\right) + O(t^2). \quad (3.21)$$

The theorem is proved by recognizing that $\tilde{\mathbf{c}} = \mathbf{x}(t)$, $\tilde{\boldsymbol{\lambda}} = \boldsymbol{\mu}(t)$, $\tilde{A}_k = B_k(t)$, $k = 0, 1, \dots,$ for $t \in \mathcal{I}$ and hence $O(\delta^2) = O(t^2)$. □

At first glance, Theorem 3.6 seems to suggest that the solution of a symmetric LiPIEP depends continuously on the problem data $\{A_k\}_{k=1}^n$ and $\{\lambda_k\}_{k=1}^n$. We must caution, nevertheless, that the solution $\mathbf{x}(t)$ implied in the theorem to a perturbed problem is only one possibly solution. While the theory guarantees that there is a perturbed solution $\mathbf{x}(t)$ near $\mathbf{c}$, we want to stress from our experience that often a numerical procedure will locate another solution that is far away from $\mathbf{c}$. This is not to say that the underlying method is necessarily bad, rather we should be more concerned about how closely $\mathbf{x}(t)$ has solved the problem. In Sun (1999), the notion of a *backward stable algorithm* is introduced to characterize any numerical procedure by which the resulting approximation solution does solve a nearby LiPIEP2 of the original problem.

Theorem 3.6 has another very important implication. The condition number $\kappa_\infty(J(\mathbf{c}))$ of the matrix $J(\mathbf{c})$, which is composed of eigenvectors of $A(\mathbf{c})$, indicates how the perturbation in the original problem data would be propagated in the solution with respect to the inverse problem. For this reason, $\kappa_\infty(J(\mathbf{c}))$ is referred to as the *condition number* of the real-valued and symmetric LiPIEP2. The larger the condition number $\kappa_\infty(J(\mathbf{c}))$, the more sensitive is the solution subject to changes. One challenge in the numerical computation is not only to find a solution for the LiPIEP2, but also to find a robust solution for which the condition number $\kappa_\infty(J(\mathbf{c}))$ is minimized with respect to $\mathbf{c}$.

3.2.3 *Ideas of computation*

Solving an IEP can generally be regarded as solving a nonlinear algebraic system. It is not surprising that many existing computational techniques can be applied to solve an IEP. However, because IEPs also involve matrices, there are more matrix structures and properties that can be and should be exploited. We have found from the literature that current numerical procedures for solving IEPs can roughly be classified into three categories. These are:

1. direct approach,
2. iterative approach,
3. continuous approach.

Thus far, we know of no superlative method whose performance is clearly superior to all others. Rather, it is often the case that a certain method should be specifically tailored to fit a certain problem. Throughout this book, we shall try to consolidate the methods available as much as possible into some general forms. Still, it will become necessary from time to time to examine some special modifications in greater detail. To help readers grasp the larger picture, we shall briefly outline the general ideas behind these methods.

Direct approach By a direct method, we mean a procedure through which a solution to an IEP can be found in finitely many arithmetic operations. It might be surprising that, for nonlinear systems such as IEPs, such a method could ever exist. However, we shall see in Section 4.2 that most IEPs with Jacobi structure can be solved by the Lanczos method that terminates in finitely many steps. The paper by Boley and Golub (1987) offers a good survey of this method.

In fact, it is also the primary concern in the paper by Ikramov and Chugunov (2000) who search for upper bounds on the number of prescribed entries so that a matrix with these prescribed entries and a prescribed spectrum could be completed by finite rational methods. By a rational method, we refer to any procedure that uses only arithmetic operations of the field over which the problem is considered. As such, a rational algorithm can employ exact arithmetic, say, using the computer algebra package MAPLE, to complete the construction of a matrix. At present, the bound on the maximal allowable number of prescribed entries in such a PEIEP appears to have been set by the Hershkowitz theorem (Hershkowitz, 1978), beyond which no existence theory is available and no direct method is known to work. However, this lack of an existence theory does not mean that no other numerical methods can be employed to find a solution. We shall discuss more of this completion problem and its historic development in Section 4.7.

Many classical results in matrix theory proved by mathematical induction may be regarded as constructible by a direct method. With advance in today's programming languages that allow a subprogram to invoke itself recursively, a traditional induction proof can be transformed into a recursive algorithm which terminates in finitely many steps. Such an approach will be demonstrated in Section 4.9.

Iterative approach To solve nonlinear problems, iterative methods seem natural. There are at least two ways to conduct the iteration for IEPs: Newton iterations and the orthogonal reduction method.

The Newton iterations are conventional Newton methods modified by taking advantage of matrix properties. We cite the article by Friedland et al. (1986) as a typical representative work in this area. There are many variations. This class of methods is fast but converges only locally and requires initial approximations. In addition, the methods have to be further adapted in the presence of multiple eigenvalues. Because this is such an important class of method, we shall give special attention of its application to real symmetric LiPIEPs in Section 3.2.4.

Orthogonal reduction methods include the Lanczos method. The basic idea is to iteratively approximate eigenvectors by employing some QR-like decompositions and then construct an approximate matrix. For the Jacobi inverse eigenvalue problem, this iteration terminates in a finite number of steps. For other structural constraints, such as the Toeplitz inverse eigenvalue problem, this iteration is terminated only after some stopping criteria are met. We shall illustrate this approach in Section 4.3. On the other hand, since this class of methods works directly with eigenvectors, the case of multiple eigenvalues usually can be easily handled.

Continuous approach By a continuous method, we mean a procedure that connects an initial value to the desired solution through a "continuous path" defined by a certain differential equation. The problem is then unravelled by tracing a specific integral curve of this differential equation. There are at least three ways to exploit this idea: the homotopy method, the projected gradient method, and the ASVD flow method.

The homotopy method builds the continuous path between two problems in a topological space with the hope that the artificial but simpler problem can be deformed in a continuous (homotopic) way so that at the end it coincides with the original but harder problem. In this process, the trivial solution to the simpler problem is also mathematically deformed into a solution to the harder problem. This method has been attractive for solving general nonlinear systems mainly because of its globally convergent property. When the homotopy is efficacious, it provides both an existence proof and a numerical method. References and applications are too numerous to list here. We shall point to the package HOMPACK (Morgan et al., 1989) as a beginning point for further reading in general and the paper (Chu, 1988) for eigenvalue computation. We shall demonstrate this method for an AIEP in Section 3.3.

The projected gradient method, on the other hand, builds the continuous path between a starting point and the desired solution on the basis of systematically reducing the difference between the current position and the target position. For IEPs, we have found this approach particularly advantageous and versatile in that the objective function is easy to recognize and that the gradient flow is easy to construct. This method also provides a general least squares setting that finds a least squares solutions of an IEP if the spectral constraints and the structural constraints cannot be satisfied simultaneously. The paper by Chu and Driessel (1990) is an appropriate introductory reference. We shall discuss its application to symmetric LiPIEP in Section 3.2.5 and its general settings in Chapters 6 and 7.

The ASVD method involves analytic singular value decomposition of a family of matrices. Its main purpose is to maintain stable coordinate transformations for nonsymmetric matrices during the integration. The method is applied to solve the nonnegative inverse eigenvalue problems and the stochastic inverse eigenvalue problem in Section 4.5.

3.2.4 *Newton's method (for LiPIEP2)*

In this section we shall concentrate exclusively on the Newton method applied to the LiPIEP2. This is the case where all matrices involved in (3.3) are in $\mathcal{S}(n)$. We choose to single out this method for consideration because, while Newton's iteration is typically regarded as the normal means to solve a nonlinear system, we shall demonstrate how the iteration can be carried out by taking into account the matrix structure. This setting sheds light on how other types of IEPs should be handled if an iterative method similar to the Newton scheme is to be used.

We first consider the case when all eigenvalues $\boldsymbol{\lambda} = \{\lambda_k\}_{k=1}^n$ are distinct and are arranged in ascending order. Consider the affine subspace

$$\mathcal{A} := \{A(\mathbf{c}) | \mathbf{c} \in \mathbb{R}^n\} \tag{3.22}$$

and the isospectral surface

$$\mathcal{M}(\Lambda) := \{Q\Lambda Q^\top | Q \in \mathcal{O}(n)\}, \tag{3.23}$$

where

$$\Lambda := \operatorname{diag}(\boldsymbol{\lambda}).$$

Using the fact that $Q(t)Q(t)^\top = I$, it follows that $Q(t)$ is a differentiable path embedded in $\mathcal{O}(n)$ if and only if

$$\dot{Q}(t) = K(t)Q(t), \quad Q(0) \in \mathcal{O}(n) \tag{3.24}$$

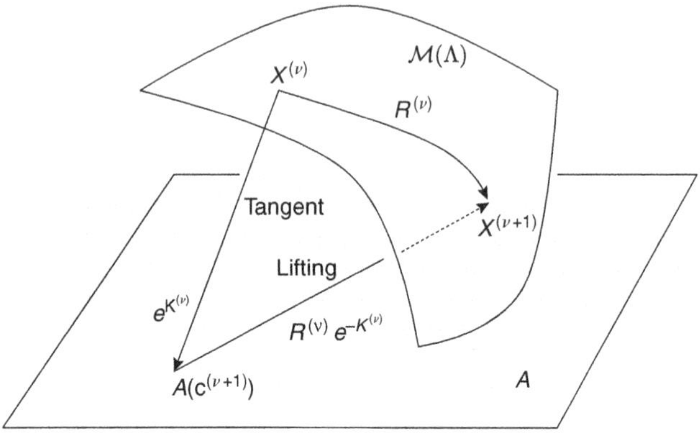

Figure 3.1. *Geometry of the Newton method for PIEP*

for some family of skew-symmetric matrices $K(t)$. Translating this to the manifold $\mathcal{M}(\Lambda)$, it follows that any tangent vector $T(X)$ to $\mathcal{M}(\Lambda)$ at a point $X \in \mathcal{M}(\Lambda)$ must be of the form

$$T(X) = XK - KX$$

for some skew-symmetric matrix $K \in \mathbb{R}^{n \times n}$.

Geometric setting The Newton method we are about to describe is analogous to the classical Newton method for finding roots of a one-variable differentiable function. Recall that one Newton step for the function $f : \mathbb{R} \to \mathbb{R}$ is

$$x^{(\nu+1)} = x^{(\nu)} - (f'(x^{(\nu)}))^{-1} f(x^{(\nu)}).$$

The new iterate $x^{(\nu+1)}$ in the scheme represents the x-intercept of the line that is tangent to the graph of f at $(x^{(\nu)}, f(x^{(\nu)}))$. The point $(x^{(\nu+1)}, f(x^{(\nu+1)}))$ represents a natural "lift" of the intercept along the y-axis to the graph of f from which the next tangent line will begin.

For the LiPIEP2, we shall think of the isospectral surface $\mathcal{M}(\Lambda)$ as playing the role of the graph of f while the affine subspace $\mathcal{A}$ plays the role of the x-axis. The mission in front of us is to find the intersection of these two geometric entities. Given $X^{(\nu)} \in \mathcal{M}(\Lambda)$, there exists a $Q^{(\nu)} \in \mathcal{O}(n)$ such that

$${Q^{(\nu)}}^{\top} X^{(\nu)} Q^{(\nu)} = \Lambda. \tag{3.25}$$

The matrix $X^{(\nu)} + KX^{(\nu)} - X^{(\nu)}K$ with any skew-symmetric matrix K represents a tangent vector to the surface $\mathcal{M}(\Lambda)$ emanating from $X^{(\nu)}$. One Newton step there should comprise of the task of seeking an $\mathcal{A}$-intercept $A(\mathbf{c}^{(\nu+1)})$ of such a vector with the affine subspace $\mathcal{A}$ and the task of lifting up the point $A(\mathbf{c}^{(\nu+1)}) \in \mathcal{A}$ to a point $X^{(\nu+1)} \in \mathcal{M}(\Lambda)$. The geometry of this idea is depicted in Figure 3.1.

Find the intercept To find the $\mathcal{A}$-intercept, we need to find a skew-symmetric matrix $K^{(\nu)}$ and a vector $\mathbf{c}^{(\nu+1)}$ such that

$$X^{(\nu)} + K^{(\nu)}X^{(\nu)} - X^{(\nu)}K^{(\nu)} = A(\mathbf{c}^{(\nu+1)}). \tag{3.26}$$

The unknowns $K^{(\nu)}$ and $\mathbf{c}^{(\nu+1)}$ in equation (3.26) can be solved separately as follows. Using (3.25), we transform (3.26) into the system

$$\Lambda + \tilde{K}^{(\nu)}\Lambda - \Lambda\tilde{K}^{(\nu)} = Q^{(\nu)^{\top}}A(\mathbf{c}^{(\nu+1)})Q^{(\nu)}, \tag{3.27}$$

where

$$\tilde{K}^{(\nu)} := Q^{(\nu)^{\top}}K^{(\nu)}Q^{(\nu)} \tag{3.28}$$

remains skew-symmetric and, hence, always has zero diagonal. Neither the pre-multiplication nor the post-multiplication of $\tilde{K}^{(\nu)}$ by Λ changes the zero diagonal entries. This observation effectively separates $\mathbf{c}^{(\nu+1)}$ from $K^{(\nu)}$. More precisely, if we write $Q^{(\nu)} = [\mathbf{q}_1^{(\nu)}, \ldots, \mathbf{q}_n^{(\nu)}]$ in columns, then a comparison of the diagonals on both sides of (3.27) gives rise to the linear system of n equations:

$$J^{(\nu)}\mathbf{c}^{(\nu+1)} = \boldsymbol{\lambda} - \mathbf{b}^{(\nu)}, \tag{3.29}$$

where

$$J^{(\nu)} := \left[\mathbf{q}_i^{(\nu)^{\top}} A_j \mathbf{q}_i^{(\nu)}\right]_{i,j=1}^{n}, \tag{3.30}$$

$$\mathbf{b}^{(\nu)} := [\mathbf{q}_1^{(\nu)^{\top}} A_0 \mathbf{q}_1^{(\nu)}, \ldots, \mathbf{q}_n^{(\nu)^{\top}} A_0 \mathbf{q}_n^{(\nu)}]^{\top}. \tag{3.31}$$

We have seen this equation before in (3.20) except that this time both $J^{(\nu)}$ and $\mathbf{b}^{(\nu)}$ are calculated from (3.30) and (3.31), respectively. The vector $\mathbf{c}^{(\nu+1)}$, therefore, can be solved from (3.29), provided $J^{(\nu)}$ is nonsingular. Note that $K^{(\nu)}$ is not involved in (3.29). Also note that we have used the fact that $A(\mathbf{c})$ is linear in $\mathbf{c}$. In this way, the $\mathcal{A}$-intercept $A(\mathbf{c}^{(\nu+1)})$ is completely determined.

To prepare for the lift, however, we still need to know the skew-symmetric matrix $\tilde{K}^{(\nu)}$ (and, hence, the matrix $K^{(\nu)}$) that creates this $\mathcal{A}$-intercept. Such a matrix can be determined from the off-diagonal equations of (3.27), once $\mathbf{c}^{(\nu+1)}$ is obtained. In fact,

$$\tilde{K}_{ij}^{(\nu)} = \frac{q_i^{(\nu)^{\top}} A(\mathbf{c}^{(\nu+1)}) q_j^{(\nu)}}{\lambda_j - \lambda_i}, \quad \text{for } 1 \leq i < j \leq n. \tag{3.32}$$

In this way, equation (3.26) is completely solved.

Note that the method can break down for two resources: when $J^{(\nu)}$ is singular or when $\{\lambda_k\}_{k=1}^{n}$ have multiple eigenvalues.

Find the lift-up In the classical Newton method the new iterate $x^{(\nu+1)}$ is "lifted up" naturally along the y-axis to the the point $(x^{(\nu+1)}, f(x^{(\nu+1)}))$ from which the next tangent line will begin. The analogy of lifting the point $A(\mathbf{c}^{(\nu+1)}) \in \mathcal{A}$ to a point $X^{(\nu+1)} \in \mathcal{M}(\Lambda)$ is somewhat more difficult due to the lack of an obvious coordinate axis to follow.

One possible way of this lifting can be motivated as follows. Recall that solving the LiPIEP2 is equivalent to finding an intersection of the two sets $\mathcal{M}(\Lambda)$ and $\mathcal{A}$. Suppose all the iterations are taking place near a point of intersection. Then we have

$$X^{(\nu+1)} \approx A(\mathbf{c}^{(\nu+1)}). \tag{3.33}$$

From (3.26), we also have

$$A(\mathbf{c}^{(\nu+1)}) \approx e^{K^{(\nu)}} X^{(\nu)} e^{-K^{(\nu)}}, \tag{3.34}$$

if the matrix exponential $e^{K^{(\nu)}}$ is expanded. High accuracy calculation of the exponential matrix $e^{K^{(\nu)}}$ in (3.34) is expensive and is not needed. So, instead, we define the Cayley transform

$$R^{(\nu)} := \left(I + \frac{K^{(\nu)}}{2}\right)\left(I - \frac{K^{(\nu)}}{2}\right)^{-1} \tag{3.35}$$

which is the (1, 1) Padé approximation of the matrix $e^{K^{(\nu)}}$. It is well known that $R^{(\nu)} \in \mathcal{O}(n)$, and that

$$R^{(\nu)} \approx e^{K^{(\nu)}} \tag{3.36}$$

if $\|K^{(\nu)}\|$ is small. Motivated by (3.33) and (3.34), we now define

$$X^{(\nu+1)} := R^{(\nu)} X^{(\nu)} R^{(\nu)^{\top}} \in \mathcal{M}_e(\Lambda) \tag{3.37}$$

and we begin the next iteration.

It is interesting to note that

$$X^{(\nu+1)} \approx R^{(\nu)} e^{-K^{(\nu)}} A(\mathbf{c}^{(\nu+1)}) e^{K^{(\nu)}} R^{(\nu)} \approx A(c^{(\nu+1)}) \tag{3.38}$$

represents a lift of the matrix $A(\mathbf{c}^{(\nu+1)})$ from the affine subspace $\mathcal{A}$ to the surface $\mathcal{M}(\Lambda)$.

The lift can also be computed by using the Wielandt–Hoffman theorem (see Sections 4.3, 7.3; Chu and Driessel (1990) and Horn and Johnson (1991)). More specifically, we may take the lift to be the nearest point on $\mathcal{M}_e(\Lambda)$ to $A(c^{(\nu+1)})$. It can be shown that $X^{(\nu+1)}$ must be given by

$$X^{(\nu+1)} := \hat{Q}^{(\nu+1)} \tilde{\Lambda}^{(\nu+1)} \hat{Q}^{(\nu+1)^{\top}}, \tag{3.39}$$

provided that

$$A(\mathbf{c}^{(\nu+1)}) = \hat{Q}^{(\nu+1)}\Sigma^{(\nu+1)}\hat{Q}^{(\nu+1)^{\top}}$$

is the spectral decomposition of $A(\mathbf{c}^{(\nu+1)})$ and that $\tilde{\Lambda}^{(\nu+1)}$ is the diagonal matrix whose elements are a rearrangement of those of Λ in the same ordering as those in $\Sigma^{(\nu+1)}$.

The above description offers a geometrical interpretation of Method III developed by Friedland et al. (1987). It can be shown that the rate of convergence for this case is quadratic (Chan et al., 1999).

Handling multiple eigenvalues The formula of the tangent vector (3.24) reaffirms that the set $\mathcal{O}(n)$ of orthogonal matrices forms a manifold of dimension $n(n-1)/2$. This is equivalent to saying that (at a local chart) orthogonal matrices can be characterized by $n(n-1)/2$ parameters. With this in mind, the relationship

$$A(\mathbf{c}) = Q\Lambda Q^{\top} \tag{3.40}$$

that must be held at the intersection of $\mathcal{A}$ and $\mathcal{M}(\Lambda)$ can be interpreted as a system of $n(n+1)/2$ equations in the $n(n+1)/2$ unknowns $\mathbf{c}$ and Q. Suppose now that there are multiple eigenvalues. For example, suppose that

$$\lambda_1 = \cdots = \lambda_\ell < \lambda_{\ell+1} < \cdots < \lambda_n.$$

Then columns of $Q_1 = [\mathbf{q}_1, \ldots, \mathbf{q}_\ell]$ span the eigenspace corresponding to λ_1 if and only if columns of Q_1U with $U \in \mathcal{O}(\ell)$ span the same eigenspace. In other words, a total of $\ell(\ell-1)/2$ many parameters in the characterization of Q are redundant. The LiPIEP2 as stated in Problem 3.4 would be over-determined unless we either reduce the number of equations or introduce an additional $\ell(\ell-1)/2$ parameters into the problem. For that reason, Friedland et al. (1987) proposed the following modified LiPIEP2.

Problem 3.8 *(Modified LiPIEP2)*
Find the parameters $c_1, \ldots, c_n$ in (3.3) so that $n - \frac{\ell(\ell-1)}{2}$ eigenvalues of $A(\mathbf{c})$ agree with the specified values

$$\lambda_1 = \cdots = \lambda_\ell < \lambda_{\ell+1} < \lambda_{n-\frac{\ell(\ell-1)}{2}}.$$

Going back to re-examine (3.27), we find that the upper-left $\frac{\ell(\ell-1)}{2} \times \frac{\ell(\ell-1)}{2}$ block of $\Lambda\tilde{K}^{(\nu)} - \tilde{K}^{(\nu)}\Lambda$ is identically zero, contributing no information to the matrix $\tilde{K}(\nu)$ which, without loss of generality, can be set to zero. It only remains to determine $\mathbf{c}^{(\nu+1)}$ from the first $n - \frac{\ell(\ell-1)}{2}$ diagonal entries,

$$\sum_{k=1}^{n} \left(\mathbf{q}_i^{(\nu)^\top} A_k \mathbf{q}_i^{(\nu)} \right) c_k^{(\nu+1)} = \lambda_i - \mathbf{q}_i^{(\nu)^\top} A_0 \mathbf{q}_i^{(\nu)}, \quad i = 1, \ldots, n - \frac{\ell(\ell-1)}{2},$$

and the off-diagonal entries of the upper-left $\ell \times \ell$ corner,

$$\sum_{k=1}^{n} \left(\mathbf{q}_i^{(\nu)^\top} A_k \mathbf{q}_j^{(\nu)} \right) c_k^{(\nu+1)} = -\mathbf{q}_i^{(\nu)^\top} A_0 \mathbf{q}_j^{(\nu)}, \quad 1 \leq i < j \leq \ell,$$

on both side of (3.27). The remaining portion of $\tilde{K}(\nu)$ can then be determined by using (3.32) with $\boldsymbol{\lambda}$ defined by

$$\boldsymbol{\lambda} = \operatorname{diag}\left(Q^{(\nu)^\top} A(\mathbf{c}^{(\nu+1)}) Q^{(\nu)} \right). \tag{3.41}$$

3.2.5 *Projected gradient method (for LiPIEP2)*

The notion of projected gradient methods is particularly applicable to IEPs in that the objective function is easy to recognize and that the projected gradient can often be explicitly calculated. Recall that an IEP usually involves two constraints. The basic idea of a project gradient method is to minimize any discrepancy between these two constraints. That is, suppose a matrix X already satisfies the spectral constraint. Let $P(X)$ denote its nearest matrix approximation satisfying the structural constraint. It is then desired to minimize the unwanted portion $\|X - P(X)\|$. We shall demonstrate how this idea works for LiPIEP2 in this section, but the idea works so long as the proximity map $P(X)$ can be defined. The general setting and other applications will be discussed in Chapter 7.

For any two matrices $M = [m_{ij}]$ and $N = [n_{ij}]$ of the same size, a commonly used metric is the Frobenuis norm induced by the Frobenius inner product

$$\langle M, N \rangle := \sum_{i,j} m_{ij} n_{ij}. \tag{3.42}$$

With respect to the Frobenuis norm, the proximity map $P(X)$ of any given $X \in \mathcal{M}(\Lambda)$ is its projection onto the affine subspace $\mathcal{A}$ defined in (3.22). Denote

$$P(X) = A_0 + \sum_{k=1}^{n} c_k(X) A_k. \tag{3.43}$$

It is easy to see that the parameters $c_1(X), \dots, c_n(X)$ can be determined from the system

$$\begin{bmatrix} \langle A_1, A_1 \rangle & \langle A_1, A_2 \rangle & \dots & \langle A_1, A_n \rangle \\ \langle A_2, A_1 \rangle & & & \\ \vdots & & & \\ \langle A_n, A_1 \rangle & & & \langle A_n, A_n \rangle \end{bmatrix} \begin{bmatrix} c_1(X) \\ c_2(X) \\ \vdots \\ c_n(X) \end{bmatrix} = \begin{bmatrix} \langle X - A_0, A_1 \rangle \\ \\ \vdots \\ \langle X - A_0, A_n \rangle \end{bmatrix}.$$

Because every $X \in \mathcal{M}(\Lambda)$ is of the form $X = Q\Lambda Q^\top$ for some $Q \in \mathcal{O}(n)$, we can rewrite the objective function in terms of the variable Q. That is, the objective is to minimize the function,

$$F(Q) := \frac{1}{2} \left\langle Q\Lambda Q^\top - P(Q\Lambda Q^\top), Q\Lambda Q^\top - P(Q\Lambda Q^\top) \right\rangle,$$

subject to the constraint $Q^\top Q = I$. Consider the functional $F : \mathbb{R}^{n\times n} \to \mathbb{R}$. The Fréchet derivative of F for a general matrix M acting on H is calculated to be

$$F'(M) \cdot H = 2\langle (M\Lambda M^\top - P(M\Lambda M^\top))M\Lambda, H\rangle.$$

It follows from the Riesz representation theorem that, with respect to the Frobenius inner product, the gradient of F for a general matrix M can be represented as

$$\nabla F(M) = 2(M\Lambda M^\top - P(M\Lambda M^\top))M\Lambda.$$

In our application, we are only interested in F restricted to the manifold $\mathcal{O}(n)$ and the projection of ∇F onto the tangent space of $\mathcal{O}(n)$.

We have already pointed out in (3.24) that the tangent space of $\mathcal{O}(n)$ for any orthogonal matrix Q is given by

$$\mathcal{T}_Q\mathcal{O}(n) = \mathcal{K}(n)Q$$

where

$$\mathcal{K}(n) = \{\text{all skew-symmetric matrices in } \mathbb{R}^{n\times n}\}.$$

It is easy to see that the normal space of $\mathcal{O}(n)$ for any orthogonal matrix Q is given by

$$\mathcal{N}_Q\mathcal{O}(n) = \mathcal{S}(n)Q.$$

Obviously, the direct sum of the tangent space and the normal space form the ambient space, that is,

$$\mathbb{R}^{n\times n} = \mathcal{K}(n)Q \oplus \mathcal{S}(n)Q.$$

Any matrix $M \in \mathbb{R}^{n\times n}$ has a unique orthogonal splitting

$$M = \left\{\left\{\tfrac{1}{2}(MQ^\top - QM^\top)\right\} + \left\{\tfrac{1}{2}(MQ^\top + QM^\top)\right\}\right\} Q.$$

Applying this splitting to $\nabla F(Q)$, in particular, we obtain the projection of $\nabla F(Q)$ into the tangent space as

$$\begin{aligned} g(Q) &= \left\{\tfrac{1}{2}(\nabla F(Q)Q^\top - Q\nabla F(Q)^\top)\right\} Q \\ &= [Q\Lambda Q^\top, P(Q\Lambda Q^\top)]Q, \end{aligned}$$

where $[M, N] := MN - NM$ denotes the Lie bracket operation. We thus have derived a steepest descent flow on the manifold $\mathcal{O}(n)$:

$$\frac{dQ}{dt} = [P(Q\Lambda Q^\top), Q\Lambda Q^\top]Q, \tag{3.44}$$

or equivalently a descent flow on the manifold $\mathcal{M}(\Lambda)$:

$$\begin{aligned} \frac{dX}{dt} &= \frac{dQ}{dt}\Lambda Q^\top + Q\Lambda\frac{dQ^\top}{dt} \\ &= [X, [X, P(X)]]. \end{aligned} \tag{3.45}$$

The projected gradient flow defined by (3.44) can now be integrated numerically. Though conventional ODE integrators are workable, we strongly suggest that orthogonal integrators be used. Research in geometric integration for dynamical systems on manifolds has been very active and fruitful recently. See, for example, Dieci et al. (1994), Hairer et al. (2002), Iserles et al. (2000).

3.3 Additive inverse eigenvalue problems

In the description of Problem 3.6 that defines a generic AIEP, the set $\mathcal{N}$ is used to specify the structure of the add-on matrix X. Suppose that the class $\mathcal{N}$ forms a linear subspace; then an AIEP is a special case of LiPIEP. The add-on matrix X plays precisely the role of the parameters $\mathbf{c}$. The class $\mathcal{N}$ from which X is selected certainly is important here. The location of free parameters and the number of free parameters imposed by $\mathcal{N}$ on X affect whether an AIEP is solvable. That avenue of thinking leads to what we call matrix completion problems with prescribed spectrum, also known as the PEIEP in this book, which will be discussed in greater detail in Section 4.7.

Problem 3.9 *(Matrix completion problem)*
Given complex $\{\lambda_k\}_{k=1}^n$ that is closed under conjugation, find $X \in \mathbb{R}^{n\times n}$ that has prescribed entries at specific locations and has eigenvalues $\{\lambda_k\}_{k=1}^n$.

Indeed, the matrix completion problem defined above is complementary to yet another interesting task, called the *eigenvalue completion problem*, by Gohberg et al. (1995) in their book.

Problem 3.10 *(Eigenvalue completion problem)*
Given a part of a matrix, describe all possible eigenvalues and their multiplicities of the matrices, called completions, that are obtained by filling in the unknown entries with arbitrary and independent complex numbers.

The cases that have been studied for the eigenvalue completion problem include those where the given part is at a submatrix in various positions (principal submatrix, full width or full length, off-diagonal), along the diagonal, or at the triangular part of the matrix. We certainly can impose similar structures on $\mathcal{N}$ for the matrix completion problem. In that case, the existence question for Problem 3.9 would have been answered by what is known for Problem 3.10. Unfortunately, there are many other cases whose eigenvalue completion problems are equally hard and no answers are in sight yet.

We have seen already in Problem 2.7 how a discrete Sturm–Liouville problem would give rise to an AIEP with X being diagonal. It is not surprising that the case of X being diagonal has attracted a lot of the attention in the literature and that traditionally the term "AIEP" refers only to the case where X is diagonal.

Even in the case that the add-on matrix X is merely diagonal, there are a few variations:

Problem 3.11 *(AIEP1)*
Given $A_0 \in \mathbb{R}^{n\times n}$ and real $\{\lambda_k\}_{k=1}^n$, solve the AIEP with $X \in \mathcal{D}_{\mathcal{R}}(n)$.

Problem 3.12 *(AIEP2)*
Given $A_0 \in \mathcal{S}(n)$ and real $\{\lambda_k\}_{k=1}^n$, solve the AIEP with $X \in \mathcal{D}_{\mathcal{R}}(n)$.

Problem 3.13 *(AIEP3)*
Given $A_0 \in \mathbb{C}^{n\times n}$ and complex $\{\lambda_k\}_{k=1}^n$, solve the AIEP with $X \in \mathcal{D}_{\mathcal{C}}(n)$.

As a reminder that the add-on matrix X need not be diagonal, recall the PEIEP arising from the Gauss–Kronrod quadrature rule described in Problem 2.13. The prescribed entries are the zeros outside the tridiagonal band of the matrix as well as the given values (associated with the Gaussian quadrature) that sit along the first $n-1$ entries in the upper-left corner. The solvability and computational schemes of this problem have been discussed in Laurie (1997), Calvetti et al. (2000) independently of other AIEPs considered in the literature.

3.3.1 *Solvability*

The issue of solvability of general PEIEPs has a long history dotted with many interesting and important classical results. The course of its development is quite intriguing and is worthy to be documented separately. We shall put off the discussion of PEIEPs to Section 4.7.

As to the case when $X = \operatorname{diag}(x_1, \dots, x_n)$ is diagonal, we may recast the AIEP as a special LiPIEP via

$$A_0 + X = A_0 + \sum_{k=1}^{n} x_k \underbrace{\mathbf{e}_k \mathbf{e}_k^\top}_{A_k}.$$

Note that each basis matrix A_i is clearly real and symmetric. Any conclusions developed earlier for general LiPIEPs, therefore, are readily applicable to the AIEP. In retrospect, however, the following results have been derived separately and specifically for AIEPs without the general theory of LiPIEPs.

Complex solvability The complex solvability of AIEP3 was first settled by Friedland (1977) using some powerful tools from algebraic geometry. Later, the same result was proved by Alexander (1978) using degree theory. The construction by Chu (1990) using homotopy can serve as a numerical method for finding all solutions. Theorem 3.1, of course, includes the following as a special case.

Theorem 3.7. (Friedland, 1977) Given any specified eigenvalues $\{\lambda_k\}_{k=1}^n \subset \mathbb{C}$,

(i) The AIEP3 is solvable.
(ii) The number of solutions is finite and there are at most $n!$ solutions.
(iii) For almost all $\{\lambda_k\}_{k=1}^n$, there are exactly $n!$ solutions.

Proof We have pointed out earlier that the Bézout theorem for polynomial systems can be used as a tool to prove Theorem 3.1 for general LiPIEP. Its application to the AIEP3 is relatively simpler, so we outline the proof below. We also suggest Section 3.3.3 for related discussion.

Suppose $X = \operatorname{diag}(\mathbf{x})$. Let $\omega_k(\mathbf{x})$ denote the k-th elementary symmetric function of $\mathbf{c}$, that is,

$$\omega_k(\mathbf{x}) = (-1)^k \sum_{i_1 < \cdots < i_k} x_{i_1} \cdots x_{i_k}, \quad k = 1, \dots, n. \tag{3.46}$$

Using the fact that X is diagonal, it is not hard to see by induction that the characteristic polynomial $p(\xi) = \det(\xi I - (A_0 + X))$ of $A_0 + X$ is given by

$$\begin{aligned} p(\xi) = \xi^n &+ (\omega_1(\mathbf{x}) + r_1(\mathbf{x}))\, \xi^{n-1} + \cdots + (\omega_{n-1}(\mathbf{x}) + r_{n-1}(\mathbf{x}))\xi \\ &+ (\omega_n(\mathbf{x}) + r_n(\mathbf{x})), \end{aligned}$$

where $r_k(\mathbf{x})$ is a polynomial of degree strictly less than k. Comparing the coefficients of $p(\xi)$ with those of $\prod_{i=1}^{n}(\xi-\lambda_i)$, we see that solving AIEP3 is equivalent to solving the polynomial system

$$\varphi_k(\mathbf{x}) = \omega_k(\mathbf{x}) + r_k(\mathbf{x}) - \omega_k(\boldsymbol{\lambda}), \quad k = 1, \ldots, n. \tag{3.47}$$

Note that the highest degree term in $\varphi_k(\mathbf{x})$ is $\omega_k(\mathbf{x})$ while we know that

$$\det\left(\frac{\partial(\omega_1, \ldots, \omega_n)}{\partial(x_1, \ldots, x_n)}\right) = (-1)^{n-1}\prod_{i<j}(x_i - x_j).$$

By the Bézout theorem, we know that the polynomial system (3.47) has at most $n!$ and has exactly $n!$ solutions over $\mathbb{C}$ for a generic $\{\lambda_k\}_{k=1}^{n}$. □

Real solvability The fundamental theorem of algebra asserts roots of a polynomial only in the complex field $\mathbb{C}$, so does the Bézout theorem when applied to the polynomial system (3.47). The following example demonstrates that some necessary conditions must be satisfied for an AIEP to be real solvable.

Example 3.2. Consider the case when $A_0 = \begin{bmatrix} 0 & 1 \\ 1 & 0 \end{bmatrix}$. The system (3.47) becomes

$$\begin{aligned} x_1 + x_2 &= \lambda_1 + \lambda_2, \\ x_1 x_2 &= \lambda_1\lambda_2 + 1. \end{aligned}$$

Assume that both prescribed eigenvalues λ_1 and λ_2 are positive. Then for the hyperbola defined by the second equation to intersect with the line defined by the first equation, it is necessary to have that

$$\sqrt{\lambda_1\lambda_2 + 1} \leq \frac{\lambda_1 + \lambda_2}{2}.$$

The case $\lambda = \{1, 2\}$, for example, does not satisfy the inequality. The corresponding AIEP, therefore, is not real solvable.

It is easy to see that the necessary condition in the above 2×2 example is equivalent to $(\lambda_1 - \lambda_2)^2 \geq 4$, implying that the gap between the two eigenvalues must be sufficiently large. Actually, it is the *total* gap that must be large enough, as is seen in the following theorem which is but one of the many necessary conditions that have been derived in the literature.

Theorem 3.8. Let entries of A_0 be denoted by $A_0 = [a_{ij}]_{i,j=1}^{n}$. If AIEP1 is solvable, then it is necessary that

$$\sum_{i\neq j}(\lambda_i - \lambda_j)^2 \geq 2n\sum_{i\neq j} a_{ij}a_{ji}. \tag{3.48}$$

Proof Without loss of generality, we may assume $\operatorname{diag}(A_0) = \mathbf{0}$. If $X = \operatorname{diag}(x)$ is a solution, then

$$\begin{aligned}
\sum_{i \neq j} (\lambda_i - \lambda_j)^2 - 2n \sum_{i \neq j} a_{ij} a_{ji} &= 2n \left(\sum_{i=1}^{n} \lambda_i^2 - \frac{\left(\sum_{i=1}^{n} \lambda_i\right)^2}{n} - \langle A_0, A_0^\top \rangle \right) \\
&= 2n \left(\sum_{i=1}^{n} x_i^2 - \frac{\left(\sum_{i=1}^{n} \lambda_i\right)^2}{n} \right) \\
&\geq 2n \left(\frac{\left(\sum_{i=1}^{n} x_i\right)^2}{n} - \frac{\left(\sum_{i=1}^{n} \lambda_i\right)^2}{n} \right) \\
&= 0,
\end{aligned}$$

because $\sum_{i=1}^{n} \lambda_i^2 = \operatorname{trace}(A_0 + X)^2 = \langle A_0 + X, A_0^\top + X \rangle$. □

On the other hand, there are also several sufficient conditions under which an AIEP is real solvable. Recall, for example, the result developed earlier by Biegler-König (1981b) that if the *minimal* gap $d(\boldsymbol{\lambda})$ between the prescribed eigenvalues (see (3.5)) is wide enough, then the LiPIEP is real solvable. The gap has something to do with the size of off-diagonal entries of A_0, which is measured by the function $\pi(M)$ defined in (3.4). We can imply directly from Theorem 3.3 the following results, although they have been proved independently in much earlier years.

Theorem 3.9. Given any specified eigenvalues $\{\lambda_k\}_{k=1}^{n} \subset \mathbb{R}$,

(i) (de Oliveira, 1970) If $d(\boldsymbol{\lambda}) > 4\pi(A_0)$, then AIEP1 is solvable.

(ii) (Hadeler, 1968) If $d(\boldsymbol{\lambda}) > 2\sqrt{3}\,(\pi(A_0 \circ A_0))^{1/2}$, then AIEP2 is solvable.

The fact that a necessary condition, such as (3.48), is too loose in that it measures the total gap (of eigenvalues) and that a sufficient condition is too restrictive in that it measures the minimal gap leaves a vacuum of knowledge for AIEPs in between.

Example 3.3. Consider the case when $A_0 = \begin{bmatrix} 0 & 4 \\ 1 & 0 \end{bmatrix}$. The eigenvalues of the matrix $A_0 + \operatorname{diag}(x_1, x_2)$ are always real and

$$d(\boldsymbol{\lambda}) = \sqrt{(x_1 - x_2)^2 + 16}.$$

Clearly, the AIEP with

$$\boldsymbol{\lambda} = \left\{ \frac{(x_1 + x_2) + \sqrt{(x_1 - x_2)^2 + 16}}{2}, \frac{(x_1 + x_2) - \sqrt{(x_1 - x_2)^2 + 16}}{2} \right\}$$

is always real solvable with solution $\{x_1, x_2\}$, even if $d(\boldsymbol{\lambda}) < 16$. The example illustrates the discrepancy between the actual $d(\boldsymbol{\lambda})$ and the estimate by de Oliveira (1970). The sufficient condition apparently is too restrictive.

Finally, from Theorem 3.4, we conclude this section with the following result on insolvability.

Theorem 3.10. Both AIEP1 and AIEP2 are unsolvable almost everywhere if a multiple eigenvalue is present in $\{\lambda_k\}_{k=1}^n$.

3.3.2 *Sensitivity and stability (for AIEP2)*

With $A_k = \mathbf{e}_k\mathbf{e}_k^\top$, $k = 1, \ldots, n$, the sensitivity analysis for the AIEP2 follows directly from Theorem 3.6.

Theorem 3.11. Suppose that the AIEP2 has a solution at $X = \operatorname{diag}(x_1, \ldots, x_n)$. With the spectral decomposition $A(X) := A_0 + X = Q(X)\Lambda Q(X)^\top$, assume the matrix

$$J(X) := [q_{ij}^2(X)]_{i,j=1}^n \tag{3.49}$$

is nonsingular and define

$$\mathbf{b}(X) := [\mathbf{q}_1(X)^\top A\mathbf{q}_1(X), \ldots, \mathbf{q}_n(X)^\top A\mathbf{q}_n(X)]^\top.$$

If the perturbation

$$\delta = \|\boldsymbol{\lambda} - \tilde{\boldsymbol{\lambda}}\|_\infty + \|A_0 - \tilde{A}_0\|_2$$

is small enough, then:

(i) The AIEP2 associated with $\tilde{A}$ and $\tilde{\boldsymbol{\lambda}}$ is solvable.

(ii) There is a solution $\tilde{X}$ near to X, that is,

$$\frac{\|X - \tilde{X}\|_\infty}{\|X\|_\infty} \leq \kappa_\infty(J(X)) \left(\frac{\|\boldsymbol{\lambda} - \tilde{\boldsymbol{\lambda}}\|_\infty + \|A_0 - \tilde{A}_0\|_2}{\|\boldsymbol{\lambda} - \mathbf{b}\|_\infty} \right) + O(\delta^2).$$

Be aware that by emphasizing the AIEP2 in the above theorem, we have deliberately assumed that the structure of the basis matrices A_k is not altered. In other words, the theorem asserts that there exists a diagonal matrix $\tilde{X}$ such that $\sigma(\tilde{A}_0 + \tilde{X}) = \tilde{\boldsymbol{\lambda}}$.

The sensitivity of an AIEP2 to perturbations can be weighted from another point of view. Suppose that $\operatorname{diag}(\hat{\mathbf{c}})$ represents an approximate solution to a given AIEP2. Recall that the backward error analysis of most numerical methods leading to $\hat{\mathbf{c}}$ would generally result in a relationship in the form

$$\sigma\left((A_0 + \Delta A_0) + \operatorname{diag}(\hat{\mathbf{c}})\right) = \boldsymbol{\lambda} + \Delta\boldsymbol{\lambda}, \tag{3.50}$$

where $\Delta A_0 \in \mathcal{S}(n)$ and $\Delta \boldsymbol{\lambda} \in \mathbb{R}^n$ represent perturbations of A_0 and $\boldsymbol{\lambda}$, respectively. Consider the quantity

$$\zeta(\Delta A_0, \Delta \boldsymbol{\lambda}; \hat{\mathbf{c}}) := \sqrt{\left(\frac{\|\Delta A_0\|}{\|A_0\|}\right)^2 + \left(\frac{\|\Delta \boldsymbol{\lambda}\|}{\|\boldsymbol{\lambda}\|}\right)^2}, \tag{3.51}$$

which measures the closeness of the inverse problem that has been exactly solved by $\hat{\mathbf{c}}$ to the original problem. Different algorithms used to produce $\hat{\mathbf{c}}$ usually will end up with different quantities of ΔA_0 and $\Delta \boldsymbol{\lambda}$. Define

$$\eta(\hat{\mathbf{c}}) := \min_{\Delta A_0, \Delta \boldsymbol{\lambda}} \zeta(\Delta A_0, \Delta \boldsymbol{\lambda}; \hat{\mathbf{c}}). \tag{3.52}$$

We shall say that $\hat{\mathbf{c}}$ is a *backward stable solution* and, hence, the algorithm producing it is backward stable if $\eta(\hat{\mathbf{c}})$ is small. Given $\hat{\mathbf{c}}$, denote

$$\sigma(A_0 + \operatorname{diag}(\hat{\mathbf{c}})) = \hat{\boldsymbol{\lambda}}. \tag{3.53}$$

In the paper (Sun, 1999), it is proved that

$$\eta(\hat{\mathbf{c}}) = \frac{\|\boldsymbol{\lambda} - \hat{\boldsymbol{\lambda}}\|_\infty}{\sqrt{\|A_0\|_2 + \|\boldsymbol{\lambda} - \hat{\boldsymbol{\lambda}}\|_\infty}}. \tag{3.54}$$

In other words, (3.54) shows that so long as the spectrum $\hat{\boldsymbol{\lambda}}$ of $A_0 + \operatorname{diag}(\hat{\mathbf{c}})$, with $\hat{\mathbf{c}}$ being the approximate solution to the original AIEP, is close to the prescribed $\{\lambda_k\}_{k=1}^n$, we may assume that $\hat{\mathbf{c}}$ is a backward stable solution. The common practice of checking the difference between $\hat{\boldsymbol{\lambda}}$ and $\boldsymbol{\lambda}$ to verify whether an AIEP2 is solved numerically is therefore justified in this context.

3.3.3 *Numerical methods*

AIEPs, particular the subclass where the add-on matrix X is diagonal, can be tackled numerically by many effective algorithms. We shall set out only two of them in this section. Something specific about these methods should be noted first.

We have observed that most methods for symmetric or Hermitian IEPs, including LiPIEP2 and AIEP2, depend heavily on the fact that the eigenvalues are real and can be totally ordered. That is, each eigenvalue λ_k can be identified individually. Because of this ordering, the corresponding eigenvectors can be arranged accordingly and, hence, the IEP can be conveniently transformed as an algebraic system of nonlinear equations. Standard techniques, with suitable modifications if necessary, can be utilized to resolve this system of equations.

For problems involving general matrices, including AIEP1 and AIEP3, spectral information is usually expected to be complex-valued. Since complex numbers do not form an ordered field, we cannot immediately identify eigenvalues in an orderly manner. Suppose that these eigenvalues evolve in the course of computation by an iterative scheme. Then what we will see is a set of n points in the complex plane which evolves into another set of n points in the complex plane. Tracking how each eigenvalue has evolved into the next stage requires some kind of matching mechanism. Under such a circumstance, it might be more effective to trail the evolution continually. The homotopy method offers a natural way to track each individual eigenvalue curve as they are predetermined by initial values.

Newton's method (for AIEP2) We have already detailed the Newton method for LiPIEP2 in Section 3.2.4. The AIEP2 is just a special case of the LiPIEP2 with basis matrices $A_k = \mathbf{e}_k\mathbf{e}_k^\top$. For reference, we assume at the ν-th iterate that the spectral constraint is satisfied by the matrix $Z^{(\nu)} \in \mathcal{M}_e(\Lambda)$ with spectral decomposition $Z^{(\nu)} = Q^{(\nu)}\Lambda {Q^{(\nu)}}^\top$ and that the structural constraint is satisfied by the matrix $A(X^{(\nu)}) := A_0 + X^{(\nu)}$ where $X^{(\nu)} = \operatorname{diag}\left(x_1^{(\nu)}, \ldots, x_n^{(\nu)}\right)$. Starting with an initial value $Z^{(0)}$, we iterate as follows:

Algorithm 3.1 (Newton's method for AIEP2)
Assume that $\{\lambda_k\}_{k=1}^n$ are all distinct. For $\nu = 0, 1, \ldots$ until convergence, do

(i) Define

$$J^{(\nu)} := \left[{q_{ij}^{(\nu)}}^2\right]_{i,j=1}^n,$$

$$\mathbf{b}^{(\nu)} := \left[{\mathbf{q}_1^{(\nu)}}^\top A_0 \mathbf{q}_1^{(\nu)}, \ldots, {\mathbf{q}_n^{(\nu)}}^\top A_0 \mathbf{q}_n^{(\nu)}\right]^\top.$$

(ii) Solve $J^{(\nu)}[x_1^{(\nu+1)}, \ldots, x_n^{(\nu+1)}]^\top = \boldsymbol{\lambda} - \mathbf{b}^{(\nu)}$ for $X^{(\nu+1)}$.

(iii) Define the skew-symmetric matrix

$$\tilde{K}^{(\nu)} := \left[\frac{{\mathbf{q}_i^{(\nu)}}^\top A(X^{(\nu+1)})\mathbf{q}_j^{(\nu)}}{\lambda_j - \lambda_i}\right], \quad 1 \le i < j \le n.$$

(iv) Update the lift by

$$Q^{(\nu+1)} := Q^{(\nu)}\left(I + \frac{\tilde{K}^{(\nu)}}{2}\right)\left(I - \frac{\tilde{K}^{(\nu)}}{2}\right)^{-1}.$$

Note that the above algorithm performs iterations on the manifold $\mathcal{O}(n)$. The sequence $\{Q^{(\nu)}\}$ of orthogonal matrices can be generated without reference to either the sequence $\{A(X^{(\nu)})\}$ of structured matrices or the sequence $\{Z^{(\nu)}\}$ of isospectral matrices.

Homotopy method (for AIEP3) The homotopy method often gives rise to both an existence proof and a numerical method for finding all solutions. It has been used as a global method for solving nonlinear algebraic equations.

We first outline the general ideas behind a homotopy method. Suppose that the roots, if there are any, of a nonlinear function $\mathbf{f} : \mathbb{R}^n \to \mathbb{R}^n$ are to be found. We establish a homotopy $\mathbf{h} : \mathbb{R}^n \times \mathbb{R} \to \mathbb{R}^n$ via, for example,

$$\mathbf{h}(\mathbf{x}, t) = (1 - t)\mathbf{g}(\mathbf{x}) + t\mathbf{f}(\mathbf{x}), \tag{3.55}$$

where $\mathbf{g} : \mathbb{R}^n \to \mathbb{R}^n$ is some function whose zeros are known. One trivial choice, for instance, is $g(\mathbf{x}) = f(\mathbf{x}) - \mathbf{x}_0$ with an arbitrary $\mathbf{x}_0 \in \mathbb{R}^n$. Under some mild assumptions such as $\mathbf{0}$ being a *regular value* for $\mathbf{h}$, the zero set

$$\mathbf{h}^{-1}(0) = \{(\mathbf{x}, t) \in \mathbb{R}^n \times \mathbb{R} | \mathbf{h}(\mathbf{x}, t) = 0\}$$

forms a one-dimensional manifold. This zero curve can be traced via the implicitly defined differential system with respect to the arc length parameter s:

$$\left[\frac{\partial \mathbf{h}}{\partial \mathbf{x}} \frac{\partial \mathbf{h}}{\partial t}\right] \begin{bmatrix} \frac{d\mathbf{x}}{ds} \\ \\ \frac{dt}{ds} \end{bmatrix} = \mathbf{0}, \tag{3.56}$$

$$x(0) = \text{any known zero of } g(\mathbf{x}),$$

$$t(0) = 0,$$

$$\left\| \frac{d\mathbf{x}}{ds} \right\|^2 + \left| \frac{dt}{ds} \right|^2 = 1. \tag{3.57}$$

The idea is that when the integration goes along and $t(s)$ hits the value 1 for some parameter s, the corresponding $\mathbf{x}(s)$ is a root of $\mathbf{f}$.

Two main concerns in a typical homotopy application are that somehow we have to ensure that the zero curve will not escape to infinity and that the curve indeed makes a connection between $t = 0$ and $t = 1$. The former can be addressed if we can show that the zero remains bounded throughout the homotopy. The latter can be addressed if, for example, we can show that the matrix $\partial \mathbf{h}/\partial \mathbf{x}$ is of full rank. If both concerns can be properly addressed, then we effectively have established the degree theory in that every zero of $\mathbf{g}(\mathbf{x})$ is guaranteed to be connected to a zero of $\mathbf{f}(\mathbf{x})$.

The application of the homotopy method to the AIEP3 can be modelled as follows. First, the ideas of homotopy (3.55) outlined above can be extended equally well to the complex field $\mathbb{C}$. We have already argued in the proof of Theorem 3.7 that solving the AIEP3 is equivalent to solving the polynomial system $\mathbf{f}(\mathbf{x}) = [\varphi_1(\mathbf{x}), \ldots, \varphi_n(\mathbf{x})]^\top$ defined in (3.47). It thus remains to select an appropriate system $\mathbf{g}(\mathbf{x})$ to establish the homotopy. We suggest the employment of

$$g_k(\mathbf{x}) = \omega_k(\mathbf{x}) - \omega_k(\hat{\mathbf{x}}), \quad k = 1, \ldots, n, \tag{3.58}$$

where $\hat{\mathbf{x}}$ is a randomly selected point in $\mathbb{C}^n$. The following theorem uses the so-called parameterized Sard theorem to prove that the two concerns mentioned above are completely dealt with. The details are given in Chu (1990). A similar but somewhat more sophisticated approach can be found in Alexander (1978).

Theorem 3.12. (Chu, 1990) For almost all $\hat{\mathbf{x}} \in \mathbb{C}^n$ with distinct components, the zero set $\mathbf{h}^{-1}(0)$ contains $n!$ analytic curves parameterized by t. Each of these curves connects a point $(\mathbf{x}^0, 0)$ where the components of $\mathbf{x}^0$ are a permutation of entries $\hat{x}_1, \ldots, \hat{x}_n$ of $\hat{\mathbf{x}}$ to a point $(\mathbf{x}^*, 1)$ where $\mathbf{x}^*$ is a solution to the AIEP3.

Be aware that the homotopy method is proposed over the complex field even if $A_0 \in \mathbb{R}^{n\times n}$. It is possible that the AIEP1 has no real solution at all. We have no knowledge of homotopy techniques for finding real-valued solutions of an AIEP.

The homotopy method offers a globally convergent method for finding *all* solutions. Since $n!$ grows rapidly, it is often the case that not all solutions are needed. Even so, the approach offers a systematic way of searching for different solutions simply by permuting the components of the vector $\hat{\mathbf{x}}$. The theory guarantees that with exactly $n!$ trials, all solutions to the AIEP3 would have been found. Although this number of trials still appears too big, it is better than an indefinite and often unpredictable number of restarts required by a locally convergent method to find all solutions. This feature can also be employed to build a hybrid method in that low order of accuracy is used in a homotopy to get into the proximity of a solution and then a faster and locally convergent method is turned on to improve the accuracy.

3.4 Multiplicative inverse eigenvalue problems

In exactly the same way as in the AIEPs, the set $\mathcal{N}$ in Problem 3.7 can be used to specify the structure of the pre-multiplier X in an MIEP. In the preconditioning application, for instance, we can go as far as taking $X = A_0^{-1}$ to perfectly condition the matrix A_0, which of course assumes no predesignated structure and whose computation is generally impractical for preconditioning purposes. Some simpler structure inherent in X therefore is expected. We do not think that this interesting idea has been fully exploited yet. Questions such as what are all the possible eigenvalues achievable by a pre-multiplication of matrices from a fixed set $\mathcal{N}$ to a fixed matrix A_0 certainly are of theoretical interest in their own right (Gohberg et al., 1995).

Example 3.4. Suppose the pre-multiplier matrix X is of the form

$$\begin{bmatrix} x_1 & x_2 & x_3 & \dots & x_n \\ 0 & 1 & 0 & & 0 \\ 0 & 0 & 1 & & 0 \\ \vdots & & & \ddots & \vdots \\ 0 & & & \dots & 1 \end{bmatrix},$$

that is, suppose that only the first row of A_0 is allowed to be modified as linear combinations of all rows. What are the achievable eigenvalues by XA_0? Consider the case $A_0 = \begin{bmatrix} 1 & 0 \\ 0 & 1 \end{bmatrix}$. Then it is clear that eigenvalues of the matrix $XA_0 = \begin{bmatrix} x_1 & x_2 \\ 0 & 1 \end{bmatrix}$ can only be $\{x_1, 1\}$. On the other hand, with $A_0 = \begin{bmatrix} 1 & 4 \\ 3 & 1 \end{bmatrix}$, the eigenvalues of the matrix $XA_0 = \begin{bmatrix} x_1 + 3x_2 & 4x_1 + x_2 \\ 3 & 1 \end{bmatrix}$ satisfy the system

$$\begin{cases} \lambda_1 \lambda_2 = 11x_1, \\ \lambda_1 + \lambda_2 = 1 + x_1 + 3x_2. \end{cases}$$

Given any $\boldsymbol{\lambda} \in \mathbb{R}^2$, we can always find a pair (x_1, x_2) of real numbers that solves the MIEP. Indeed, the solution is unique in this case.

In many other applications, the matrix X often assumes the diagonal structure as we have seen in Problem 2.3. Even so, the following variations distinguish themselves from others.

Problem 3.14 *(MIEP1)*
Solve the MIEP when $A_0 \in \mathbb{R}^{n\times n}$, $X \in \mathcal{D}_{\mathbb{R}}(n)$, and $\{\lambda_k\}_{k=1}^n \subset \mathbb{R}$.

Problem 3.15 *(MIEP2)*
Solve the MIEP when $A_0 \in \mathcal{S}(n)$ and is positive definite, $X \in \mathcal{D}_{\mathbb{R}}(n)$ and is nonnegative, and $\{\lambda_k\}_{k=1}^n \subset \mathbb{R}$.

Example 3.5. Consider the case $A_0 = \begin{bmatrix} 2 & -1 \\ -1 & 2 \end{bmatrix}$. Then the MIEP2 is algebraically equivalent to finding the inverse function of the map $\mathbf{f} : \mathbb{R}_+^2 \to \mathbb{R}_+^2$ where

$$\mathbf{f}(x_1, x_2) = \begin{bmatrix} x_1 + x_2 + \sqrt{x_1^2 - x_1x_2 + x_2^2} \\ x_1 + x_2 - \sqrt{x_1^2 - x_1x_2 + x_2^2} \end{bmatrix}.$$

The range $\mathbf{f}(\mathbb{R}_+^2)$ is shown in Figure 3.2 where the slanted boundaries have slopes 3 and $\frac{1}{3}$, respectively. Only for those eigenvalues $\boldsymbol{\lambda}$ from the shaded region will the MIEP2 be solvable.

Problem 3.16 *(MIEP3)*
Solve the MIEP when $A_0 \in \mathbb{C}^{n\times n}$, $X \in \mathcal{D}_{\mathbb{C}}(n)$, and $\{\lambda_k\}_{k=1}^n \subset \mathbb{C}$.

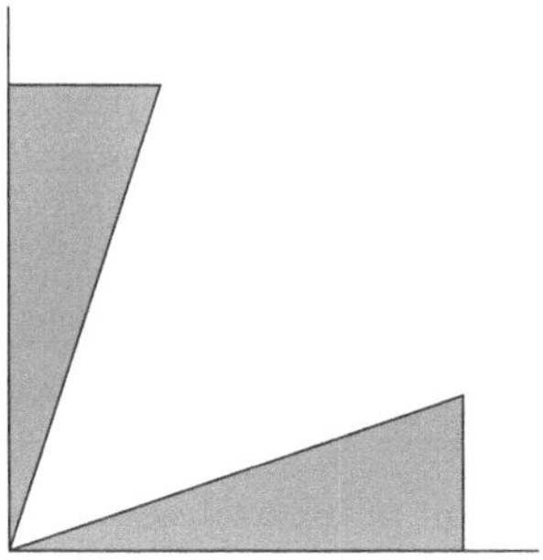

Figure 3.2. *Range of solvable* $\{\lambda_1, \lambda_2\}$ *for MIEP2*

Problem 3.17 *(MIEP4)*
Given $A_0 \in \mathcal{H}(n)$ and $\{\lambda_k\}_{k=1}^n \subset \mathbb{R}$, find $X \in \mathcal{D}_{\mathbb{R}}(n)$ such that $\sigma(X^{-1}A_0X^{-1}) = \{\lambda_k\}_{k=1}^n$.

Problem 3.17 is perhaps one of the earliest IEPs in matrix form first proposed by Downing and Householder (1956). The problem is closely related to the question of determining a diagonal matrix X such that $X^{-1}A_0X^{-1}$ is best conditioned in the sense that the ratio of the maximal eigenvalue to the minimal eigenvalue of $X^{-1}A_0X^{-1}$ is a minimum. See Theorem 3.15.

3.4.1 *Solvability*

Most of the discussion for MIEPs is concentrated on the case when X is diagonal. We have seen in the formulation (3.2) that such an MIEP is a special case of the LiPIEP with basis matrices $A_k = \mathbf{e}_k\mathbf{a}_k^\top$ where $\mathbf{a}_k^\top$ is the k-th row of A_0. While an AIEP can be cast in the larger context to become a PEIEP and has enjoyed considerable attention in the literature, we are not aware of as much discussion for MIEPs as for AIEPs. This is not because MIEPs are easier. Rather, many interesting questions remain to be answered for MIEPs even with modest variations.

Complex solvability It is truly amazing to see that, even with the very limited structure that X is diagonal, a matrix A_0 can be pre-multiplied to retain any prescribed spectrum in the complex plane. This is in contrast to Example 3.5 we have just examined. More precisely, the complex solvability is completely answered in the following theorem.

Theorem 3.13. (Friedland, 1975) Assume that all principal minors of A_0 are distinct from zero. Then:

(i) For any specified $\{\lambda_k\}_{k=1}^n \in \mathbb{C}$, the MIEP3 is solvable.
(ii) There are at most $n!$ solutions.

We stress again by Theorem 3.13 that a diagonal pre-multiplier can reposition eigenvalues of XA_0 anywhere in the complex plane. In particular, it asserts a matrix A_0 can be perfectly conditioned by a diagonal matrix. It has to be noted, however, that this powerful and elegant result does have its limitation. That is, the cost of repositioning eigenvalues of XA_0 may be too expensive to be of practical usage. It would be desirable to have an efficient algorithm to implement Theorem 3.13, but we are not aware of any numerical procedure to do this.

Real solvability To apply Theorem 3.3, we will have to assume that the diagonal elements of A_0 are normalized to 1, that is, $a_{ii} = 1$. This is easy to do because, otherwise, we simply need to pre-multiply A_0 by the diagonal matrix $\operatorname{diag}(\operatorname{diag}(A_0))^{-1}$.

Theorem 3.14. Assume that $\operatorname{diag}(A_0) = [1, \ldots, 1]^\top$ and $\pi(A_0) < 1$. If

$$d(\boldsymbol{\lambda}) \geq \frac{4\pi(A_0)\|\boldsymbol{\lambda}\|_\infty}{1 - \pi(A_0)},$$

then the MIEP1 is solvable.

Some obvious necessary conditions for the MIEP1 to be solvable include, for example, that $\sum_{i=1}^n x_i = \sum_{i=1}^n \lambda_i$ and $\det(A_0)\prod_{i=1}^n x_i = \prod_{i=1}^n \lambda_i$.

Optimal conditioning by diagonal matrices Related to the IEPs is an interesting question of finding the optimal preconditioner with a specified sparsity pattern (Greenbaum, 1992). This question has not been satisfactorily answered even to this date. In passing we mention one classical result by Forsythe and Straus (1955) in this regard.

Theorem 3.15. (Forsythe and Straus, 1955) Suppose A_0 is symmetric, positive definite, and has property $\mathcal{A}$, that is, suppose A_0 can be symmetrically permuted into

$$\begin{bmatrix} D_1 & B \\ B^\top & D_2 \end{bmatrix},$$

where D_1 and D_2 are diagonal. Let $D = \operatorname{diag}(\operatorname{diag}(A_0))$. Then

$$\kappa_2(D^{-1/2}AD^{-1/2}) = \min_{\widehat{D}>0,\widehat{D}=\text{diagonal}} \kappa_2(\widehat{D}A\widehat{D}).$$

Theorem 3.15 has been generalized to various extents. For instance, Bauer (1963) considered the problem of minimizing $\kappa(D_1AD_2)$ with respect to a given norm by nonsingular diagonal matrices D_1 and D_2. One of his results shows that the minimum of $\kappa_\infty(D_1AD_2)$ is equal to the Perron root of $|A|\,|A^{-1}|$. When the Frobenius norm or any one of the Hölder norms are used in defining $\kappa(A)$, van der Sluis (1969/1970a) showed that the minimizing diagonal matrices D_1 and D_2 in either cases of $\kappa(AD_1)$ or $\kappa(D_2A)$ are unique. See also Golub and Varah (1974) for a characterization of the best ℓ_2-scaling and Eisenstat et al. (1982) for a generalization to the block diagonal scaling.

3.4.2 *Sensitivity (for MIEP2)*

The result in Theorem 3.6 cannot be applied directly because the basis matrices $A_k = \mathbf{e}_k\mathbf{a}_k^\top$ are not symmetric. Observe that the MIEP2 is equivalent to the symmetrized problem

$$X^{-1/2}(XA_0)X^{1/2} = X^{1/2}A_0X^{1/2},$$

but the latter is not linear in X.

Theorem 3.16. Suppose that the MIEP2 has a solution at $X = \operatorname{diag}(x_1,\dots,x_n)$ with $x_i > 0$ for all i. With the spectral decomposition $X^{1/2}A_0X^{1/2} = U(X)\Lambda U(X)^\top$, define

$$W(X) := [u_{ji}^2(X)]_{i,j=1}^n,$$

where $U(X) = [u_{ij}(X)]_{i,j=1}^n$. Assume that $W(X)$ is nonsingular and that the perturbation

$$\delta = \|\boldsymbol{\lambda} - \tilde{\boldsymbol{\lambda}}\|_\infty + \|A_0 - \tilde{A}_0\|_2$$

is small enough. Then:

(i) The MIEP2 associated with $\tilde{A}_0$ and $\tilde{\boldsymbol{\lambda}}$ is solvable.
(ii) There is a solution $\tilde{X}$ in a neighborhood of X, that is,

$$\frac{\|X-\tilde{X}\|_\infty}{\|X\|_\infty} \le \frac{\lambda_n}{\lambda_1}\|W(X)^{-1}\|_\infty \left(\frac{\|\boldsymbol{\lambda}-\tilde{\boldsymbol{\lambda}}\|_\infty}{\|\boldsymbol{\lambda}\|_\infty} + \|A_0 - \tilde{A}_0\|_2\right) + O(\delta^2).$$

Proof Using (3.12), we find that the Jacobian matrix of $\boldsymbol{\lambda}$ with respect to $\mathbf{x} = [x_1,\dots,x_n]^\top$ can be conveniently summarized as

$$\left[\frac{\partial \boldsymbol{\lambda}}{\partial \mathbf{x}}\right] = U(X)^\top X^{-1/2} \circ \left(A_0X^{1/2}U(X)^\top\right)^\top, \tag{3.59}$$

where $\circ$ stands for the element-to-element Hadamard product between two matrices. With the modification of (3.16) to

$$B(t,\mathbf{y}) = \operatorname{diag}(\operatorname{diag}(\mathbf{y}))(A_0 + tE_0)$$

and the fact that (3.18) becomes

$$\left.\frac{\partial \mathbf{f}}{\partial \mathbf{x}}\right|_{(0,\mathbf{x})} = \Lambda W X^{-1}, \tag{3.60}$$

the proof now is essentially the same as that for Theorem 3.6. □

Converting MIEP2 to LiPIEP2 Another way to convert the MIEP2 into a symmetric LiPIEP is worth mentioning. This approach does not deal with the product XA_0 directly, but rather changes to work on a new LiPIEP2 which has the same solution X as the original MIEP2.

Suppose $A_0 = LL^\top$ is the Cholesky decomposition of A_0. Note that the matrix XA_0 is similar to $X^{1/2}A_0DX^{1/2} = (X^{1/2}L)(X^{1/2})L)^\top$ which, in turn, has exactly the same eigenvalues as the matrix $(X^{1/2}L)^\top(X^{1/2}L) = L^\top XL$. The same diagonal matrix X and eigenvalues are involved in both XA_0 and $L^\top XL$. The latter may be considered as solving the LiPIEP with $A_0 = 0$ and basis matrices

$$A_k = L^\top \text{diag}(\mathbf{e}_k)L, \quad k = 1, \ldots, n. \tag{3.61}$$

Note that each A_k is now in $\mathcal{S}(n)$.

Though it requires an extra Choleskey decomposition, the conversion has the advantage that the sensitivity result in Theorem 3.6 can now be applied with this new definition (3.61) as basis matrices. Also the numerical methods developed earlier for the LiPIEP2 can be applied directly.

3.4.3 *Numerical methods*

If the purpose is solely for the preconditioning of a given matrix, many effective techniques for selecting a preconditioner are available. There is no need to solve the MIEP precisely, which would be too expensive to be practical.

On the other hand, be aware that even though the MIEP is linear in X, it is not a symmetric LiPIEP even if A is symmetric. The numerical methods developed for LiPIEP2 need to be modified. In particular, if A is a Jacobi matrix, the problem can be solved by direct methods, which will be discussed in Chapter 4.

Reformulate MIEP1 as nonlinear equations The most straightforward way to solve an inverse problem is by considering it as an algebraic system of nonlinear equations. To solve the MIEP1, for example, we can consider solving $\mathbf{f}(\mathbf{x}) = 0$ where $\mathbf{f} : \mathbb{R}^n \to \mathbb{R}^n$. Some obvious ways of formulating $f(x)$ include, if we write $X = \text{diag}(\text{diag}(\mathbf{x}))$,

$$f_i(\mathbf{x}) := \det(XA_0 - \lambda_i I), \quad i = 1, \ldots, n, \tag{3.62}$$

or simply that

$$\mathbf{f}(\mathbf{x}) := \sigma(XA_0) - \boldsymbol{\lambda}. \tag{3.63}$$

In formulation such as (3.63), be aware that we have assumed an ordering among the eigenvalues. If XA_0 has complex eigenvalues, it is not a straightforward task to pair the computed $\sigma(XA_0)$ and the original $\boldsymbol{\lambda}$ together. Some additional matching mechanism might be needed in defining $\mathbf{f}$.

Newton's method (for MIEP2) For MIEP2s, $\sigma(XA_0) = \sigma(X^{1/2}A_0X^{1/2})$. It is guaranteed that XA_0 has a real spectrum. In formulating the MIEP2 as the nonlinear system (3.63), note that we have already calculated the Jacobian matrix,

$$\mathbf{f}'(\mathbf{x}) = \Lambda W X^{-1},$$

as in (3.60). The conventional Newton iteration therefore can be carried out in the usual way. Be cautioned that the solution requires that $\mathbf{x}$ be non-negative.

Alternatively, a modified Newton iteration has been proposed by Joseph (1992) to maintain the symmetry of the problem. The idea is to recast the MIEP2 as a symmetric-definite generalized eigenvalue problem

$$A_0\mathbf{q}_i = \lambda_i Y\mathbf{q}_i, \quad i = 1, \ldots, n,$$

with $Y = X^ck{-}1$. Iterations then take place to correct the value of Y with the desire of driving $\sigma(A_0 - \lambda Y)$ to $\boldsymbol{\lambda}$.

We remark that this approach can be generalized equally well to the case where Y is symmetric and positive definite, but not necessarily diagonal, making it useful for solving MIEPs arising from applied mechanics. See, for example, the system (2.11) arising in Section 2.3.

Without elaborating on more details, we give a summary of the iterative scheme below.

Algorithm 3.2 (Joseph–Newton method for MIEP2)
Given an initial guess $Y^{(0)}$, do the following iteration for $\nu = 0, 1, \ldots$ until convergence:

1. Solve the generalized eigenvalue problem

$$\left(A_0 - \lambda^{(\nu)}Y^{(\nu)}\right)\mathbf{q}^{(\nu)} = 0.$$

2. Normalize eigenvectors $\mathbf{q}_i^{(\nu)}$, $i = 1, \ldots, n$, in the sense that $\mathbf{q}_i^{(\nu)\top} Y^{(\nu)} \times \mathbf{q}_i^{(\nu)} = 1$.
3. Denote $Q^{(\nu)} = [\mathbf{q}_1^{(\nu)}, \ldots, \mathbf{q}_n^{(\nu)}] = [q_{ij}^{(\nu)}]_{i,j=1}^n$.
4. Define the Jacobian matrix for eigenvalues of the matrix pencil $A_0 - \lambda Y$,

$$J(Y^{(\nu)}) := \left[-\lambda_i^{(\nu)}\left(q_{ij}^{(\nu)}\right)^2\right]_{i,j=1}^n.$$

5. Solve $J(Y^{(\nu)})\mathbf{d}^{(\nu)} = \boldsymbol{\lambda} - \boldsymbol{\lambda}^{(\nu)}$.
6. Update $Y^{(\nu+1)} := Y^{(\nu)} + \mathrm{diag}(\mathrm{diag}(\mathbf{d}^{(\nu)}))$.

3.5 Summary

In this chapter, we have directed our attention toward one single object – that the structural constraint in an IEP is regulated by a set of parameters. At present, most of our understanding has been limited to the LiPIEPs which include the intensively studied AIEPs and MIEPs as special cases. Our development is far from being complete in many respects. We mention a few:

1. Most of our discussion is for the case where the number m of parameters used to regulate the structure is the same as the dimension n of the underlying matrices. Many of the fundamental questions such as solvability and computability need to be addressed when $m \neq n$.
2. Most of our discussion is for the case where the entire spectrum $\{\lambda_k\}_{k=1}^n$ is specified. This is not necessarily the situation in practice. In structure design, for example, often we will be satisfied with just a few low-order natural frequencies being matched. In fact, for engineering applications involving large and complicated structures, it is infeasible to expect that all frequencies or modes are available. How to formulate and solve the problem when only a few low-order frequencies, that is, the first few small eigenvalues, are given is a very practical and important question.
3. Given a set of basis matrices in an LiPIEP, what are the reachable eigenvalues by varying $\mathbf{c}$? For example, given the standard Jacobi matrix A_0 with nonzero row entries $[-1, 2, -1]$, what is the set of all reachable spectra of XA_0 via a nonnegative diagonal matrix X? This question, raised in Section 2.3, generalizes the eigenvalue completion problems considered in Gohberg et al. (1995).
4. Most of the Newton methods described thus far are for symmetric matrices only. How to generalize these methods effectively for general matrices?
5. Most of our discussion is for the case when no eigenvector information is involved. Redevelop the theory and algorithms when eigenvectors are part of the spectral constraints.

4

STRUCTURED INVERSE EIGENVALUE PROBLEMS

4.1 Overview

We have attributed IEPs to many applications where parameters of a certain physical system are to be determined from the knowledge or expectation of its dynamical behavior. We usually impose two constraints on an IEP. Spectral information is entailed because the dynamical behavior often is governed by the underlying natural frequencies and normal modes. Structural stipulation is designated because the physical system is often subject to some feasibility constraints. The spectral data involved may consist of complete or only partial information on eigenvalues or eigenvectors. The structure the matrices embodied can take many forms.

In the previous chapter, our emphasis has been on how the parameters regulate the problem. The objective of this chapter is to emphasize the structure. That is, we want to consider the following structured IEP (**SIEP**).

> **Problem 4.1** *(Generic SIEP)*
> Given a set of scalars $\Omega \subset \mathbf{F}$, find $X \in \mathcal{N}$ which consists of specially structured matrices such that $\sigma(X) \subset \Omega$.

Note that Problem 4.1 is very analogous to Problem 3.1. Indeed, these two definitions can imply each other. Given a characterization of $A(\mathbf{c})$, we can use it to define a special structure, that is, $\mathcal{N} = \{A(\mathbf{c}) \mid \mathbf{c} \in \mathbb{F}^m\}$. Conversely, given a characterization of $\mathcal{N}$, we might be able to specify, for example, how its members can be expressed in terms of a set of basis matrices and, hence, a LiPIEP. Insofar as every IEP could be regarded as an SIEP, that view certainly is too broad to be apprehended by prose of finite length. Thus, our scope of "structure" is limited to those structures delineated in this chapter. In particular, we shall focus on nine selected special structures. These are the IEPs for Jacobi matrices, Toeplitz matrices, nonnegative matrices, stochastic matrices, unitary matrices, matrices with prescribed entries, matrices with prescribed singular values, matrices with prescribed singular values and eigenvalues, and matrices with equality constraints. Our criteria of selection are simply that these nine problems are representative of a variety of structural constraints and are slightly better studied in the literature.

Some of these structures, such as Jacobi or Toeplitz, result in matrices forming linear subspaces; some structures, such as nonnegative or stochastic, limit entries of matrices in a certain range; while others, such as matrices with prescribed entries or with prescribed singular values, lead to some implicitly defined structural constraints. We shall touch upon a variety of SIEPs by describing their formulations, highlighting some theories or numerical procedures, and suggesting some pertinent references. The SIEPs enjoy quite a few classical results in matrix theory. From time to time, we shall point out some open questions.

Let it be noted that, as in the previous chapter, while we sometimes seem to be concentrating on one particular numerical method applied to one particular problem, often the method has enough generality that with some suitable modifications it can also be applied to other types of problems. We choose not to encumber readers with the details.

4.2 Jacobi inverse eigenvalue problems

By a Jacobi structure, we mean a symmetric, tridiagonal matrix of the form

$$J = \begin{bmatrix} a_1 & b_1 & 0 & & & 0 \\ b_1 & a_2 & b_2 & & & 0 \\ 0 & b_2 & a_3 & & & 0 \\ \vdots & & & \ddots & & \\ & & & & a_{n-1} & b_{n-1} \\ 0 & & & & b_{n-1} & a_n \end{bmatrix}, \tag{4.1}$$

with positive subdiagonal elements $b_i > 0$, $i = 1, \dots, n-1$. We have already seen in Chapter 2 that this structure arises in many important areas of applications, including oscillatory mass–spring systems, composite pendulum, and Sturm–Liouville problems. The eigenvalues of a Jacobi matrix are necessarily real and distinct.

Jacobi matrices enjoy many nice properties. These properties make the study of JIEPs the most complete and fruitful among other IEPs. For that reason, we shall provide somewhat more detail on the theory and development of JIEPs. We shall touch upon all four fundamental questions raised in Section 1.2.2 for JIEPs. We hope that this exertion can serve as a study guide for further developments of other IEPs.

Observe that J is characterized by the $2n-1$ unknown entries, $\{a_i\}_{i=1}^{n}$ and $\{b_j\}_{j=1}^{n-1}$. Thus it is intuitively true that $2n-1$ pieces of information are needed to solve the inverse problems. In other words, to fully describe a Jacobi inverse eigenvalue problem (**JIEP**), we need additional information other than just the spectrum of J. This additional information comes from different sources and defines additional structures for JIEPs. We shall survey a few JIEPs in this section. One unique and important feature for the JIEPs is that often the inverse problem can be solved by direct methods in finitely many steps.

Before we move on, we should emphasize that the JIEPs under discussion here are of tridiagonal structure only. The generalization to band matrices is possible. Some initial studies on the IEP for band matrices can be found in the paper by Biegler-König (1981a). In the review paper by Boley and Golub (1987) it was argued that some of the numerical methods for JIEPs could be generalized to the banded problems. However, be aware that there are some fundamental differences in the generalization. For instance, generally two sets of eigenvalues determine a tridiagonal matrix uniquely (see Theorem 4.1 in Section 4.2.3) whereas three sets of eigenvalues do not give a pentadiagonal matrix uniquely and, in fact, sometimes there is a continuum of solutions (Boley and Golub, 1987; Dai, 1988).

4.2.1 *Variations*

There are several variations in formulating a JIEP. Each formulation can be associated with a mass–spring system. Because the literature of JIEPs is very rich, we think it is appropriate to devote this section only to give some background information. We shall describe first the setup, a brief history and some relevant references on the original settings for each problem. Topics on physical interpretation, mathematical theory, and computational methods will be discussed in the next few sections.

In the following, J_k denotes the $k \times k$ principal submatrix of J and J_{n-1} is abbreviated as $\bar{J}$. Whenever it is possible, we refer to each variation by the same identification name as that used in Chu (1998).

Problem 4.2 *(SIEP6a)*
Given real scalars $n\{\lambda_k\}_{k=1}^n$ and $\{\mu_1, \ldots, \mu_{n-1}\}$ satisfying the interlacing property:

$$\lambda_i < \mu_i < \lambda_{i+1}, \quad i = 1, \ldots, n-1, \tag{4.2}$$

find a Jacobi matrix J such that

$$\begin{cases} \sigma(J) = \{\lambda_k\}_{k=1}^n, \\ \sigma(\bar{J}) = \{\mu_1, \ldots, \mu_{n-1}\}. \end{cases}$$

This problem SIEP6a is perhaps the most fundamental and extensively studied IEP in the literature. It appears that the problem was originally proposed by Hochstadt (1967), although Downing and Householder (1956) had formulated a more general inverse characteristic value problem much earlier. Much of the existence theory and continuous dependence of the solution on data were developed later in Hochstadt (1974), Gray and Wilson (1976), and Hald (1972, 1976). Dangerously many numerical methods are available! We use the word "dangerous" because some of the methods are stable while many others are subtly unstable. We shall discuss some of these theories and methods later.

For the time being, it is sufficient to mention some representative work, including Boley and Golub (1987), de Boor and Golub (1978), Erra and Philippe (1997), Gragg and Harrod (1984), Hochstadt (1979), Parlett (1998), in this regard.

Problem 4.3 *(SIEP2)*
Given real scalars $\{\lambda_k\}_{k=1}^n$, find a Jacobi matrix J such that

$$\begin{cases} \sigma(J) = \{\lambda_k\}_{k=1}^n, \\ a_i = a_{n+1-i}, \\ b_i = b_{n+2-i}. \end{cases}$$

A matrix that is symmetric with respect to the unit perdiagonal matrix $\Xi = [\xi_{ij}]$, defined by

$$\xi_{ij} = \begin{cases} 1, & \text{if } i = n+1-j, \\ 0, & \text{otherwise}, \end{cases}$$

is said to be persymmetric. The structure required in SIEP2 is persymmetric Jacobi. A persymmetric Jacobi matrix involves only n independent entries. The spectral constraint therefore requires the spectrum information only. This problem was first considered in Hochstadt (1967) and then in Hald (1976). Numerical methods for the SIEP2 usually come along with those for the SIEP6a with appropriate modifications (de Boor and Golub, 1978; Parlett, 1998).

Problem 4.4 *(SIEP7)*
Given real scalars $\{\lambda_k\}_{k=1}^n$ and $\{\mu_1, \dots, \mu_{n-1}\}$ satisfying the interlacing property

$$\begin{cases} \lambda_i \le \mu_i \le \lambda_{i+1}, \\ \mu_i < \mu_{i+1}, \end{cases} \qquad i = 1, \dots, n-1 \tag{4.3}$$

and a positive number β, find a periodic Jacobi matrix J of the form

$$J = \begin{bmatrix} a_1 & b_1 & & & & b_n \\ b_1 & a_2 & b_2 & & & 0 \\ 0 & b_2 & a_3 & & & 0 \\ \vdots & & & \ddots & & \\ & & & & a_{n-1} & b_{n-1} \\ b_n & & & & b_{n-1} & a_n \end{bmatrix}$$

such that

$$\begin{cases} \sigma(J) = \{\lambda_k\}_{k=1}^n, \\ \sigma(\bar{J}) = \{\mu_1, \dots, \mu_{n-1}\}, \\ \prod_1^n b_i = \beta. \end{cases}$$

A periodic Jacobi matrix differs from a Jacobi matrix in that its eigenvalues need not be strictly separated. The interlacing property (4.3) therefore differs from (4.2) in that equalities are allowed. The notion of periodic Jacobi matrices arises in applications such as periodic Toda lattices (Adler et al., 1993) or continued fractions (Andrea and Berry, 1992). Spectral properties of the periodic Jacobi matrices were first analyzed by Ferguson (1980) using a discrete version of Floquet theory, but numerical methods had been proposed even earlier in Boley and Golub (1978). See also discussions in Booley and Golub (1984, 1987).

Problem 4.5 *(SIEP8)*
Given real scalars $\{\lambda_k\}_{k=1}^{n}$ and $\{\mu_1, \ldots, \mu_n\}$ satisfying the interlacing property

$$\lambda_i < \mu_i < \lambda_{i+1}, \quad i = 1, \ldots, n,$$

with $\lambda_{n+1} = \infty$, find Jacobi matrices J and $\tilde{J}$ so that

$$\begin{cases} \sigma(J) = \{\lambda_k\}_{k=1}^{n}, \\ \sigma(\tilde{J}) = \{\mu_1, \ldots, \mu_n\}, \\ J - \tilde{J} \neq 0, \quad \text{only at the } (n,n) \text{ position.} \end{cases}$$

This problem originally appeared in de Boor and Golub (1978). Note that $\tilde{J}$ is a special rank one update of J. This problem is closely related to the SIEP6a in that the theory and numerical methods for the SIEP6a will work almost identically for the SIEP8. A similar problem involving the preconditioning of a matrix by a rank one matrix is Problem 2.11 mentioned earlier in Section 2.6. An application of rank one updating involving the inverse quadratic eigenvalue problem can be found in Datta et al. (1997), Ram (1995).

Problem 4.6 *(SIEP9)*
Given distinct real scalars $\{\lambda_k\}_{k=1}^{2n}$ and an $n \times n$ Jacobi matrix $\tilde{J}$, find a $2n \times 2n$ Jacobi matrix J so that

$$\begin{cases} \sigma(J) = \{\lambda_k\}_{k=1}^{2n}, \\ J_n = \tilde{J}. \end{cases}$$

The SIEP9, first discussed in Hochstadt (1979), corresponds exactly to the problem of computing the Gaussian quadrature of order $2n$ with degree of precision $4n-1$, given the Gaussian quadrature of order n with degree of precision $2n-1$. Several numerical algorithms are available. See Boley and Golub (1987). An IEP as such is actually a special case of the more general PEIEPs which, in turn, is a subset of the so-called completion problems in the literature. We shall discuss the completion problems in Section 4.7. The prescribed entries need not to be in a diagonal block as in SIEP9.

Observe in the SIEP9 that there are n^2 entries specified in the upper-left corner of J. We stress the number and the location of the specified entries. An interesting question related to the general PEIEP is to find the largest permissible cardinality of the prescribed entries so that the completed matrix has a prescribed spectrum. This problem has aroused considerable interest in the field starting with the work by London and Minc (1972), followed by the series of articles (de Oliveira, 1973a,b, 1975). A recent survey on this particular subject by Ikramov and Chugunov (2000) suggests that thus far the strongest result in this class of problem would be that contained in the thesis by Hershkowitz (1983). We shall chronicle an interesting history of development in Section 4.7. Also presented in Ikramov and Chugunov (2000) is a careful treatment showing how the completion problems can be solved by finite rational algorithms.

There are several other formulations with features similar to the SIEP9. Constructing matrices with prescribed entries and characteristic polynomial, for example, has been considered by Dias da Silva (1974). Inverse problems for matrices with prescribed characteristic polynomial and principal submatrices have been studied by Silva (1987a); and for matrices with prescribed spectrum and principal submatrices in Silva (1987b).

Problem 4.7 *(SIEP6b)*
Given complex scalars $\{\lambda_1, \dots, \lambda_{2n}\}$ and $\{\mu_1, \dots, \mu_{2n-2}\}$, distinct and closed under complex conjugation, find tridiagonal symmetric matrices C and K for the λ-matrix $Q(\lambda) = \lambda^2 I + \lambda C + K$ so that

$$\begin{cases} \sigma(Q) = \{\lambda_1, \dots, \lambda_{2n}\}, \\ \sigma(\tilde{Q}) = \{\mu_1, \dots, \mu_{2n-2}\}. \end{cases}$$

Clearly, the SIEP6b is an analogy of the SIEP6a applied to a damped system. Strictly speaking, to maintain the physical feasibility, a practical solution imposes additional conditions on K and C, that is, both matrices are supposed to have positive diagonal entries, negative off-diagonal entries, and be weakly diagonally dominant. The setup of SIEP6b, where two sets of eigenvalues are given, was considered by Ram and Elhay (1996). There are other types of settings, such as the partial eigenstructure assignment problem described in Problem 2.6. Similar inverse problems for quadratic pencils with prescribed eigenvalues and eigenvectors have been studied in Starek et al. (1992), Starek and Inman (1997, 2001), in a series of work by Ram (2003), and most recently in Chu et al. (2003).

4.2.2 *Physical interpretations*

The JIEPs described above can be related to various physical systems, for example, a vibrating beam or rod (Gladwell, 1986b), a composite pendulum (Hald, 1976), or a string with beads (Hochstadt, 1967). Correspondingly, the quantities to be determined in a JIEP represent different physical parameters, for instance, the stress, the mass, the length, and so on. In this section, we shall use a serially linked, undamped mass–spring system with n particles to demonstrate the physical interpretation of JIEPs. The physical system is depicted in Figure 4.1.

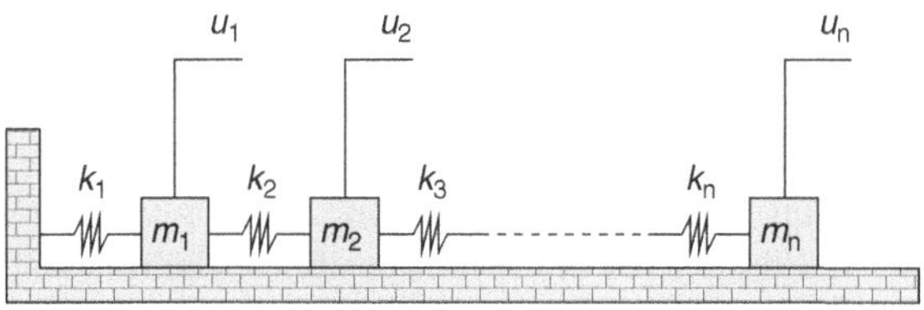

Figure 4.1. *Mass–spring system*

Suppose that the i-th particle has mass m_i. Suppose that the springs satisfy Hooke's law and that the i-th spring has spring constant k_i. Let $u_i(t)$ denote the horizontal displacement of the i-th particle at time t. Then it is easy to see that the equation of motion is given by

$$\begin{aligned} m_1 \frac{d^2 u_1}{dt} &= -k_1 u_1 + k_2(u_2 - u_1), \\ m_i \frac{d^2 u_i}{dt} &= -k_i(u_i - u_{i-1}) + k_{i+1}(u_{i+1} - u_i), \quad i = 2, \dots, n-1, \\ m_n \frac{d^2 u_n}{dt} &= -k_n(u_n - u_{n-1}). \end{aligned}$$

In matrix form, we have

$$M \frac{d^2 \mathbf{u}}{dt} = -K\mathbf{u}, \tag{4.4}$$

where $\mathbf{u} = [u_1, \dots, u_n]^\top$, $M = \operatorname{diag}(m_1, \dots, m_n)$, and K is the Jacobi matrix given by

$$K = \begin{bmatrix} k_1 + k_2 & -k_2 & 0 & \dots & 0 & 0 \\ -k_2 & k_2 + k_3 & -k_3 & & & 0 \\ 0 & -k_3 & k_3 + k_4 & & & 0 \\ \vdots & & & \ddots & & \vdots \\ 0 & & & & & -k_n \\ 0 & & & & -k_n & k_n \end{bmatrix}.$$

A fundamental solution of the form $\mathbf{u}(t) = e^{i\omega t}\mathbf{v}$ leads to the generalized eigenvalue problem that we have already seen in (2.11).

At this point, if we perform transformations $J = M^{-1/2}KM^{-1}$, $\mathbf{z} = M^{1/2}\mathbf{v}$, and $\lambda = \omega^2$, then (4.4) leads to the Jacobi eigenvalue problem (2.12). The direct problem calculates the natural frequencies and modes of the mass–spring system from given values of m_i and k_k. The inverse problem requires calculating quantities such as $(k_i + k_{i+1})/m_i$ and $k_{i+1}/\sqrt{m_i m_{i+1}}$ from the spectral data. Based on this model, we make the following observations.

If the last mass m_n is fastened to the floor, then the motion of mass m_{n-1} is governed by

$$m_{n-1}\frac{d^2u_{n-1}}{dt} = -k_{n-1}(u_{n-1} - u_{n-2}) + k_n(-u_{n-1}),$$

instead. In matrix form the equation of motion for the first $n-1$ particles corresponds to exactly that of deleting the last row and the last column from (4.4). Thus, solving the SIEP6a is equivalent to identifying the mass–spring system in Figure 4.1 from its spectrum and from the spectrum of the reduced system where the last mass is held to have no motion. The recovery of the spring stiffness and the masses from the matrix J was carefully analyzed in Gladwell (1986b).

Likewise, if another spring from m_n is attached to a wall on the far right side of the system, then the equation of motion for m_n is modified to become

$$m_n\frac{d^2u_n}{dt} = -k_n(u_n - u_{n-1}) + k_{n+1}(-u_n).$$

The SIEP2 corresponds to the construction of such a mass–spring system from its spectrum if all parameters m_i and k_i are known *a priori* to be symmetric with respect to the center of the system.

It is a little bit more complicated to sketch a diagram for the physical layout of SIEP7. Basically, we imagine that masses m_1 and m_n are somehow connected by another spring mechanism so as to form a loop (such as the periodic Toda lattice discussed in the literature). Any displacement in either particle of m_1 or m_n will affect each other via that mechanism, contributing nonzero but equal entries at the $(1, n)$ and $(n, 1)$ positions of K, respectively. Apart from this extra connection, the meaning of SIEP7 now is the same as that of SIEP6a.

We can also identify a mass–spring system from its spectrum and from the spectrum of a new system by replacing the last mass and spring with new parameters $\tilde{m}_n$ and $\tilde{k}_n$ satisfying the relationship

$$\frac{k_n^2}{m_n} = \frac{\tilde{k}_n^2}{\tilde{m}_n}.$$

The resulting inverse problem is precisely the SIEP8.

The interpretation of SIEP9 is straightforward. It means to complete the construction of a mass–spring system of size $2n$ from its spectrum and from existing physical parameters m_i, k_i of the first half of the particles.

Up to this point, we have assumed that the system in Figure 4.1 has no friction. For a damped system, the damping matrix C will be part of the parameters and we shall face a quadratic eigenvalue problem (2.10). Other than this, the physical interpretation for each of the JIEPs described above can be extended to a damped system. The SIEP6b, for example, is to identify the damped system, including its damper configurations, from its spectrum and from the spectrum of the reduced system where the last mass is held immobile. This problem is still open. The principal difficulty is to find the conditions on the (complex) spectra which ensure a realistic solution. A number of other variants of JIEPs for quadratic pencils may be found in Nylen and Uhlig (1997a,b). The corresponding damped problems can be found in Nylen (1999), Gladwell (2001) and Foltete et al. (2001).

It is important to point out again that thus far we have considered using only eigenvalues to construct Jacobi matrices. For large and complex systems, often it is practically impossible to gather the entire spectrum information for reconstruction. Partial information with some from eigenvalues and some from eigenvectors can also be used to determine a Jacobi or other structured matrices. This type of problem, referred to as PDIEP, has not been thoroughly understood yet. Some will be discussed in Chapter 5. An interesting question related to the PDIEP is how much eigenvector information is needed to determine such a Jacobi (or any structured) matrix. Gladwell (1996) has offered an account from engineering perspectives on why using low-frequency normal modes is important in practice. Some related discussions can be found in the books by Zhou and Dai (1991) and Xu (1998).

Applications of inverse problem with given spectral and modal data were studied in Gladwell (1986a), Starek et al. (1992), Starek and Inman (2001) and the many references cited in the review paper by Gladwell (1996).

4.2.3 *Existence theory*

Among all the IEPs, the class of JIEPs probably enjoys the most satisfactory theory on its solvability. Most of the existence proofs are based on a recurrence relationship among the characteristic polynomials. More precisely, let $p_k(t) = \det(tI - J_k)$ denote the characteristic polynomial of the leading $k \times k$ principal submatrix J_k. Then

$$p_k(t) = (t - a_k)p_{k-1}(t) - b_{k-1}^2 p_{k-2}(t), \quad k = 2, \ldots, n, \tag{4.5}$$

if $p_0 \equiv 1$. Such a recurrence relationship in fact gives rise to a constructive proof that, in turn, can be implemented as a numerical algorithm. Because there is an extensive literature in this regard, and because some of the constructions will be discussed as numerical methods, we shall only state the existence theorems without repeating any of the proofs in this section. We give references for each of the results where the original thoughts that led to these conclusions can be found.

Theorem 4.1. Suppose that all the given eigenvalues are distinct. Then

(i) (Hald, 1976) The SIEP6a has a unique solution.
(ii) (Hald, 1976) The SIEP2 has a unique solution.
(iii) (de Boor and Golub, 1978) The SIEP8 has a unique solution.

It should be noted that the MIEP (of uniformly spaced beads on a taut string) described in Problem 2.3 is very different from the JIEPs described in this chapter in several respects. Problem 2.3 involves only one *single spectrum* while most of the problems in this chapter involve *two spectra*. In the former, we have only one set of parameters (the masses) to adjust. In the latter, we have two sets (the m_i's and the k_i's) to combine. The solution for the latter is often unique while the former is a much harder problem.

Theorem 4.2. (Ram and Elhay, 1996) Over the complex field $\mathbb{C}$, suppose that all the given eigenvalues are distinct. Then the SIEP6b is solvable and has at most $2^n(2n-3)!/(n-2)!$ different solutions. In the event that there are common eigenvalues, then there are infinitely many solutions for the SIEP6b.

Be aware that the solvability of SIEP6b established in the above theorem is over the algebraically closed field $\mathbb{C}$. It is not known when the problem will be realistically solvable with positive masses, springs, and dampers (Gladwell, 2001).

Theorem 4.3. (Xu, 1998) The SIEP7 is solvable if and only if

$$\prod_{k=1}^{n} |\mu_j - \lambda_k| \geq 2\beta(1 + (-1)^{n-j+1}),$$

for all $j = 1, \ldots, n-1$. Even in the case of existence, no uniqueness can be ascertained.

It is worth mentioning that while trying to characterize periodic Jacob matrices, Ferguson (1980) developed a notion of "compatible" data that can be turned into a numerical algorithm. Each set of compatible data uniquely determines a periodic Jacobi matrix.

Example 4.1. No periodic Jacobi matrices can have eigenvalues $\lambda = \{1, 3, 5\}$ and $\mu = \{2, 4\}$ with $\beta = 1$. This is a counterexample showing that the SIEP7 is not always solvable (Xu, 1998).

Theorem 4.4. (Xu, 1998) Assume that all eigenvalues are distinct. Define

$$\Delta_k = \det\left(\begin{bmatrix} 1 & \ldots & 1 & 1 & \ldots & 1 \\ \lambda_1 & \ldots & \mathbf{e}_1^\top \tilde{J} \mathbf{e}_1 & \lambda_{k+1} & \ldots & \lambda_{2n} \\ \vdots & \vdots & \vdots & \vdots & \vdots & \vdots \\ \lambda_1^{2n-1} & \ldots & \mathbf{e}_1^\top \tilde{J}^{2n-1} \mathbf{e}_1 & \lambda_{k+1}^{2n-1} & \ldots & \lambda_{2n}^{2n-1} \end{bmatrix}\right).$$

Then the SIEP9 has a unique solution if and only if

$$\Delta_k > 0$$

for all $k = 1, \ldots, 2n$.

Example 4.2. A simple counterexample showing that the SIEP9 is not always solvable is as follows: No 2×2 symmetric matrix J can have a fixed $(1, 1)$ entry a_1 and eigenvalues satisfying either $a_1 < \lambda_1 < \lambda_2$ or $\lambda_1 < \lambda_2 < a_1$.

4.2.4 *Sensitivity issues*

If the numerical computation is to be done using finite precision arithmetic, it is critical to understand the perturbation behavior of the underlying mathematical problem. The notion of conditioning is normally used as an indication of the sensitivity dependence.

For IEPs, partially because inverse problems are harder to analyze than the direct problems by nature, partially because most of the IEPs have multiple solutions, not many results on sensitivity analysis have been performed. We believe that this is an important yet wide open area for further research. We mentioned earlier that such a study could have the application of finding a robust solution that is least sensitive to perturbations. Despite the fact that considerable effort has been devoted to the development of numerical algorithms for many of the IEPs discussed here, we should make it clear that thus far very little attention has been paid to this direction. The analysis of either the conditioning of IEPs or the stability of the associated numerical methods is lacking. For that reason, we can address only partially the sensitivity issues by demonstrating known results for the SIEP6a in this section. Clearly, more effort needs to be done here.

For the direct problem, it is easy to see that the function $\mathbf{f}: \mathbb{R}^n \times \mathbb{R}_+^{n-1} \rightarrow \mathbb{R}^n \times \mathbb{R}^{n-1}$ where

$$\mathbf{f}(a_1, \ldots, a_n, b_1, \ldots, b_{n-1}) = (\sigma(J), \sigma(\bar{J}))$$

is differentiable. The well-posedness of the inverse problem was initially established by Hochstadt (1974).

Theorem 4.5. (Hochstadt, 1974) The unique solution J to the SIEP6a depends continuously on the given data $\{\lambda_k\}_{k=1}^n$ and $\{\mu_1, \ldots, \mu_{n-1}\}$.

Mere continuous dependence is not enough for numerical computation. We need to quantify how the solution is changed subject to the change in problem data. Using the implicit function theorem, Hald (1976) refined this dependence and provided the following sensitivity dependence.

Theorem 4.6. (Hald, 1976) Suppose J and $\tilde{J}$ are the solutions to the SIEP6a with data

$$\lambda_1 < \mu_1 < \lambda_2 < \cdots < \mu_{n-1} < \lambda_n,$$
$$\tilde{\lambda}_1 < \tilde{\mu}_1 < \tilde{\lambda}_2 < \cdots < \tilde{\mu}_{n-1} < \tilde{\lambda}_n,$$

respectively. Then there exists a constant K such that

$$\|J - \tilde{J}\|_F \leq K \left(\sum_{i=1}^{n} |\lambda_i - \tilde{\lambda}_i|^2 + \sum_{i=1}^{n-1} |\mu_i - \tilde{\mu}_i|^2 \right)^{1/2}, \tag{4.6}$$

where the constant K depends on the quantities

$$d = \max\{\lambda_n, \tilde{\lambda}_n\} - \min\{\lambda_1, \tilde{\lambda}_1\},$$
$$\epsilon_0 = \frac{1}{d} \min_{j \neq k}\{|\lambda_j - \mu_k|, |\tilde{\lambda}_j - \tilde{\mu}_k|\},$$
$$\delta_0 = \frac{1}{2d} \min_{j \neq k}\{|\lambda_j - \lambda_k|, |\mu_j - \mu_k|, |\tilde{\lambda}_j - \tilde{\lambda}_k|, |\tilde{\mu}_j - \tilde{\mu}_k|\},$$

which measure the separation of the given data.

The constant K in (4.6) is critical in that it determines how the perturbation in the given data would be amplified. Its actual quantity, however, remains opaque because the implicit function theorem warrants only its existence but not its content. To remedy this deficiency, a notion of condition number that could be explicitly estimated, at least for the SIEP6a, has been introduced by Xu (1993). So as not to be sidetracked into another interesting area that is wide open for further research, we shall not elaborate on details of the condition number but merely suggest that, as a general rule, the smaller the separation of the given data are, the more ill conditioned the JIEP6a becomes.

At present, we do not know much about the sensitivity dependence of other types of JIEPs.

4.2.5 *Numerical methods*

Numerical algorithms for the solution of JIEPs often follow directly from constructive proofs of its existence. Consequently, there is no lack of numerical methods for JIEPs. Nevertheless, we have to point out that some of the obvious procedures are subtly unstable. It is of no point to evaluate each known method in this book. Rather, we shall concentrate on the basic ideas of two popular approaches: the Lanczos method and the orthogonal reduction method (Boley and Golub, 1987; Parlett, 1998).

The Lanczos method We first recall the following well-known theorem that is the basis of the Lanczos approach.

Theorem 4.7. The orthogonal matrix Q and the upper Hessenberg matrix H with positive subdiagonal entries can be completely determined by a given matrix A and the last (or any) column of Q if the relationship $Q^\top AQ = H$ holds.

In our application, we want to construct the symmetric tridiagonal matrix $J = Q^\top \Lambda Q$ with $\Lambda = \mathrm{diag}(\lambda_1, \ldots, \lambda_n)$. The Lanczos method simply carries out Theorem 4.7 in the following constructive manner.

Algorithm 4.1 (Generic Lanczos method for JIEP)
Suppose that the last column $\mathbf{q}_n$ is known. Then the entries a_i and b_i of the Jacobi matrix J with eigenvalue $\{\lambda_k\}_{k=1}^n$ can be constructed via finitely many steps as follows:

1. Define

$$
\begin{aligned}
a_n &:= \mathbf{q}_n^\top \Lambda \mathbf{q}_n, \\
b_{n-1} &:= \|\Lambda \mathbf{q}_n - a_n \mathbf{q}_n\|, \\
\mathbf{q}_{n-1} &:= (\Lambda \mathbf{q}_n - a_n \mathbf{q}_n)/b_{n-1}.
\end{aligned}
$$

2. For $i = 1, \ldots, n-2$, compute recursively the quantities:

$$
\begin{aligned}
a_{n-i} &:= \mathbf{q}_{n-i}^\top \Lambda \mathbf{q}_{n-i}, \\
b_{n-i-1} &:= \|\Lambda \mathbf{q}_{n-i} - a_{n-i}\mathbf{q}_{n-i} - b_{n-i}\mathbf{q}_{n-i+1}\|, \\
q_{n-i-1} &:= (\Lambda \mathbf{q}_{n-i} - a_{n-i}\mathbf{q}_{n-i} - b_{n-i}\mathbf{q}_{n-i+1})/b_{n-i-1},
\end{aligned}
$$

3. Define $a_1 := \mathbf{q}_1^\top \Lambda \mathbf{q}_1$.

It only remains to calculate the column vector $\mathbf{q}_n$. It is important to note that in most of the JIEP described thus far there is a set of conditions. In the above algorithm, however, that second set of conditions has not been employed yet. Indeed, that piece of information will now be needed in discriminating $\mathbf{q}_n$. Toward that end, we demonstrate how the second set of eigenvalues is used to determine the unique solution of the SIEP6a. This construction would fulfill what we have promised in Section 4.2.3. We first recall a classical result by Thompson and McEnteggert (1968).

Theorem 4.8. (Thompson and McEnteggert, 1968) Let $(\lambda_i, \mathbf{x}_i)$, $i = 1, \ldots, n$, be orthonormal eigenpairs that form the spectrum decomposition of a given symmetric matrix A. Then

$$
\mathrm{adj}(\lambda_i I - A) = \prod_{\substack{k=1 \\ k \neq i}}^{n} (\lambda_i - \lambda_k)\mathbf{x}_i \mathbf{x}_i^\top. \tag{4.7}
$$

Evaluating both sides of (4.7) at the (n,n) position, we obtain

$$\det(\lambda_i I_{n-1} - A_{n-1}) = x_{ni}^2 \prod_{\substack{k=1 \\ k \neq i}}^{n} (\lambda_i - \lambda_k),$$

where x_{ni} is the last entry of $\mathbf{x}_i$ and recall that A_{n-1} denotes the principal sub-matrix of size $n-1$. In our application, the last column $\mathbf{q}_n$ is precisely $[x_{n1}, \dots, x_{nn}]^\top$, if A is replaced by J. It follows that

$$x_{ni}^2 = \frac{\prod_{k=1}^{n-1} (\lambda_i - \mu_k)}{\prod_{\substack{k=1 \\ k \neq i}}^{n} (\lambda_i - \lambda_k)}. \tag{4.8}$$

In other words, the last column $\mathbf{q}_n$ for J can be expressed from the spectral data $\{\lambda_k\}_{k=1}^n$ and $\{\mu_1, \dots, \mu_{n-1}\}$. The Lanczos algorithm kicks in and the SIEP6a is solved in finite steps.

We remark that other types of JIEPs can be solved in similar ways with appropriate modifications. We shall refrain from examining them here. Readers are referred to the review paper by Boley and Golub (1987) and the book by Xu (1998) for more details.

We caution that the method by de Boor and Golub (1978) using the orthogonal polynomial approach is entirely equivalent to the above Lanczos approach, but is less stable in the face of roundoff error. We suggest that an reorthogonalization process should take place even along the Lanczos steps to ensure stability.

Orthogonal reduction method In the orthogonal reduction method, the given data are used first to construct a bordered diagonal matrix A of the form

$$A = \begin{bmatrix} \alpha & \beta_1 & \dots & \beta_{n-1} \\ \beta_1 & \mu_1 & & 0 \\ \vdots & & \ddots & \\ \beta_{n-1} & 0 & \dots & \mu_{n-1} \end{bmatrix}$$

so that $\sigma(A) = \{\lambda_k\}_{k=1}^n$. Such a construction is entirely possible. First, α is trivially determined as $\alpha = \sum_{i=1}^n \lambda_i - \sum_{i=1}^{n-1} \mu_i$. Second, the characteristic polynomial is given by

$$\det(\lambda I - A) = (\lambda - \alpha) \prod_{k=1}^{n-1} (\lambda - \mu_k) - \sum_{i=1}^{n-1} \beta_i^2 \left(\prod_{\substack{k=1 \\ k \neq i}}^{n-1} (\lambda - \mu_k) \right). \tag{4.9}$$

Thus, border elements β_i are given by

$$\beta_i^2 = -\frac{\prod_{k=1}^{n} (\mu_i - \lambda_k)}{\prod_{\substack{k=1 \\ k \neq i}}^{n-1} (\mu_i - \mu_k)}.$$

Let $\boldsymbol{\beta} = [\beta_1, \ldots, \beta_{n-1}]^\top$. The next step is to construct an orthogonal matrix Q efficiently so that the product

$$\begin{bmatrix} 1 & \mathbf{0}^\top \\ \mathbf{0} & Q^\top \end{bmatrix} A \begin{bmatrix} 1 & \mathbf{0}^\top \\ 0 & Q \end{bmatrix} = \begin{bmatrix} \alpha & b_1\mathbf{e}_1^\top \\ b_1\mathbf{e}_1 & \bar{J} \end{bmatrix} \tag{4.10}$$

becomes a Jacobi matrix. For this to happen, we must have $Q^\top \boldsymbol{\beta} = b_1\mathbf{e}_1$ where $\mathbf{e}_1$ is the first standard unit vector in $\mathbb{R}^{n-1}$. It follows that $b_1 = \|\boldsymbol{\beta}\|$ and that the first column of Q is given by $\boldsymbol{\beta}/b_1$. The Lanczos procedure can now be employed to complete the construction $Q^\top \mathrm{diag}(\mu_1, \ldots, \mu_{n-1})Q = \bar{J}$ and the SIEP6a is solved in finite steps.

We conclude this section by mentioning that other tridiagonalization processes, including Householder transformation, Givens rotations, the Rutishauser method (Gragg and Harrod, 1984), and so on, may also be used effectively to explore the bordered diagonal structure (Boley and Golub, 1987).

4.3 Toeplitz inverse eigenvalue problems

Given a column vector $\mathbf{r} = [r_1, \ldots, r_n]^\top$, a matrix $T = T(\mathbf{r})$ of the form

$$T := \begin{bmatrix} r_1 & r_2 & \ldots & r_{n-1} & r_n \\ r_2 & r_1 & & r_{n-2} & r_{n-1} \\ \vdots & \ddots & \ddots & & \vdots \\ r_{n-1} & & & r_1 & r_2 \\ r_n & r_{n-1} & & r_2 & r_1 \end{bmatrix}$$

is called a symmetric Toeplitz matrix. Toeplitz matrices play important roles in many applications, including the trigonometric moment problem, Szegö theory, and signal processing. We recommend an interesting article (Gray, 2002) for additional reading on this subject. A simple example might illustrate how Toeplitz structure comes into play.

Example 4.3. (Haykin, 1996) In linear discrete time series analysis, a "basic" filtering equation takes the form

$$\mathbf{u} = F\mathbf{x} + \mathbf{v}, \tag{4.11}$$

where $\mathbf{x}$ stands for transmitted signal vectors, F stands for the overall impulse response and is termed a filtering matrix. The filter matrix has a Toeplitz structure of the form

$$F := \begin{bmatrix} f_0 & f_1 & \ldots & f_{n-1} & f_n & 0 & \ldots & 0 \\ 0 & f_0 & \ldots & & f_{n-1} & f_n & \ldots & 0 \\ \vdots & \ddots & \ddots & & \vdots & & \ddots & \\ 0 & & & & f_0 & & & f_n \end{bmatrix} \tag{4.12}$$

by the design of the linear model. In trying to recover the signals from the observation vector $\mathbf{u}$, the least squares procedure must explicitly or implicitly find the inverse of the covariance matrix $R = FF^\top$ which itself is a symmetric Toeplitz matrix.

Toeplitz matrices have been a well studied subject over the years. Efficient algorithms for solve a Toeplitz system of equations (Golub and Van Loan, 1996, Section 2.1) or Toeplitz eigensystem (Hu and Kung, 1985) in $O(n^2)$, for example, are available in the literature.

One of the fundamental issues in communication theory is to determine a finite impulse response (FIR) filter that is optimal in the sense that the output signal-to-noise ratio (SNR) is maximized. This filter optimization is closely related to the eigensystem of the covariance matrix R (Haykin, 1996). In our context, we are interested in the inverse problem. An inverse Toeplitz eigenvalue problem (**ToIEP**) concerns the following:

Problem 4.8 *(ToIEP)*
Given a set of real numbers $\{\lambda_1, \dots, \lambda_n\}$, construct a symmetric Toeplitz matrix $T \in \mathbb{R}^{n\times n}$ such that

$$\sigma(T) = \{\lambda_k\}_{k=1}^n.$$

Problem 4.9 *(ToIEP1)*
Given a set of positive numbers $\{\lambda_1, \dots, \lambda_n\}$, construct a symmetric and positive definite Toeplitz matrix $R \in \mathbb{R}^{n\times n}$ such that

$$\sigma(R) = \{\lambda_k\}_{k=1}^n.$$

We mention in passing that in view of Example 4.3, we may modify ToIEP1 to the problem of finding F in the form of a filtering matrix (4.12) so that F has a prescribed set of singular values. This is a specially structured inverse singular value problem. Also, a similar IEP could be asked for a Hankel matrix $H(\mathbf{r})$ that is related to the Toeplitz matrix $T(\mathbf{r})$ via $H(\mathbf{r}) = \Xi T(\mathbf{r})$ and $T(\mathbf{r}) = \Xi H(\mathbf{r})$ where $\Xi = [\xi_{ij}]$ is the unit perdiagonal matrix (of appropriate size) defined earlier.

The set $\mathcal{T}(n)$ of symmetric Toeplitz matrices forms a subset of a larger class

$$\mathcal{C}(n) := \{M \in \mathbb{R}^{n\times n} | M = M^\top, M = \Xi M \Xi\}$$

of centrosymmetric matrices. A vector $\mathbf{v}$ is said to be even if $\Xi\mathbf{v} = \mathbf{v}$, and odd if $\Xi\mathbf{v} = -\mathbf{v}$. It is known that eigenvectors of centrosymmetric matrices and, hence, of symmetric Toeplitz matrices, are necessarily even or odd vectors (Cantoni and Bulter, 1976; R. D. Hill and Waters, 1990). The prescribed eigenvalues $\{\lambda_k\}_{k=1}^n$ in a ToIEP, therefore, should also carry a corresponding parity assignment as we shall explore in the next section.

4.3.1 *Symmetry and parity*

Summarized in Table 4.1 are some characteristics of centrosymmetric matrices (Cantoni and Bulter, 1976). Depending on whether n is even or odd, any centrosymmetric matrix M must assume the symmetry as indicated in Table 4.1, where $A, C, \Xi \in \mathbb{R}^{\lfloor \frac{n}{2} \rfloor \times \lfloor \frac{n}{2} \rfloor}$, $\mathbf{x} \in \mathbb{R}^{\lfloor \frac{n}{2} \rfloor}$, $q \in \mathbb{R}$, and $A = A^\top$. Let K be the orthogonal matrix defined in the table. Then M can be decomposed into 2×2 diagonal blocks via orthogonal similarity transformation by K. The blocks assume the forms shown in the last row of Table 4.1.

Effectively the spectral decomposition of M is reduced to that of two submatrices with about half size. If Z_1 denotes the $\lfloor \frac{n}{2} \rfloor \times \lfloor \frac{n}{2} \rfloor$ matrix of orthonormal eigenvectors for $A - \Xi C$, then columns from the matrix $K^\top \begin{bmatrix} Z_1 \\ 0 \end{bmatrix}$ will be eigenvectors of M. These eigenvectors are odd vectors. Similarly, there are $\lceil \frac{n}{2} \rceil$ even eigenvectors of M computable from those of $A + \Xi C$ or $\begin{bmatrix} q & \sqrt{2}x^\top \\ \sqrt{2}x & A + \Xi C \end{bmatrix}$. It is interesting to ask whether a symmetric Toeplitz matrix can have arbitrary spectrum with arbitrary parity. Can the parity be arbitrarily assigned to the prescribed eigenvalues in a ToIEP?

We consider the 3×3 case to further explore this question. Any matrix $M \in \mathcal{C}(3)$ is of the form

$$M = \begin{bmatrix} m_{11} & m_{12} & m_{13} \\ m_{12} & m_{22} & m_{12} \\ m_{13} & m_{12} & m_{11} \end{bmatrix}.$$

Without loss of generality, we may assume that the trace of M is zero. In this way, the parameters are reduced to m_{11}, m_{12} and m_{13}. Let $\mathcal{M}_\mathcal{C} = \mathcal{M}_\mathcal{C}(\lambda_1, \lambda_2, \lambda_3)$ denote the subset of centrosymmetric matrices that are isospectral to eigenvalues

Table 4.1. *Structure of centrosymmetric matrices*

n	Even	Odd
M	$\begin{bmatrix} A & C^\top \\ C & \Xi A \Xi \end{bmatrix}$	$\begin{bmatrix} A & x & C^\top \\ x^\top & q & x^\top \Xi \\ C & \Xi x & \Xi A \Xi \end{bmatrix}$
$\sqrt{2}K$	$\begin{bmatrix} I & -\Xi \\ I & \Xi \end{bmatrix}$	$\begin{bmatrix} I & 0 & -\Xi \\ 0 & \sqrt{2} & 0 \\ I & 0 & \Xi \end{bmatrix}$
$KMK^\top$	$\begin{bmatrix} A - \Xi C & 0 \\ 0 & A + \Xi C \end{bmatrix}$	$\begin{bmatrix} A - \Xi C & 0 & 0 \\ 0 & q & \sqrt{2}x^\top \\ 0 & \sqrt{2}x & A + \Xi C \end{bmatrix}$

$\{\lambda_1, \lambda_2, \lambda_3\}$. Assuming $\sum_{i=1}^{3} \lambda_i = 0$, elements in $\mathcal{M}_\mathcal{C}$ must satisfy the equations

$$\left(m_{11} - \frac{\lambda_{\varrho_1}}{4}\right)^2 + \frac{1}{2}m_{12}^2 = \frac{(\lambda_{\varrho_2} - \lambda_{\varrho_3})^2}{16},$$
$$m_{13} = m_{11} - \lambda_{\varrho_1},$$

where ϱ denotes any of the six permutation of integers $\{1, 2, 3\}$. Thus $\mathcal{M}_\mathcal{C}$ consists of three ellipses in $\mathbb{R}^3$. It suffices to plot these ellipses in the (m_{11}, m_{12})-plane only, since m_{13} is simply a shift of m_{11}. Several plots with qualitatively different eigenvalues are depicted in Figure 4.2. Observe that it is always the case that one ellipse circumscribes the other two.

Example 4.4. A 3×3 centrosymmetric matrix is a solution to the ToIEP only if $m_{11} = 0$. By counting the number of m_{12}-intercepts, we should be able to know the number of solutions to the ToIEP. Specifically, we find that there are four solutions if eigenvalues are distinct and two solutions if one eigenvalue has multiplicity 2. We note further that the parities of the prescribed eigenvalues in the ToIEP cannot be arbitrary. Each of the ellipses corresponds to one particular parity assignment among the eigenvalues. A "wrong" parity assignment, such as

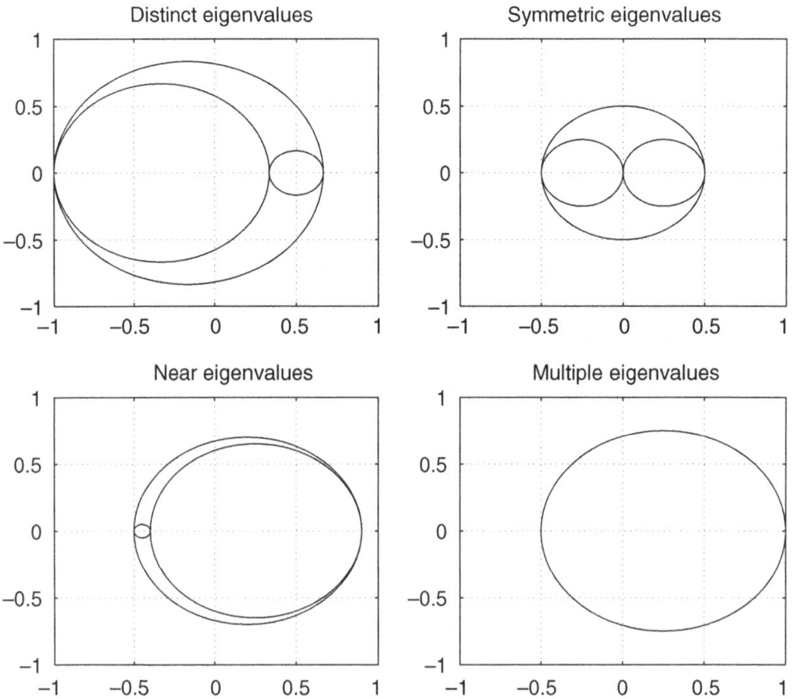

Figure 4.2. *Plots of $\mathcal{M}_\mathcal{C}$ in the (m_{11}, m_{12})-plane*

the two smallest ellipses in the left column of Figure 4.2, implies that there is no m_{12}-intercept and, hence, no isospectral Toeplitz matrix.

As far as the ToIEP is concerned, parity assignment is not explicitly given as part of the constraint. As a safeguard for ensuring existence, it has been suggested in the literature that the ordered eigenvalues should have alternating parity, partially because of the following existence theory.

4.3.2 *Existence*

Despite the simplicity of its appearance of an ToIEP, the issue of its solvability has been quite a challenge. Delsarte and Genin (1984) argued that the problem would be analytically intractable if $n \geq 5$. Eventually, using a topological degree argument, Landau (1994) settled the following theorem with a nonconstructive proof.

Theorem 4.9. (Landau, 1994) Every set of n real numbers is the spectrum of an $n \times n$ real symmetric Toeplitz matrix.

It might be useful to briefly outline the proof as it shows the existence of an even more restricted class of Toeplitz matrices. A Toeplitz matrix $T(r_1, \ldots, r_n)$ is said to be *regular* if every principal submatrix $T(c_1, \ldots, c_k)$, $1 \leq k \leq n$, has distinct eigenvalues that, when arranged in ascending order, alternate parity with the largest one even. Consider the map $\varphi : \mathbb{R}^{n-2} \to \mathbb{R}^{n-2}$ defined by

$$\varphi(t_3, \ldots, t_n) = (y_2, \ldots, y_{n-1}),$$

where $y_i := -\mu_i/\mu_1$, $i = 2, \ldots, n-1$, if $\mu_1 \leq \cdots \leq \mu_n$ are eigenvalues of $T(0, 1, t_3, \ldots, t_n)$. Note that since $\sum_{i=1}^{n} \mu_i = 0$, it is necessary that $\mu_1 < 0$. The range of φ is the simplex

$$\Delta := \left\{ (y_2, \ldots, y_{n-1}) \in \mathbb{R}^{n-2} \middle| \begin{array}{c} -1 \leq y_2 \leq \cdots \leq y_{n-1} \\ y_2 + \cdots + y_{n-2} + 2y_{n-1} \leq 1 \end{array} \right\}.$$

The key components in the proof by Landau (1994) are as follows. First, the set $\mathcal{F}$ of regular Toeplitz matrices of the form $T(0, 1, t_3, \ldots, t_n)$ is not empty. Second, the map φ restricted to those points $(t_3, \ldots, t_n) \in \mathbb{R}^{n-2}$ such that $T(0, 1, t_3, \ldots, t_n) \in \mathcal{F}$ is a *surjective* map onto the interior of Δ. Finally, any given $\lambda_1 \leq \cdots \leq \lambda_n$ can be *shifted* and *scaled* to a unique point in Δ whose preimage(s) can then be scaled and shifted backward to a symmetric Toeplitz matrix with eigenvalues $\{\lambda_1, \ldots, \lambda_n\}$.

4.3.3 *Numerical methods*

The lack of understanding does not necessarily preclude the development of effective numerical algorithms for the ToIEP. There are two basic approaches to tackle the ToIEP numerically. One is by iteration and the other by continuation. We briefly describe the basic ideas for each approach in this section.

Restrictive Newton iteration Regarding the ToIEP as a nonlinear system of n equations in n unknowns, a natural tactic would be the Newton-type iteration. The schemes in Friedland et al. (1987) originally proposed for the more general class of PIEPs are in this class and certainly applicable to the ToIEP. The inverse Rayleigh quotient algorithm in Laurie (1988, 1991) is also equivalent to a Newton variation. These methods do not exploit the Toeplitz structure and can suffer from local convergence. The iterative scheme proposed by Trench (1997) seems to have more robust performance, but still no global convergence can be proved. The following discussion is another Newton-type iteration by Chu (1994). The iterations are confined in $\mathcal{C}(n)$. Since the centrosymmetric structure is preserved, the cost is substantially reduced and the case of double eigenvalues can be handled effectively.

Recall, as we have already suggested in Section 3.2.4, that the classical Newton method

$$x^{(\nu+1)} = x^{(\nu)} - (f'(x^{(\nu)}))^{-1} f(x^{(\nu)})$$

for a scalar function $f : \mathbb{R} \rightarrow \mathbb{R}$ can be thought of as two steps: the tangent step defines $x^{(\nu+1)}$ by the x-intercept of the tangent line emanating from the point $(x^{(\nu)}, f(x^{(\nu)}))$ on the graph of f and the lift step lifts the intercept to the point $(x^{(\nu+1)}, f(x^{(\nu+1)}))$ on the graph of f along the y-axis. We shall mimic similar procedures for the ToIEP. Such an idea has been proposed earlier in Section 3.2.4 for the LiPIEP2. The exception in the following approach is that we further limit the iteration to a specially selected subspace.

The parity of the prescribed eigenvalues must be taken into account. Let the given eigenvalues $\{\lambda_k\}_{k=1}^{n}$ be arranged as $\boldsymbol{\lambda} = [\phi_1, \ldots, \phi_{\lfloor \frac{n}{2} \rfloor}, \psi_1, \ldots, \psi_{\lceil \frac{n}{2} \rceil}]^\top$ where ϕ_k and ψ_k are of even and odd parity, respectively. Let $\Lambda = \operatorname{diag}(\boldsymbol{\lambda})$. An analogue of this idea applied to the ToIEP is to think of the isospectral subset $\mathcal{M}_{\mathcal{C}} = \mathcal{M}_{\mathcal{C}}(\Lambda)$ of centrosymmetric matrices as the graph of some unknown f and the subspace $\mathcal{T}(n)$ of symmetric Toeplitz matrices as the x-axis. We want to do the tangent and lift iterations between these to entities.

From Section 4.3.1, we see that every element $M \in \mathcal{M}_{\mathcal{C}}$ can be characterized by the parameter $Z \in \mathcal{O}(\lfloor \frac{n}{2} \rfloor) \times \mathcal{O}(\lceil \frac{n}{2} \rceil)$ where $M = Q\Lambda Q^\top$ and $Z = KQ$. It follows that tangent vectors of $\mathcal{M}_{\mathcal{C}}$ at M are of the form

$$T_M(\mathcal{M}_{\mathcal{C}}) = \tilde{S}M - M\tilde{S}, \tag{4.13}$$

with $\tilde{S} := Q\operatorname{diag}(S_1, S_2)Q^\top$ where S_1 and S_2 are arbitrary skew-symmetric matrices in $\mathbb{R}^{\lfloor \frac{n}{2} \rfloor \times \lfloor \frac{n}{2} \rfloor}$ and $\mathbb{R}^{\lceil \frac{n}{2} \rceil \times \lceil \frac{n}{2} \rceil}$, respectively. Thus a tangent step from a given $M^{(\nu)} \in \mathcal{M}_{\mathcal{C}}(\Lambda)$ amounts to finding a skew-symmetric matrix $\tilde{S}^{(\nu)}$ and a vector $\mathbf{r}^{(\nu+1)}$ so that

$$M^{(\nu)} + \tilde{S}^{(\nu)}M^{(\nu)} - M^{(\nu)}\tilde{S}^{(\nu)} = T(\mathbf{r}^{(\nu+1)}). \tag{4.14}$$

Assume the spectral decomposition $M^{(\nu)} = Q^{(\nu)}\Lambda Q^{(\nu)^\top}$ and define $Z^{(\nu)} = KQ^{(\nu)}$. The tangent equation is reduced to

$$\Lambda + S^{(\nu)}\Lambda - \Lambda S^{(\nu)} = Z^{(\nu)^\top}\left(KT(\mathbf{r}^{(\nu+1)})K^\top\right)Z^{(\nu)}, \tag{4.15}$$

where $S^{(\nu)} = Q^{(\nu)^\top}\tilde{S}^{(\nu)}Q^{(\nu)}$ remains to be skew-symmetric. Observe that the product $KT(\mathbf{r}^{(\nu+1)})K^\top$ is a 2×2 diagonal block matrix, denoted by $\operatorname{diag}(T_1^{(\nu+1)}, T_2^{(\nu+1)})$, because $T(\mathbf{r}^{(\nu+1)})$ is centrosymmetric. We also know from the discussion in Section 4.3.1 that $Z^{(\nu)} = \operatorname{diag}(Z_1^{(\nu)}, Z_2^{(\nu)})$. Thus the system (4.15) is effectively split in half.

We first retrieve the vector $\mathbf{r}^{(\nu+1)}$. It suffices to compare the diagonal elements on both sides without reference to $S^{(\nu)}$. Note that the right-hand side of (4.15) is linear in $\mathbf{r}^{(\nu+1)}$. This linear relationship can be expressed as

$$\Omega^{(\nu)}\mathbf{r}^{(\nu+1)} = \boldsymbol{\lambda}$$

for $r^{(\nu+1)}$, where the entries in the matrix $\Omega^{(\nu)} = [\Omega_{ij}^{(\nu)}]$ are defined by

$$\Omega_{ij}^{(\nu)} := \begin{cases} (Z_1^{(\nu)})_{*i}^\top E_1^{[j]}(Z_1^{(\nu)})_{*i}, & \text{if } 1 \le i \le \left\lfloor \frac{n}{2} \right\rfloor, \\ (Z_2^{(\nu)})_{*i}^\top E_2^{[j]}(Z_2^{(\nu)})_{*i}, & \text{if } \left\lfloor \frac{n}{2} \right\rfloor < i \le n. \end{cases}$$

In the above, $E_1^{[j]}$ and $E_2^{[j]}$ are the diagonal blocks in the 2×2 diagonal block matrix $KT(\mathbf{e}_j)K^\top$, $\mathbf{e}_j$ is the j-th standard unit vector, and $(Z_k^{(\nu)})_{*i}$ denotes the i-th column of the matrix $Z_k^{(\nu)}$. Throughout all the calculations, we need only to multiply vectors or matrices of lengths $\lfloor \frac{n}{2} \rfloor$ or $\lceil \frac{n}{2} \rceil$. Once $T(\mathbf{r}^{(\nu+1)})$ is determined, off-diagonal elements in (4.15) determine $S^{(\nu)}$. Specifically, if eigenvalues within each parity group are distinct, then it is easy to see that

$$(S_1^{(\nu)})_{ij} = \frac{(Z_1^{(\nu)})_{*i}^\top T_1^{(\nu+1)}(Z_1^{(\nu)})_{*j}}{\phi_i - \phi_j}, \quad 1 \le i < j \le \left\lfloor \frac{n}{2} \right\rfloor,$$

$$(S_2^{(\nu)})_{ij} = \frac{(Z_2^{(\nu)})_{*i}^\top T_2^{(\nu+1)}(Z_2^{(\nu)})_{*j}}{\psi_i - \psi_j}, \quad 1 \le i < j \le \left\lceil \frac{n}{2} \right\rceil.$$

This completes the calculation for the tangent step. We remark that the scheme is capable of handling the case of double eigenvalues because such eigenvalues have to be split into one even and one odd (Delsarte and Genin, 1984).

To obtain a lift from $T(r^{(\nu+1)})$ to $\mathcal{M}_\mathcal{C}$, we look for a matrix $M^{(\nu+1)} \in \mathcal{M}_\mathcal{C}$ that is nearest to $T(r^{(\nu+1)})$. The idea is sketched in Figure 4.3. Such a nearest approximation can be obtained by the Wielandt–Hoffman theorem. That is, suppose the spectral decomposition of $T(r^{(\nu+1)})$ is given by

$$\overline{Z}^{(\nu+1)^\top} KT(r^{(\nu+1)})K^\top \overline{Z}^{(\nu+1)} = \begin{bmatrix} \overline{\Lambda}_1^{(\nu+1)} & 0 \\ 0 & \overline{\Lambda}_2^{(\nu+1)} \end{bmatrix}.$$

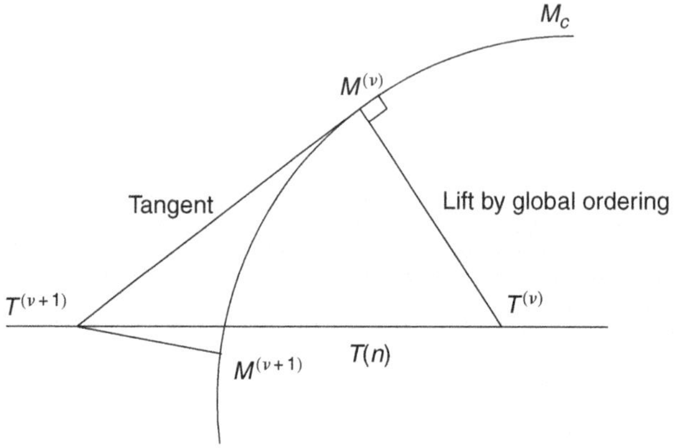

Figure 4.3. *Geometry of the lift by the Wielandt–Hoffman theorem*

Rearrange $\{\lambda_1, \dots, \lambda_n\}$ in the same ordering as in $\overline{\Lambda}_1^{(\nu+1)}$ and $\overline{\Lambda}_2^{(\nu+1)}$ to obtain $\tilde{\Lambda}_1^{(\nu+1)}$ and $\tilde{\Lambda}_2^{(\nu+1)}$. Then the nearest approximation $M^{(\nu+1)}$ is given by

$$M^{(\nu+1)} := K^\top \overline{Z}^{(\nu+1)} \operatorname{diag}(\tilde{\Lambda}_1^{(\nu+1)}, \tilde{\Lambda}_2^{(\nu+1)}) \overline{Z}^{(\nu+1)^\top} K.$$

For computational purpose, this $M^{(\nu+1)}$ never needs to be calculated. We only need to repeat the tangent step with the new parity assignment $\Lambda = \operatorname{diag}(\tilde{\Lambda}_1^{(\nu+1)}, \tilde{\Lambda}_2^{(\nu+1)})$ and the new parameter matrix $Z^{(\nu+1)} := \overline{Z}^{(\nu+1)}$. It can be proved that this method converges quadratically.

Continuation method An alternative approach fundamentally different from the iterative methods is to solve the ToIEP by tracing curves defined by differential equations. Similar to the homotopy method in spirit, but different in setup, the idea is to continually transform the matrix Λ until a Toeplitz matrix is found. This approach actually finds its roots in a more profound framework whose theory will be established in Chapter 7. For the time being, we simply give an example. We shall revisit the method in Section 7.3.

Example 4.5. Consider the initial value problem

$$\begin{cases} \dfrac{dX}{dt} = [X, k(X)], \\ X(0) = \Lambda, \end{cases} \tag{4.16}$$

where $[A, B] := AB - BA$ denotes the Lie bracket and $k(X) = [k_{ij}(X)]$ is a skew-symmetric matrix depending on X. The skew-symmetry of $k(X)$ guarantees that the solution $X(t)$ to (4.16) exists for all t and enjoys a continuous spectral decomposition $X(t) = Q(t)^\top \Lambda Q(t)$. The orthogonal transformation $Q(t)$ is determined by the initial value problem

$$\begin{cases} \frac{dQ}{dt} = Qk(Q^\top \Lambda Q), \\ Q(0) = I. \end{cases} \tag{4.17}$$

The limiting behavior of $Q(t)$ determines the limiting behavior of $X(t)$ and vice versa.

One possible way of defining $k(X)$ is via the formulation

$$k_{ij}(X) := \begin{cases} x_{i+1,j} - x_{i,j-1}, & \text{if } 1 \le i < j \le n, \\ 0, & \text{if } 1 \le i = j \le n, \\ x_{i,j-1} - x_{i+1,j}, & \text{if } 1 \le j < i \le n. \end{cases} \tag{4.18}$$

The hope of getting a solution to the ToIEP hinges upon the fact that $k(X) = 0$ if and only if X is a Toeplitz matrix. For that reason, $k(X)$ is called a Toeplitz annihilator in Chu (1993).

To maintain orthogonality, differential systems such as (4.17) can be integrated effectively by available geometric integrators (Calvo et al., 1997; Dieci et al., 1994; Iserles et al., 2000). Numerical experiences seem to suggest that the flows always converge (Diele and Sgura, 1999), but a rigorous proof of the global convergence is still missing.

4.4 Nonnegative inverse eigenvalue problems

The notion of nonnegative matrices is so important that voluminous literature has been devoted to its study. Examples of applications include the Leontief input–output analysis in economics, finite Markov chains, linear complementary problems, and so on (Berman and Plemmons, 1979). One of the most elegant and important results in this field is the Perron–Frobenius theorem which characterizes properties of eigenvalues and eigenvectors of a nonnegative matrix. Perhaps it is for the same reason that the inverse problem has drawn considerable interest in the literature as well. The nonnegative inverse eigenvalue problem (**NIEP**) is concerned with the following construction of a matrix:

Problem 4.10 *(NIEP)*
Given a set $\{\lambda_k\}_{k=1}^n$ of numbers that are closed under complex conjugation, find an entry-wise nonnegative matrix $A \in \mathbb{R}^{n\times n}$ such that

$$\sigma(A) = \{\lambda_k\}_{k=1}^n.$$

We think that the earliest study on the subject of NIEP was perhaps due to the Russian mathematician Suleĭmanova (1949) on stochastic matrices, followed by Perfect (1953, 1955). The first systematic treatment of eigenvalues of symmetric nonnegative matrices can probably be attributed to Fiedler (1974).

A more comprehensive study was conducted by Boyle and Handelman (1991) using the notion of symbolic dynamics to characterize the conditions under which a given set is a portion of the spectrum of a nonnegative matrix or primitive matrix. General treatises on nonnegative matrices and applications include the classics by Berman and Plemmons (1979) and Minc (1988). Both books devote extensive discussion to the NIEPs as well.

Most of the discussions in the literature center around finding conditions to qualify a given set of values as the spectrum of some nonnegative matrices. A short list of references giving various necessary or sufficient conditions includes Barrett and Johnson (1984), Boyle and Handelman (1991), Friedland (1978), Friedland and Melkman (1979), Loewy and London (1978), de Oliveira (1983), and Reams (1996). The difficulty is that the necessary condition is usually too general and the sufficient condition too specific. Under a few special sufficient conditions, the nonnegative matrices can be constructed numerically (Soules, 1983). A general numerical treatment for the NIEP, even knowing the existence of a solution, is not available at the time of writing of this treatise.

A further refinement in the posing of the NIEP has also attracted some attention. We specify only two variants. The real-valued problem (RNIEP) is concerned with the problem of determining the set of real numbers $\{\lambda_k\}_{k=1}^n$ which can be the spectrum of a nonnegative matrix. The symmetric problem (SNIEP) is concerned with which set of real numbers $\{\lambda_k\}_{k=1}^n$ can occur as the spectrum of a symmetric nonnegative matrix. It was proved that there exist real numbers $\{\lambda_k\}_{k=1}^n$ that solve the RNIEP but *not* the SNIEP (Guo, 1996; Johnson et al., 1996). Of course, both RNIEP and SNIEP also concern the construction of such a nonnegative matrix once the spectrum $\{\lambda_k\}_{k=1}^n$ is known to be feasible.

4.4.1 *Some existence results*

The solvability of the NIEP has been the major issue of discussion in the literature. Existence results, either necessary or sufficient, are too numerous to be listed here. We shall mention only two results that in some sense provide the most distinct criteria in this regard.

Given a matrix A, the moments of A are defined to be the sequence of numbers $s_p = \text{trace}(A^p)$. Recall that if $\sigma(A) = \{\lambda_k\}_{k=1}^n$, then

$$s_p = \sum_{k=1}^{n} \lambda_k^p.$$

For nonnegative matrices, the moments are always nonnegative. The following necessary condition is due to Loewy and London (1978).

Theorem 4.10. (Loewy and London, 1978) Suppose $\{\lambda_k\}_{k=1}^n$ are eigenvalues of an $n \times n$ nonnegative matrix. Then the inequalities

$$s_p^m \leq n^{m-1} s_{pm} \tag{4.19}$$

hold for all $p, m = 1, 2, \ldots$.

Note also that the inequalities in (4.19) are sharp because equalities hold in them for the identity matrix. Note also that by taking $p = 1$, the inequalities in Theorem 4.10 imply the obvious necessary condition that for all $m = 1, 2, \ldots,$

$$s_m \geq \frac{s_1^m}{n^{m-1}} \geq 0, \tag{4.20}$$

provided $s_1 \geq 0$.

If we further limit the inverse problem to positive matrices, i.e., every entry exceeds zero, it turns out that the eigenvalues can be completely characterized. The following necessary and sufficient condition appeared at the end of the long treatise by Boyle and Handelman (1991, p. 313).

Theorem 4.11. (Boyle and Handelman, 1991) The set $\{\lambda_k\}_{k=1}^n \subset \mathbb{C}$ with $\lambda_1 = \max_{1 \leq k \leq n} |\lambda_k|$ is the nonzero spectrum of a positive matrix of size $m \geq n$ if and only if

(i) $\lambda_1 > |\lambda_i|$ for all $i > 1$,
(ii) $s_p > 0$ for all $p = 1, 2, \ldots,$ and
(iii) the polynomial $\prod_{k=1}^n (t - \lambda_k)$ has real coefficients in the indeterminate t.

It is exciting to see that Theorem 4.11 gives rise to the necessary and sufficient conditions of the existence of a nonnegative matrix with specified spectrum. However, one must read the statement carefully. What is not specified is the size m of the positive matrix. That is to say, we only know the existence of a larger size nonnegative matrix with a smaller size of eigenvalues, but the theorem does not indicate how large the size m is to be. This setting is somewhat different from the usual posing of NIEPs where the size of the matrix is the same as the cardinality of the prescribed spectrum.

4.4.2 *Symmetric nonnegative inverse eigenvalue problem*

We shall discuss a least squares approach for solving the general NIEP in the next section. At present, we touch upon the SNIEP with few more comments.

First, we remark that the solvability of a SNIEP remains open. Some sufficient conditions are listed in (Berman and Plemmons, 1979, Chapter 4), but these results are very limited.

Example 4.6. It can be checked that the set $\lambda = \{\sqrt[3]{51} + \epsilon, 1, 1, 1, -3, -3\}$ with $\epsilon > 0$ does not satisfy any of those conditions mentioned in Berman and Plemmons (1979, Chapter 4). In fact, it cannot be the nonzero spectrum of any symmetric nonnegative matrix (Johnson et al., 1996).

SNIEP for tridiagonal matrices One idea of solving the NIEPs is to reduce the number of free variables. By limiting the consideration of an NIEP to symmetric tridiagonal structure, for example, Friedland and Melkman (1979) have established a fairly elegant result that answers both RNIEP and SNIEP under a special circumstance.

Theorem 4.12. (Friedland and Melkman, 1979) A set of real numbers $\lambda_1 \geq \lambda_2 \geq \cdots \geq \lambda_n$ is the spectrum of an $n \times n$ nonnegative tridiagonal matrix if and only if $\lambda_i + \lambda_{n-i+1} = 0$ for all i. In this case, the matrix is a block diagonal matrix given by $J = \text{diag}(A_1, \cdots, A_{[(n+1)/2]})$, where

$$A_i = \begin{cases} \dfrac{1}{2}\begin{bmatrix} \lambda_i + \lambda_{n-i+1} & \lambda_i - \lambda_{n-i+1} \\ \lambda_i - \lambda_{n-i+1} & \lambda_i + \lambda_{n-i+1} \end{bmatrix}, & \text{if } 1 \leq i < (n+1)/2, \\ [\lambda_i], & \text{if } i = (n+1)/2 \text{ and } n \text{ is odd.} \end{cases}$$

Note that the matrix J constructed above is reducible when $n > 2$. The problem becomes harder if J is required further to be Jacobi, that is, a symmetric tridiagonal matrix with positive subdiagonal elements. One sufficient but not necessary condition for this particular JIEP is as follows.

Theorem 4.13. If $\lambda_1 > \lambda_2 > \cdots > \lambda_n$ and if $\lambda_i + \lambda_{n-i+1} > 0$ for all i, then there exists a positive Jacobi matrix with spectrum $\{\lambda_k\}_{k=1}^n$.

It would be very difficult to achieve a nearly as simple characterization of solutions to a general SNIEP.

Least squares SNIEP We could formulate the SNIEP as a constrained optimization problem of minimizing the objective function

$$F(Q, R) := \tfrac{1}{2}\|Q^\top \Lambda Q - R \circ R\|^2,$$

subject to the constraint $(Q, R) \in \mathcal{O}(n) \times \mathcal{S}(n)$. Recall that $\circ$ stands for the Hadamard product and $\mathcal{S}(n)$ stands for the subspace of $n \times n$ symmetric matrices. The idea is to parameterize any symmetric matrix $X = Q^\top \Lambda Q$ that is isospectral to Λ by the orthogonal matrix Q and to parameterize any symmetric nonnegative matrix $Y = R \circ R$ by the symmetric matrix R via entry-wise squares. The SNIEP is solvable if and only if $F(Q, R) = 0$ for some Q and R. Such a formulation offers a handle for numerical computation by optimization techniques. In Chu and Driessel (1991), the dynamical system

$$\begin{cases} \dfrac{dX}{dt} = [X, [X, Y]], \\ \dfrac{dY}{dt} = 4Y \circ (X - Y), \end{cases} \tag{4.21}$$

resulting from projected gradient flow with $X = Q^\top \Lambda Q$ and $Y = R \circ R$, has been proposed as a possible numerical means for solving the SNIEP. It is interesting to note that even if the SNIEP is not solvable, the limit point of the gradient flow gives rise to a least squares solution. We shall discuss an analogous approach of (4.21) for general NIEPs in Section 4.5.2.

4.4.3 *Minimum realizable spectral radius*

In many situations, satisfying the necessary conditions has been used as an exclusion procedure to eliminate unwanted situations. For instance, finding critical points to satisfy the first-order optimality condition is usually the first step in most optimization techniques. Clearly, other conditions such as the second-order derivative information need to be brought in to further determine whether a critical point is a maximizer or minimizer. In a similar manner, we could use necessary conditions to "exclude" eigenvalues that are not suitable for NIEPs. Neither of the criteria mentioned in the preceding section is practical for computation. There are infinitely many inequalities involved. Indeed, most other existing criteria can be of little help for such a task. Thus the issue of constructing a nonnegative matrix with a prescribed spectrum, even the very first step of verifying whether a given set $\{\lambda_k\}_{k=1}^n$ is feasible or not, has long been an open question.

In this section, we offer yet another avenue to tackle this challenging problem. Our idea is based on the Perron–Frobenius theorem asserting that every nonnegative matrix necessarily has at least one positive eigenvalue λ_1, called the *Perron root*, which is equal to the spectral radius of the matrix. Given $\{\lambda_k\}_{k=1}^n \subset \mathbb{C}$ that is closed under complex conjugation, we seek conditions on λ_1 in terms of $\{\lambda_2, \ldots, \lambda_n\}$ so that $\{\lambda_k\}_{k=1}^n$ is the spectrum of some nonnegative matrix.

It turns out that the notion of minimal realizable spectral radius precisely suits this purpose. Our main objective in this section is to introduce this simple yet momentous concept and to offer a simple numerical means to approximately locate the minimal realizable spectral radius. Though the significance of this approximation is correlated to the necessary conditions used to establish it, the computation is much more effective than using the necessary conditions themselves. Furthermore, our numerical experiments seem to suggest that the approximation in general is quite robust and close to the true minimal realizable spectral radius.

The notion of minimal realizable spectral radius for nonnegative matrices was first introduced in a 1991 M.S. Thesis at Peking University, which later appeared in Guo (1997). Following the notation adopted by Xu (1998, Chapter 5), we briefly outline the basic idea below which leads to a computational means.

For convenience, we shall denote the spectral radius of a matrix A by $\rho(A)$. Let

$$\mathcal{E}_{n-1} := \{\mathcal{L} = \{\lambda_2, \ldots, \lambda_n\} \subset \mathbb{C} \mid \lambda_i \in \mathcal{L} \text{ if and only if } \bar{\lambda}_i \in \mathcal{L}\} \tag{4.22}$$

denote the set of all $(n-1)$-tuples that are closed under complex conjugation. We say that the set $\{\lambda_1; \mathcal{L}\}$ with $\mathcal{L} = \{\lambda_2, \ldots, \lambda_n\} \in \mathcal{E}_{n-1}$ is *realizable* if and only if there exists an $n \times n$ nonnegative matrix A such that $\sigma(A) = \{\lambda_k\}_{k=1}^n$ and $\rho(A) = \lambda_1$.

Given any prescribed set $\{\lambda_k\}_{k=1}^n$ as possible eigenvalues in the formulation of an NIEP, the Perron–Frobenius theorem makes it easy to identify the Perron root, say, λ_1. The question of whether the underlying NIEP is solvable, therefore, is reduced to whether the remaining $\mathcal{L} = \{\lambda_2, \ldots, \lambda_n\}$ can be realizable via its association with λ_1. To address this question, we make a slightly different inquiry and recast the NIEP as the following problem:

> **Problem 4.11** *(Realizable spectrum problem)*
> Given any $\mathcal{L} = \{\lambda_2, \ldots, \lambda_n\} \in \mathcal{E}_{n-1}$, determine whether it can be associated with *some* λ_1 to become realizable. If so, characterize values of λ_1.

The following observations made in (Guo, 1997; Xu, 1998) provide critical insights into issues raised in Problem 4.11. First, it is true that any given $\mathcal{L} \in \mathcal{E}_{n-1}$ is realizable for some $\lambda_1 > 0$. Indeed, assume that $\mathcal{L} = \{\lambda_2, \ldots, \lambda_n\}$ where $\lambda_{2k} = \bar{\lambda}_{2k+1} = \mu_k + i\nu_k$, $k = 1, \ldots, s$, are complex-valued and $\lambda_{2s+2}, \ldots, \lambda_n$ are real-valued. Then it is straightforward to see that the product

$$P \begin{pmatrix} 2n\delta & 2\delta & 2\delta & \cdots & & & 2\delta & \cdots & 2\delta \\ 0 & \mu_1 & \nu_1 & & & & 0 & \cdots & 0 \\ 0 & -\nu_1 & \mu_1 & & & & & & \\ \vdots & & & \ddots & & & & & \vdots \\ & & & & \mu_s & \nu_s & & & 0 \\ & & & & -\nu_s & \mu_s & 0 & & 0 \\ & & & & & & \lambda_{2s+2} & & 0 \\ & & & & & & & \ddots & \\ 0 & 0 & & & & & 0 & \cdots & \lambda_n \end{pmatrix} P^{-1}, \tag{4.23}$$

with $\delta = \delta(\mathcal{L}) := \max_{2 \le i \le n} |\lambda_i|$ and $P = [\mathbf{1}, \mathbf{e}_2, \ldots, \mathbf{e}_n]$ where $\mathbf{1} = [1, \ldots, 1]^\top$ and $\mathbf{e}_j$ is the standard j-th unit vector, is a positive matrix with spectrum $\{2n\delta; \lambda_2, \ldots, \lambda_n\}$.

Secondly, it is clear that the map $R : \mathcal{E}_{n-1} \to \mathbb{R}$ via

$$R(\mathcal{L}) := \inf\{\lambda_1 \in \mathbb{R} | \{\lambda_1; \mathcal{L}\} \text{ is realizable}\} \tag{4.24}$$

is well defined. Furthermore, from the above example, we see that

$$\delta(\mathcal{L}) \le R(\mathcal{L}) \le 2n\delta(\mathcal{L}). \tag{4.25}$$

The quantity $R(\mathcal{L})$, called the *minimal realizable radius* of $\mathcal{L}$, plays a decisive role in the solvability of an NIEP because of the following result (Guo, 1997; Xu, 1998).

Theorem 4.14. (Guo, 1997) A given set $\{\lambda_k\}_{k=1}^n$ with $\lambda_1 \geq |\lambda_i|$, $i = 2, \ldots, n$, and $\{\lambda_2, \ldots, \lambda_n\} \in \mathcal{E}_{n-1}$ is precisely the spectrum of a certain $n \times n$ nonnegative matrix with λ_1 as its spectral radius if and only if $\lambda_1 \geq \mathcal{R}(\mathcal{L})$. The nonnegativity can be replaced by strick positivity if the inequality is replaced by strick inequality.

This necessary and sufficient criterion $\lambda_1 \geq R(\{\lambda_2, \ldots, \lambda_n\})$ is all we need to determine whether the NIEP with a prescribed set of eigenvalue $\{\lambda_k\}_{k=1}^n$ is solvable, provided the minimal realizable radius $R(\{\lambda_2, \ldots, \lambda_n\})$ is known. It thus becomes an interesting and worthy task to compute $R(\mathcal{L})$ for any given $\mathcal{L}$. As far as we know, no effort has been taken to carry out this work.

Bisection method We now outline an idea for the computation of $\mathcal{R}(\mathcal{L})$ by Chu and Xu (2004), followed by an illustrative algorithm. Our approach is quite straightforward, but it offers an effective way to judge whether an NIEP is solvable.

For convenience, denote

$$\omega_p = \omega_p(\mathcal{L}) := \sum_{k=2}^{n} \lambda_k^p, \quad p = 1, 2, \ldots \tag{4.26}$$

for each given $\mathcal{L}$. To estimate $R(\mathcal{L})$, it suffices to consider the sequence of (moment) functions $s_p : [\delta, 2n\delta] \to \mathbb{R}$, $p = 1, 2, \ldots$, where

$$s_p(t) := t^p + \omega_p. \tag{4.27}$$

We propose a bisection strategy by using the necessary conditions (4.19) and (4.20) to adjust the variable t. In principle, we seek after the minimal value of t so that $s_p(t)$ satisfies all necessary conditions. Note that this is a one-dimensional computation.

It should be pointed out right away that even if *all* necessary conditions are satisfied, we can at most assert that the estimate we obtained serves as a *lower bound* of $\mathcal{R}(\mathcal{L})$. Any λ_1 that is less than the estimated $\mathcal{R}(\mathcal{L})$ is assured to be unfeasible. For practicality, we shall check in fact only a predetermined set $\mathcal{N}$ of *finitely many* necessary conditions in the form of either (4.19) or (4.20).

Beginning with the interval $[\delta, 2n\delta]$, we take the midpoint $t = (2n + 1)\delta/2$. If $s_p(t)$ violates any of these necessary conditions in $\mathcal{N}$, then it must be that $t < R(\mathcal{L})$. In this case, the left endpoint could be adjusted safely since we know that $\mathcal{R}(\mathcal{L}) \in [t, 2n\delta]$. We then proceed to the next midpoint and repeat the procedure. On the other hand, if all the given conditions in $\mathcal{N}$ are satisfied, then we have to face two choices: one is to increase the level of complexity in $\mathcal{N}$ by adding more necessary conditions and repeat the procedure; the other is to *presume* that $\mathcal{R}(\mathcal{L}) < t$, adjust the right endpoint, and proceed to work with the next interval $[\delta, t]$. The latter strategy obviously has the danger of greatly underestimate the location of $\mathcal{R}(\mathcal{L})$.

The following algorithm demonstrates one of the tactics in selecting $\mathcal{N}$. We use the parameter ℓ to prepare the "level" of table lookup for values of ω_p and the parameter d to decide the "depth" of checkup on necessary conditions.

Algorithm 4.2 (Bisection method for minimal realizable spectral radius) Given $\mathcal{L} \in \mathcal{E}_{n-1}$ and a tolerance $\epsilon > 0$, the following iterations converge to a lower estimate of $\mathcal{R}(\mathcal{L})$.

1. Define $L := \delta$ and $R := 2n\delta$.
2. Select an integer level parameter ℓ and an integer depth parameter d.
3. Generate $\omega_p(\mathcal{L})$ for $k = 1, 2, \ldots, \ell!$
4. While $(R - L)/2 > \epsilon$, do
   (a) $t = (R + L)/2$;
   (b) for $k = 1 : \ell!/d$
         if $t^k + \omega_p \leq 0$, go to (4d).
         else
           for $m = 2 : \lfloor \ell!/k \rfloor$
             if $(t^k + \omega_p)^m > n^{m-1}(t^{km} + \omega_{km})$, go to (4d).
           end
         end
       end
   (c) go to (4f). (No left-end is reset. Might need more depth.)
   (d) set $L = t$.
   (e) go to (4).
   (f) set $R = t$.

We iterate again that the decision made at Step (4c) to cut back the right endpoint has the danger of underestimating $\mathcal{R}(\mathcal{L})$. We also emphasize that Step (4b) is fully flexible in that any additional necessary conditions can easily be embraced, if necessary, into the set $\mathcal{N}$ to provide a mechanism of multilayer filtering before we have to make a decision at (4c).

The advantage of determining $\mathcal{R}(\mathcal{L})$ is quite obvious. For each given $\mathcal{L}$, it settles the issue of feasibility of all potential $\{\lambda_1; \mathcal{L}\}$. Once the spectrum $\{\lambda_1; \mathcal{L}\}$ is determined to be feasible, what remains to be done is to construct such a nonnegative matrix with the prescribed spectrum. We have mentioned earlier that techniques for the construction of nonnegative matrices are in need of further research.

Numerical experiment It is highly dubious that Algorithm 4.2, as simple as it is, can do much in answering the long standing inverse problem for non-negative matrices. We think that it might be more convincing to throw in a few examples to demonstrate its performance in estimating $\mathcal{R}(\mathcal{L})$. We choose $\epsilon = 10^{-12}$, $\ell = 4$ and $d = 1$. The level parameter and the depth parameter can certainly be enlarged to increase the complexity of $\mathcal{N}$ and possibly the degree of confidence of the computed result.

Example 4.7. We begin with a pathological example demonstrating the case when the algorithm fails! The algorithm returns $\mathcal{R}(\{\sqrt{2}, i, -i\}) = \sqrt{2}$, suggesting that $\{\sqrt{2}, \sqrt{2}, i, -i\}$ should be the spectrum of a 4×4 nonnegative matrix. By the Perron–Frobenius theorem, such a matrix must be reducible. It follows that $\{\sqrt{2}, i, -i\}$ must be the spectrum of a 3×3 nonnegative matrix, which is impossible because the necessary condition (4.19) is clearly violated. Indeed, the algorithm returns $\mathcal{R}(\{i, -i\}) = \sqrt{3}$, assuring that $\{\sqrt{2}, i, -i\}$ cannot be spectrum of any 3×3 nonnegative matrix.

The problem with the wrongly calculated minimal realizable radius, that is, $\mathcal{R}(\{\sqrt{2}, i, -i\}) = \sqrt{2}$, is because the set $\{\sqrt{2}, \sqrt{2}, i, -i\}$ happens to satisfy necessary conditions (4.19) and (4.20) for all p and m. That is, the necessary conditions we have used in (4b) are never enough to capture a correct upper bound. This extreme example shows that the checking mechanism (4b) as it is now fails, regardless of the values of level ℓ and d. Remedies might come only if some other necessary conditions independent of (4.19) and (4.20) are implemented in Step (4b) as additional filters.

For 3×3 NIEPs, the solutions are known in closed form. It is known that with $\nu > 0$, the exact minimal realizable radius is given by $\mathcal{R}(\{\mu + i\nu, \mu - i\nu\}) = \max\{-2\mu, \mu + \sqrt{3}\nu\}$. We test our algorithm against this closed form solution.

Example 4.8. The Algorithm 4.2 is run on a grid of size 0.05 over the range $-5 \leq \mu \leq 5$ and $0 < \nu \leq 5$. We compare the computed minimal realizable spectral radii with the theoretical ones point by point on the grid. The maximal error is approximately 5.9×10^{-11}, clearly indicating that the algorithm works reasonably well.

A well-known result by Suleimanova (see Theorem 4.20) asserts that if $\mathcal{L} = \{\lambda_2, \ldots, \lambda_n\} \subset \mathbb{R}$ and if all $\lambda_i \in \mathcal{L}$ are nonpositive, then $\mathcal{R}(\mathcal{L}) = -\lambda_2 - \cdots - \lambda_n$. This knowledge provides us with a chance to test Algorithm 4.2 against relatively large-sized problems.

Example 4.9. We generate 100 sets of random numbers with sizes ranging from 2 to 101. The random numbers come from a uniform distribution over the interval $[-10, 0]$. Using each set as the prescribed $\mathcal{L}$, we apply our algorithm and compare the computed minimal realizable spectral radii with the Suleimanova bound. Errors in Figure 4.4 clearly indicates that the algorithm correctly (up to the stopping criterion) computes the minimal realizable spectral radius in each case.

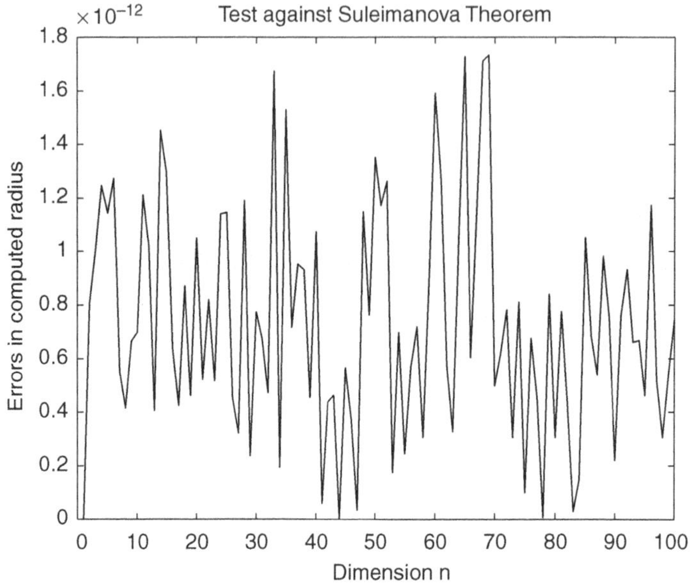

Figure 4.4. *Errors in predicting the Suleimanova bound*

Consider the example by Johnson et al. (1996) where it is shown that the spectrum with $\lambda_1 = \sqrt[3]{51} + \epsilon$, $\lambda_2 = \lambda_3 = \lambda_4 = 1$, $\lambda_5 = \lambda_6 = -3$, and $\lambda_7 = \cdots = \lambda_m = 0$ provides an example that solves the RNIEP but not the SNIEP.

Example 4.10. We apply Algorithm 4.2 to the set $\mathcal{L}_m = \{1, 1, 1, -3, -3, 0, \ldots, 0\}$ with m ranging from 7 to 250. We are curious to know how Johnson's λ_1 would differ from the computed $\mathcal{R}(\mathcal{L}_m)$. Depicted in Figure 4.5 is the difference $\mathcal{R}(\mathcal{L}_m) - \sqrt[3]{51}$. We notice an interesting relationship. The result affirms that $\mathcal{R}(\mathcal{L}_m) < \lambda_1$ and suggests that $\mathcal{R}(\mathcal{L}_m)$ converges to the critical value of $\sqrt[3]{51}$ as m goes to infinity. It also seems to provide a bound on the size of ϵ needed to substantiate Johnson's RNIEP result.

We conclude this section by reiterating that the minimal realizable spectral radius of any given $\mathcal{L} = \{\lambda_2, \ldots, \lambda_n\}$, closed under complex conjugation, serves as the sole critical threshold for determining whether the set $\{\lambda_k\}_{k=1}^n$ with the additional real number λ_1 can be the spectrum of an $n \times n$ nonnegative matrix. We stress that the more necessary conditions we can implement into the filtering mechanism $\mathcal{N}$ for the bisection method, the more accurate we can estimate the location of $\mathcal{R}(\mathcal{L})$ which, in turn, offers an effective way to determine whether an NIEP is solvable. The notion does not offer a way to construct the nonnegative matrix, but it does dictate whether such a construction is possible.

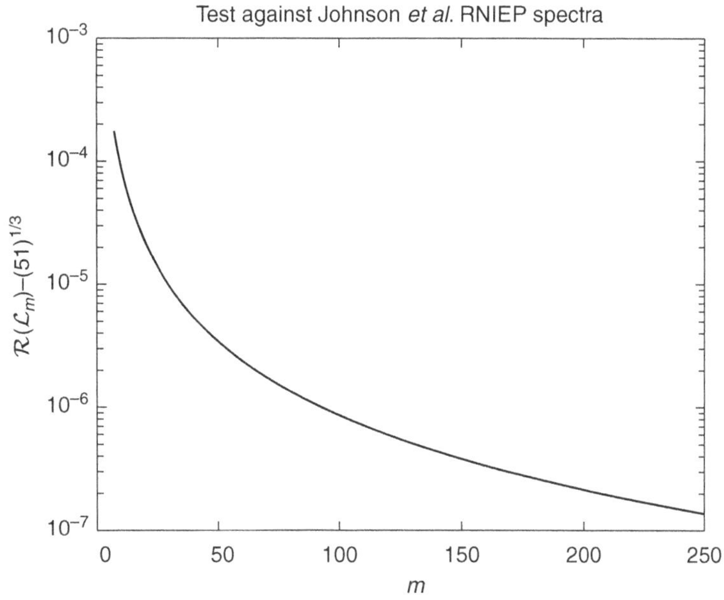

Figure 4.5. *Computed $R(\mathcal{L}_m)$ for the Johnson et al. spectra*

4.5 Stochastic inverse eigenvalue problems

An $n \times n$ nonnegative matrix $C = [c_{ij}]_{i,j=1}^n$ is a (row) stochastic matrix if all its row sums are 1. That is, $\sum_{j=1}^n c_{ij} = 1$, for all $i = 1, \ldots, n$. Each entry c_{ij} represents the probability to transit from state i to state j. The spectral information of C, particularly the left dominant eigenvector satisfying $\boldsymbol{\pi}^\top C = \boldsymbol{\pi}^\top$, called the *stationary distribution vector*, plays a critical role in Markov analysis. The stochastic inverse eigenvalue problem (**StIEP**) concerns whether a Markov chain can be built with the desirable spectral property. One kind of inverse problem can be stated as follows.

Problem 4.12 *(StIEP)*
Given a set of numbers $\{\lambda_k\}_{k=1}^n$ that is closed under complex conjugation and $\lambda_1 = 1$, construct a stochastic matrix C so that

$$\sigma(C) = \{\lambda_k\}_{k=1}^n.$$

Another kind of inverse problem is to satisfy a specified stationary vector. This is more in the context of PDIEP that we shall discuss in Chapter 5.

Problem 4.13 *(StIEP2)*
Given a positive vector $\boldsymbol{\pi} \in \mathbb{R}^n$, find a stochastic matrix C so that

$$\boldsymbol{\pi}^\top C = \boldsymbol{\pi}^\top.$$

Clearly the StIEP is a special NIEP with the additional row sum structure. It should be noted that, in contrast to the linearly constrained IEPs discussed thus far, the structure involved in the StIEP is nonlinear in the sense that the sum of two (stochastic) structured matrices does not have the same (stochastic) structure.

The Perron–Frobenius theorem asserts that the spectral radius $\rho(A)$ of an irreducible nonnegative matrix A is a positive maximal eigenvalue of A. The corresponding maximal eigenvector can be chosen to be all positive. Recall also that the set of reducible matrices forms a subset of measure zero. With this in mind, the spectral properties for stochastic matrices do not differ much from those of other nonnegative matrices because of the following connection (Minc, 1988).

Theorem 4.15. Suppose A is a (generic) nonnegative matrix with positive maximal eigenvalue $\rho(A)$ and a positive maximal eigenvector $\mathbf{x}$. Let $D = \operatorname{diag}(\mathbf{x})$. Then $\frac{1}{\rho(A)} D^{-1} A D$ is a stochastic matrix.

Theorem 4.15 suggests that an StIEP could be solved by first constructing a nonnegative solution, followed by a diagonal similarity transformation. We shall pursue this avenue as a numerical means for both the NIEP and the StIEP.

4.5.1 *Existence*

The StIEP is a hard problem. We note that the StIEP formulated in Problem 4.12 is called an *inverse spectrum problem* by Minc (1988) to carefully distinguish it from the problem of determining conditions under which one *single* complex number is an eigenvalue of a stochastic matrix. For the latter problem, the set Θ_n of points in the complex plane that are eigenvalues of any $n \times n$ stochastic matrices has been completely characterized by Karpelevič (1951). The complete statement of Karpelevič's theorem is rather lengthy (Minc, 1988, Theorem 1.8), so we shall only highlight the main points below.

Theorem 4.16. (Karpelevič, 1951) The region Θ_n is contained in the unit disk and is symmetric with respect to the real axis. It intersects the unit circle at points $e^{2\pi i a/b}$ where a and b range over all integers such that $0 \leq a < b \leq n$. The boundary of Θ_n consists of curvilinear arcs connecting these points in circular

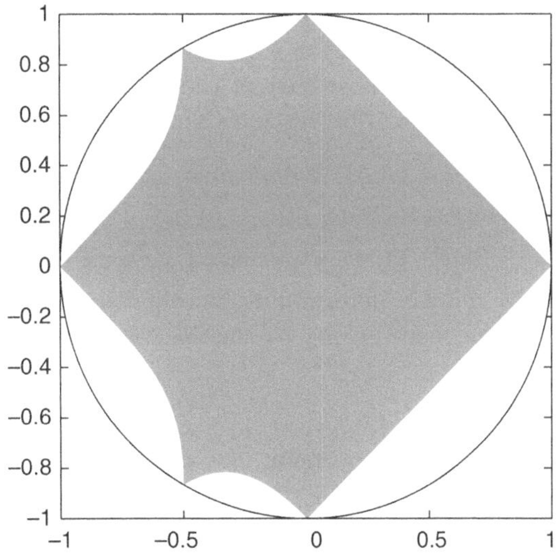

Figure 4.6. Θ_4 *by the Karpelevič theorem*

order. Any λ on these arcs must satisfy one of these parametric equations

$$\lambda^q(\lambda^p - t)^r = (1-t)^r,$$
$$(\lambda^b - t)^d = (1-t)^d\lambda^q,$$

where $0 \leq t \leq 1$, and b, d, p, q, r are natural integers determined from certain specific rules (explicitly given in Karpelevič (1951), Minc (1988)).

Example 4.11. It might help to better understand Theorem 4.16 by examining the set Θ_4 is sketched in Figure 4.6.

It has to be stressed that the Karpelevič theorem characterizes only one complex value at a time. The theorem does not provide further insights into when two or more points in Θ_n are eigenvalues belonging to the *same* stochastic matrix. It provides only a necessary condition for the StIEP. Determining necessary and sufficient conditions for a complex n-tuple to be a spectrum of a row stochastic matrix is a hard problem.

Example 4.12. Each element of the set $\mathcal{L} = \{-0.90, 0.88+0.10i, -0.88-0.10i\}$ belongs to Θ_4, but it can be shown that these points cannot belong to the same spectrum of any 4×4 stochastic matrix.

The minimal realizable spectral radius discussed in the early sections has an immediate application to the stochastic inverse spectrum problem (Minc, 1988). Recall what we have suggested earlier to first construct a generic nonnegative matrix which then is modified into a stochastic matrix via a diagonal similarity transformation. Such an approach requires that the maximal eigenvector of a nonnegative matrix associated with the Perron root be positive. We now point out that the assumption on the positivity of the maximal eigenvector is in fact not needed, due to the following observation (Xu, 1998, Lemma 5.3.2).

Theorem 4.17. If the set $\{\lambda_1; \mathcal{L}\}$ is realizable, then the set $\{1; \mathcal{L}/\lambda_1\}$ is the spectrum of a row stochastic matrix.

More significantly, with the aid of Theorem 4.14, we can extend the Karpelevič theorem by quickly determining whether any given $n-1$ points in the domain Θ_n simultaneously belong to the spectrum of a certain stochastic matrix or not.

Corollary 4.18. Given $\mathcal{L} = \{\lambda_2, \dots, \lambda_n\} \in \mathcal{E}_{n-1}$ whose elements are from within Θ_n, the set $\{1; \mathcal{L}\}$ is the spectrum of a certain row stochastic matrix if and only if $\mathcal{R}(\mathcal{L}) \leq 1$.

Once we have identified one set of $n-1$ points in Θ_n to be realizable by a stochastic matrix, we can identify a "segment" of sets in Θ_n, each of which is realizable by some other stochastic matrices.

Corollary 4.19. If the set $\{1; \mathcal{L}\}$ is the spectrum of a row stochastic matrix, then every set $\{1; \alpha\mathcal{L}\}$, $1 \leq \alpha \leq 1/\mathcal{R}(\mathcal{L})$, is also the spectrum of a certain row stochastic matrix.

Most of the sufficient conditions for StIEPs are imposed upon the case when the desirable spectrum is wholly made of real numbers (Suleĭmanova, 1949; Perfect, 1953, 1955). By Theorem 4.15, we now see that any sufficient conditions for NIEPs will also be applicable for StIEPs, provided that the maximal eigenvector of the resulting nonnegative matrix is positive.

We conclude this section with a sufficient condition due to Suleĭmanova (1949). This is perhaps one of the earliest results for StIEP.

Theorem 4.20. (Suleĭmanova, 1949) Any n given real numbers $1, \lambda_1, \dots, \lambda_{n-1}$ with $|\lambda_j| < 1$ are the spectrum of some $n \times n$ positive stochastic matrix if the sum of all $|\lambda_j|$ over those $\lambda_j < 0$ is less than 1. If the λ_j's are all negative the condition is also necessary.

4.5.2 *Numerical method*

From the fact that there are many sufficient conditions in the literature, it might be surprising that, except for Soules (1983), we have found none of the proofs thus far is constructive. There are no satisfactory numerical algorithms available either, even if a sufficient condition is met. Even with Soules (1983), the construction is quite limited in that the components of the Perron eigenvector must satisfy some additional inequalities. Recently, Chu and Guo (1998) proposed the following least squares approach which might be employed to solve the NIEP and the StIEP for generally prescribed eigenvalues. The idea is parallel to that of (4.21) for SNIEP, but the method is capable of handling general NIEPs.

The diagonal matrix $\mathrm{diag}(\lambda_1, \ldots, \lambda_n)$ of the prescribed eigenvalues can be transformed similarly, if necessary, into a diagonal block matrix Λ with 2×2 real blocks if some of the given values appear in complex conjugate pairs. The set

$$\mathcal{M}(\Lambda) = \{P\Lambda P^{-1} \in \mathbb{R}^{n\times n} | P \in \mathbb{R}^{n\times n} \text{ is nonsingular}\}$$

denotes isospectral matrices parameterized by nonsingular matrices P. The cone of nonnegative matrices

$$\pi(\mathbb{R}_+^n) := \{B \circ B | B \in \mathbb{R}^{n\times n}\}$$

is characterized by the Hadamard product of general square matrices. A solution to the NIEP must be at the intersection of $\mathcal{M}(\Lambda)$ and $\pi(R^n+)$, if there is any. If such a nonnegative matrix has a positive maximal eigenvector, it can be reduced to a stochastic matrix by a diagonal similarity transformation. We thus formulate the constrained optimization problem

$$\begin{array}{ll} \text{Minimize} & F(P,R) := \frac{1}{2}\|P\Lambda P^{-1} - R \circ R\|^2, \\ \text{subject to} & P \subset \mathcal{G}l(n), \; R \in \mathbb{R}^{n\times n}, \end{array}$$

where $\mathcal{G}l(n)$ denotes the general linear group of invertible matrices in $\mathbb{R}^{n\times n}$. We use P and R as variables to maneuver elements in $\mathcal{M}(\Lambda)$ and $\pi(R_+^n)$ to reduce the objective value. This is different from the SNIEP; note that the feasible sets are open and a minimum may not exist. With respect to the induced inner product,

$$\langle (X_1, Y_1), (X_2, Y_2)\rangle := \langle X_1, X_2\rangle + \langle Y_1, Y_2\rangle,$$

in the product topology of $\mathbb{R}^{n\times n} \times \mathbb{R}^{n\times n}$, the gradient of F can be expressed as the pair of matrices

$$\nabla F(P,R) = \left([\Delta(P,R), \Gamma(P)^\top]P^{-\top}, -2\Delta(P,R) \circ R\right), \tag{4.28}$$

where we recall that $[M,N] = MN - NM$ is the Lie bracket and we abbreviate

$$\Gamma(P) := P\Lambda P^{-1},$$
$$\Delta(P,R) := \Gamma(P) - R \circ R.$$

The differential system

$$\frac{dP}{dt} = [\Gamma(P)^\top, \Delta(P,R)]P^{-\top}, \tag{4.29}$$

$$\frac{dR}{dt} = 2\Delta(P,R) \circ R, \tag{4.30}$$

thus provides a steepest descent flow on the feasible set $\mathcal{G}l(n) \times \mathbb{R}^{n\times n}$ for the objective function $F(P,R)$.

There is an unexpected advantage that deserves notice. Observe that the zero structure, if there is any, in the original matrix $R(0)$ is preserved throughout the integration due to the Hadamard product. This feature may be exploited to construct a Markov chain with both prescribed linkages and spectrum. In other words, if it is desirable that state i is not allowed to transit into state j, we simply assign $r_{ij} = 0$ in the initial value $R(0)$. Then that zero transit status is maintained throughout the evolution.

On the other hand, the solution flow $P(t)$ is susceptible of becoming singular and the involvement of P^{-1} is somewhat worrisome. A remedy is to monitor the analytic singular value decomposition (ASVD) (Bunse-Gerstner et al., 1991),

$$P(t) = U(t)\Sigma(t)V(t)^{\top}, \tag{4.31}$$

of the path of matrices $P(t)$. In (4.31), $U(t)$ and $V(t)$ are orthogonal matrices of singular vectors, $\Sigma(t)$ is the diagonal of singular values, and all are analytic in t. Such an ASVD flow exists because the solution $P(t)$ is analytic by the Cauchy–Kovalevskaya theorem. We propose to replace the flow $(P(t), R(t))$ by the flow $(U(t), \Sigma(t), V(t), R(t))$ where the differential equations governing $U(t), \Sigma(t)$, and $V(t)$ can be obtained as follows (Wright, 1992).

Upon differentiating (4.31), we have

$$\underbrace{U^{\top}\frac{dP}{dt}V}_{Q} = \underbrace{U^{\top}\frac{dU}{dt}}_{Z}\Sigma + \frac{d\Sigma}{dt} + \Sigma\underbrace{\frac{dV^{\top}}{dt}V}_{W}, \tag{4.32}$$

where both Z and W are skew-symmetric matrices. Note that

$$Q := U^{\top}\frac{dP}{dt}V$$

is known because the vector field dP/dt has already been specified in (4.29) . Note also that the inverse of $P(t)$ can be calculated from $P^{-1} = V\Sigma^{-1}U^{\top}$ whereas the diagonal entries of $\Sigma = \text{diag}\{s_1, \ldots, s_n\}$ provide us with information about the proximity of $P(t)$ to singularity.

The diagonals on both sides of (4.32) lead to the equation

$$\frac{d\Sigma}{dt} = \text{diag}(Q)$$

for $\Sigma(t)$. The off-diagonals on both sides of (4.32) give rise to

$$\frac{dU}{dt} = UZ,$$
$$\frac{dV}{dt} = VW,$$

where, if $s_k^2 \neq s_j^2$, then entries of W and Z are obtained from

$$z_{jk} = \frac{s_k q_{jk} + s_j q_{kj}}{s_k^2 - s_j^2},$$

$$w_{jk} = \frac{s_j q_{jk} + s_k q_{kj}}{s_j^2 - s_k^2}$$

for all $j > k$. We note that in case $s_k^2 = s_j^2$, then W and Z can still be obtained with some modifications (Wright, 1992). The flow is now ready to be integrated by available geometric integrators (Hairer et al., 2002; Iserles et al., 2000).

Example 4.13. Suppose we want to find a stochastic matrix with eigenvalues $\{1.0000, -0.2608, 0.5046, 0.6438, -0.4483\}$. Applying Algorithm 4.2, we estimate that the minimum realizable spectral radius is approximately 0.6438. Thus the StIEP is solvable. Suppose we further want the Markov chain to be of a "ring" structure where each state is linked at most to its two immediate neighbors. We begin with the initial matrices

$$P_0 = \begin{bmatrix} 0.1825 & 0.7922 & 0.2567 & 0.9260 & 0.9063 \\ 0.1967 & 0.5737 & 0.7206 & 0.5153 & 0.0186 \\ 0.5281 & 0.2994 & 0.9550 & 0.6994 & 0.1383 \\ 0.7948 & 0.6379 & 0.5787 & 0.1005 & 0.9024 \\ 0.5094 & 0.8956 & 0.3954 & 0.6125 & 0.4410 \end{bmatrix}$$

and

$$R_0 = 0.9210 \begin{bmatrix} 1 & 1 & 0 & 0 & 1 \\ 1 & 1 & 1 & 0 & 0 \\ 0 & 1 & 1 & 1 & 0 \\ 0 & 0 & 1 & 1 & 1 \\ 1 & 0 & 0 & 1 & 1 \end{bmatrix}.$$

We have noted earlier that the zeros in R_0 are invariant under the integration of (4.29) and (4.30). Thus we are maintaining the ring structure while searching for the one Markov chain with matched spectrum. It turns out that the stochastic matrix

$$D = \begin{bmatrix} 0.0000 & 0.3094 & 0 & 0 & 0.6906 \\ 0.0040 & 0.5063 & 0.4896 & 0 & 0 \\ 0 & 0 & 0.5134 & 0.4866 & 0 \\ 0 & 0 & 0.7733 & 0.2246 & 0.0021 \\ 0.4149 & 0 & 0 & 0.3900 & 0.1951 \end{bmatrix},$$

with decimals rounded to 4 digits, is the limit point of the solution flow and possesses the desirable spectrum.

It is easy to check that along the solution curve $(P(t), R(t))$, we always have

$$\frac{dF(P(t),R(t))}{dt} = -\|\nabla F(P(t), R(t))\|^2 \leq 0.$$

Thus the method fails to solve the NIEP only in two situations: either $P(t)$ becomes singular in finite time or $F(P(t), R(t))$ converges to a least squares local solution. In the former case, a restart might avoid the problem. In the latter case, either the NIEP has no solution at all or the algorithm needs a restart.

4.6 Unitary Hessenberg inverse eigenvalue problems

Eigenvalues of unitary matrices are on the unit circle. At first glance, the unitary Hessenberg structure looks peculiar. But the associated eigenvalue problem arises naturally in several signal processing applications including the frequency estimation procedure and the harmonic retrieval problem in radar or sonar navigation. The characteristic polynomials of unitary Hessenberg matrices are the well known Szegö polynomials. The Szegö polynomials are orthogonal with respect to a certain measure on the unit circle in the same way as the characteristic polynomials of the Jacobi matrices are orthogonal with respect to a certain weight on an interval. Most of the discussion in this section are results adapted from Ammar et al. (1991), Ammar and He (1995), Faßbender (1997).

Let $H = [h_{ij}]_{i,j=1}^n$ denote an upper Hessenberg matrix with positive subdiagonal entries. Set $\eta_1 = -h_{11}$ and $\zeta_1 = h_{21}$. Then clearly $|\eta_1|^2 + \zeta_1^2 = 1$ and

$$G_1^*(\eta_1)H = \begin{bmatrix} 1 & 0 & \dots & 0 \\ 0 & \vdots & \hat{H} & \\ 0 & & & \end{bmatrix},$$

where $\hat{H}$ is an upper Hessenberg matrix of size $n-1$. Proceeding in this manner, we see that any upper Hessenberg unitary matrix H with positive subdiagonal entries can be uniquely expressed as the product

$$H = G_1(\eta_1) \cdots G_{n-1}(\eta_{n-1})\tilde{G}_n(\eta_n), \tag{4.33}$$

where η_k are complex numbers with $|\eta_k| < 1$ for $1 \leq k < n$ and $|\eta_n| = 1$, each $G_k(\eta_k)$, $k = 1, \dots, n-1$ is a Givens rotation,

$$G_k(\eta_k) = \begin{bmatrix} I_{k-1} & & & \\ & -\eta_k & \zeta_k & \\ & \zeta_k & \bar{\eta}_k & \\ & & & I_{n-k+1} \end{bmatrix}$$

with $\zeta_k := \sqrt{1 - |\eta_k|^2}$ and $\tilde{G}_n(\eta_n) = \operatorname{diag}[I_{n-1}, -\eta_n]$. In other words, each unitary upper Hessenberg matrix is determined by $2n-1$ *real* parameters, another feature analogous to the Jacobi matrices. For convenience, this dependence is denoted as $H = H(\eta_1, \dots, \eta_n)$. The decomposition (4.33), referred to as the Schur parameterization, plays a fundamental role in efficient algorithms for upper Hessenberg unitary matrices.

In many ways, theory and algorithms for the unitary Hessenberg inverse eigenvalue problem (**UHIEP**) are similar to those for the JIEP. So we state only a generic formulation.

> **Problem 4.14** *(Generic UHIEP)*
> Construct a unitary Hessenberg matrix with prescribed points on the unit circle.

Similar to the JIEPs, often a second set of spectral constraints is needed to specify a UHIEP. The following two UHIEPs, for example, are analogous to the SIEP8 and the SIEP6a, respectively.

Theorem 4.21. (Ammar and He, 1995) Given two sets $\{\lambda_k\}_{k-1}^n$ and $\{\mu_i\}_{i=1}^n$ of strictly interlaced points on the unit circle, there exist a unique unitary upper Hessenberg matrix $H = H(\eta_1, \dots, \eta_n)$ and a unique complex number α of unit modulus such that

$$\begin{cases} \sigma(H) = \{\lambda_k\}_{k=1}^n, \\ \sigma(H(\alpha\eta_1, \dots, \alpha\eta_n)) = \{\mu_i\}_{i=1}^n. \end{cases} \tag{4.34}$$

Note that the matrix $\tilde{H} = H(\alpha\eta_1, \dots, \alpha\eta_n)$ is a rank one perturbation of $H(\eta_1, \dots, \eta_n)$ in that

$$\tilde{H} = (I - (1 - \alpha)\mathbf{e}_1\mathbf{e}_1^\top)H.$$

However, this perturbation is different from that in SIEP8 in that its added effect is not limited only at the (n, n) entry, rather, it changes every entry of H.

The leading principal submatrix H_{n-1} of a unitary matrix is not unitary, and its eigenvalues do not lie on the unit circle. One way to modify the notion of submatrix matrix is as follows.

Theorem 4.22. (Ammar and He, 1995) Given two sets of strictly interlaced points $\{\lambda_k\}_{k=1}^n$ and $\{\mu_0, \mu_1, \dots, \mu_{n-1}\}$ on the unit circle, there exist a unique unitary upper Hessenberg matrix $H = H(\eta_1, \dots, \eta_n)$ such that

$$\begin{cases} \sigma(H) = \{\lambda_k\}_{k=1}^n, \\ \sigma(H(\eta_1, \dots, \eta_{n-2}, \rho_{n-1})) = \{\mu_1, \dots, \mu_{n-1}\}, \end{cases} \tag{4.35}$$

with $\rho_{n-1} = (\eta_{n-1} + \bar{\mu}_0\eta_n)/(1 + \bar{\mu}_0\bar{\eta}_{n-1}\eta_n)$.

Just like the JIEPs, the proofs of existence for the above results can be turned into numerical methods in the same spirit as the Lanczos or Arnoldi algorithm.

We shall not repeat the details here which can be found in Ammar et al., 1991; Ammar and He, 1995.

4.7 Inverse eigenvalue problems with prescribed entries

Thus far, we have not formally set forth the definition, but we have seen already that a very large class of IEPs can be described as inverse eigenvalue problems with prescribed entries (**PEIEP**). The prescribed entries are used to characterize the underlying structure. The most general setting can be delineated as follows (Ikramov and Chugunov, 2000).

Problem 4.15 *(PEIEP)*
Given a certain subset $\mathcal{L} = \{(i_t, j_t)\}_{t=1}^{\ell}$ of double subscripts, a certain set of values $\{a_t\}_{t=1}^{\ell}$ over a field $\mathbb{F}$, and a set of n values $\{\lambda_k\}_{k=1}^{n}$, find a matrix $X = [x_{ij}]_{i,j=1}^{n} \in \mathbb{F}^{n\times n}$ such that

$$\begin{cases} \sigma(X) = \{\lambda_k\}_{k=1}^{n}, \\ x_{i_t,j_t} = a_t, \qquad \text{for } t = 1, \ldots, \ell. \end{cases}$$

Let $|\mathcal{L}|$ denote the cardinality ℓ of a general index set $\mathcal{L}$. The PEIEP is to determine (complete) values for the remaining $n^2 - |\mathcal{L}|$ positions that do not belong to $\mathcal{L}$ so as to satisfy the spectral constraint. The Jacobi structure can be considered as a special case of PEIEP where, in addition to the desired symmetry of the band, elements outside the tridiagonal band are required to be zero. The PEIEP has another interesting variation in which the completion problem requires only a one-to-one correspondence between the ℓ positions in $\mathcal{L}$ and the ℓ prescribed values $\{a_1, \ldots, a_\ell\}$, but not in any specific order. Such a problem is more general than (4.15), but will not be considered in this presentation.

PEIEPs have a long history of development in the literature. We shall chronicle several in this section. Generally speaking, attention has been centered around two focal points: to determine the cardinality $|\mathcal{L}|$ so that the problem makes sense and to study the effect of the locations in $\mathcal{L}$.

4.7.1 *Prescribed entries along the diagonal*

Perhaps a natural place to begin the discussion of PEIEP is the construction of a Hermitian matrix with prescribed diagonal entries and eigenvalues. Recall that a vector $\mathbf{a} \in \mathbb{R}^n$ is said to majorize $\boldsymbol{\lambda} \in \mathbb{R}^n$ if, assuming the ordering $a_{j_1} \leq \cdots \leq a_{j_n}$ and $\lambda_{m_1} \leq \cdots \leq \lambda_{m_n}$ of their elements, the following relationships hold:

$$\begin{cases} \sum_{i=1}^{k} \lambda_{m_i} \leq \sum_{i=1}^{k} a_{j_i}, \quad k = 1, \ldots, n, \\ \sum_{i=1}^{n} \lambda_{m_i} = \sum_{i=1}^{n} a_{j_i}. \end{cases} \tag{4.36}$$

The necessary and sufficient relationship between the diagonal entries and the eigenvalues of a Hermitian matrix is completely characterized by the Schur–Horn theorem.

Theorem 4.23. (Horn, 1954a) A Hermitian matrix H with eigenvalues $\boldsymbol{\lambda}$ and diagonal entries $\mathbf{a}$ exists if and only if $\mathbf{a}$ *majorizes* $\boldsymbol{\lambda}$.

The notion of majorization has many applications, including matrix theory and statistics. Through the Schur–Horn theorem, for instance, the total least squares problem can be seen equivalent to a linear programming problem (Bloch et al., 1990). There are extensive research results on this subject. See, for example, the books by Arnold (1987) and Marshall and Olkin (1979) and the references contained therein.

The Schur–Horn theorem expresses an important relation of two seemingly unrelated notions in $\mathbb{R}^n$: majorization and convexity. In 1923 Schur proved that the diagonal $\mathbf{a} = \operatorname{diag}(H)$ of an $n \times n$ Hermitian matrix H with eigenvalues $\boldsymbol{\lambda} = \{\lambda_k\}_{k=1}^{n}$ is contained in the convex hull of symmetric group actions on $\boldsymbol{\lambda}$. It took 31 years before Horn proved that each point in the convex hull can be obtained in the same way, that is, as the diagonal of some Hermitian matrix. Equivalently, $\mathbf{a}$ majorizes $\boldsymbol{\lambda}$ if and only $\boldsymbol{\lambda} = P\mathbf{a}$ for some doubly stochastic matrix P. Apparently, the sufficient condition is the harder part of the proof and that is precisely the heart of the Schur–Horn inverse eigenvalue problem (**SHIEP**):

> **Problem 4.16** *(SHIEP)*
> Given two vectors $\mathbf{a}$ and $\boldsymbol{\lambda}$ where $\mathbf{a}$ majorizes $\boldsymbol{\lambda}$, construct a Hermitian matrix H such that
> $$\begin{cases} \operatorname{diag}(H) = \mathbf{a}, \\ \sigma(H) = \boldsymbol{\lambda}. \end{cases}$$

The original proof of the Schur–Horn theorem was done by mathematical induction. A constructive proof by a continuous method which can be used as a numerical means for solving the SHIEP was first proposed by Chu (1995). Shortly after, Zha and Zhang (1995) argued that the induction proof actually can be converted into an appropriate iterative method that enjoyes the property of termination in finitely many steps.

Without the Hermitian structure, the relationship between eigenvalues and diagonal entries of a general matrix is much simpler.

Theorem 4.24. (Mirsky, 1958) A matrix with eigenvalues $\lambda_1, \dots, \lambda_n$ and main diagonal elements $a_1, \dots, a_n$ exists if and only if

$$\sum_{i=1}^{n} a_i = \sum_{i=1}^{n} \lambda_i. \tag{4.37}$$

Again, the sufficient condition in the Mirsky theorem constitutes an inverse eigenvalue problem with prescribed entries along the diagonal. So straightforward and expected is the relationship (4.37) that one might think that an inverse problem as such would be of little interest. It is *not* so. It can be shown that such an inverse problem has a closed form solution by a rational algorithm for all n (Ikramov and Chugunov, 2000), but the algebraic expressions involved are quite complicated.

Example 4.14. Consider the problem of constructing a 5×5 matrix A with symbolic, and hence arbitrary, diagonal elements $\{a_{11}, \dots, a_{55}\}$ and eigenvalues $\{\lambda_1, \dots, \lambda_5\}$ where necessarily $\lambda_5 = \sum_{i=1}^{5} a_{ii} - \sum_{j=1}^{4} \lambda_j$. It can be verified that the matrix

$$A = \begin{bmatrix} a_{11} & 1 & 0 & -1 & 0 \\ a_{21} & a_{22} & a_{23} & a_{24} & 1 \\ 0 & 0 & a_{33} & 1 & 0 \\ 0 & a_{42} & a_{43} & a_{44} & 1 \\ a_{51} & a_{52} & a_{53} & a_{54} & a_{55} \end{bmatrix},$$

where

$$\begin{aligned}
a_{21} &= (\lambda_1 - a_{11})\,\lambda_1 - (\lambda_1 - a_{11})\,(\lambda_2 + \lambda_1 - a_{11})\,, \\
a_{23} &= 1 + (\lambda_3 - a_{33})\,\lambda_3 - (\lambda_3 - a_{33})\,(\lambda_4 + \lambda_3 - a_{33})\,, \\
a_{24} &= -\lambda_2 - \lambda_1 + a_{11} + a_{44}, \\
a_{42} &= -\lambda_2 - \lambda_1 + a_{11} + a_{22}, \\
a_{43} &= (\lambda_3 - a_{33})\,\lambda_3 - (\lambda_3 - a_{33})\,(\lambda_4 + \lambda_3 - a_{33})\,, \\
a_{51} &= (\lambda_2 + \lambda_1 - a_{11} - a_{22})\,((\lambda_1 - a_{11})\,\lambda_1 - (\lambda_1 - a_{11})\,(\lambda_2 + \lambda_1 - a_{11}))\,, \\
a_{52} &= (\lambda_2 + \lambda_1 - a_{11} - a_{22})\,(\lambda_2 + \lambda_1 - a_{11}) - (\lambda_2 + \lambda_1 - a_{11} - a_{22}) \\
&\quad \times (a_{33} + a_{44} + a_5 - \lambda_3 - \lambda_4) + (\lambda_4 + \lambda_3 - a_{33} - a_{44}) \\
&\quad \times (-\lambda_2 - \lambda_1 + a_{11} + a_{22})\,, \\
a_{53} &= (\lambda_2 + \lambda_1 - a_{11} - a_{22})\,(1 + (\lambda_3 - a_{33})\,\lambda_3 - (\lambda_3 - a_{33})\,(\lambda_4 + \lambda_3 - a_{33})) \\
&\quad + (\lambda_4 + \lambda_3 - a_{33} - a_{44})\,((\lambda_3 - a_{33})\,\lambda_3 - (\lambda_3 - a_{33})\,(\lambda_4 + \lambda_3 - a_{33}))\,, \\
a_{54} &= (\lambda_2 + \lambda_1 - a_{11} - a_{22})\,(\lambda_4 + \lambda_3 - a_{33} - \lambda_2 - \lambda_1 + a_{11}) \\
&\quad + (\lambda_4 + \lambda_3 - a_{33} - a_{44})\,(\lambda_4 + \lambda_3 - a_{33}) - (\lambda_4 + \lambda_3 - a_{33} - a_{44})\,a_{55},
\end{aligned}$$

has spectrum $\{\lambda_1, \dots, \lambda_5\}$.

The prescribed entries in both theorems above are located along the diagonal. The generalization of the Mirsky theorem to the case of nonprincipal diagonals was carried out by de Oliveira (1973a) followed by de Oliveira (1973b, 1975). Given a permutation ϱ, the positions in a matrix corresponding to the index set $\mathcal{L} = \{(i, \varrho(i))\}_{i=1}^{n}$ is called a ϱ-diagonal of that matrix.

Theorem 4.25. (de Oliveira, 1973b) Let $\{\lambda_k\}_{k=1}^n$ and $\{a_i\}_{i=1}^n$ be two sets of arbitrary numbers over a field $\mathbb{F}$. Suppose that at least one of the disjoint cycles in the product representation $\varrho = \varrho_1 \dots \varrho_s$ has length > 2. Then there exists a matrix $X = [x_{ij}]_{i,j=1}^n \in \mathbb{F}^{n\times n}$ such that $\sigma(X) = \{\lambda_k\}_{k=1}^n$ and $x_{i,\varrho(i)} = a_i$ for $i = 1, \dots, n$.

The assumption in the de Oliveira theorem that at least one cycle has length great than 2 precludes the case of ϱ being the identity and, hence, the equality (4.37) is not needed. If no cycle is of length > 2, then a similar result holds under some additional restrictions (de Oliveira, 1973b, Theorem 2). Using the so-called L-transform, the proof of the de Oliveira theorem is constructive and can be converted into a finite step numerical algorithm. We shall comment more on this in the next few sections.

Note that all three theorems, Schur–Horn, Mirsky, and de Oliveira, give rise to an IEP with prescribed entries along the diagonal. We shall continue to study other PEIEPs where the prescribed entries are at general locations. At present, however, it might be natural to bring forth another class of inverse problems similar to the PEIEPs except that singular values instead of eigenvalues are involved as the spectral constraint. Wc propose the following inverse singular value problem with prescribed entries (**PEISVP**).

Problem 4.17 *(PEISVP)*
Given a certain subset $\mathcal{L} = \{(i_t, j_t)\}_{t=1}^{\ell}$ of double subscripts, a certain set of real numbers $\{a_t\}_{t=1}^{\ell}$, and a set of m positive number $\{s_i\}_{i=1}^n$, find a matrix $X = [x_{ij}] \in \mathbb{R}^{m\times n}$, $m \geq n$, such that

$$\begin{cases} \varpi(X) = \{s_i\}_{i=1}^n, \\ x_{i_t,j_t} = a_t, \qquad \text{for } t = 1, \dots, \ell. \end{cases}$$

We are not aware if much work done has been done on this problem, but we think there are lots of interesting questions here to be explored. For instance, analogous to the Schur–Horn theorem, the following Sing–Thompson theorem characterizes the relationship between singular values and diagonal entries of a general matrix.

Theorem 4.26. (Sing, 1976; Thompson, 1977) Assume that elements in two given vectors $\mathbf{d}, \mathbf{s} \in \mathbb{R}^n$ satisfy $s_1 \geq s_2 \geq \dots \geq s_n$ and $|d_1| \geq |d_2| \geq \dots \geq |d_n|$. Then a real matrix with singular values $\mathbf{s}$ and main diagonal entries $\mathbf{d}$ (possibly in different order) exists if and only if

$$\begin{cases} \sum_{i=1}^k |d_i| \leq \sum_{i=1}^k s_i, \quad \text{for } k = 1, \dots, n, \\ \left(\sum_{i=1}^{n-1} |d_i|\right) - |d_n| \leq \left(\sum_{i=1}^{n-1} s_i\right) - s_n. \end{cases} \tag{4.38}$$

Once again, the sufficient condition in the Sing–Thompson theorem gives rise to a special inverse singular value problem (**STISVP**).

> **Problem 4.18** *(STISVP)*
> Given two vectors $\mathbf{d}$ and $\mathbf{s}$ in $\mathbb{R}^n$ satisfying (4.38), construct a matrix $X \in \mathbb{R}^{n\times n}$ such that
> $$\begin{cases} \varpi(X) = \mathbf{s}, \\ \operatorname{diag}(X) = \mathbf{d}. \end{cases}$$

The original proof of the Sing–Thompson theorem was done by mathematical induction. Chu (1999) developed it into a divide-and-conquer algorithm that can be implemented in any programming environment that supports recursion. We shall describe one similar algorithm in Section 4.9.

Finally, we note that a PEISVP can be converted into a PEIEP because eigenvalues of the *structured* symmetric matrix

$$C = \begin{bmatrix} 0 & X \\ X^\top & 0 \end{bmatrix} \tag{4.39}$$

are precisely the pluses and minuses of singular values of X. The PEIEP for C has the fixed structure of zero diagonal blocks plus whatever prescribed entries that are inherited from X. The PEISVP for a structured X is solvable if and only if the PEIEP for C with structure defined in (4.39) is solvable.

4.7.2 *Prescribed entries at arbitrary locations*

Strictly speaking, the PEIEP in the Mirsky theorem really involves only $n-1$ prescribed entries $a_1, \ldots, a_{n-1}$ because a_n is determined from (4.37). It is quite remarkable that London and Minc (1972) showed that such a restriction of the $n-1$ prescribed entries to the main diagonal is entirely unnecessary.

Theorem 4.27. (London and Minc, 1972) Given two sets $\{\lambda_k\}_{k=1}^n$ and $\{a_i\}_{i=1}^{n-1}$ of arbitrary numbers over a field $\mathbb{F}$, suppose $\mathcal{L} = \{(i_t, j_t)\}_{t=1}^{n-1}$ is a set of arbitrary but distinct positions. Then there exists a matrix $X \in \mathbb{F}^{n\times n}$ such that $\sigma(X) = \{\lambda_k\}_{k=1}^n$ and $X_{i_t,j_t} = a_t$ for $t = 1, \ldots, n-1$.

An alternative proof was given in de Oliveira, 1973b. Both proofs used mathematical induction. In principle, we think a fast recursive algorithm similar to those for the SHIEP and the STISVP could be devised. We have not seen a numerical implementation of it, yet we expect no difficulties. Similar inverse problems of constructing matrices with $n-1$ arbitrary prescribed entries and prescribed characteristic polynomials are considered in Dias da Silva (1974), Ikramov and Chugunov (2000).

An interesting follow-up question to the London–Minc theorem is how many more entries of a matrix can be specified while the associated PEIEP is still solvable. Obviously, as we have learned, the locations of these prescribed entries also have some effect on the solvability. To help better grasp the scope of this complicated issue, we turn our attention to a special subclass of PEIEPs before we return to this question in Section 4.7.4.

4.7.3 *Additive inverse eigenvalue problem revisit*

Thus far, we have considered several cases of PEIEPs with small $|\mathcal{L}|$. With $|\mathcal{L}| = n - 1$, the London and Minc theorem asserts that the PEIEP is always solvable with no other constraints. With $|\mathcal{L}| = n$, the PEIEP is solvable under some constraints. Indeed, Ikramov and Chugunov (2000, Section 3b) argued meticulously through various cases to draw the most general conclusion.

Theorem 4.28. (Ikramov and Chugunov, 2000) Suppose that the field $\mathbb{F}$ is algebraically closed and that $|\mathcal{L}| = n$. Assume that the following two conditions are met:

$$\begin{cases} \text{that (4.37) is satisfied,} & \text{if } \mathcal{L} = \{(i,i)\}_{i=1}^{n}, \text{or} \\ \text{that } a_i = \lambda_j \text{ for some } j, & \text{if } \mathcal{L} = \{(i,j_t)\}_{t=1}^{n} \text{ and } a_t = 0 \text{ for all } j_t \neq i. \end{cases}$$

Then the PEIEP is solvable via rational algorithms in $\mathbb{F}$.

To emphasize that the PEIEP with $|\mathcal{L}| = n$ can be solved via a rational algorithm, Chugunov (2001) has developed a MAPLE code that generates a solution in closed-form. Indeed, the relationships in Example 4.14 were generated by using Chugunov's code.

In both Theorems 4.27 and 4.28, there is much room, that is, $n^2 - |\mathcal{L}|$ free locations, for constructing such a matrix. In contrast, the classical AIEP (see Section 3.3) is another type of PEIEP with much less room for free locations. Recall that a classical AIEP concerns adding a diagonal matrix D to a given matrix A so that $\sigma(A + D)$ has a prescribed spectrum. (Note that in a more general context, the matrix D need not be diagonal but rather may be defined by the *complement* to any given index set $\mathcal{L}$.) In the classical AIEP, the prescribed entries consist of all off-diagonal elements, that is, $|\mathcal{L}| = n^2 - n$. In this case, the following result by Friedland (1972), which is slightly more general than what we have stated previously in Theorem 3.7, shows the solvability of the AIEP over *any* algebraically closed field.

Theorem 4.29. (Friedland, 1972) The AIEP over any algebraically closed field is always solvable. If n is the order of the problem, then there exist at most $n!$ solutions. For almost all given $\{\lambda_k\}_{k=1}^{n}$, there are exactly $n!$ solutions.

It is critical to observe that the solvability assured in both Theorem 4.28 and Theorem 4.29 requires that the underlying field $\mathbb{F}$ is algebraically closed. We have already argued in Section 3.3 that the AIEP is not always solvable over $\mathbb{R}$ from two perspectives: that only adequate separation of prescribed eigenvalues relative to the size of the (prescribed) off-diagonal entries of A renders some sufficient conditions for real solvability (Theorem 3.9) and that coalescent eigenvalues almost always render no solution at all (Theorem 3.10).

It is perhaps fitting to point out that the PEIEPs are a special case of the LiPIEP (see 3.1) by identifying $|\mathcal{L}| = \ell$ and $A_t = \mathbf{e}_{i_t}\mathbf{e}_{j_t}^\top$ where $\mathbf{e}_k$ denotes the k-th standard unit vector in $\mathbb{R}^n$ and (i_t, j_t) is the t-th pair of indices in $\mathcal{L}$. The existence theory, sensitivity analysis, and numerical methods developed earlier for LiPIEPs can therefore be applied to a PEIEP in general.

There is a fundamental difference between Theorems 4.28 and 4.29. Somewhere between $|\mathcal{L}| = n$ and $\mathcal{L} = n^2 - n$ there is a threshold on the cardinality of prescribed entries that changes the PEIEP from finitely solvable to finitely unsolvable. It is known that the classical AIEPs, with $|\mathcal{L}| = n^2 - n$ being too large, in general cannot be solved in finitely many steps. The AIEP generally has to be solved by other means of numerical methods such as those discussed in Section 3.2.

Example 4.15. The AIEP in which all off-diagonal entries are 1 is not solvable in radicals for $n \geq 5$. The AIEP for a Jacobi matrix with subdiagonal (and superdiagonal) entries 1 is not solvable in radicals even for $n = 4$ (Ikramov and Chugunov, 2000).

4.7.4 *Cardinality and locations*

We have seen two extreme cases thus far. The prescribed entries in the SHIEP and the STISVP, on one hand, are required to be on the diagonal. It follows that certain equalities or inequalities (Theorems 4.23 and 4.26) involving the prescribed eigenvalues and entries must be satisfied. The prescribed entries in an AIEP, on the other hand, are required to be on the off-diagonal. The complex solvability was answered in Theorem 4.29, but its real solvability is only partially understood. In all these cases, the prescribed entries are located at special positions.

Theorem 4.28 is welcome in that it relaxes the specification to arbitrary locations and, under very mild conditions, asserts the existence of a solution to the PEIEP when $|\mathcal{L}| = n$. It is natural to ask what is the interplay between cardinality and locations so that a PEIEP is solvable. In other words, atop Problems 3.9 or 4.15 is the theoretical issue of how the prescribed eigenvalues, the cardinality, and the locations of prescribed entries intertwine to ensure the solvability.

Problem 4.19 *(Cardinality and location issue of PEIEP)*
Specify the conditions on the locations and the cardinality of $\mathcal{L}$ under which a PEIEP is solvable over an algebraically closed field.

Starting with the conditions in Theorem 4.28 and onward until the conditions in Theorem 4.29, we believe that to answer Problem 4.19 for almost every kind of

$\mathcal{L}$ in between is a challenge. As a partial answer, we describe what is possibly the strongest result on $|\mathcal{L}|$ in the class of PEIEPs at arbitrary locations. The original work was presented in the M.Sc. thesis by Hershkowitz (1978). We restate the result from Hershkowitz (1983).

Theorem 4.30. (Hershkowitz, 1983) Suppose that the field $\mathbb{F}$ is algebraically closed and that $|\mathcal{L}| = 2n - 3$. Assume that the following two conditions, if they occur, are met:

$$\begin{cases} \text{that (4.37) is satisfied,} & \text{if } \mathcal{L} \supseteq \{(i,i)\}_{i=1}^{n}, \text{or} \\ \text{that } a_i = \lambda_j \text{ for some } j, & \text{if } \mathcal{L} \supseteq \{(i,j_t)\}_{t=1}^{n} \text{ and } a_t = 0 \text{ for all } j_t \neq i. \end{cases}$$

Then the PEIEP is solvable in $\mathbb{F}$.

Note that the effect of locations of positions in $\mathcal{L}$ is limited to the two necessary conditions stated in the theorem. These conditions are remarkably general. The proof of the Hershkowitz theorem is established by induction. It was claimed in Ikramov and Chugunov (2000) that, in principle, the construction could be done by a rational algorithm. However, we find that there are seven basic cases plus the many subcases of analysis in the 15-page proof. Such complexity might make its computer implementation quite a challenge. So far as we know, no software for constructing the Hershkowitz result has been accomplished yet. It would be interesting to see if other numerical algorithms could be developed.

4.7.5 *Numerical methods*

Research thus far indicates that constructive proofs for the solution of a PEIEP exist to a certain point beyond which very few theories or numerical algorithms are available. In this section we propose to recast the completion problem as one of minimizing the distance between the isospectral matrices with the prescribed eigenvalues and the affine matrices with the prescribed entries. This approach is general enough that it can be used to explore the existence question when the prescribed entries are at arbitrary locations with arbitrary cardinalities (Chu et al., 2004). The approach proposed below can easily be generalized to the complex case, but we shall limit our discussion to real matrices. Consequently, the prescribed eigenvalues $\lambda_1, \ldots, \lambda_n$ are necessarily closed under complex conjugation.

Let $\Lambda \in \mathbb{R}^{n\times n}$ denote a matrix with eigenvalues $\lambda_1, \ldots, \lambda_n$. If necessary, we could consider Λ as the (real) Jordan canonical form or the Schur form to accommodate the geometric multiplicities. The set

$$\mathcal{M}(\Lambda) = \{V\Lambda V^{-1} | V \in \mathcal{G}l(n)\} \tag{4.40}$$

consists of all matrices that are isospectral to (and with the same kind of geometric multiplicity as) Λ. Given an index subset of locations $\mathcal{L} = \{(i_\nu, j_\nu)\}_{\nu=1}^{\ell}$

and the prescribed values $\mathbf{a} = \{a_1, \ldots, a_\ell\}$, the set

$$\mathcal{S}(\mathcal{L}, \mathbf{a}) = \{A \in \mathbb{R}^{n\times n} | A_{i_\nu j_\nu} = a_\nu, \nu = 1, \ldots, \ell\} \tag{4.41}$$

contains all matrices with the prescribed entries at the desired locations. Solving the PEIEP is equivalent to finding the intersection of $\mathcal{M}(\Lambda)$ and $\mathcal{S}(\mathcal{L}, \mathbf{a})$. Toward that end, we propose a least squares approximation.

For convenience, split any given matrix X as the sum

$$X = X_{\mathcal{L}} + X_{\mathcal{L}^c}, \tag{4.42}$$

where entries in $X_{\mathcal{L}}$ are the same as X, except that those that do not correspond to positions in $\mathcal{L}$ are set identically zero and $\mathcal{L}^c$ is simply the index subset complementary to $\mathcal{L}$. The drawing in Figure 4.7, though only symbolic, indicates the various situations in our approach. With respect to the Frobenius inner product, the projection $P(X)$ of any matrix X onto the affine subspace $\mathcal{S}(\mathcal{L}, \mathbf{a})$ is given by

$$P(X) = A_{\mathcal{L}} + X_{\mathcal{L}^c}, \tag{4.43}$$

where $A_{\mathcal{L}}$ is the constant matrix in $\mathcal{S}(\mathcal{L}, \mathbf{a})$ with zero entries at all locations corresponding to $\mathcal{L}^c$. For each given $X \in \mathcal{M}(\Lambda)$, we intend to minimize the distance between X and $\mathcal{S}(\mathcal{L}, \mathbf{a})$. Equivalently, we want to minimize the function defined by

$$f(X) = \frac{1}{2}\langle X - P(X), X - P(X)\rangle. \tag{4.44}$$

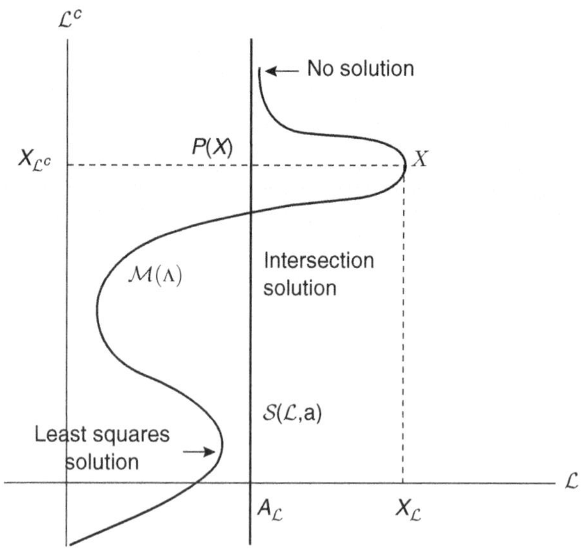

Figure 4.7. *Representation of splitting, intersection, and least squares solution*

We can rewrite the minimization more conveniently as an unconstrained optimization problem in terms of V in the open set $\mathcal{G}l(n)$. Let $X = V\Lambda V^{-1}$. The objective function $f(X)$ can be written as

$$g(V) = \frac{1}{2}\langle V\Lambda V^{-1} - P(V\Lambda V^{-1}), V\Lambda V^{-1} - P(V\Lambda V^{-1})\rangle. \tag{4.45}$$

Since the action of the derivative of P is perpendicular to the residual $X - P(X) = X_{\mathcal{L}} - A_{\mathcal{L}}$, the Fréchet derivative of g at $V \in \mathcal{G}l(n)$ acting on $H \in \mathbb{R}^{n\times n}$ is given by

$$g'(V)\cdot H = \langle H\Lambda V^{-1} - V\Lambda V^{-1}HV^{-1}, V\Lambda V^{-1} - P(V\Lambda V^{-1})\rangle. \tag{4.46}$$

By the Riesz representation theorem and the fact that

$$\langle A, BC\rangle = \langle B^{\top}A, C\rangle = \langle AC^{\top}, B\rangle,$$

we find that the gradient ∇g of the objective function g is given by

$$\begin{aligned}\nabla g(V) = &\left(V\Lambda V^{-1} - P(V\Lambda V^{-1})\right) V^{-\top}\Lambda \\ &- (V\Lambda V^{-1})^{\top}\left(V\Lambda V^{-1} - P(V\Lambda V^{-1})\right) V^{-\top}.\end{aligned} \tag{4.47}$$

Equivalently, we may write

$$\nabla g(V)V^{\top} = \left[X - P(X), X^{\top}\right]. \tag{4.48}$$

It follows that the vector field

$$\frac{dV}{dt} = k(X)V^{-\top}, \tag{4.49}$$

where

$$k(X) = [X^{\top}, X - P(X)] \tag{4.50}$$

defines a flow in the open set $\mathcal{G}l(n)$ and moves in the steepest descent direction to reduce the value of $g(V)$. Likewise, the vector field

$$\frac{dX}{dt} = [h_V(X), X] \tag{4.51}$$

with

$$h_V(X) = k(X)V^{-\top}V^{-1} \tag{4.52}$$

defines the steepest descent flow on the manifold $\mathcal{M}(\Lambda)$ for $f(X)$. The system (4.51) is not particularly important in practice since it is known that its solution is $X(t) = V(t)\Lambda V(t)^{-1}$.

We can rewrite the differential equation (4.49) as a self-sustaining system

$$\frac{dV}{dt} = k(V\Lambda V^{-1})V^{-\mathsf{T}}, \tag{4.53}$$

where we recall that Λ is a constant real matrix with prescribed eigenvalues $\{\lambda_k\}_{k=1}^n$. Since $V(t)$ generally has no additional structure, the system can be integrated by any available ODE solver starting with initial value $V(0) = I$. This naturally constitutes a reasonable numerical method for solving the PEIEP.

At this moment, it is appropriate to point out that the framework of our gradient flow (4.53) is applicable to general PEIEPs with any kind of index subset $\mathcal{L}$. Different specifications of $\mathcal{L}$ simply mean different projections $P(X)$. Our differential equation offers a continuous approach that has no limitation on either the locations $\mathcal{L}$ or the cardinality $|\mathcal{L}|$. If a PEIEP is not solvable, our approach finds a least squares solution.

Also, we point out that the formulation above is for general matrices in $\mathbb{R}^{n\times n}$. If we are interested only in symmetric matrices, the group action by $\mathcal{G}l(n)$ is replaced by the group $\mathcal{O}(n)$ of $n \times n$ orthogonal matrices, $V^{-1} = V^{\mathsf{T}}$, and many of the expressions can be simplified.

We outline below some additional strategies that can be adopted, not only for (4.53), but for the numerical computation of any continuous method.

Control robustness The notion of robustness was first introduced by Kautsky et al. (1985) in the context of the pole assignment problem. Since then, it has been extensively studied, including a recent generalization to the quadratic eigenstructure assignment problem (Nichols and Kautsky, 2001a). However, unlike the robust eigenstructure assignment problems where entries in either the gain matrix or the output matrix are freely subject to control, the robust PEIEP imposes additional challenges in that the solution matrix has a fixed structure specified by the prescribed entries. Finding a solution for the PEIEP is already a hard problem by itself; finding a solution that is robust is even harder.

We need to quantify the meaning of robustness more discerningly. Because the multiple solutions often appear in continuum form, it does not seem sensible to compare the nearness of solutions with one another. Rather, the robustness of an approximate solution should be measured by the sensitivity of its eigenvalues to perturbations. Toward that end, we recall the well-known Bauer–Fike theorem that characterizes the sensitivity of eigenvalues to perturbations.

Theorem 4.31. If μ is an eigenvalue of $X + E \in \mathbb{C}^{n\times n}$ and $V^{-1}XV = \operatorname{diag}\{\lambda_k\}_{k=1}^n$, then

$$\min_{\lambda\in\sigma(X)} |\lambda - \mu| \le \kappa_p(V)\|E\|_p, \tag{4.54}$$

where $\|\cdot\|_p$ denotes any of the p-norms and $\kappa_p(V)$ denotes the condition number of V with respect to the p-norm.

It is therefore intuitively true that a solution X_1 to the PEIEP is relatively more robust than another solution X_2 if the corresponding matrices V_1 and V_2 of eigenvectors satisfy the inequality $\kappa_p(V_1) < \kappa_p(V_2)$. Note that eigenvectors are necessarily involved in the determination of robustness.

The quantity $\kappa_p(V)$ sometime serves only as an outlier estimate that does not necessarily reflect the texture of the underlying eigenstructure. A more refined way to assess the robustness is the notion of *condition number* of each individual eigenvalue introduced by Wilkinson (1965). Let the left and right eigenvectors corresponding to the eigenvalue λ of a matrix X be denoted by $\mathbf{u}^\top$ and $\mathbf{v}$, respectively. Then the condition number $c(\lambda)$ of λ is defined to be the rate of change of λ relative to change of X. It is a well-known fact that $c(\lambda)$ is given by the formula

$$c(\lambda) = \frac{\|\mathbf{u}^\top\|_2 \|\mathbf{v}\|_2}{|\mathbf{u}^\top \mathbf{v}|}. \tag{4.55}$$

Let $V = [\mathbf{v}_1, \ldots, \mathbf{v}_n]$ denote the matrix of eigenvectors. If we assume that $U = V^{-\top}$ and $U^\top X V = \operatorname{diag}\{\lambda_k\}_{k=1}^n$, then the sum

$$\nu^2(V) = \sum_{i=1}^{n} \|\mathbf{u}_i^\top\|_2^2 \|\mathbf{v}_i\|_2^2, \tag{4.56}$$

can be considered as a total measure of robustness of X (and hence of V). It is easy to see that $\nu(V) \leq \kappa_F(V)$.

Example 4.16. A simple illustration should manifest the notion of robust completion. Consider the PEIEP with $n=2$, $\mathcal{L}=\{(1,2)\}$, $a_1=4$, and $\sigma(X)=\{1,2\}$. With three positions of X to be completed, the solutions constitute a one-parameter family of matrices,

$$X^{[a]} = \begin{bmatrix} a & 4 \\ \frac{a(3-a)-2}{4} & 3-a \end{bmatrix},$$

where a is arbitrary in $\mathbb{R}$. The condition numbers of $X^{[a]}$ as well as the corresponding matrix $V^{[a]}$ of eigenvectors are depicted in Figure 4.8.

Each $X^{[a]}$ is a solution to the PEIEP, but our ultimate goal is to minimize this condition number among all possible solutions. Such a quest for the absolute robustness in general would be too expensive to be practical. Thus, strictly speaking, in what follows we seek only to lower this condition number while constructing a solution to the PEIEP.

We not only would like to integrate (4.53) to its equilibrium, but also to somehow control the condition number of $V(t)$. Obviously, as it stands now, $\kappa_p(V(t))$ is an innate property of $V(t)$ which is already determined by (4.53) and there is not much we can do. To control the robustness, we must provide a mechanism to redirect the course of integration.

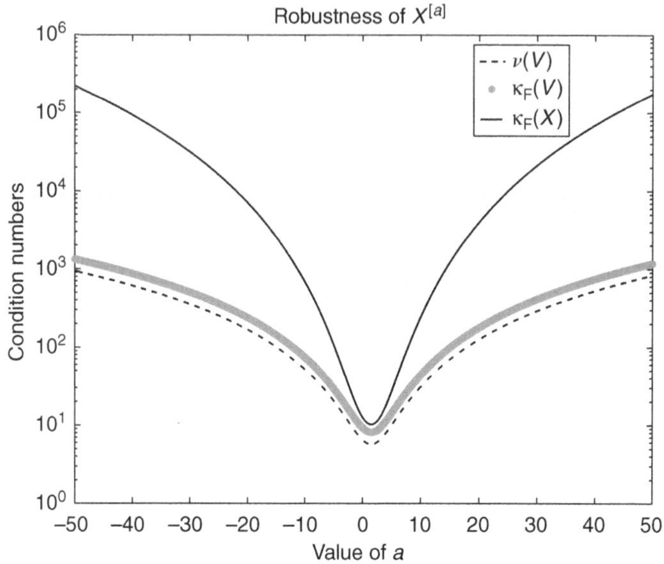

Figure 4.8. *Condition numbers of PEIEP solutions when* $n = 2$, $\mathcal{L} = \{(1,2)\}$, $a_1 = 4$, *and* $\sigma(X^{[a]}) = \{1, 2\}$

One strategy to control the robustness is to build the condition number of V into the objective function to be minimized. That is, instead of (4.45), we can consider one of the new objective functions,

$$\xi(V) = g(V) + \frac{\alpha}{2}\langle V, V\rangle\langle V^{-1}, V^{-1}\rangle, \tag{4.57}$$

or

$$\zeta(V) = g(V) + \frac{\beta}{2}\nu^2(V). \tag{4.58}$$

The new terms added to $g(V)$ stand for a *penalty* on either $\kappa_F^2(V)$ or $\nu^2(V)$. The multipliers α and β are positive numbers selected to reflect how much weight we want to emphasize the penalty. It remains to calculate the gradient of $\xi(V)$ or $\zeta(V)$, and then everything described above about gradient flows should follow through. Towards that end, we claim after some manipulation of calculus that the gradients give rise to the following two dynamical systems.

Theorem 4.32. (Chu et al., 2003) If $\xi(V)$ is used as the objective function, then the steepest descent flow in the same spirit of the system (4.53) should be defined by the modified equation

$$\frac{dV}{dt} = \left(k(V\Lambda V^{-1}) - \alpha(\langle V^{-1}, V^{-1}\rangle VV^{\mathsf{T}} - \langle V, V\rangle V^{-\mathsf{T}}V^{-1})\right)V^{-\mathsf{T}}. \tag{4.59}$$

Similarly, if $\zeta(V)$ is used, then the steepest descent flow becomes

$$\frac{dV}{dt} = \left(k(V\Lambda V^{-1}) - \beta\left(V\Theta(V^{-\top})V^{\top} - V^{-\top}\Theta(V)V^{-1}\right)\right)V^{-\top}, \tag{4.60}$$

where

$$\Theta(V) = \text{diag}\{\|\mathbf{v}_1\|_2^2, \ldots, \|\mathbf{v}_n\|_2^2\}, \tag{4.61}$$

if $V = [\mathbf{v}_1, \ldots, \mathbf{v}_n]$.

Numerical experiments indicate that the weights α and β are fairly difficult to manage. Too little penalty does not improve the condition number significantly. On the other hand, too much penalty would compromise the real purpose of finding an exact or nearly exact solution to the PEIEP and, instead, land at a least squares solution. It seems from our numerical experience that often only a small fraction of the condition number built into the dynamical system will suffice to tame the behavior. In some cases, we see that α or β should be in the range of 10^{-10} to avoid an inadvertent least squares solution. We speculate that a more sophisticated scheme in selecting α and β adaptively, such as those used in the interior point methods, might help to improve the situation, but that investigation will have to be further extended.

Restart Observe that the initial value $V(0) = I$ is perfectly conditioned. By continuity, the conditioning of $V(t)$ will remain reasonable well for at least small values of t. As the integration continues, it is possible that $\kappa_F(V(t))$ (or $\nu(V(t))$) will grow, particularly when $V(t)$ converges to singularity or becomes ill-conditioned. Is it possible that, when this occurs, we restart the integration so as to maintain the computational stability? We exploit this idea more carefully in the sequel.

First of all, it is important to note that all three differential systems (4.53), (4.59), and (4.60) assume the same kind of format as

$$\frac{dV}{dt} = \varphi(V; \Lambda)V^{-\top}, \tag{4.62}$$

but with different definitions of $\varphi(V; \Lambda)$. Let $t_0, t_1, \ldots,$ be a sequence of positive numbers whose values will be defined later. Let $V(t)$ denote the solution to (4.62) with initial value $V(0) = I$. For each $i = 0, 1, \ldots,$ let $V_i(\tau)$ denote the solution to the *temporal* system

$$\frac{dV_i}{d\tau} = \varphi(V_i; X_i)V_i^{-\top}, \quad V_i(0) = I, \quad \tau \in [0, t_i], \tag{4.63}$$

where, starting with $X_0 = \Lambda$, we recursively define

$$X_{i+1} = V_i(t_i)X_iV_i^{-1}(t_i). \tag{4.64}$$

The following theorem concerning the factorization of $V(t)$ is a consequence of the fact that (4.62) is an autonomous differential system. A proof for the case that $\varphi(V;\Lambda) = k(V\Lambda V^{-1})$ has been done in Chu et al. (2004), and the generalization to other types of $\varphi(V;\Lambda)$ is not difficult.

Theorem 4.33. (Chu et al., 2004) At each point $t = s + \sum_{j=0}^{i-1} t_j$ in the interval of existence for the initial value problem (4.62) with $V(0) = I$, the solution $V(t)$ can be factorized as

$$V(t) = V_i(s)V_{i-1}(t_{i-1}) \cdots V_0(t_0), \tag{4.65}$$

where each $V_j(\tau)$ is the solution of the temporal system (4.63).

Other than the obvious fact that the value of t_i for each $i = 0, 1, \ldots,$ must be within the maximal interval of existence for each system (4.63), the choice of t_i can be quite flexible. One strategy is to continue integrating (4.63) within that interval until $\kappa_F(V_i(\tau))$ reaches some predesignated threshold. The value of τ in reaching that threshold is the *maximal* t_i we can define. We then update X_{i+1} according to (4.64) and switch to solve a new initial value problem (4.63). We call this process a *restart*. Restart is expensive because we have to update X_{i+1}. In the extreme (and not recommended) case, we can restart at every single integration step. By restarting, however, we effectively improve the computational stability because the condition number of $V_i(\tau)$ is kept under some upper bound.

Thus far, the restart strategy does not seem to have provided any hope of improving $\kappa_p(V(t))$. The theory implies that the solution $V(t)$ from the continuation scheme (4.62) and the product (4.65) from the restart scheme (4.63) are identical. However, we have made an interesting and somewhat surprising remark – we have observed frequently in our numerical experiments that the computed results of these two schemes behave very differently. More candidly, they converge to two entirely different limit points. The reason for this is not completely understood. We surmise that this convergence behavior is partially due to the "curse" that the PEIEP is ill-posed in two respects: that the problem has multiple (and often continuum) solutions and that a small perturbation of the vector field in (4.63) by the restart changes the course of integration dramatically. Computations for ill-posed problems such as this generally would have been disastrous, but notice that we have built the descent property into our vector field by which the objective function is allowed only to go downhill. For instance, if (4.53) is used, then the distance between $\mathcal{M}(X_i)$ and $\mathcal{S}(\mathcal{L}, \mathbf{a})$ can only be reduced. That is, it is always true that

$$\|X_{i+1} - P(X_{i+1})\|_F \leq \|V_i(t)X_iV_i(t)^{-1} - P(V_i(t)X_iV_i(t)^{-1})\|_F \leq \|X_i - P(X_i)\|_F$$

for all $t \in [0, t_i)$. Changing the course of integration does not affect this descent property. Rather, the restart together with the built-in descent property offers an opportunity to exploit the ill-posedness and turn the disadvantages into a possible advantage of improving the stability and conditioning.

Numerical schemes We have now described several ways to tackle the robust PEIEPs numerically. We categorize them into two classes of schemes.

Algorithm 4.3 (Gradient method for PEIEP)
Given an index set $\mathcal{L}$ of locations with prescribed values $\mathbf{a}$, a least squares approximation to the solution of PEIEP can be obtained as follows:

1. Define the projection map $P(X)$ onto the affine space $\mathcal{S}(\mathcal{L}, \mathbf{a})$.
2. Starting with $V(0) = I$ and $X_0 = \Lambda$, integrate any of the three differential equations (4.53), (4.59), or (4.60) by any available ODE integrator until convergence.

Two comments on the application of Algorithm 4.3 deserve attention. First, *modifying the initial value X_0 by choosing a different Λ can make a substantial difference.* The difference is not only that different equilibrium points may emerge, but also that these equilibrium points may vary significantly in their robustness. For instance, suppose that the given spectrum $\{\lambda_k\}_{k=1}^n$ is all real. We find that starting with $\Lambda = \Lambda_0 = \operatorname{diag}\{\lambda_k\}_{k=1}^n$ generally is less desirable than starting with $\Lambda = \Lambda_0 + T_0$ where T_0 is an arbitrary strictly upper triangular matrix. The flow starting from a triangular starting value usually converges faster to a more robust equilibrium point than that starting from a diagonal Λ_0. One possible explanation is as follows. Recall that the Schur triangular form (or the real-valued block triangular form with 2×2 blocks along the diagonal for complex conjugate eigenvalues) is reachable via orthogonal similarity transformations. Recall also that the isospectral orbits by $\mathcal{O}(n)$ group action are closed and bounded. It seems intuitively true that points reachable by $\mathcal{O}(n)$ actions should be easier and more "stable" to locate than, say, the Jordan canonical form (which, under the above assumption, is precisely Λ_0.) Since our integration process in a sense is a "reversal of the reduction procedure" used to find canonical forms, that is, since we want to start with some points in the "reduced form" and travel to some other points in the "full form" while maintaining the spectrum, it should be expected that starting with isospectral triangular matrices performs better than starting with the more restricted diagonal Λ_0.

Secondly, in case that we want to employ the dynamical systems (4.59) or (4.60) to further a solution with improved robustness, *we have to choose the weight α or β appropriately.* Too much emphasis on the penalty term translates into too much restriction on the condition number which, in turn, may jeopardize the original intention of finding an approximate solution to PEIEP and result in a least squares solution instead. Employing (4.59) or (4.60) will improve the condition number in general, but might slow down the convergence.

Algorithm 4.4 (Gradient method with restart for PEIEP)
Given an index set $\mathcal{L}$ of locations with prescribed values $\mathbf{a}$, a least squares approximation to the solution of PEIEP can be obtained as follows:

1. Define the projection map $P(X)$ onto the affine space $\mathcal{S}(\mathcal{L}, \mathbf{a})$.
2. Integrate the system (4.63) using any of the three aforementioned dynamical systems until the condition number of $V(t)$ violates a given threshold.
3. Automatically apply a restart and repeat (2).

This hybrid method has the ability to cap the condition number of the factors of $V(t)$ under a given threshold. The product of the condition numbers of these factors provides only an upper bound on the condition number of $V(t)$, so this method does not control the robustness directly. However, this restart scheme does have quite an impressive impact on the condition number after all. Additionally, the solution flow $V_i(t)$ in each segment $[0, t_i)$ being well conditioned (otherwise, a restart will be activated), it also improves the computational stability. Again, be warned that if the threshold for restart is set too low, then there will be frequent (expensive) restarts and more factors involved in the product (4.65). The overall performance may be degraded due to this frequent restart.

4.8 Inverse singular value problems

The notion of IEPs can naturally be extended to the inverse singular value problems (**ISVP**). An ISVP concerns the construction of a structured matrix with prescribed singular values. Just like the IEPs, we emphasize once again that an ISVP should also satisfy a certain structural constraint. For symmetric matrices, the IEPs and the ISVPs are essentially the same. But for general matrices, it does appear that the class of ISVPs is a whole new territory that has barely been explored in the literature. Adding to the complication is that the underlying matrix could be rectangular. We have already seen a few special types of ISVPs, including Problem 2.11 in Section 2.6 where a given matrix was to be conditioned by a rank one perturbation, the PEISVP (Problem 4.17), and the STISVP (Problem 4.18) in Section 4.7.1 where a matrix is to be constructed with prescribed diagonal entries and singular values. Clearly, every other type of IEP, except for the symmetric problems, has a counterpart under the context of ISVP.

We have already mentioned that an ISVP can be converted into an IEP. Recall that eigenvalues of the structured symmetric matrix C defined in (4.39) are precisely the pluses and minuses of singular values of X. The IEP for C has the fixed structure of zero diagonal blocks plus whatever structure is inherited from X. An ISVP for a structured B is solvable if and only if an IEP for C with structure defined in (4.39) is solvable. Because of this additional zero block structure in C, to establish conditions on the solvability of a structured ISVP should be an interesting question in its own right.

To introduce the notion of ISVP's, we shall limit our discussion to a special class of parameterized ISVPs.

> **Problem 4.20** *(PISVP)*
> Given general matrices $B_0, B_1, \ldots, B_n \in \mathbb{R}^{m\times n}$, $m \geq n$, and nonnegative real numbers $s_1 \geq \cdots \geq s_n$, find values of $\mathbf{c} := (c_1, \ldots, c_n)^\top \in \mathbb{R}^n$ such that the singular values of the matrix
>
> $$B(\mathbf{c}) := B_0 + \sum_{i=1}^{n} c_i B_i \tag{4.66}$$
>
> are precisely $\{s_1, \ldots, s_n\}$.

Analogous to the LiPIEP (3.1), the matrices B_i can be used to delineate certain quite general structures. We have already discussed a Newton-type iterative procedure for the LiPIEP2. We now demonstrate how the ISVP can be handled in an almost parallel but more subtle way. The subtlety comes from the fact that for ISVPs we have to deal with the left and the right singular vectors at the same time. The following discussion summarizes major points from the discussion in Chu (1992b).

4.8.1 *Distinct singular values*

Consider first the case that all prescribed singular values $s_1, \ldots, s_n$ are positive and distinct. Let $\Sigma \in \mathbb{R}^{m\times n}$ denote the "diagonal" matrix with diagonal elements $\{s_1, \ldots, s_n\}$. Define the affine subspace

$$\mathcal{B} := \{B(\mathbf{c}) | \mathbf{c} \in \mathbb{R}^n\},$$

and the surface

$$\mathcal{M}_s(\Sigma) := \{U\Sigma V^\top | U \in \mathcal{O}(m), V \in \mathcal{O}(n)\}$$

of all matrices with singular values $\{s_1, \ldots, s_n\}$. Solving the ISVP is equivalent to finding an intersection of the two sets $\mathcal{M}_s(\Sigma)$ and $\mathcal{B}$.

Any tangent vector $T(X)$ to $\mathcal{M}_s(\Sigma)$ at a point $X \in \mathcal{M}_s(\Sigma)$ about which a local chart can be defined must be of the form

$$T(X) = XK - HX$$

for some skew-symmetric matrices $H \in \mathbb{R}^{m\times m}$ and $K \in \mathbb{R}^{n\times n}$. Any given $X^{(\nu)} \in \mathcal{M}_s(\Sigma)$ can be factorized as

$${U^{(\nu)}}^\top X^{(\nu)} V^{(\nu)} = \Sigma$$

with $U^{(\nu)} \in \mathcal{O}(m)$ and $V^{(\nu)} \in \mathcal{O}(n)$. Mimicking what we have done for the LiPIEP2 in Section 3.2.4, our goal is to, first, seek a $\mathcal{B}$-intercept $B(c^{(\nu+1)})$ from a line that is tangent to the manifold $\mathcal{M}_s(\Sigma)$ at $X^{(\nu)}$ and, then, to seek a way to lift the matrix $B(c^{(\nu+1)}) \in \mathcal{B}$ to a point $X^{(\nu+1)} \in \mathcal{M}_s(\Sigma)$.

To determine the intercept, we need to calculate skew-symmetric matrices $H^{(\nu)} \in \mathbb{R}^{m\times m}$ and $K^{(\nu)} \in \mathbb{R}^{n\times n}$, and a vector $c^{(\nu+1)} \in \mathbb{R}^n$ so that the equation

$$X^{(\nu)} + X^{(\nu)}K^{(\nu)} - H^{(\nu)}X^{(\nu)} = B(c^{(\nu+1)}) \tag{4.67}$$

is satisfied. Equivalently, the skew-symmetric matrices $\tilde{H}^{(\nu)} := {U^{(\nu)}}^\top H^{(\nu)}U^{(\nu)}$ and $\tilde{K}^{(\nu)} := {V^{(\nu)}}^\top K^{(\nu)}V^{(\nu)}$ satisfying the equation

$$\Sigma + \Sigma\tilde{K}^{(\nu)} - \tilde{H}^{(\nu)}\Sigma = \underbrace{{U^{(\nu)}}^\top B(c^{(\nu+1)})V^{(\nu)}}_{W^{(\nu)}} \tag{4.68}$$

are needed. The values for $c^{(\nu+1)}$, $H^{(\nu)}$, and $K^{(\nu)}$ can be determined separately.

Observe that in total there are $m(m-1)/2 + n(n-1)/2 + n$ unknowns and mn equations involved in (4.68). A closer examination of (4.68) shows that the lower-right corner of size $(m-n)\times(m-n)$ in $\tilde{H}^{(\nu)}$ can be arbitrary. For simplicity, we set this part identically zero. Then it suffices to consider the mn equations

$$W_{ij}^{(\nu)} = \Sigma_{ij} + \Sigma_{ii}\tilde{K}_{ij}^{(\nu)} - \tilde{H}_{ij}^{(\nu)}\Sigma_{jj}, \quad 1 \le i \le m,\ 1 \le j \le n, \tag{4.69}$$

where $\tilde{K}_{ij}^{(\nu)}$ is understood to be zero if $i \ge n+1$, for the remaining quantities.

For $1 \le i = j \le n$, we obtain the equations for $c^{(\nu+1)}$

$$\Omega^{(\nu)}c^{(\nu+1)} = \mathbf{s} - \mathbf{b}^{(\nu)}, \tag{4.70}$$

where

$$\begin{aligned}
\Omega_{st}^{(\nu)} &:= {\mathbf{u}_s^{(\nu)}}^\top B_t \mathbf{v}_s^{(\nu)}, \quad 1 \le s,t \le n,\\
\mathbf{s} &:= [s_1,\dots,s_n]^\top,\\
b_s^{(\nu)} &:= {\mathbf{u}_s^{(\nu)}}^\top B_0 \mathbf{v}_s^{(\nu)}, \quad 1 \le s \le n,
\end{aligned}$$

if $\mathbf{u}_s^{(\nu)}$ and $\mathbf{v}_s^{(\nu)}$ denote column vectors of $U^{(\nu)}$ and $V^{(\nu)}$, respectively. Under mild assumptions, the matrix $\Omega^{(\nu)}$ is nonsingular. The vector $c^{(\nu+1)}$ and, hence, the matrix $W^{(\nu)}$ are thus obtained.

The skew-symmetric matrices $H^{(\nu)}$ and $K^{(\nu)}$ can be obtained by comparing the "off-diagonal" entries in (4.68). For $n+1 \le i \le m$ and $1 \le j \le n$, it is clear that

$$\tilde{H}_{ij}^{(\nu)} = -\tilde{H}_{ji}^{(\nu)} = -\frac{W_{ij}^{(\nu)}}{s_j}. \tag{4.71}$$

For $1 \leq i < j \leq n$, we have from (4.69)

$$W_{ij}^{(\nu)} = \Sigma_{ii}\tilde{K}_{ij}^{(\nu)} - \tilde{H}_{ij}^{(\nu)}\Sigma_{jj}, \tag{4.72}$$

$$W_{ji}^{(\nu)} = \Sigma_{jj}\tilde{K}_{ji}^{(\nu)} - \tilde{H}_{ji}^{(\nu)}\Sigma_{ii} = -\Sigma_{jj}\tilde{K}_{ij}^{(\nu)} + \tilde{H}_{ij}^{(\nu)}\Sigma_{ii}. \tag{4.73}$$

Solving for $\tilde{H}_{ij}^{(\nu)}$ and $\tilde{K}_{ij}^{(\nu)}$ yields

$$\tilde{H}_{ij}^{(\nu)} = -\tilde{H}_{ji}^{(\nu)} = \frac{s_i W_{ji}^{(\nu)} + s_j W_{ij}^{(\nu)}}{s_i^2 - s_j^2}, \tag{4.74}$$

$$\tilde{K}_{ij}^{(\nu)} = -\tilde{K}_{ji}^{(\nu)} = \frac{s_i W_{ij}^{(\nu)} + s_j W_{ji}^{(\nu)}}{s_i^2 - s_j^2}. \tag{4.75}$$

The intercept is now completely determined.

It only remains to lift the intercept $B(c^{(\nu+1)})$ back to $\mathcal{M}_s(\Sigma)$. Toward that end, one possible way is to define the lift as

$$X^{(\nu+1)} := R^\top X^{(\nu)} S,$$

where R and S are the Cayley transforms

$$R := \left(I + \frac{H^{(\nu)}}{2}\right)\left(I - \frac{H^{(\nu)}}{2}\right)^{-1},$$

$$S := \left(I + \frac{K^{(\nu)}}{2}\right)\left(I - \frac{K^{(\nu)}}{2}\right)^{-1}.$$

Note also that the matrix $X^{(\nu+1)}$ need not to be formed explicitly in the computation. Rather, only the orthogonal matrices

$$U^{(\nu+1)} := R^\top U^{(\nu)} \tag{4.76}$$

and

$$V^{(\nu+1)} := S^\top V^{(\nu)} \tag{4.77}$$

are needed in the computation. This completes one cycle of the Newton step and the iteration repeats until convergence, if it converges at all. Regarding the efficiency of this algorithm, Chu (1992b) proved the following result on the rate of convergence.

Theorem 4.34. Suppose that the ISVP (4.66) has an exact solution at c^* and that $B(c^*) = \hat{U}\Sigma\hat{V}^\top$ is the corresponding singular value decomposition. Define the error matrix $E := (E_1, E_2) := (U - \hat{U}, V - \hat{V})$. Suppose that the matrix $\Omega^{(\nu)}$ is nonsingular. Then $\|E^{(\nu+1)}\| = O(\|E^{(\nu)}\|^2)$.

4.8.2 *Multiple singular values*

In searching for the $\mathcal{B}$-intercept of a tangent line of $\mathcal{M}_s(\Sigma)$ the definition (4.71) allows no zero singular values. Similarly, the definitions (4.74) and (4.75) require all of the singular values to be distinct. In this section, we consider the case when multiple singular values are present. For clarity, we shall continue to assume that all singular values are positive. To simplify the notation, we shall also assume that only the first singular value σ_1^* is multiple, with multiplicity p.

We observe first that all the formulas (4.70) and (4.71) are still well defined. For $1 \le i < j \le p$, however, we can conclude from (4.72) and (4.73) only that

$$W_{ij}^{(\nu)} + W_{ji}^{(\nu)} = 0. \tag{4.78}$$

Instead of determining values for $\tilde{H}_{ij}^{(\nu)}$ and $\tilde{K}_{ij}^{(\nu)}$, the system (4.78) gives rise to an additional $q := p(p-1)/2$ equations for the vector $c^{(\nu+1)}$. That is, multiple singular values gives rise to an over-determined system for $c^{(\nu+1)}$, a situation analogous to that discussed in Section 3.2.4 for LiPIEP2. Geometrically, the case implies that maybe no tangent line from $\mathcal{M}_s(\Sigma)$ will intercept the affine subspace $\mathcal{B}$ at all. To remedy this, we follow a strategy of Friedland et al. to modify the ISVP as:

Problem 4.21 *(PISVP2)*
Given positive values $\sigma_1^* = \cdots = \sigma_p^* > \sigma_{p+1}^* > \cdots > \sigma_{n-q}^*$, find real values of $c_1, \ldots, c_n$ such that the $n-q$ largest singular values of the matrix $B(c)$ are $\sigma_1^*, \ldots, \sigma_{n-q}^*$.

Now that we have q degrees of freedom in choosing the remaining singular values, so we shall use the equation (compare with (4.68))

$$\hat{\Sigma} + \hat{\Sigma}\tilde{K}^{(\nu)} - \tilde{H}^{(\nu)}\hat{\Sigma} = {U^{(\nu)}}^\top B(c^{(\nu+1)})V^{(\nu)} \tag{4.79}$$

to find the $\mathcal{B}$-intercept, where

$$\hat{\Sigma} := \operatorname{diag}\{\sigma_1^*, \ldots, \sigma_{n-q}^*, \hat{\sigma}_{n-q+1}, \ldots, \hat{\sigma}_n\} \tag{4.80}$$

and $\hat{\sigma}_{n-q+1}, \ldots, \hat{\sigma}_n$ are free parameters. We summarize a procedure for solving the ISVP2 in the following algorithm:

Algorithm 4.5 (Newton method for PISVP2)
Given $U^{(0)} \in \mathcal{O}(m)$ and $V^{(0)} \in \mathcal{O}(n)$, repeat for $\nu = 0, 1, \ldots$, until convergence:

1. Solve for $c^{(\nu+1)}$ from the system of equations:

$$\sum_{k=1}^{n} \left(u_i^{(\nu)\top} B_k v_i^{(\nu)} \right) c_k^{(\nu+1)}$$
$$= \sigma_i^* - u_i^{(\nu)\top} B_0 v_i^{(\nu)}, \quad \text{for } i = 1, \ldots, n-q;$$
$$\sum_{k=1}^{n} \left(u_s^{(\nu)\top} B_k v_t^{(\nu)} + u_t^{(\nu)\top} B_k v_s^{(\nu)} \right) c_k^{(\nu+1)}$$
$$= -u_s^{(\nu)\top} B_0 v_t^{(\nu)} - u_t^{(\nu)\top} B_0 v_s^{(\nu)}, \qquad \text{for } 1 \le s < t \le p.$$

2. Define $\hat{\sigma}_k^{(\nu)}$ by

$$\hat{\sigma}_k^{(\nu)} := \begin{cases} \sigma_k^*, & \text{if } 1 \le k \le n-q; \\ u_k^{(\nu)\top} B(c^{(\nu+1)}) v_k^{(\nu)}, & \text{if } n-q < k \le n. \end{cases}$$

3. Once $c^{(\nu+1)}$ is determined, calculate $W^{(\nu)} = U^{(\nu)\top} B(c^{(\nu+1)}) V^{(\nu)}$.
4. Define the skew-symmetric matrices $\tilde{K}^{(\nu)}$ and $\tilde{H}^{(\nu)}$ by

$$\tilde{K}_{ij}^{(\nu)} := \begin{cases} \dfrac{\hat{\sigma}_i^{(\nu)} W_{ij}^{(\nu)} + \hat{\sigma}_j^{(\nu)} W_{ji}^{(\nu)}}{(\hat{\sigma}_i^{(\nu)})^2 - (\hat{\sigma}_j^{(\nu)})^2}, & \text{if } 1 \le i < j \le n, \text{ and } p < j; \\ 0, & \text{if } 1 \le i < j \le p, \end{cases}$$

and

$$\tilde{H}_{ij}^{(\nu)} := \begin{cases} -\dfrac{W_{ij}^{(\nu)}}{\hat{\sigma}_j^{(\nu)}}, & \text{if } 1 \le i < j \le p; \\ -\dfrac{W_{ij}^{(\nu)}}{\hat{\sigma}_j^{(\nu)}}, & \text{if } n+1 \le i \le m, \text{ and } 1 \le j \le n; \\ \dfrac{\hat{\sigma}_i^{(\nu)} W_{ji}^{(\nu)} + \hat{\sigma}_j^{(\nu)} W_{ij}^{(\nu)}}{(\hat{\sigma}_i^{(\nu)})^2 - (\hat{\sigma}_j^{(\nu)})^2}, & \text{if } 1 \le i < j \le n, \text{ and } p < j; \\ 0, & \text{if } n+1 \le i \ne j \le m. \end{cases}$$

5. Once $\tilde{H}^{(\nu)}$ and $\tilde{K}^{(\nu)}$ are determined, proceed with the lifting in the same way as for the PISVP.

It is important to keep in mind that we are no longer on a fixed manifold $\mathcal{M}_s(\Sigma)$, since the portion $\hat{\Sigma}$ is changed at each step. We believe a proof of convergence similar to that given by Friedland et al. (1986) for the LiPIEP2 can easily be carried out. Numerical experiments seem to support the conclusion that the above algorithm for the case of multiple singular value converges quadratically (Chu, 1992a).

4.8.3 *Rank deficiency*

A zero singular value indicates rank deficiency. To find a lower rank matrix in a generic affine subspace $\mathcal{B}$ is intuitively a more difficult problem. In this case, it is most likely that the PISVP does not have a solution. We shall study low rank approximation problems in Chapter 8. At present, we simply make comment on the solvability of low rank PISVP.

To demonstrate what might happen to the algorithm proposed earlier, we consider the simplest case where the rank deficiency is merely 1.

Example 4.17. Assume that $\sigma_1^* > \cdots > \sigma_{n-1}^* > \sigma_n^* = 0$. Equation (4.69) indicates that except for $\tilde{H}_{in}$ (and $\tilde{H}_{ni}$), $i = n+1, \ldots, m$, all other quantities including $c^{(\nu+1)}$ are well defined. Furthermore, in order that equation (4.69) be valid, it is necessary that

$$W_{in}^{(\nu)} = 0, \quad \text{for } i = n+1, \ldots, m. \tag{4.81}$$

If condition (4.81) fails, then no tangent line of $\mathcal{M}_s(\Sigma)$ from the current iterate $X^{(\nu)}$ will intersect the affine subspace $\mathcal{B}$. The iteration, therefore, cannot be continued.

4.9 Inverse singular/eigenvalue problems

The structure involved in the SIEPs can be quite general. Thus far, we have seen structures including Jacobi, Toeplitz, nonnegative, stochastic, unitary Hessenberg, prescribed entries, and the special form (4.39). Most of these structural constraints are explicitly or, at least, semi-explicitly given in terms of the appearance of the underlying matrix. It is possible that the structure is described implicitly as the solution set of some nonlinear functions. In this section, we discuss one particular class of SIEPs where the "structure" is implicitly defined by the singular values. The problem can also be considered as an ISVP where the structure is characterized implicitly by prescribed eigenvalues.

Recall that the Schur–Horn theorem identifies the connection between diagonal entries and eigenvalues of a Hermitian matrix. The Mirsky theorem gives the connection between diagonal entries and eigenvalues of a general matrix. The Sing–Thompson theorem characterizes the connection between diagonal entries and singular values of a general matrix. It is natural to ask whether there is any connection between singular values and eigenvalues of a matrix. This is an important question because, indeed, eigenvalue and singular values are two of the most distinguished attributes that characterize many important properties of a given general square matrix.

For Hermitian matrices, the singular values are simply the absolute values of eigenvalues. But for general square matrices, the connection is much more involved. Fortunately, this relation can be elegantly described by the Weyl–Horn theorem.

Theorem 4.35. (Weyl, 1949; Horn, 1954b) Given vectors $\boldsymbol{\lambda} \in \mathbb{C}^n$ and $\mathbf{s} \in \mathbb{R}^n$, suppose the entries are arranged in the ordering that $|\lambda_1| \geq \cdots \geq |\lambda_n|$ and $s_1 \geq \cdots \geq s_n$. Then a matrix with eigenvalues $\lambda_1, \ldots, \lambda_n$ and singular values $s_1, \ldots, s_n$ exists if and only if

$$\begin{cases} \prod_{j=1}^{k} |\lambda_j| \leq \prod_{j=1}^{k} s_j, & k = 1, \ldots, n-1, \\ \prod_{j=1}^{n} |\lambda_j| = \prod_{j=1}^{n} s_j. \end{cases} \tag{4.82}$$

It is interesting to note that for nonsingular matrices the Weyl–Horn condition is equivalent to the statement that the vector $\log(\mathbf{s})$ majorizes the vector $\log|\boldsymbol{\lambda}|$.

The inverse problem we are concerned about, denoted by **ISEP**, is to construct a matrix with prescribed eigenvalues and singular values. Such a construction, if possible, would allow us to create test matrices with desirable spectral properties.

Problem 4.22 *(ISEP)*
Given $\boldsymbol{\lambda} \in \mathbb{C}^n$ and $\mathbf{s} \in \mathbb{R}^n$ satisfying the Weyl–Horn conditions (4.82), construct a matrix A so that

$$\begin{cases} \sigma(A) = \boldsymbol{\lambda}, \\ \varpi(A) = \mathbf{s}. \end{cases}$$

The Weyl–Horn theorem was originally proved by induction. Recently, Chu (2000) modified the proof to avoid "triangularization" and converted it into a fast divide-and-conquer recursive algorithm. We shall outline the proof and the algorithm in subsequent sections. In passing we mention that Li and Mathias (2001) extended the Weyl–Horn condition to the case when only $m(\leq n)$ eigenvalues are given. Also, the task of constructing unit triangular matrices with prescribed singular values discussed in Kosowski and Smoktunowicz (2000) can be considered as a special case of our discussion.

We point out that the constructed matrix obtained by the Weyl–Horn theorem is usually complex-valued, if $\boldsymbol{\lambda}$ contains complex eigenvalues. It might be desirable to construct a real-valued solution, if all eigenvalues are complete in conjugation. A stable algorithm using diagonal unitary matrices, permutations, and rotation matrices to construct a real matrix if the specified eigenvalues are closed in complex conjugation can be found in Li and Mathias (2001), but will not be reviewed here.

4.9.1 *The 2×2 building block*

The basic ideas in Weyl and Horn's proof contains three major components:

- that the original problem can be reduced to two problems of smaller sizes;
- that problems of smaller sizes are guaranteed to be solvable by the induction hypothesis; and
- that the subproblems can be affixed together by working on a suitable 2×2 corner that has an explicit solution.

If we repeatedly apply these principles, then the original inverse problem is *divided* into subproblems of size 2×2 or 1×1 that can eventually be *conquered* to build up the original size. We begin with a description of how a 2×2 matrix with such eigenvalues and singular values can be constructed in closed form. Such a construction will serve as the building block in our recursive algorithm.

First, observe that the 2×2 triangular matrix

$$A = \begin{bmatrix} \lambda_1 & \mu \\ 0 & \lambda_2 \end{bmatrix}$$

has singular value $\{s_1, s_2\}$ if and only if

$$\mu = \pm\sqrt{s_1^2 + s_2^2 - |\lambda_1|^2 - |\lambda_2|^2}.$$

The fact that μ is well defined follows from the Weyl–Horn conditions that $|\lambda_1| \leq s_1$ and $|\lambda_1||\lambda_2| = s_1 s_2$. It is interesting to note that μ^2 is precisely the so-called departure of A from normality. For the sake of better computational stability, we suggest replacing μ by the definition

$$\mu = \begin{cases} 0, & \text{if } |(s_1 - s_2)^2 - (|\lambda_1| - |\lambda_2|)^2| \leq \epsilon, \\ \sqrt{|(s_1 - s_2)^2 - (|\lambda_1| - |\lambda_2|)^2|}, & \text{otherwise,} \end{cases}$$

in practice to stabilize the calculation. The 2×2 matrix plays a key role in the recursion formulation.

4.9.2 *Divide and conquer*

The original idea on how the problems could be divided is quite intriguing. Since this is an entirely new approach different from either the iterative methods or the continuous methods we have discussed thus far for other types of IEPs, we outline the proof in slightly more detail as follows.

For simplicity, we assume that $s_i > 0$ for all $i = 1, \ldots, n$. It follows that $\lambda_i \neq 0$ for all i. The case of zero singular values can be handled in a similar way (Chu, 2000). Starting with $\gamma_1 := s_1$, define the sequence

$$\gamma_i := \gamma_{i-1} \frac{s_i}{|\lambda_i|}, \quad i = 2, \ldots, n-1. \tag{4.83}$$

Assume that the maximum $\gamma := \min_{1\le i\le n-1}\gamma_i$ is attained at the index j. Define

$$\theta := \frac{|\lambda_1\lambda_n|}{\gamma}. \tag{4.84}$$

Then the following three sets of inequalities are true:

$$\begin{cases} |\lambda_1| \ge |\lambda_n|, \\ \gamma \ \ge \ \theta; \end{cases} \tag{4.85}$$

$$\begin{cases} \gamma \ge |\lambda_2| \ge \cdots \ge |\lambda_j|, \\ s_1 \ge \ s_2 \ \ge \cdots \ge \ s_j; \end{cases} \tag{4.86}$$

$$\begin{cases} |\lambda_{j+1}| \ge \cdots \ge |\lambda_{n-1}| \ge \theta, \\ s_{j+1} \ \ge \cdots \ge \ s_{n-1} \ \ge s_n. \end{cases} \tag{4.87}$$

More importantly, the numbers in each of the above sets satisfy the Weyl–Horn conditions, respectively, with the first row playing the role of eigenvalues and the second row playing the singular values in each set. Since these are problems of *smaller* sizes, by the induction hypothesis, the ISEPs associated with (4.86) and (4.87) are solvable. In particular, there exist unitary matrices U_1, $V_1 \in \mathbb{C}^{j\times j}$ and *triangular* matrices A_1 such that

$$U_1 \begin{bmatrix} s_1 & 0 & \dots & 0 \\ 0 & s_2 & & 0 \\ \vdots & \ddots & & \\ 0 & 0 & \dots & s_j \end{bmatrix} V_1^* = A_1 = \begin{bmatrix} \gamma & \times & \times & \dots & \times \\ 0 & \lambda_2 & & & \times \\ & & & & \\ \vdots & & & \ddots & \\ 0 & 0 & & & \lambda_j \end{bmatrix},$$

and unitary matrices U_2, $V_2 \in \mathbb{C}^{(n-j)\times(n-j)}$, and *triangular* matrices A_2 such that

$$U_2 \begin{bmatrix} s_{j+1} & 0 & \dots & 0 \\ 0 & s_{j+2} & & 0 \\ \vdots & & \ddots & \\ 0 & 0 & \dots & s_n \end{bmatrix} V_2^* = A_2 = \begin{bmatrix} \lambda_{j+1} & \times & \dots & \times & \times \\ 0 & \lambda_{j+2} & & & \times \\ \vdots & & \ddots & & \vdots \\ & & & \lambda_{n-1} & \times \\ 0 & 0 & \dots & 0 & \theta \end{bmatrix}.$$

Note the positions of γ and θ in the matrices. If we augment A_1 and A_2 into

$$\begin{bmatrix} A_1 & \bigcirc \\ \bigcirc & A_2 \end{bmatrix}, \tag{4.88}$$

the γ and θ reside, respectively, in the $(1,1)$ and the (n,n) positions. In his original proof, Horn claimed that the block matrix could be *permuted* to the triangular matrix

$$\begin{bmatrix}
\lambda_2 & \times & \cdots & \times & \times & & & & & \\
0 & & & & \times & & & & & \\
\vdots & & \ddots & & \vdots & & & \bigcirc & & \\
 & & & \lambda_j & \times & & & & & \\
0 & & \cdots & 0 & \gamma & 0 & & & & \\
0 & 0 & \cdots & 0 & 0 & \theta & \times & \times & \cdots & \times \\
 & & & & & & \lambda_{j+1} & & & \times \\
 & & \bigcirc & & & & & & & \\
 & & & & & \vdots & & \ddots & & \\
 & & \cdots & & & 0 & 0 & & & \lambda_{n-1}
\end{bmatrix}.$$

If this claim were true, it is obvious that the resulting matrix would have singular values $\{s_1, \ldots, s_n\}$ and miss only the eigenvalues $\{\lambda_1, \lambda_n\}$. The next step is to glue the 2×2 corner adjacent to the two blocks together by an equivalence transformation

$$U_0 \begin{bmatrix} \gamma & 0 \\ 0 & \theta \end{bmatrix} V_0^* = A_0 = \begin{bmatrix} \lambda_1 & \mu \\ 0 & \lambda_n \end{bmatrix}$$

that does not tamper with the eigenvalues $\{\lambda_2, \ldots, \lambda_{n-1}\}$.

In Horn's proof, the ordering of diagonal entries is important and the resulting matrix is upper triangular. While the final result in the Weyl–Horn theorem remains true, it is unfortunate that it takes more than permutation to rearrange the diagonals of a triangular matrix while maintaining the singular values. Such a rearrangement is needed at every conquering step, but it requires a new Schur decomposition and is expensive to compute in general.

It was proved in Chu (2000) that the triangular structure was entirely unnecessary and the rearrangement of the diagonal entries was not needed. It can be shown that modifying the first and the last rows and columns of the block diagonal matrix in (4.88) is enough to solve the ISEP and the resulting matrix is permutation similar to a triangular matrix. This advance of understanding makes it possible to effectively implement the induction proof as a numerical algorithm.

More precisely, denote the 2×2 orthogonal matrices as $U_0 = [u_{st}^{(0)}]_{s,t=1}^2$ and $V_0 = [v_{st}^{(0)}]_{s,t=1}^2$. Then the matrix

$$A = \begin{bmatrix} u_{11}^{(0)} & 0 & u_{12}^{(0)} \\ 0 & I_{n-1} & 0 \\ u_{21}^{(0)} & 0 & u_{22}^{(0)} \end{bmatrix} \begin{bmatrix} A_1 & \bigcirc \\ \bigcirc & A_2 \end{bmatrix} \begin{bmatrix} v_{11}^{(0)} & 0 & v_{12}^{(0)} \\ 0 & I_{n-1} & 0 \\ v_{21}^{(0)} & 0 & v_{22}^{(0)} \end{bmatrix}^*$$

has eigenvalues $\{\lambda_k\}_{k=1}^n$ and singular values $\{s_1, \ldots, s_n\}$. The resulting matrix A has the structure

$$A = \begin{bmatrix} \lambda_1 & \otimes & \ldots & \otimes & \otimes & * & * & & & \mu \\ \otimes & \lambda_2 & & & \times & 0 & 0 & & & * \\ \vdots & & \ddots & & \vdots & & & \bigcirc & & \\ & & & \lambda_{j-1} & \times & & & & & \\ \otimes & \times & \ldots & \times & \lambda_j & & & & & * \\ * & 0 & \ldots & 0 & 0 & \lambda_{j+1} & \times & \times & \ldots & \otimes \\ * & & & & 0 & \times & \lambda_{j+2} & & & \otimes \\ & & \bigcirc & & & & & & & \\ & & & & & \vdots & & & \ddots & \\ 0 & * & \ldots & * & * & \otimes & \otimes & & & \lambda_n \end{bmatrix},$$

where $\times$ stands for unchanged, original entries from A_1 or A_2, $\otimes$ stands for entries of A_1 or A_2 that are modified by scalar multiplications, and $*$ denotes possible new entries that were originally zero. This pattern repeats itself when going through the cycle of recursion. Note that diagonal entries of A_1 and A_2 are in fixed orders, $\gamma, \lambda_2, \ldots, \lambda_j$ and $\lambda_{j+1}, \ldots, \lambda_{n-1}, \theta$, respectively. Each A_i is similar through permutations, which need not be known, to a lower triangular matrix whose diagonal entries constitute the same set as the diagonal entries of A_i. Thus, each A_i has precisely its own diagonal entries as its eigenvalues. The first row and the last row have the same zero pattern except that the lower-left corner is always zero. The first column and the last column have the same zero pattern except that the lower-left corner is always zero. Using graph theory (we shall omit the details), it can be shown that the affixed matrix A has exactly the same properties (Chu, 2000).

With this realization, the entire induction process can easily be implemented in any programming language that supports a routine calling itself recursively. The main feature in the routine should be a single divide and conquer mechanism as we have just described. As the routine is calling itself recursively, the problem is divided down and conquered up accordingly. A sample MATLAB program is listed below.

Algorithm 4.6 (Recursive method for inverse singular/eigenvalue problem) Given arrays `lambda` of eigenvalues and `alpha` of singular values satisfying Theorem 4.35, the following code generates a matrix A with the prescribed eigenvalues and singular values.

```
function [A]=svd_eig(alpha,lambda);
n = length(alpha);
if n == 1                                  % The 1 by 1 case
   A = [lambda(1)];
elseif n == 2                              % The 2 by 2 case
   [U,V,A] = two_by_two(alpha,lambda);
else                                       % Check zero singular values
   tol = n*alpha(1)*eps;
   k = sum(alpha > tol); m = sum(abs(lambda) > tol);
   if k == n                               % Case 1
      j = 1; s = alpha(1); temp = s;
      for i = 2:n-1
         temp = temp*alpha(i)/abs(lambda(i));
         if temp < s, j = i; s = temp; end
      end
      rho = abs(lambda(1)*lambda(n))/s;
      [U0,V0,A0] = two_by_two([s;rho],[lambda(1);lambda(n)]);
      [A1] = svd_eig(alpha(1:j),[s;lambda(2:j)]);
      [A2] = svd_eig(alpha(j+1:n),[lambda(j+1:n-1);rho]);
      A = [A1,zeros(j,n-j);zeros(n-j,j),A2];
      Temp = A;
         A(1,:)=U0(1,1)*Temp(1,:)+U0(1,2)*Temp(n,:);
         A(n,:)=U0(2,1)*Temp(1,:)+U0(2,2)*Temp(n,:);
      Temp = A;
         A(:,1)=V0(1,1)*Temp(:,1)+V0(1,2)*Temp(:,n);
         A(:,n)=V0(2,1)*Temp(:,1)+V0(2,2)*Temp(:,n);
   else                                    % Case 2
      beta = prod(abs(lambda(1:m)))/prod(alpha(1:m-1));
      [A3] = svd_eig([alpha(1:m-1);beta],lambda(1:m));
      A = zeros(n); [U3,S3,V3] = svd(A3);
      A(1:m,1:m) = V3'*A3*V3;
      for i = m+1:k, A(i,i+1) = alpha(i); end
      A(m,m+1) = sqrt(abs(alpha(m)^2-beta^2));
   end
end
```

Be aware of how the function `svd_eig` calls itself, which results in further splitting. At the end of the splitting, the self-calling begins to build up.

The program returns a matrix A which has eigenvalues $\{\lambda_k\}_{k=1}^n$ and singular values $\{\alpha_i\}_{i=1}^n$. Once A is constructed, any similarity transformation QAQ^* by a unitary matrix Q will maintain the same eigenvalues and singular values.

4.9.3 *A symbolic example*

It might be helpful to illustrate how the divide-and-conquer algorithm actually works by a 6×6 symbolic example. In the following, the integers j_ℓ, selected randomly only for demonstration, indicate where the problem should be divided.

The dividing process along with the corresponding eigenvalues and singular values for each subproblems is depicted in the boxed frames in Figure 4.9. A blank framed box indicates that the division immediately above it has reached its bottom. In this example, the original 6×6 problem is divided into two 1×1 problems and two 2×2 problems. Each of these small problems can trivially be solved. The pair of numbers beside j_ℓ and in between rows of framed boxes are the eigenvalues and singular values for the 2×2 matrix used to fasten the smaller matrices together in the conquering process.

The conquering process that uses the small matrices to build larger matrices is depicted in Figure 4.10. The matrices beside j_ℓ and in between rows of framed boxes are the augmented matrices (4.88) with the wrong eigenvalues. After fixing by some appropriated 2×2 matrices, we see in Figure 4.10 that some rows and columns must be modified. The symbols $\times, \otimes$ and $*$, indicating how the values have been changed during the conquering process, have the same meaning as defined before. The final 6×6 matrix with the desirable eigenvalues and singular values has the structure indicated at the top of Figure 4.10.

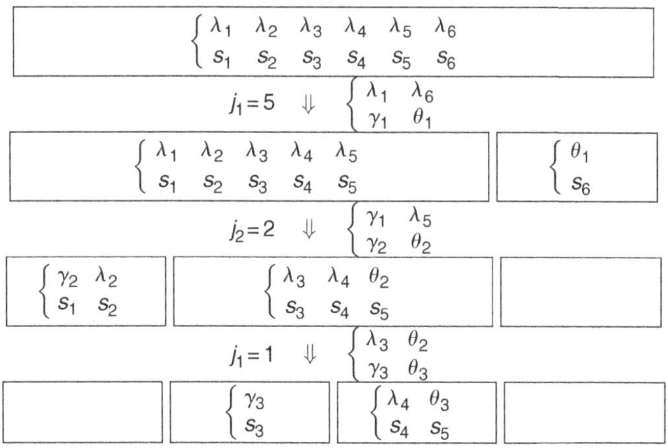

Figure 4.9. *An illustration of the dividing process*

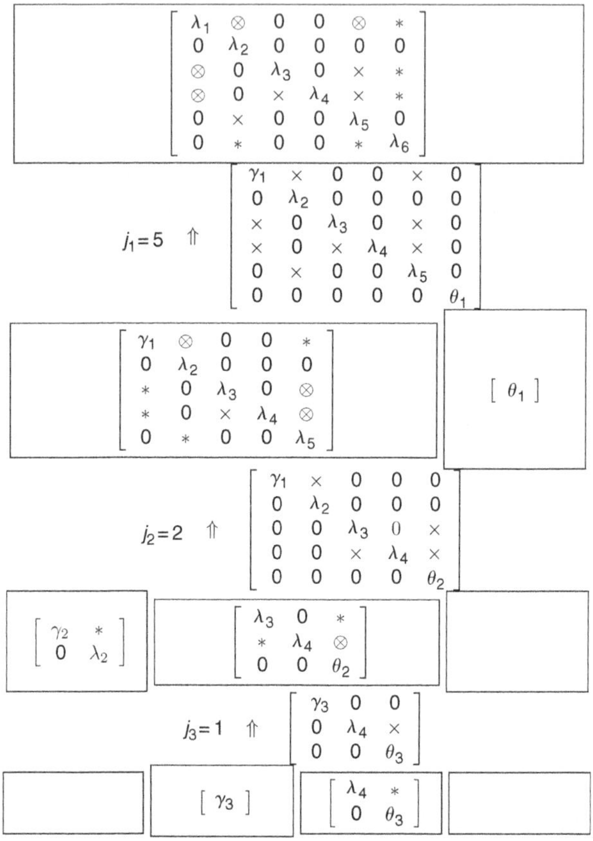

Figure 4.10. *An illustration of the conquering process*

Finally, we remark that the divide-and-conquer feature brings on fast computation. Numerical experiments seem to suggest that the overall cost in solving an n-dimensional ISEP is of the order of $O(n^2)$.

4.9.4 *A numerical example*

We know give a numerical example to illustrate how the constructed matrix might look. The 8×8 Rosser matrix R with integer elements,

$$R = \begin{bmatrix} 611 & 196 & -192 & 407 & -8 & -52 & -49 & 29 \\ 196 & 899 & 113 & -192 & -71 & -43 & -8 & -44 \\ -192 & 113 & 899 & 196 & 61 & 49 & 8 & 52 \\ 407 & -192 & 196 & 611 & 8 & 44 & 59 & -23 \\ -8 & -71 & 61 & 8 & 411 & -599 & 208 & 208 \\ -52 & -43 & 49 & 44 & -599 & 411 & 208 & 208 \\ -49 & -8 & 8 & 59 & 208 & 208 & 99 & -911 \\ 29 & -44 & 52 & -23 & 208 & 208 & -911 & 99 \end{bmatrix},$$

is a classical challenge for many eigenvalue algorithms. The Rosser matrix is symmetric, but has a double eigenvalue, three nearly equal eigenvalues, a zero eigenvalue, two dominant eigenvalues of opposite sign and a small nonzero eigenvalue. From MATLAB, the *computed* eigenvalues and singular values of R are

$$\boldsymbol{\lambda} = \begin{bmatrix} -1.020049018429997e{+}03 \\ 1.020049018429997e{+}03 \\ 1.020000000000000e{+}03 \\ 1.019901951359278e{+}03 \\ 1.000000000000001e{+}03 \\ 9.999999999999998e{+}02 \\ 9.804864072152601e{-}02 \\ 4.851119506099622e{-}13 \end{bmatrix}, \quad \boldsymbol{\alpha} = \begin{bmatrix} 1.020049018429997e{+}03 \\ 1.020049018429996e{+}03 \\ 1.020000000000000e{+}03 \\ 1.019901951359279e{+}03 \\ 1.000000000000000e{+}03 \\ 9.999999999999998e{+}02 \\ 9.804864072162672e{-}02 \\ 1.054603342667098e{-}14 \end{bmatrix},$$

respectively. We notice that MATLAB estimates the rank of R to be 7 due to the nearly zero singular value. The machine zero corresponding to a matrix A of size n is defined to be $n\epsilon\|A\|_2$ where the floating point relative accuracy ϵ is approximately 2.2204×10^{-16}.

Using our recursive algorithm for the above $\boldsymbol{\lambda}$ and $\boldsymbol{\alpha}$, we obtain the following matrix A which, for the convenience of running text, is displayed to only five digits:

$$A = \begin{bmatrix} 1.0200e{+}03 & 0 & 0 & 0 & 0 & 0 & 0 & 0 \\ 0 & -1.0200e{+}03 & 0 & 0 & 0 & 0 & 0 & 0 \\ 0 & 0 & 1.0200e{+}03 & 0 & 0 & 0 & 0 & 0 \\ 0 & 0 & 0 & 1.0199e{+}03 & 0 & 0 & 1.4668e{-}09 & 0 \\ 0 & 0 & 0 & 0 & 1.0000e{+}03 & 0 & 0 & 0 \\ 0 & 0 & 0 & 0 & 0 & 1.0000e{+}03 & 0 & 0 \\ 0 & 0 & 0 & -1.5257e{-}05 & 0 & 0 & 9.8049e{-}02 & 0 \\ 0 & 0 & 0 & 0 & 0 & 0 & 1.4045e{-}07 & 0 \end{bmatrix}.$$

The computed eigenvalues and singular values of A are

$$\hat{\boldsymbol{\lambda}} = \begin{bmatrix} -1.020049018429997e{+}03 \\ 1.020049018429997e{+}03 \\ 1.020000000000000e{+}03 \\ 1.019901951359278e{+}03 \\ 1.000000000000001e{+}03 \\ 9.999999999999998e{+}02 \\ 9.80486407215721e{-}02 \\ 0 \end{bmatrix}, \quad \hat{\boldsymbol{\alpha}} = \begin{bmatrix} 1.020049018429997e{+}03 \\ 1.020049018429997e{+}03 \\ 1.020000000000000e{+}03 \\ 1.019901951359279e{+}03 \\ 1.000000000000001e{+}03 \\ 9.999999999999998e{+}02 \\ 9.804864072162672e{-}02 \\ 0 \end{bmatrix},$$

respectively.

It is interesting to note that the resulting A is *not* symmetric, yet the eigenvalues and singular values of A agree with those of R up to the machine accuracy. Note also the sparsity and the relatively small off-diagonal entries of A.

4.10 Equality constrained inverse eigenvalue problems

Recall the inverse problem ECIEP defined in Problem 2.9 in which a matrix $\Upsilon \in \mathbb{R}^{n\times n}$ is to be found so as to satisfy the conditions

$$\begin{aligned} \Upsilon \mathbf{x}_i &= \mathbf{y}_i, \quad i = 1, \dots, p, \\ \sigma(\Upsilon) &= \{\lambda_k\}_{k=1}^n, \end{aligned} \tag{4.89}$$

where $\{\mathbf{x}_1, \dots, \mathbf{x}_p\}$ and $\{\mathbf{y}_1, \dots \mathbf{y}_p\}$, $p \leq n$, are two sets of given vectors in $\mathbb{R}^n$, and $\{\lambda_k\}_{k=1}^n$ is a set of given complex numbers closed under complex conjugation. This problem may be considered as a variation of the generic SIEP defined in Problem 4.1 where the structure $\mathcal{N}$ is characterized by the linear equations (4.89).

4.10.1 *Existence and equivalence to PAPs*

There is a close relationship between the ECIEP and the PAP. By exploiting that relationship, we can establish existence theory and numerical methods for the ECIEP from what is already known for the PAP.

Rewrite (4.89) as

$$\Upsilon X = Y,$$

with $X = [\mathbf{x}_1, \dots, \mathbf{x}_p]$ and $Y = [\mathbf{y}_1, \dots, \mathbf{y}_p]$ as matrices in $\mathbb{R}^{n\times p}$. Let the QR decomposition of X be denoted by

$$X = [Q_1, Q_2] \begin{bmatrix} R \\ 0 \end{bmatrix} \tag{4.90}$$

where $R \in \mathbb{R}^{p\times p}$ is nonsingular and the columns of the orthogonal matrix $[Q_1, Q_2]$ span, respectively, the range space and its orthogonal complement of X. Then a feasible solution A must be of the form

$$\Upsilon = YR^{-1}Q_1^\top + ZQ_2^\top, \tag{4.91}$$

with arbitrary $Z \in \mathbb{R}^{n\times(n-p)}$. The ECIEP now is reduced to finding the matrix Z so that $\sigma(YR^{-1}Q_1^\top + ZQ_2^\top) = \{\lambda_k\}_{k=1}^n$, which is equivalent to the state feedback PAP discussed in Problem 2.1 (in reversed order).

A standard result addressing the solvability of Problem 2.1 is as follows (Kautsky et al., 1985).

Theorem 4.36. Given $A \in \mathbb{R}^{n\times n}$, $B \in \mathbb{R}^{n\times m}$, and a set of n complex numbers $\{\lambda_k\}_{k=1}^n$ closed under complex conjugation, then there exist a matrix $F \in \mathbb{R}^{m\times n}$

such that $\sigma(A + BF) = \{\lambda_k\}_{k=1}^n$ if and only if the pair (A, B) is controllable, that is, if and only if the following condition holds:

$$\{y^\top A = \mu y^\top \text{ and } y^\top B = 0\} \Longleftrightarrow y^\top = 0. \tag{4.92}$$

Applied to our problem in (4.91), the spectral condition can be satisfied by some Z if and only the pair of matrices $(YR^{-1}Q_1^\top, Q_2^\top)$ is observable. That is, the ECIEP is solvable if and only if the $(2n-p) \times n$ matrix

$$\begin{bmatrix} YR^{-1}Q_1^\top - sI \\ Q_2^\top \end{bmatrix} \tag{4.93}$$

is of full column rank for every eigenvalue s of the square matrix $YR^{-1}Q_1^\top$. Observe that

$$\begin{bmatrix} YR^{-1}Q_1^\top - sI \\ Q_2^\top \end{bmatrix} [Q_1 R, Q_2] = \begin{bmatrix} Y - sX & -sQ_2 \\ 0 & I_{n-p} \end{bmatrix}.$$

We thus conclude the following existence theorem (Li, 1997b).

Theorem 4.37. Suppose that $X \in \mathbb{R}^{n \times p}$ is of full column rank. Then the ECIEP is solvable for any $\{\lambda_k\}_{k=1}^n$ if and only if the matrix $Y - sX$ is of full rank for every $s \in \sigma(YX^\dagger)$, where $X^\dagger$ is the Moore–Penrose generalized inverse of X.

The solution Z is not unique. Being regarded as a PAP, algorithms for robust pole assignment such as those developed in Byers and Nash (1989), Kautsky et al. (1985) are readily applicable.

4.11 Summary

In a general sense, every IEP is a structured problem. The structure the problem embodies can be characterized in many different ways, ranging from straightforward linear dependence on some basis matrices to sophisticated implicitly defined qualification criteria. In this chapter, we demonstrate the eight different structures. This choice of study represents only a small segment of the full scope of structured inverse eigenvalue problems. We have noted that some of the problems have enjoyed extensive attention in the literature because of their connections to other important applications. We have also noted many open problems that need further study.

Once again, be aware that our consideration has been limited to the setting that the entire spectrum is known and that the structural constraint must be satisfied. We have not discussed the structured problems where only partial eigenvalues and eigenvectors are given. Neither have we examined the case where a least squares solution with approximate spectrum or approximate structure is sufficient for practical purposes.

5

PARTIALLY DESCRIBED INVERSE EIGENVALUE PROBLEMS

5.1 Overview

For most of the IEPs we have discussed thus far, the imposed spectral constraint involves only eigenvalues. In practice, the spectral information furnished for an IEP often is from empirical data. Through vibration tests where the excitation and the response of the structure at selected points are measured experimentally, there are identification techniques to extract a portion of eigenpair information from the measurements. However, the physical system might be too complicated or the size might be too large that it is not always possible to attain knowledge of the entire spectrum. On one hand, there is no reasonable analytical tool available to evaluate the entire spectral information. On the other hand, it is simply unwise to use experimental values of high natural frequencies to reconstruct a model. A general rule of thumb is that a discrete model with n degrees of freedom can provide accurate values for only about $n/3$ natural frequencies. Furthermore, it is often demanded, especially in structural design, that certain eigenvectors should also satisfy some specific conditions. The spectral constraints involved in an IEP, therefore, may consist of only partial information on eigenvalues or eigenvectors. This kind of inverse problem will be referred to as the partially described inverse eigenvalue problem (PDIEP).

A generic PDIEP can be stated as follows.

Problem 5.1 *(Generic PDIEP)*
Given vectors $\{\mathbf{v}^{(1)}, \ldots, \mathbf{v}^{(k)}\} \subset \mathbb{F}^n$, values $\{\lambda_1, \ldots \lambda_k\} \subset \mathbb{F}$, and a set $\mathcal{N}$ of structured matrices, find $X \in \mathcal{N}$ such that

$$X\mathbf{v}^{(i)} = \lambda_i \mathbf{v}^{(i)}, \quad i = 1, \ldots, k.$$

PDIEPs certainly are not new. The famous paper and filmed lecture by Kac (1966) with the inquisitive title "Can one hear the shape of a drum?" is a special kind of PDIEP. This subject sometimes appears under different names, including the inverse mode problem (Gladwell, 1986a; Ram, 1994a), inverse eigenvector problem, inverse shape problem (Lai and Ananthasuresh, 2003; Scascighini and Troxler, 2001), inverse nodal problem (for DEs) (Hald and McLaughlin, 1989, 1996; Yang, 1997), model updating problem (Datta, 2002), and so on.

It should be noted that eigenpair information sometimes provides additional intrinsic relationships. A question immediately arising is that, for a structured matrix, how many pairs of information are needed to determine such a matrix. Gladwell (1986a), for example, derived the necessary and sufficient conditions applicable to the eigenpairs to permit the construction of a realizable beam. A simply connected spring–mass system, fixed at one end and free at the other, may be reconstructed uniquely from two eigenvalues, two eigenvectors and the total mass of the system. Two sets of eigenvalues and eigenvectors for the fixed–fixed and fixed–free boundary conditions can be used to reconstruct the mass and stiffness matrices of a discrete finite difference model of the nonuniform axially vibrating bar uniquely. The inverse mode problem for the continuous model of an axially vibrating bar from two eigenvalues, the corresponding eigenvectors and the total mass of the system was solved by Ram (1994a). Ram and Gladwell (1994) proposed a way to reconstruct a finite element model of a vibrating bar from a single eigenvalue and two eigenvectors based on the fact that both the mass and stiffness matrices of the finite element model are tridiagonal. Ram (1994b) also derived a method of reconstructing a finite difference model of a vibrating beam from three eigenvectors, one eigenvalue and the total mass of the beam.

It is quite a relief to know, from an application point of view, that in many cases a few eigenpairs can determine much of the desirable reconstruction. We shall illustrate this point by concentrating on the Toeplitz structure and the quadratic pencil in this chapter. We shall not review those existing results mentioned above because it will take considerable length to introduce the background. A least squares formulation of PDIEP will be studied in the next chapter. Currently it does not appear that there are any general systematic studies for PDIEPs. We think PDIEPs for most other matrix structures are still widely open for further studies.

5.2 PDIEP for Toeplitz matrices

Earlier in Section 4.3 we discussed the Toeplitz inverse eigenvalue problem where the entire spectrum is prescribed. In this section, we are more interested in the inverse problem of constructing a real symmetric Toeplitz matrix from prescribed eigenpairs.

> **Problem 5.2** *(PDIEP1)*
> Given a set of real orthonormal vectors, $\{\mathbf{v}^{(1)}, \ldots, \mathbf{v}^{(k)}\}$, $k \geq 1$, each symmetric or skew-symmetric, and a set of real numbers $\{\lambda_1, \ldots, \lambda_k\}$, find a symmetric Toeplitz matrix T (other than a scalar matrix) such that
> $$T\mathbf{v}^{(i)} = \lambda_i \mathbf{v}^{(i)}, \quad i = 1, \ldots, k. \tag{5.1}$$

We shall consider in particular the construction based on *two* prescribed eigenpairs. We shall show that, similar to those mass–spring systems mentioned earlier, two prescribed eigenpairs are almost sufficient in determining a unique symmetric Toeplitz matrix.

Recall from Section 4.3.1 that as a symmetric centrosymmetric matrix (Cantoni and Bulter, 1976, Theorem 2), any symmetric Toeplitz matrix of order n has $\lceil n/2 \rceil$ even and $\lfloor n/2 \rfloor$ odd eigenvectors. For eigenvalues of multiplicity greater than one, the corresponding eigenspace has an orthonormal basis which splits as evenly as possible between even and odd eigenvectors (Delsarte and Genin, 1984, Theorem 8). Thus it is sensible to say that the eigenvectors of a symmetric Toeplitz matrix can be split into two classes. For convenience, we shall use $\sigma^+(T)$ and $\sigma^-(T)$ to denote, respectively, the spectrum of eigenvalues corresponding to even and odd eigenvectors. A few other discussions about eigenvectors of Toeplitz matrices can be found in Makhoul (1981) and Trench (1989). For the moment, we note that the description of the given eigenpairs in Problem 5.2 cannot be totally arbitrary. For instance, it is improper to request that all vectors be even if $k > \lceil n/2 \rceil$. A notion of *universal* distribution of eigenvalues for Toeplitz matrices was mentioned in Delsarte and Genin (1984) suggesting that $\sigma^+(T)$ and $\sigma^-(T)$ should interlace. We have illustrated in Section 4.3.1 that a Toeplitz matrix whose spectrum does not satisfy the interlaced distribution is perhaps more difficult to find (Laurie, 1988). On the other hand, if the number k of specified eigenpairs in Problem 5.2 is not too large, there might be enough room for the remaining unspecified eigenpairs to shuffle so that the spectrum of a solution eventually satisfies the interlaced condition.

It is noted in Cantoni and Bulter (1976, Theorem 3) that any real $n \times n$ matrix with a set of n real orthonormal eigenvectors, each being either even or odd, is necessarily both symmetric and centrosymmetric. In contrast, it is a much harder problem to identify an orthogonal matrix so that its columns form the eigenvectors of some Toeplitz matrix.

For a single vector alone, Cybenko (1984) proved that its being even or odd is sufficient to be an eigenvector of a Toeplitz matrix. In fact, let

$$\mathcal{R}_0(\mathbf{v}) := \{\mathbf{r} \in \mathbb{R}^n | T(\mathbf{r})\mathbf{v} = 0\}, \tag{5.2}$$

denote the collection of (the first columns of) all symmetric Toeplitz matrices for which $\mathbf{v}$ is an eigenvector corresponding to the eigenvalue 0. The following result is proved in Cybenko (1984, Corollary 1).

Theorem 5.1. (Cybenko, 1984) The set $\mathcal{R}_0(\mathbf{v})$ is a linear subspace with dimension

$$\dim(\mathcal{R}_0(\mathbf{v})) = n - \pi(\mathbf{v}), \tag{5.3}$$

where

$$\pi(\mathbf{v}) := \begin{cases} \lceil n/2 \rceil, & \text{if } \mathbf{v} \text{ is even,} \\ \lfloor n/2 \rfloor, & \text{if } \mathbf{v} \text{ is odd.} \end{cases} \tag{5.4}$$

Clearly $T(\mathbf{r})\mathbf{v} = \lambda\mathbf{v}$ if and only if $\mathbf{r} - \lambda\mathbf{e}_1 \in \mathcal{R}_0(\mathbf{v})$, where $\mathbf{e}_1 = [1, 0, \ldots, 0]^\top$ is the first standard basis vector in $\mathbb{R}^n$. Thus the set

$$\mathcal{R}(\mathbf{v}) := \{\mathbf{r} \in \mathbb{R}^n | T(\mathbf{r})\mathbf{v} = \lambda\mathbf{v} \text{ for some } \lambda \in \mathbb{R}\}, \tag{5.5}$$

is precisely the direct sum $\text{span}\{\mathbf{e}_1\} \oplus \mathcal{R}_0(\mathbf{v})$.

Suppose now $\{\mathbf{v}^{(1)}, \ldots, \mathbf{v}^{(k)}\}$, $k \geq 1$, is a set of real orthonormal vectors, each being even or odd. Then $\cap_{i=1}^k \mathcal{R}(\mathbf{v}^{(i)})$ contains all symmetric Toeplitz matrices for which each $\mathbf{v}_i$ is an eigenvector. Evidently, $\mathbf{e}_1 \in \mathcal{R}(\mathbf{v}^{(i)})$ for all i. So $\cap_{i=1}^k \mathcal{R}(\mathbf{v}^{(i)})$ is at least of dimension 1. An interesting question then is to obtain a non-trivial lower bound on the dimension of $\cap_{i=1}^k \mathcal{R}(\mathbf{v}^{(i)})$.

Toward this end, we show in the sequel that for the case $k = 2$, the dimension of $\cap_{i=1}^2 \mathcal{R}(\mathbf{v}^{(i)})$ is almost always independent of the size of the problem, and in fact is either two, three, or four, depending upon whether the eigenvectors are even or odd. We show further that in each direction in the subspace $\cap_{i=1}^2 \mathcal{R}(\mathbf{v}^{(i)})$ there is one and only one Toeplitz matrix for Problem 5.2. In particular, we show that if n is odd and if at least one of the given eigenvectors is even, or if n is even and one eigenvector is even and the other is odd, then the Toeplitz matrix is uniquely determined.

5.2.1 *An example*

As we shall consider only the case $k = 2$ throughout the discussion, it is more convenient to denote, henceforth, the eigenvectors $\mathbf{v}^{(1)}$ and $\mathbf{v}^{(2)}$ by $\mathbf{u}$ and $\mathbf{v}$, respectively.

We begin our study of the set $\mathcal{R}(\mathbf{u}) \cap \mathcal{R}(\mathbf{v})$ with the special case where $n = 3$. The example should give some insights on the higher dimensional case.

Due to the special eigenstructure of symmetric Toeplitz matrices, it is necessary that one of the two given eigenvectors, say $\mathbf{u}$, must be even. Denote $\mathbf{u} = [u_1, u_2, u_1]^\top$ where $2{u_1}^2 + u_2^2 = 1$. It can also be proved that the odd vector $\hat{\mathbf{u}} = [1/\sqrt{2}, 0, -1/\sqrt{2}]^\top$ is always an eigenvector for every symmetric Toeplitz matrix of order 3. Thus, given $\mathbf{u}$, we imply from the orthogonality condition that the second prescribed eigenvector $\mathbf{v}$ must be either the second or the third column of the matrix

$$Q = \begin{bmatrix} u_1 & \dfrac{1}{\sqrt{2}} & -\dfrac{u_2}{\sqrt{2}} \\ u_2 & 0 & u_1\sqrt{2} \\ u_1 & -\dfrac{1}{\sqrt{2}} & -\dfrac{u_2}{\sqrt{2}} \end{bmatrix}. \tag{5.6}$$

In other words, one even eigenvector completely determines all three orthonormal eigenvectors (up to a $\pm$ sign). It follows, from Cybenko (1984), that $\dim \mathcal{R}(\mathbf{u}) \cap \mathcal{R}(\mathbf{v}) = 2$. (Note that if the orthogonality condition is violated, then trivially $\mathcal{R}(\mathbf{u}) \cap \mathcal{R}(\mathbf{v}) = \langle \mathbf{e}_1 \rangle$.)

Let $\Lambda = \operatorname{diag}\{\lambda_1, \lambda_2, \lambda_3\}$. We already know $Q\Lambda Q^\top$ is a centrosymmetric matrix. It is not difficult to see that $Q\Lambda Q^\top$ is Toeplitz if and only if

$$(3{u_1}^2 - 1)\lambda_1 + \frac{\lambda_2}{2} + \left(\frac{1}{2} - 3{u_1}^2\right)\lambda_3 = 0. \tag{5.7}$$

From (5.7), the following facts can easily be observed.

Example 5.1. Suppose $2u_1^2 + u_2^2 = 1$. Then:

1. If the second given eigenvector is even, then corresponding to any given scalars λ_1 and λ_3 there is a unique symmetric Toeplitz matrix T such that

$$T\begin{bmatrix} u_1 & -\dfrac{u_2}{\sqrt{2}} \\ u_2 & u_1\sqrt{2} \\ u_1 & -\dfrac{u_2}{\sqrt{2}} \end{bmatrix} = \begin{bmatrix} u_1 & -\dfrac{u_2}{\sqrt{2}} \\ u_2 & u_1\sqrt{2} \\ u_1 & -\dfrac{u_2}{\sqrt{2}} \end{bmatrix} \begin{bmatrix} \lambda_1 & 0 \\ 0 & \lambda_3 \end{bmatrix}. \tag{5.8}$$

2. If the second given eigenvector is odd and if $u_1 \neq \pm\sqrt{1/6}$, then corresponding to any given scalars λ_1 and λ_2 there is a unique symmetric Toeplitz matrix T such that

$$T\begin{bmatrix} u_1 & \dfrac{1}{\sqrt{2}} \\ u_2 & 0 \\ u_1 & -\dfrac{1}{\sqrt{2}} \end{bmatrix} = \begin{bmatrix} u_1 & \dfrac{1}{\sqrt{2}} \\ u_2 & 0 \\ u_1 & -\dfrac{1}{\sqrt{2}} \end{bmatrix} \begin{bmatrix} \lambda_1 & 0 \\ 0 & \lambda_2 \end{bmatrix}. \tag{5.9}$$

3. If $u_1 = \pm\sqrt{1/6}$ then there are infinitely many symmetric Toeplitz matrices T which satisfy (5.9) if $\lambda_1 = \lambda_2$; however, if $\lambda_1 \neq \lambda_2$, then (5.9) does not hold for any symmetric Toeplitz matrix T.

5.2.2 *General consideration*

We now consider the case for general n. When $\mathbf{v}$ is an eigenvector, the idea of rewriting the linear relationship in the matrix–vector product (Cybenko, 1984)

$$T(\mathbf{r})\mathbf{v} = M(\mathbf{v})\mathbf{r}, \tag{5.10}$$

where $M(\mathbf{v})$ is some matrix depending on $\mathbf{v}$, can be very useful. It is easy to see that the columns of $M(\mathbf{v})$ have the same symmetry as the vector $\mathbf{v}$ has. That is, $\Xi M(\mathbf{v}) = \pm M(\mathbf{v})$ if and only if $\Xi\mathbf{v} = \pm\mathbf{v}$. Thus only the first $\pi(\mathbf{v})$ rows of $M(\mathbf{v})$ need to be considered.

We now take a closer examination of the matrix $M(\mathbf{v})$. For convenience, let $p := \lceil n/2 \rceil$ and let $N(\mathbf{v})$ denote the $p \times n$ submatrix of the first p rows of $M(\mathbf{v})$.

It is easy to verify that $N(\mathbf{v})$ can be decomposed into blocks:

$$N(\mathbf{v}) = \underbrace{[h(\mathbf{v}), H(\mathbf{v})}_{\tilde{H}(\mathbf{v})}, 0] + \underbrace{[0, L(\mathbf{v})}_{\tilde{L}(\mathbf{v})}, 0] + [0, 0, U(\mathbf{v})], \tag{5.11}$$

where $h(\mathbf{v})$ is a $p \times 1$ column vector, $\tilde{H}(\mathbf{v}) := [h(\mathbf{v}), H(\mathbf{v})] = [\tilde{h}_{ij}(\mathbf{v})]$ is the $p \times p$ Hankel matrix

$$\tilde{h}_{ij}(\mathbf{v}) := v_{i+j-1}, \tag{5.12}$$

$\tilde{L}(\mathbf{v}) := [0, L] = [\tilde{l}_{ij}(\mathbf{v})]$ is the $p \times p$ lower triangular matrix

$$\tilde{l}_{ij}(\mathbf{v}) := \begin{cases} v_{i-j+1}, & \text{if } 1 < j \le i, \\ 0, & \text{otherwise,} \end{cases} \tag{5.13}$$

and $U(\mathbf{v}) = [u_{ij}(\mathbf{v})]$ is the $p \times (n-p)$ triangular matrix

$$u_{ij}(\mathbf{v}) := \begin{cases} v_{p+i+j-1} & \text{if } i + j \le n - p + 1, \\ 0 & \text{otherwise.} \end{cases} \tag{5.14}$$

We note that the last row of $N(\mathbf{v})$ is identically zero when n is odd and $\mathbf{v}$ is skew-symmetric. The rows of $N(\mathbf{u})$ and $N(\mathbf{v})$ will be used to construct a larger matrix.

Suppose that $\mathbf{u}$ and $\mathbf{v}$ are two given eigenvectors of a certain Toeplitz matrix $T(\mathbf{r})$, that is, suppose that we have

$$\begin{aligned} T(\mathbf{r})\mathbf{u} &= \lambda_1 \mathbf{u}, \\ T(\mathbf{r})\mathbf{v} &= \lambda_2 \mathbf{v}. \end{aligned} \tag{5.15}$$

Then the vector $\mathbf{r}$ must be such that the linear equations

$$\begin{aligned} N(\mathbf{u})(\mathbf{r} - \lambda_1 \mathbf{e}_1) &= 0, \\ N(\mathbf{v})(\mathbf{r} - \lambda_2 \mathbf{e}_1) &= 0, \end{aligned} \tag{5.16}$$

are satisfied. If we write

$$\mathbf{x} := [r_1 - \lambda_2, r_1 - \lambda_1, r_2, \ldots, r_n]^\top \in \mathbb{R}^{n+1}, \tag{5.17}$$

then the system (5.16) is equivalent to

$$\tilde{M}(\mathbf{u}, \mathbf{v})\mathbf{x} = 0, \tag{5.18}$$

where $\tilde{M}(\mathbf{u}, \mathbf{v})$ is the $(2p) \times (n+1)$ matrix defined by

$$\tilde{M}(\mathbf{u}, \mathbf{v}) := \begin{bmatrix} 0 & h(\mathbf{u}) & H(\mathbf{u}) + L(\mathbf{u}) & U(\mathbf{u}) \\ h(\mathbf{v}) & 0 & H(\mathbf{v}) + L(\mathbf{v}) & U(\mathbf{v}) \end{bmatrix}. \tag{5.19}$$

Given even or odd vectors $\mathbf{u}$ and $\mathbf{v}$, a solution to (5.18) can be used to construct a Toeplitz matrix in the following way, whose proof is by direct observation.

Theorem 5.2. Suppose $[x_0, x_1, \ldots, x_n]^\top$ is a solution to (5.18). For arbitrary real numbers λ_1 and α, define $\mathbf{r} = [r_1, \ldots, r_n]^\top$ where

$$\begin{aligned} r_1 &:= \alpha x_1 + \lambda_1, \\ r_i &:= \alpha x_i, \qquad \text{for } i = 2, \ldots, n. \end{aligned} \tag{5.20}$$

Then $\mathbf{u}$ and $\mathbf{v}$ are eigenvectors of the Toeplitz matrix $T(\mathbf{r})$ with corresponding eigenvalues λ_1 and

$$\lambda_2 := \alpha(x_1 - x_0) + \lambda_1. \tag{5.21}$$

In other words, $\mathcal{R}(\mathbf{u}) \cap \mathcal{R}(\mathbf{v})$ is the direct sum of the subspace spanned by $\mathbf{e}_1$ and the subspace obtained by deleting the first component from $\ker(\tilde{M})$.

On the other hand, suppose that the two eigenvalues λ_1 and λ_2 are prescribed. Then the equation (5.21) implies that the constant α in (5.20) must be

$$\alpha = \frac{\lambda_1 - \lambda_2}{x_0 - x_1}, \tag{5.22}$$

provided $x_0 \neq x_1$. We conclude, therefore,

Theorem 5.3. Suppose $\mathbf{x}$ is a nontrivial solution of (5.18) satisfying $x_0 \neq x_1$. Then corresponding to the direction of $\mathbf{x}$, there is a unique solution to Problem 5.2 when $k = 2$.

To construct a solution to Problem 5.2, it only remains to determine the null space of $\tilde{M}(\mathbf{u}, \mathbf{v})$. It turns out that the dimension depends upon whether n is even or odd and whether the two eigenvectors are even or odd. In all cases, we shall show that $\tilde{M}$ has a nontrivial null space. It is most interesting to note that the dimension does not depend upon the size of n.

For clarity, we shall divide the discussion into six different cases. Only one proof is detailed, but the rest can be done in a very similar way. We abbreviate $\tilde{M}(\mathbf{u}, \mathbf{v})$ to $\tilde{M}$.

Case 1: n is odd and both eigenvectors are even When n is odd, the Hankel matrix $\tilde{H}(\mathbf{v})$ for an even vector $\mathbf{v}$ takes the form:

$$\tilde{H}(\mathbf{v}) = \begin{bmatrix} v_1 & v_2 & \cdots & v_{p-1} & v_p \\ v_2 & v_3 & \cdots & v_p & v_{p-1} \\ \vdots & & & & \vdots \\ v_{p-1} & v_p & \cdots & v_3 & v_2 \\ v_p & v_{p-1} & \cdots & v_2 & v_1 \end{bmatrix}. \tag{5.23}$$

The corresponding $U(\mathbf{v})$ becomes

$$U(\mathbf{v}) = \Xi L(\mathbf{v}) = \begin{bmatrix} v_{p-1} & v_{p-2} & \dots & v_2 & v_1 \\ v_{p-2} & & \dots & v_1 & 0 \\ \vdots & & & & \vdots \\ v_1 & 0 & \dots & & 0 \\ 0 & & \dots & & 0 \end{bmatrix}. \tag{5.24}$$

The basic structure of $\tilde{M}$ can be seen from the example when $n = 5$,

$$\tilde{M} = \begin{bmatrix} 0 & u_1 & u_2 & u_3 & u_2 & u_1 \\ 0 & u_2 & u_3 + u_1 & u_2 & u_1 & 0 \\ 0 & u_3 & 2u_2 & 2\,u_1 & 0 & 0 \\ v_1 & 0 & v_2 & v_3 & v_2 & v_1 \\ v_2 & 0 & v_3 + v_1 & v_2 & v_1 & 0 \\ v_3 & 0 & 2v_2 & 2v_1 & 0 & 0 \end{bmatrix}. \tag{5.25}$$

The matrix $\tilde{M}$ in general is a square matrix of order $n+1$. The determinant of $\tilde{M}$ is an algebraic expression involving independent variables $v_1, \dots, v_p, u_1, \dots, u_p$. We need to show that, under the additional condition that $\mathbf{u}$ and $\mathbf{v}$ are perpendicular to each other, the matrix $\tilde{M}$ is rank deficient.

Given even vectors $\mathbf{u}$ and $\mathbf{v}$, define the $2p \times 2p$ matrix

$$\tilde{G} = \tilde{G}(\mathbf{u}, \mathbf{v}) := \begin{bmatrix} I & 0 \\ -G(\mathbf{v}) & G(\mathbf{u}) \end{bmatrix}, \tag{5.26}$$

where $G(\mathbf{v})$ is the $p \times p$ upper triangular matrix given by

$$G(\mathbf{v}) := \begin{bmatrix} 2v_1 & 2v_2 & \dots & 2v_{p-1} & v_p \\ 0 & 2v_1 & \dots & 2v_{p-2} & v_{p-1} \\ \vdots & 0 & \ddots & \vdots & \vdots \\ 0 & & & 2v_1 & v_2 \\ 0 & 0 & 0 & 0 & v_1 \end{bmatrix}. \tag{5.27}$$

Without loss of generality, we may assume $u_1 \neq 0$ and hence the matrix $\tilde{G}(\mathbf{u}, \mathbf{v})$ is nonsingular, since otherwise we may swap the roles of $\mathbf{u}$ and $\mathbf{v}$ if $v_1 \neq 0$, and the problem is reduced to lower dimension if both $u_1 = v_1 = 0$. It follows that the product

$$\tilde{W} := \tilde{G}\tilde{M} = \begin{bmatrix} I & 0 \\ -G(\mathbf{v}) & G(\mathbf{u}) \end{bmatrix} \begin{bmatrix} 0 & h(\mathbf{u}) & H(\mathbf{u}) + L(\mathbf{u}) & U(\mathbf{u}) \\ h(\mathbf{v}) & 0 & H(\mathbf{v}) + L(\mathbf{v}) & U(\mathbf{v}) \end{bmatrix} \tag{5.28}$$

has the same rank as $\tilde{M}$. The role of $\tilde{G}$ is simply a sequence of elementary row operations by which we produce special zero patterns in $\tilde{W}$.

Lemma 5.4. Suppose that n is odd and that the two symmetric vectors $\mathbf{u}$ and $\mathbf{v}$ are orthogonal. Then the matrix $\tilde{W}$ is rank deficient. In fact,

$$p + 1 \leq \operatorname{rank}(\tilde{\mathrm{W}}) \leq \mathrm{n}. \tag{5.29}$$

Proof The proof is tedious but straightforward. We check out the product block by block. For $i = 1, \ldots, p$, the i-th component of $G(\mathbf{u})h(\mathbf{v})$ is given by

$$2 \sum_{\substack{s=1 \\ t-s=i-1}}^{p-i} u_s v_t + u_{p-i+1} v_p.$$

The first component is trivially seen to be $2\sum_{s=1}^{p-1} u_s v_s + u_p v_p$ which is zero because $\mathbf{u}$ and $\mathbf{v}$ are perpendicular to each other. For the same reason, the first component of $-G(\mathbf{v})h(\mathbf{u})$ is zero.

The product $G(\mathbf{u})U(\mathbf{v})$ has the same triangular structure as $U(\mathbf{v})$. On the other hand, the (i, j)-component of $G(\mathbf{u})U(\mathbf{v})$ with $i + j \leq p$ is given by

$$2 \sum_{s+t=p-i-j+2} u_s v_t. \tag{5.30}$$

It is important to note that the summation (5.30) is a symmetric function of $\mathbf{u}$ and $\mathbf{v}$. It follows that the lower-right $p \times (n-p)$ block $-G(\mathbf{v})U(\mathbf{u}) + G(\mathbf{u})U(\mathbf{v})$ is identically zero.

For $i = 1, \ldots, p$ and $j = 1, \ldots, p-1$, the (i, j)-component of $G(\mathbf{u})H(\mathbf{v})$ is given by

$$\begin{cases} 2\sum_{\substack{s=1 \\ t-s=i+j-1}}^{p-i-j+1} u_s v_t + 2\sum_{\substack{s=p-i-j+2 \\ t+s=2p-i-j+1}}^{p-i} u_s v_t + u_{p-i+1} v_{p-j}, & \text{if } j < p-i+1, \\[2ex] 2\sum_{\substack{s=1 \\ t+s=2p-i-j+1}}^{p-i} u_s v_t + u_{p-i+1} v_{p-j}, & \text{if } j \geq p-i+1. \end{cases}$$

The (i, j)-component of $G(\mathbf{u})L(v)$ is given by

$$\begin{cases} 2\sum_{\substack{s=1 \\ t-s=i-j-1}}^{p-i} u_s v_t + u_{p-i+1} v_{p-j}, & \text{if } j \leq i-1, \\[2ex] 2\sum_{\substack{s=j-i+2 \\ t-s=i-j-1}}^{p-i} u_s v_t + u_{p-i+1} v_{p-j}, & \text{if } j > i-1. \end{cases}$$

It follows that the $(1, j)$-component of $G(\mathbf{u})(H(\mathbf{v}) + L(\mathbf{v}))$ is given by

$$2 \sum_{\substack{s=1 \\ t-s=j}}^{p-j} u_s v_t + 2 \sum_{\substack{s=p-j+1 \\ t+s=2p-j}}^{p-1} u_s v_t + 2 \sum_{\substack{t=1 \\ s-t=j}}^{p-j} u_s v_t. \tag{5.31}$$

The first and the last summations in (5.31) are symmetric to each other. The second summation in (5.31) is a symmetric function of $\mathbf{u}$ and $\mathbf{v}$. Over all, (5.31) is a symmetric function of $\mathbf{u}$ and $\mathbf{v}$ which will be completely canceled by its counterpart in $-G(\mathbf{v})(H(\mathbf{u}) + L(\mathbf{u}))$.

By now we have proved that the $(p+1)$-th row of $\tilde{W}$ is identically zero. If t follows that the null space of $\tilde{M}$ is at least of dimension one.

It is further observed that the $(i, p-1)$-component of $G(\mathbf{u})(H(\mathbf{v}) + L(\mathbf{v}))$ is given by

$$2 \sum_{\substack{s=1 \\ t+s=p+2-i}}^{p} u_s v_t,$$

which, once again, is a symmetric function of $\mathbf{u}$ and $\mathbf{v}$, and will be completely canceled by its counterpart in $-G(\mathbf{v})(H(\mathbf{u}) + L(\mathbf{u}))$. The zero structure of $\tilde{W}$ clearly indicates that (5.29) is true. □

The typical structure of $\tilde{W}$, particularly the pattern of zeros registered in the above proof, is illustrated in the example below where $n = 5$ and $p = 3$:

$$\tilde{W} = \begin{bmatrix} 0 & u_1 & u_2 & u_3 & u_2 & u_1 \\ 0 & u_2 & u_3 + u_1 & u_2 & u_1 & 0 \\ 0 & u_3 & 2u_2 & 2u_1 & 0 & 0 \\ 0 & 0 & 0 & 0 & 0 & 0 \\ 2v_2u_1 + v_3u_2 & -2v_1u_2 - v_2u_3 & 2v_3u_1 - 2v_1u_3 & 0 & 0 & 0 \\ v_3u_1 & -v_1u_3 & 2v_2u_1 - 2v_1u_2 & 0 & 0 & 0 \end{bmatrix}. \quad (5.32)$$

Let W denote the lower-left $(p-1) \times p$ submatrix of $\tilde{W}$. That is, let W be the matrix obtained by deleting the first row and the last column of the matrix

$$[G(\mathbf{u})h(\mathbf{v}), -G(\mathbf{v})h(\mathbf{u}), -G(\mathbf{v})(H(\mathbf{u}) + L(\mathbf{u})) + G(\mathbf{u})(H(\mathbf{v}) + L(\mathbf{v}))]. \quad (5.33)$$

The rank of $\tilde{W}$ can be less than n if and only if W is rank deficient, which will be true if and only if values of u_i and v_i are such that $\det(WW^\top) = 0$. We note that $\det(WW^\top)$ is a polynomial in the independent variables u_i and v_i. It can easily be checked that $\det(WW^\top)$ cannot be identically zero for all $\mathbf{u}$ and $\mathbf{v}$. Thus $\text{rank}(\tilde{\text{W}}) < \text{n}$ if and only if u_i and v_i come from a codimension-one surface. We conclude, therefore, that for almost all $\mathbf{u}$ and $\mathbf{v}$ satisfying $\mathbf{u}^\top\mathbf{v} = 0$ the matrix $\tilde{W}$ and, hence, the matrix $\tilde{M}$ is of rank n. In this case, $\ker(\tilde{M})$ is a one-dimensional subspace. Once a null vector $\mathbf{x}$ is found, by Theorem 5.2, the Toeplitz matrix can be constructed.

In summary, we have proved the following theorem.

Theorem 5.5. Suppose that n is odd and that $\mathbf{u}$ and $\mathbf{v}$ are two symmetric vectors satisfying $u^\top \mathbf{v} = 0$. Then

(i) The dimension of $\mathcal{R}(\mathbf{u}) \cap \mathcal{R}(\mathbf{v})$ is at least two.
(ii) For almost all $\mathbf{u}$ and $\mathbf{v}$, the dimension of $\mathcal{R}(\mathbf{u}) \cap \mathcal{R}(\mathbf{v})$ is exactly two.
(iii) For almost all $\mathbf{u}$ and $\mathbf{v}$ and for any values of λ_1 and λ_2, there exists a unique symmetric Toeplitz matrix T satisfying $T\mathbf{u} = \lambda_1 \mathbf{u}$ and $T\mathbf{v} = \lambda_2 \mathbf{v}$.

Case 2: n is odd and both eigenvectors are odd When n is odd, the Hankel matrix $\tilde{H}(\mathbf{v})$ for an odd vector $\mathbf{v}$ takes the form:

$$\tilde{H}(\mathbf{v}) = \begin{bmatrix} v_1 & v_2 & \dots & v_{p-1} & 0 \\ v_2 & v_3 & \dots & 0 & -v_{p-1} \\ \vdots & & & & \vdots \\ v_{p-1} & 0 & \dots & -v_3 & -v_2 \\ 0 & -v_{p-1} & \dots & -v_2 & -v_1 \end{bmatrix}. \tag{5.34}$$

The corresponding $U(\mathbf{v})$ becomes

$$U(\mathbf{v}) = -\Xi L(\mathbf{v}) = \begin{bmatrix} -v_{p-1} & -v_{p-2} & \dots & -v_2 & -v_1 \\ -v_{p-2} & & \dots & -v_1 & 0 \\ \vdots & & & & \vdots \\ -v_1 & 0 & \dots & & 0 \\ 0 & 0 & \dots & & 0 \end{bmatrix}. \tag{5.35}$$

It follows that the last row of $N(\mathbf{v})$ is identically zero. For the equation (5.18), it is now obvious that $\ker(\tilde{M})$ is of dimension at least two.

Lemma 5.6. Suppose that n is odd and that the skew-symmetric vectors $\mathbf{u}$ and $\mathbf{v}$ are perpendicular. Then

$$p \leq \operatorname{rank}(\tilde{\mathrm{W}}) \leq n - 2. \tag{5.36}$$

Indeed, for almost all $\mathbf{u}$ and $\mathbf{v}$, $\operatorname{rank}(\tilde{\mathrm{W}}) = n - 2$.

Proof The proof is very similar to that of Lemma 5.4. So we simply outline a recipe for constructing the transformation matrix that does the elimination. The details of the justification are omitted.

It suffices to consider the $2(p-1) \times (n+1)$ submatrix $\hat{M}$ obtained by deleting the p-th and the $2p$-th rows of $\tilde{M}$. For an odd vector $\mathbf{v}$, define the $(p-1) \times (p-1)$

matrix $G(\mathbf{v})$ by

$$G(\mathbf{v}) := \begin{bmatrix} -v_1 & -v_2 & \dots & -v_{p-2} & -v_{p-1} \\ 0 & v_1 & & v_{p-3} & v_{p-2} \\ \vdots & 0 & \ddots & \vdots & \vdots \\ 0 & & & v_1 & v_2 \\ 0 & 0 & 0 & 0 & v_1 \end{bmatrix}. \tag{5.37}$$

Then construct the $2(p-1) \times 2(p-1)$ transformation matrix $\tilde{G}(\mathbf{u}, \mathbf{v})$ in the same way as is defined in (5.26). It can be proved now that the p-th row of the product $\hat{W} := \tilde{G}\hat{M}$ is identically zero. Furthermore, the lower-right $(p-1) \times (p-1)$ submatrix of $\hat{W}$ is also identically zero. The assertion follows from these observations. □

As an example, for $n = 5$, the matrix $\hat{M}$ takes the form

$$\hat{M} = \begin{bmatrix} 0 & u_1 & u_2 & 0 & -u_2 & -u_1 \\ 0 & u_2 & u_1 & -u_2 & -u_1 & 0 \\ v_1 & 0 & v_2 & 0 & -v_2 & -v_1 \\ v_2 & 0 & v_1 & -v_2 & -v_1 & 0 \end{bmatrix}, \tag{5.38}$$

and after the transformation the matrix $\hat{W}$ looks like

$$\hat{W} = \begin{bmatrix} 0 & u_1 & u_2 & 0 & -u_2 & -u_1 \\ 0 & u_2 & u_1 & -u_2 & -u_1 & 0 \\ -v_1u_1 - v_2u_2 & v_1u_1 + v_2u_2 & 0 & 0 & 0 & 0 \\ u_1v_2 & -v_1u_2 & 0 & v_1u_2 - u_1v_2 & 0 & 0 \end{bmatrix}. \tag{5.39}$$

The third row of $\hat{W}$ is identically zero because $u^\top \mathbf{v} = 0$.

We conclude this case by the following theorem.

Theorem 5.7. Suppose that $n \geq 5$ is odd and that $\mathbf{u}$ and $\mathbf{v}$ are two odd vectors satisfying $\mathbf{u}^\top \mathbf{v} = 0$. Then

(i) The dimension of $\mathcal{R}(\mathbf{u}) \cap \mathcal{R}(\mathbf{v})$ is at least four.
(ii) For almost all $\mathbf{u}$ and $\mathbf{v}$, the dimension of $\mathcal{R}(\mathbf{u}) \cap \mathcal{R}(\mathbf{v})$ is exactly four.
(iii) For almost all $\mathbf{u}$ and $\mathbf{v}$ and for any values of λ_1 and λ_2, the symmetric Toeplitz matrices T satisfying $T\mathbf{u} = \lambda_1\mathbf{u}$ and $T\mathbf{v} = \lambda_2\mathbf{v}$ form a two-dimensional manifold.

Case 3: n is odd, one eigenvector is even and the other odd A symmetric vector is always orthogonal to a skew-symmetric vector regardless of what the values of the components are. Thus, unlike the previous two cases, the orthogonality condition $\mathbf{u}^\top \mathbf{v} = 0$ no longer helps to reduce the rank of $\tilde{M}$. Since $\tilde{M}$ does contain an identically zero row, we should have the same conclusion as in Theorem 5.5.

Theorem 5.8. Suppose that n is odd and that $\mathbf{u}$ and $\mathbf{v}$ are even and odd vectors, respectively. Then

(i) The dimension of $\mathcal{R}(\mathbf{u}) \cap \mathcal{R}(\mathbf{v})$ is at least two.
(ii) For almost all $\mathbf{u}$ and $\mathbf{v}$, the dimension of $\mathcal{R}(\mathbf{u}) \cap \mathcal{R}(\mathbf{v})$ is exactly two.
(iii) For almost all $\mathbf{u}$ and $\mathbf{v}$ and for any values of λ_1 and λ_2, there exists a unique symmetric Toeplitz matrix T satisfying $T\mathbf{u} = \lambda_1\mathbf{u}$ and $T\mathbf{v} = \lambda_2\mathbf{v}$.

Case 4: n is even and both eigenvectors are even When n is even, the Hankel matrix $\tilde{H}(\mathbf{v})$ for an even vector $\mathbf{v}$ assumes the form:

$$\tilde{H}(\mathbf{v}) = \begin{bmatrix} v_1 & v_2 & \cdots & v_{p-1} & v_p \\ v_2 & v_3 & & v_p & v_p \\ \vdots & & & & \vdots \\ v_{p-1} & v_p & \cdots & v_4 & v_3 \\ v_p & v_p & \cdots & v_3 & v_2 \end{bmatrix}. \tag{5.40}$$

Once again, we define

$$G(\mathbf{v}) := \begin{bmatrix} v_1 & v_2 & \cdots & v_{p-1} & v_p \\ 0 & v_1 & & v_{p-2} & v_{p-1} \\ \vdots & 0 & \ddots & \vdots & \vdots \\ 0 & & & v_1 & v_2 \\ 0 & 0 & 0 & 0 & v_1 \end{bmatrix} \tag{5.41}$$

and construct $\tilde{G}(\mathbf{u}, \mathbf{v})$ according to (5.26). If the two symmetric vectors $\mathbf{u}$ and $\mathbf{v}$ are orthogonal, then it can be shown that the rank of the $n \times (n+1)$ matrix $\tilde{W} := \tilde{G}\tilde{M}$ satisfies

$$p + 1 \leq \operatorname{rank}(\tilde{W}) \leq n - 1 \tag{5.42}$$

and for almost all $\mathbf{u}$ and $\mathbf{v}$, the dimension of $\ker(\tilde{M})$ is exactly two.

Theorem 5.9. Suppose that n is even and that $\mathbf{u}$ and $\mathbf{v}$ are two even vectors satisfying $\mathbf{u}^\top\mathbf{v} = 0$. Then

(i) The dimension of $\mathcal{R}(\mathbf{u}) \cap \mathcal{R}(\mathbf{v})$ is at least three.
(ii) For almost all $\mathbf{u}$ and $\mathbf{v}$, the dimension of $\mathcal{R}(\mathbf{u}) \cap \mathcal{R}(\mathbf{v})$ is exactly three.
(iii) For almost all $\mathbf{u}$ and $\mathbf{v}$ and for any values of λ_1 and λ_2, the symmetric Toeplitz matrices T satisfying $T\mathbf{u} = \lambda_1\mathbf{u}$ and $T\mathbf{v} = \lambda_2\mathbf{v}$ form a one-dimensional manifold.

Case 5: n is even and both eigenvectors are odd When n is even, the Hankel matrix $\tilde{H}(\mathbf{v})$ for an odd vector $\mathbf{v}$ assumes the form:

$$\tilde{H}(\mathbf{v}) = \begin{bmatrix} v_1 & v_2 & \dots & v_{p-1} & v_p \\ v_2 & v_3 & \dots & v_p & -v_p \\ \vdots & & & & \vdots \\ v_{p-1} & v_p & \dots & -v_4 & -v_3 \\ v_p & -v_p & \dots & -v_3 & -v_2 \end{bmatrix}. \tag{5.43}$$

The transformation matrix $\tilde{G}(\mathbf{u}, \mathbf{v})$ is constructed with $G(\mathbf{v})$ defined exactly in the same way as (5.41).

Theorem 5.10. Suppose that n is even and that $\mathbf{u}$ and $\mathbf{v}$ are two skew-symmetric vectors satisfying $\mathbf{u}^\top \mathbf{v} = 0$. Then

(i) The dimension of $\mathcal{R}(\mathbf{u}) \cap \mathcal{R}(\mathbf{v})$ is at least three.
(ii) For almost all $\mathbf{u}$ and $\mathbf{v}$, the dimension of $\mathcal{R}(\mathbf{u}) \cap \mathcal{R}(\mathbf{v})$ is exactly three.
(iii) For almost all $\mathbf{u}$ and $\mathbf{v}$ and for any values of λ_1 and λ_2, the symmetric Toeplitz matrices T satisfying $T\mathbf{u} = \lambda_1\mathbf{u}$ and $T\mathbf{v} = \lambda_2\mathbf{v}$ form a one-dimensional manifold.

Case 6: n is even, one eigenvector is even and the other odd Just like Case 3, the orthogonality condition does not help to reduce the rank of $\tilde{M}$. The $n \times (n+1)$ matrix $\tilde{M}$ in general is of full rank. The conclusion, therefore, is similar to Theorem 5.8:

Theorem 5.11. Suppose that n is even and that $\mathbf{u}$ and $\mathbf{v}$ are even and odd vectors, respectively. Then

(i) The dimension of $\mathcal{R}(\mathbf{u}) \cap \mathcal{R}(\mathbf{v})$ is at least two.
(ii) For almost all $\mathbf{u}$ and $\mathbf{v}$, the dimension of $\mathcal{R}(\mathbf{u}) \cap \mathcal{R}(\mathbf{v})$ is exactly two.
(iii) For almost all $\mathbf{u}$ and $\mathbf{v}$ and for any values of λ_1 and λ_2, there exists a unique symmetric Toeplitz matrix T satisfying $T\mathbf{u} = \lambda_1\mathbf{u}$ and $T\mathbf{v} = \lambda_2\mathbf{v}$.

The results for the above six cases are summarized in Table 5.1. The symbolic ratio p/q merely indicates the number $p = \dim(\mathcal{R}(\mathbf{u}) \cap \mathcal{R}(\mathbf{v}))$ versus the dimension q of the affine subspace of solutions to the PDIEP1 with $k = 2$. Thus, $q = 0$ means that the solution to the PEIEP1 with two prescribed eigenpairs is unique.

Table 5.1. *Solution counts for PDIEP1*

	Even	Odd
even	2/0 (3/1)	2/0 (2/0)
odd		4/2 (3/1)

The cases of odd–even or even–odd eigenvectors are the same by swapping the roles of $\mathbf{u}$ and $\mathbf{v}$, if necessary. The ratio outside the parentheses is the result for the case of odd n and that inside is for even n.

5.3 PDIEP for quadratic pencils

We have noted earlier in Section 2.3.2 that the dynamical behavior of a classical vibrating system (2.9) can be understood from the associated algebraic equation (2.10) referred to as a quadratic eigenvalue problem (**QEP**). The latter task concerns, given $n \times n$ complex matrices M, C and K, finding scalars λ and nonzero vectors $\mathbf{x}$ satisfying

$$Q(\lambda)\mathbf{x} = 0, \tag{5.44}$$

where

$$Q(\lambda) := Q(\lambda; M, C, K) = \lambda^2 M + \lambda C + K, \tag{5.45}$$

is called a *quadratic pencil*. The scalars λ and the corresponding vectors $\mathbf{x}$ are called, respectively, eigenvalues and eigenvectors of the quadratic pencil $Q(\lambda)$. Together, $(\lambda, \mathbf{x})$ is called an *eigenpair* of $Q(\lambda)$. The QEP has received much attention because its formation has repeatedly arisen in many different disciplines, including applied mechanics, electrical oscillation, vibro-acoustics, fluid mechanics, and signal processing. In a recent treatise, Tisseur and Meerbergen (2001) surveyed a good many applications, mathematical properties, and a variety of numerical techniques for the QEP. The book by Gohberg et al. (1982, Chapters 10 and 13) also offers a careful theoretical treatment under minimal assumptions. It is known that the QEP has $2n$ finite eigenvalues over the complex field, provided that the leading matrix coefficient M is nonsingular. The QEP arising in practice often entails some additional conditions on the matrices. For example, if M, C, and K represent the mass, damping, and stiffness matrices, respectively, in a mass–spring system, then it is required that all matrices be real-valued and symmetric, and that M and K be positive definite and semidefinite, respectively. It is this class of constraints on the matrix coefficients in (5.45) that underlines our main contribution in this section. Following the theme of this discourse, we are concerned with the inverse problem of the QEP.

The term *quadratic inverse eigenvalue problem* (**QIEP**) adopted in the literature usually is for general matrix coefficients. In this section we shall use it exclusively to stress the additional structure imposed upon the matrix coefficients. Two scenarios will be considered separately:

- Determine real, symmetric matrix coefficients M, C, and K with M positive definite and K positive semidefinite so that the resulting QEP has a prescribed set of k eigenpairs.
- Assuming that the symmetric and positive definite leading matrix coefficient M is known and fixed, determine real and symmetric matrix

coefficients C and K so that the resulting QEP has a prescribed set of k eigenpairs.

Other types of QIEPs have been studied under modified conditions. For instance, the QIEP studied by Ram and Ehlay (Ram, 1995) is for symmetric tridiagonal coefficients and, instead of prescribed eigenpairs, two sets of eigenvalues are given. In a series of articles, Starek and Inman (2001) studied the QIEPs associated with nonproportional underdamped systems. Recently Lancaster and Prells (2003) considered the inverse problem where all eigenvalues are simple and non-real and derived conditions under which the QIEP is solvable, particularly for symmetric and positive semidefinite damping C, with *complete* information on eigenvalues and eigenvectors. Settings for some other mechanical applications can be found at the web site (Ram, 2003). Our study in the following presentation stems from the speculation that the notion of the QIEP has the potential of leading to an important modification tool for model updating (Datta, 2002), model tuning, and model correction (Carvalho et al., 2001; Ferng et al., 2001; Sivan and Y.M. Ram, 1999; Zimmerman and Widengren, 1990), when compared with an analytical model. It might be appropriate to attribute the first technique for solving the inverse problem of QEP to a short exposition in the book (Gohberg et al., 1982, page 173). Unfortunately, the method derived from that discussion is not capable of producing symmetric C and K.

We note that in several recent works, including those by Chu and Datta (1996), Nichols and Kautsky (2001b), as well as Datta et al. (2000) and Datta and Sarkissian (2001), studies are undertaken toward a feedback design problem for a second-order control system. That consideration eventually leads to either a full or a partial eigenstructure assignment problem for the QEP. See, for example, Problem 2.6 outlined in Section 2.3.2. The proportional and derivative state feedback controller designated in these studies is capable of assigning specific eigenvalues and, additionally, making the resulting system insensitive to perturbations. Nonetheless, these results still cannot meet the basic requirement that the quadratic pencil be symmetric.

We have already explained in Section 5.1 that it is often impossible to obtain the entire spectral information in a large or complicated physical system. We further point out that quantities related to high frequency terms generally are susceptible to measurement errors due to the finite bandwidth of measuring devices. Spectral information, therefore, should not be used at its full extent. For these reasons, it might be more sensible to consider a QIEP where only a *portion* of the eigenvalues and eigenvectors is prescribed. Again, a critical question to ask is how much eigeninformation is needed to ensure that a QIEP is solvable. We mention in passing that in setting forth their work of solving a QIEP with complete eigeninformation, Lancaster and Prells (2003) argued that "dominant eigenvalues and modes could be assigned as indicated by experiment, and sub-dominant data can be assigned in a physically plausible but otherwise arbitrary fashion." What is not clear, and would be a very interesting research topic in its own right, is how this arbitrary assignment of high frequency terms would affect the final answer. More fundamentally, can the remaining eigenstructure be arbitrarily assigned? The subsequent discussion, first considered by Chu et al. (2004), offers some useful insights on this issue.

To facilitate the discussion, we shall describe the partial eigeninformation via the pair $(\Lambda, X) \in \mathbb{R}^{k\times k} \times \mathbb{R}^{n\times k}$ of matrices where

$$\Lambda = \operatorname{diag}\{\lambda_1^{[2]}, \ldots, \lambda_\ell^{[2]}, \lambda_{2\ell+1}, \ldots, \lambda_k\} \tag{5.46}$$

with

$$\lambda_j^{[2]} = \begin{bmatrix} \alpha_j & \beta_j \\ -\beta_j & \alpha_j \end{bmatrix} \in \mathbb{R}^{2\times 2}, \quad \beta_j \neq 0, \quad \text{for } j = 1, \ldots, \ell, \tag{5.47}$$

and

$$X = [\mathbf{x}_{1R}, \mathbf{x}_{1I}, \ldots, \mathbf{x}_{\ell R}, \mathbf{x}_{\ell I}, \mathbf{x}_{2\ell+1}, \ldots, \mathbf{x}_k]. \tag{5.48}$$

The true eigenvalues and eigenvectors are readily identifiable via the transformation

$$R := \operatorname{diag}\left\{\frac{1}{\sqrt{2}}\begin{bmatrix} 1 & 1 \\ i & -i \end{bmatrix}, \ldots, \frac{1}{\sqrt{2}}\begin{bmatrix} 1 & 1 \\ i & -i \end{bmatrix}, I_{k-2\ell}\right\} \tag{5.49}$$

with $i = \sqrt{-1}$. In other words, by defining

$$\tilde{\Lambda} = R^H \Lambda R = \operatorname{diag}\{\lambda_1, \lambda_2, \ldots, \lambda_{2\ell-1}, \lambda_{2\ell}, \lambda_{2\ell+1}, \ldots, \lambda_k\} \in \mathbb{C}^{k\times k}, \tag{5.50}$$

$$\tilde{X} = XR = [\mathbf{x}_1, \mathbf{x}_2, \ldots, \mathbf{x}_{2\ell-1}, \mathbf{x}_{2\ell}, \mathbf{x}_{2\ell+1}, \ldots, \mathbf{x}_k] \in \mathbb{C}^{n\times k}, \tag{5.51}$$

respectively, the QIEP is concerned with finding a *real-valued* quadratic pencil $Q(\lambda)$ (with its matrix coefficients possessing a certain specified structure) so that $Q(\lambda_j)\mathbf{x}_j = 0$ for all $j = 1, \ldots, k$. The true (complex-valued) eigenvalues and eigenvectors of the desired quadratic pencil $Q(\lambda)$ can be induced from the pair (Λ, X) of real matrices. In this case, note that $\mathbf{x}_{2j-1} = \mathbf{x}_{jR} + i\mathbf{x}_{jI}$, $\mathbf{x}_{2j} = \mathbf{x}_{jR} - i\mathbf{x}_{jI}$, $\lambda_{2j-1} = \alpha_j + i\beta_j$, and $\lambda_{2j} = \alpha_j - i\beta_j$ for $j = 1, \ldots, \ell$, whereas $\mathbf{x}_j$ and λ_j are all real-valued for $j = 2\ell + 1, \ldots, k$. For convenience, we shall denote henceforth the set of diagonal elements of $\tilde{\Lambda}$, which is precisely the spectrum of Λ, by $\sigma(\Lambda)$. We shall call (Λ, X) an *eigeninformation pair* of the quadratic pencil $Q(\lambda)$.

We shall consider two types of partially described QIEPs. The standard quadratic inverse eigenvalue problem (**SQIEP**) concerns the identification of the entire system parameters, that is, all three matrix coefficients, so as to satisfy the partially described spectral information.

Problem 5.3 *(SQIEP)*
Given an eigeninformation pair (Λ, X), find real and symmetric matrices M, C and K with M and K positive definite and semidefinite, respectively, so that the equation

$$MX\Lambda^2 + CX\Lambda + KX = 0, \tag{5.52}$$

is satisfied.

The surprisingly harder monic quadratic inverse eigenvalue problem (**MQIEP**) is where the leading coefficient M is positive definite and fixed and only the matrix coefficients C and K are to be determined. Since M is known, let $M = LL^\top$ denotes the Cholesky decomposition of M. Then

$$Q(\lambda)\mathbf{x} = 0 \;\Leftrightarrow\; \widetilde{Q}(\lambda)(L^\top \mathbf{x}) = 0, \tag{5.53}$$

where

$$\widetilde{Q}(\lambda) := \lambda^2 I_n + \lambda L^{-1}CL^{-\top} + L^{-1}KL^{-\top}. \tag{5.54}$$

Thus, without loss of generality, we may assume that the given matrix M in the MQIEP is the $n \times n$ identity matrix I_n to begin with.

Problem 5.4 *(MQIEP)*
Given an eigeninformation pair (Λ, X), find real and symmetric matrices C and K that satisfy the equation

$$X\Lambda^2 + CX\Lambda + KX = 0. \tag{5.55}$$

At first glance, one might easily mistake that the SQIEP is a special case of the MQIEP. It might be fitting, before we move into further details, to first highlight some fundamental differences between the SQIEP and the MQIEP.

1. In the MQIEP, it suffices to consider the monic quadratic pencil (5.55) for the more general case where the leading matrix coefficient M is positive definite and *fixed*. This is not the case with the SQIEP. The leading matrix coefficient M in the SQIEP is part of the unknowns to be determined.
2. The MQIEP requires only symmetry and nothing else of the two matrix coefficients C and K. The symmetry of C and K implies that there are in total $n(n+1)$ unknowns to be determined in the inverse problem. Since each eigenpair $(\lambda, \mathbf{x})$ characterizes a system of n equations, it is natural to conjecture that a monic quadratic pencil could be determined from any given $n+1$ eigenpairs that are closed under complex conjugation. We shall substantiate this conjecture by proving constructively that, after

a necessary condition is satisfied, the solution for the MQIEP is in fact unique.

3. In contrast, the positive definiteness imposed on the SQIEP is much more complicated than a mere count of the numbers of the unknowns and equations. It is an important discovery that the amount of eigeninformation cannot contain more than n eigenpairs. Given any $k \leq n$ distinct eigenvalues and linearly independent eigenvectors closed under complex conjugation, the SQIEP is always solvable but the solution often is not unique. Furthermore, contrary to what Lancaster and Prells (2003) has proposed, the remaining unspecified eigenstructure of the reconstructed quadratic pencil is in fact quite limited. In particular, at the upper end when $k = n$, that is, when the number of prescribed eigenpairs is equal to the dimension of the ambient space, every prescribed eigenvalue is a double eigenvalue and the remaining eigenstructure is completely fixed.
4. Though both problems are solved by constructive proofs, the mathematical techniques employed to derive the main results for the two problems are indispensably different. It appears counter to intuition that the MQIEP is much harder to analyze than the SQIEP.

The materials in this section are adopted from our work (Chu et al., 2003). We think the progress made is significant in four fronts. First, we give a recipe for the construction of a solution to each of the two inverse problems. Specifically, the symmetry is kept in the solution. These recipes can be turned into numerical algorithms. Secondly, we specify necessary and sufficient conditions under which the inverse problem is solvable. Thirdly, we completely characterize the eigenstructure of the reconstructed quadratic pencil. Specifically, we understand what the remaining eigenstructure should look like. Finally, we propose a way to refine the construction process so that the best approximation subject to some additional optimal conditions can be established.

5.3.1 *Recipe of construction*

In this section, we present a general theory elucidating how the SQIEP could be solved with the prescribed spectral information (Λ, X). Our proof is constructive. As a by-product, numerical algorithms can also be developed thence. We shall assume henceforth, in the formulation of an SQIEP, that the given spectral information (Λ, X) is always in the form of (5.46) and (5.48).

Basic formula Starting with the given pair of matrices (Λ, X), consider the null space $\mathcal{N}(\Omega)$ of the augmented matrix

$$\Omega := \begin{bmatrix} X^\top & \Lambda^\top X^\top \end{bmatrix} \in \mathbb{R}^{k \times 2n}.$$

Denote the dimension of $\mathcal{N}(\Omega)$ by m. If X has linearly independently columns (as we will assume later), then $m = 2n - k$. Note that $m \geq n$, if we have assumed

$k \leq n$ (for the reason to be seen later) in the formulation of the SQIEP. Let the columns of the matrix

$$\begin{bmatrix} U^\top \\ V^\top \end{bmatrix} \in \mathbb{R}^{2n \times m}$$

with $U^\top, V^\top \in \mathbb{R}^{n \times m}$ denote *any* basis of the subspace $\mathcal{N}(\Omega)$. We shall see later that the choices of U and V make a difference in the properties of a solution. The equation

$$\begin{bmatrix} X^\top & \Lambda^\top X^\top \end{bmatrix} \begin{bmatrix} U^\top \\ V^\top \end{bmatrix} = 0, \tag{5.56}$$

holds. Define the quadratic pencil $Q(\lambda)$ by the matrix coefficients

$$M = V^\top V, \tag{5.57}$$

$$C = V^\top U + U^\top V, \tag{5.58}$$

$$K = U^\top U. \tag{5.59}$$

We claim that the above definitions are sufficient for constructing a solution to the SQIEP. The theory will be established in several steps.

Theorem 5.12. Given any pair of matrices (Λ, X) in the form of (5.46) and (5.48), let U and V be an arbitrary solution to the equation (5.56). Then (Λ, X) is an eigenpair of the quadratic pencil $Q(\lambda)$ with matrix coefficients M, C, and K defined according to (5.57), (5.58), and (5.59), respectively.

Proof Upon substitution, we see that

$$\begin{aligned} MX\Lambda^2 + CX\Lambda + KX &= V^\top VX\Lambda^2 + \left(V^\top U + U^\top V\right) X\Lambda + \left(U^\top U\right) X \\ &= V^\top \left(VX\Lambda + UX\right)\Lambda + U^\top \left(VX\Lambda + UX\right) \\ &= 0. \end{aligned}$$

The last equality is due to the properties of U and V in (5.56). □

By this construction, all matrix coefficients in $Q(\lambda)$ are obviously real and symmetric. Note also that both matrices M and K are positive semidefinite. However, it is not clear whether $Q(\lambda)$ is a trivial quadratic pencil. Toward that end, we claim that the assumption that X has full column rank is sufficient and necessary for the regularity of $Q(\lambda)$.

Regularity

Theorem 5.13. The leading matrix coefficient $M = V^\top V$ is nonsingular, provided that X has full column rank. In this case, the resulting quadratic pencil $Q(\lambda)$ is regular, that is, $\det(Q(\lambda))$ is not identically zero.

Proof Suppose that $V^\top \in \mathbb{R}^{n\times m}$ is not of full row rank. There exists an orthogonal matrix $G \in \mathbb{R}^{m\times m}$ such that

$$V^\top G = \begin{bmatrix} V_1^\top & 0_{n\times m_2} \end{bmatrix},$$

where $V_1^\top \in \mathbb{R}^{n\times m_1}$ and $0_{n\times m_2}$ denotes the zero matrix of size $n \times m_2$. Note that $m_1 < n$ and $m_2 = m - m_1$. Post-multiply the same G to $U^\top$ and partition the product into

$$U^\top G = \begin{bmatrix} U_1^\top & U_2^\top \end{bmatrix},$$

where $U_1^\top \in \mathbb{R}^{n\times m_1}$ and $U_2^\top \in \mathbb{R}^{n\times m_2}$. Note that $m_2 > m - n$. On the other hand, we see from (5.56) that

$$X^\top U_2^\top = 0,$$

whereas the columns of $U_2^\top$ are necessarily linearly independent by construction. It follows that

$$n - k \geq m_2 > m - n,$$

which contradicts the fact that $m = 2n - k$. Thus, the matrix $V^\top$ must be of full row rank and then $M = V^\top V$ is nonsingular. □

Theorem 5.14. Suppose in a given pair of matrices (Λ, X) that all eigenvalues in Λ are distinct and that X is not of full column rank. Then the quadratic pencil $Q(\lambda)$ defined by (5.57), (5.58), and (5.59) is singular.

Proof It is easy to check that equation (5.56) remains true if Λ and X are replaced by $\tilde{\Lambda}$ and $\tilde{X}$ defined in (5.50) and (5.51), respectively. Let μ be an arbitrary complex number not in $\sigma(\Lambda)$. From the obvious fact that

$$\begin{bmatrix} \tilde{X}^\top & \tilde{\Lambda}^\top \tilde{X}^\top \end{bmatrix} \begin{bmatrix} I & -\mu I \\ 0 & I \end{bmatrix} \begin{bmatrix} I & \mu I \\ 0 & I \end{bmatrix} \begin{bmatrix} U^\top \\ V^\top \end{bmatrix} = 0,$$

it follows that

$$\begin{bmatrix} \tilde{X}^\top & (\tilde{\Lambda}^\top - \mu I)\tilde{X}^\top \end{bmatrix} \begin{bmatrix} \mu V^\top + U^\top \\ V^\top \end{bmatrix} = 0.$$

By assumption, $\tilde{X}$ is not of full column rank. We may therefore assume that for some $2 \leq q \leq k$, we can express the combination

$$\tilde{\mathbf{x}}_q = \sum_{j=1}^{q-1} r_j \tilde{\mathbf{x}}_j,$$

and that not all r_j, $j = 1, \ldots, q-1$, are zero. Define

$$\Gamma := \begin{bmatrix} 1 & & \gamma_{1,q} & & 0 \\ & \ddots & \vdots & & \\ & & \ddots \;\; \gamma_{q-1,q} & & \\ & & 1 & & \\ & & & \ddots & \\ 0 & & & & 1 \end{bmatrix} \in \mathbb{C}^{k\times k}$$

with

$$\gamma_{j,q} = -\frac{\lambda_q - \mu}{\lambda_j - \mu} r_j, \quad j = 1, \ldots, q-1.$$

Clearly,

$$\Gamma^\top \begin{bmatrix} \tilde{X}^\top & (\tilde{\Lambda}^\top - \mu I)\tilde{X}^\top \end{bmatrix} \begin{bmatrix} \mu V^\top + U^\top \\ V^\top \end{bmatrix} = 0. \tag{5.60}$$

Notice that, by construction, the q-th row of $\Gamma^\top(\tilde{\Lambda}^\top - \mu I)\tilde{X}^\top$ is zero. Let $\mathbf{y}(\mu)^\top$ denote the q-th row of $\Gamma^\top \tilde{X}^\top$, which cannot be identically zero because the spectrum of Λ are distinct. We thus see from (5.60) that $\mathbf{y}(\mu)^\top(\mu V^\top + U^\top) = 0$. It follows that for every $\mu \in \mathbb{C}$, we can always find a nontrivial row vector $\mathbf{y}(\mu)^\top$ such that $\mathbf{y}(\mu)^\top Q(\mu) = 0$. Since $\mu \in \mathbb{C}$ is arbitrary, the constructed pencil $Q(\lambda)$ must be singular. □

Rank condition We conclude this section with one important remark. The rank condition $k = n$ plays a pivotal role in SQIEP. It is the critical value for the regularity of the quadratic pencil $Q(\lambda)$ defined by the matrix coefficients (5.57), (5.58), and (5.59). In fact, it is clear now that corresponding to any given $\Lambda \in \mathbb{R}^{k\times k}$, $X \in \mathbb{R}^{n\times k}$ in the form of (5.46) and (5.48), the quadratic pencil $Q(\lambda)$ can always be factorized into the product

$$Q(\lambda) = (\lambda V^\top + U^\top)(\lambda V + U). \tag{5.61}$$

If $k > n$, then $\operatorname{rank}(\lambda V + U) \leq 2n - k < n$ and hence $\det(Q(\lambda)) \equiv 0$ for all λ. It is for this reason that we always assume that $k \leq n$ in the formulation of an SQIEP.

5.3.2 *Eigenstructure of $Q(\lambda)$*

Now that we know how to define the matrix coefficients so that the corresponding quadratic pencil possesses a prescribed set of k eigenvalues and eigenvectors, a compelling question to ask is what the unspecified eigenpair in the constructed pencil should look like. In this section we examine the remaining eigenstructure of the quadratic pencil $Q(\lambda)$ created from our scheme.

Remaining eigenstructure

Theorem 5.15. Let $(\Lambda, X) \in \mathbb{R}^{k\times k} \times \mathbb{R}^{n\times k}$ in the form of (5.46) and (5.48) denote the partial eigeninformation and $Q(\lambda)$ be the quadratic pencil defined by coefficients (5.57), (5.58), and (5.59). Assume that X has full column rank k.

1. If $k = n$, then $Q(\lambda)$ has double eigenvalue λ_j for each $\lambda_j \in \sigma(\Lambda)$;
2. If $k < n$, then $Q(\lambda)$ has double eigenvalue λ_j for each $\lambda_j \in \sigma(\Lambda)$. The remaining $2(n-k)$ eigenvalues of $Q(\lambda)$ are all complex conjugate with nonzero imaginary parts. In addition, if the matrices U and V in (5.56) are chosen from an orthogonal basis of the null space of Ω, then the remaining $2(n-k)$ eigenvalues are only $\pm i$ with corresponding eigenvectors $\mathbf{z} \pm i\mathbf{z}$ where $X^\top \mathbf{z} = 0$.

Proof The case $k = n$ is easy. The matrices $U^\top$ and $V^\top$ involved in (5.56) forming the null space of Ω are square matrices of size n. We also know from Theorem 5.13 that $V^\top$ is nonsingular. Observe that

$$V^{-1}U = -X\Lambda X^{-1}. \tag{5.62}$$

Using the factorization (5.61), we see that

$$\det(Q(\lambda)) = (\det(\lambda V + U))^2.$$

It is clear that $Q(\lambda)$ has double eigenvalue λ_j at every $\lambda_j \in \sigma(\Lambda)$.

We now consider the case when $k < n$. Since $X^\top \in \mathbb{R}^{k\times n}$ is of full row rank, there exists an orthogonal matrix $P_1 \in \mathbb{R}^{n\times n}$ such that

$$X^\top P_1^\top = \begin{bmatrix} X_{11}^\top & 0_{n\times(n-k)} \end{bmatrix}, \tag{5.63}$$

where $X_{11}^\top \in \mathbb{R}^{k\times k}$ is nonsingular. There also exists an orthogonal matrix $Q_1 \in \mathbb{R}^{m\times m}$ such that

$$P_1 V^\top Q_1 = \begin{bmatrix} V_{11}^\top & 0_{k\times(n-k)} & 0_{k\times(m-n)} \\ V_{21}^\top & \mathcal{A} & 0_{(n-k)\times(m-n)} \end{bmatrix} \in \mathbb{R}^{n\times m}, \tag{5.64}$$

with appropriate sizes for the other three submatrices. In particular, note that both $V_{11}^\top \in \mathbb{R}^{k\times k}$ and $\mathcal{A} \in \mathbb{R}^{(n-k)\times(n-k)}$ are nonsingular matrices, because $V^\top$ is of full row rank by Theorem 5.13. From the fact that

$$\begin{bmatrix} X^\top & \Lambda^\top X^\top \end{bmatrix} \begin{bmatrix} P_1^\top & 0 \\ 0 & P_1^\top \end{bmatrix} \begin{bmatrix} P_1 & 0 \\ 0 & P_1 \end{bmatrix} \begin{bmatrix} U^\top \\ V^\top \end{bmatrix} Q_1 = 0, \tag{5.65}$$

we conclude that the structure of $P_1 U^\top Q_1$ must be of the form

$$P_1 U^\top Q_1 = \begin{bmatrix} U_{11}^\top & 0_{k\times(n-k)} & 0_{k\times(m-n)} \\ U_{21}^\top & \Delta & \mathcal{B} \end{bmatrix} \in \mathbb{R}^{n\times m}, \tag{5.66}$$

where $\mathcal{B} \in \mathbb{R}^{(n-k)\times(n-k)}$ is nonsingular. Because $\begin{bmatrix} U^\top \\ V^\top \end{bmatrix}$ is of full column rank, together with the fact that both $\mathcal{A}$ and $\mathcal{B}$ in (5.64) and (5.66) are nonsingular, it follows that $\begin{bmatrix} U_{11}^\top \\ V_{11}^\top \end{bmatrix}$ must be of full column rank. Note that $U_{11}^\top$ is nonsingular if and only if Λ has no zero eigenvalue. Using $V_{11}^\top$ as a pivot matrix to eliminate $V_{21}^\top$ in (5.64), we may claim that there exits a nonsingular matrix P_2 such that

$$\tilde{U}^\top := P_2 P_1 U^\top Q_1 = \begin{bmatrix} U_{11}^\top & 0 & 0 \\ \tilde{U}_{21}^\top & \Delta & \mathcal{B} \end{bmatrix},$$

$$\tilde{V}^\top := P_2 P_1 V^\top Q_1 = \begin{bmatrix} V_{11}^\top & 0 & 0 \\ 0 & \mathcal{A} & 0 \end{bmatrix}.$$

Compute the three matrices

$$\tilde{M} := \tilde{V}^\top \tilde{V} = \begin{bmatrix} V_{11}^\top V_{11} & 0 \\ 0 & \mathcal{A}\mathcal{A}^\top \end{bmatrix},$$

$$\tilde{C} := \tilde{U}^\top \tilde{V} + \tilde{V}^\top \tilde{U} = \begin{bmatrix} U_{11}^\top V_{11} + V_{11}^\top U_{11} & V_{11}^\top \tilde{U}_{21} \\ \tilde{U}_{21}^\top V_{11} & \mathcal{A}\Delta^\top + \Delta\mathcal{A}^\top \end{bmatrix},$$

$$\tilde{K} := \tilde{U}^\top \tilde{U} = \begin{bmatrix} U_{11}^\top U_{11} & U_{11}^\top \tilde{U}_{21} \\ \tilde{U}_{21}^\top U_{11} & \tilde{U}_{21}^\top \tilde{U}_{21} + \mathcal{B}\mathcal{B}^\top + \Delta\Delta^\top \end{bmatrix},$$

and define the quadratic pencil $\tilde{Q}(\lambda) := \lambda^2 \tilde{M} + \lambda \tilde{C} + \tilde{K}$. By construction, it is clear that $\tilde{Q}(\lambda) = (P_2 P_1) Q(\lambda) (P_2 P_1)^\top$. This congruence relation ensures that $\tilde{Q}(\lambda)$ preserves the same eigenvalue information as $Q(\lambda)$. Define

$$Q_{11}(\lambda) := \lambda^2 (V_{11}^\top V_{11}) + \lambda (V_{11}^\top U_{11} + U_{11}^\top V_{11}) + U_{11}^\top U_{11}, \tag{5.67}$$

$$P_3 := \begin{bmatrix} I & 0 \\ -\tilde{U}_{21}^\top (\lambda V_{11} + U_{11}) Q_{11}^{-1}(\lambda) & I \end{bmatrix}. \tag{5.68}$$

It is further seen that $\tilde{Q}(\lambda)$ can be factorized as

$$P_3 \begin{bmatrix} Q_{11}(\lambda) & 0 \\ 0 & (\lambda\mathcal{A} + \Delta)(\lambda\mathcal{A}^\top + \Delta^\top) + \mathcal{B}\mathcal{B}^\top \end{bmatrix} P_3^\top. \tag{5.69}$$

We thus have effectively decomposed the quadratic pencil $\tilde{Q}(\lambda)$ into two subpencils.

By construction, we see from (5.63), (5.65), and (5.67) that the quadratic subpencil $Q_{11}(\lambda)$ in (5.67) solves exactly the SQIEP with spectral data (Λ, X_{11}). For this problem, we have already proved in the first part that $Q_{11}(\lambda)$ must have double eigenvalue λ_j for each $\lambda_j \in \sigma(\Lambda)$. It only remains to check the eigenvalues for the subpencil $(\mu\mathcal{A} + \Delta)(\mu\mathcal{A}^\top + \Delta^\top) + \mathcal{B}\mathcal{B}^\top$. Recall that the matrix $\mathcal{B}$ in (5.66) is nonsingular. The matrix $(\mu\mathcal{A} + \Delta)(\mu\mathcal{A}^\top + \Delta^\top) + \mathcal{B}\mathcal{B}^\top$ is positive definite for every $\mu \in \mathbb{R}$. In particular, its determinant cannot be zero for any real μ. Therefore, the remaining eigenvalues of $Q(\lambda)$ must be all complex conjugate with nonzero imaginary parts.

If, in addition, the columns of $\begin{bmatrix} U^\top \\ V^\top \end{bmatrix}$ in (5.65) are orthogonal to begin with, then both $\mathcal{A}$ and $\mathcal{B}$ are $(n-k)\times(n-k)$ orthogonal matrices and the submatrix Δ in (5.66) must be a zero matrix. By (5.69), the remaining eigenvalues of $Q(\lambda)$ can only be $\pm i$. Observe further that there exists a nonsingular $W \in \mathbb{R}^{k\times k}$ such that

$$\begin{bmatrix} I & 0 \\ 0 & W \end{bmatrix} \begin{bmatrix} U & V \\ X^\top & \Lambda^\top X^\top \end{bmatrix} \begin{bmatrix} U^\top & X \\ V^\top & X\Lambda \end{bmatrix} \begin{bmatrix} I & 0 \\ 0 & W^\top \end{bmatrix} = \begin{bmatrix} I_{2n-k} & 0 \\ 0 & I_k \end{bmatrix}. \tag{5.70}$$

It follows that

$$\begin{aligned} U^\top U + XW^\top W X^\top &= I_n, \\ U^\top V + XW^\top W \Lambda^\top X^\top &= 0, \\ V^\top U + X\Lambda W^\top W X^\top &= 0, \\ V^\top V + X\Lambda W^\top W \Lambda^\top X^\top &= I_n. \end{aligned} \tag{5.71}$$

For any $\mathbf{z}$ satisfying $X^\top \mathbf{z} = 0$, we see from the above equations that

$$U^\top U\mathbf{z} = \mathbf{z},$$

$$V^\top V\mathbf{z} = \mathbf{z},$$

$$U^\top V\mathbf{z} + V^\top U\mathbf{z} = 0.$$

This show that $Q(\pm i)(\mathbf{z} \pm i\mathbf{z}) = 0$. □

Theorem 5.15 is significant on several fronts. First of all, all given eigenvalues are double roots. Second, if $k = n$, then *all* eigenvalues of $Q(\lambda)$ are completely counted. Third, if $k < n$ and if the basis of the null space $\mathcal{N}(\Omega)$ is selected to be mutually orthogonal (as we normally would do by using, say, MATLAB), then again all eigenvalues of $Q(\lambda)$ are completed determined. In other words, in this construction we are *not* allowed to supplement any additional $n-k$ eigenpairs to simplify this SQIEP. This observation is in sharp contrast to the claim that "sub-dominant (eigen)data can be assigned in a physically plausible but otherwise arbitrary fashion" made in Lancaster and Prells (2003). These seemingly conflicting statements must be understood clearly: It is our special construction that has this limited eigenstructure whereas there might be other ways to solve the SQIEP, such as those in Lancaster and Prells (2003) if special conditions are met.

Our construction thus far is the most natural method for a solution to the SQIEP with $k(< n)$ prescribed pairs of eigenvalues and eigenvectors. In Section 5.3.3, we shall study how a nonorthogonal basis of $\mathcal{N}(\Omega)$ can be exploited to improve the SQIEP approximation.

Geometric multiplicity We can further calculate the geometric multiplicity of the double roots characterized in Theorem 5.16.

Theorem 5.16. Let (Λ, X) in the form of (5.46) and (5.48) denote the prescribed eigenpair of the quadratic pencil $Q(\lambda)$ defined before. Assume that Λ has distinct spectrum and X has full column rank. Then

1. Each real-valued $\lambda_j \in \sigma(\Lambda)$ has an elementary divisor of degree 2, that is, the dimension of the null space $\mathcal{N}(Q(\lambda_j))$ is 1.
2. The dimension of $\mathcal{N}(Q(\lambda_j))$ of a complex-valued eigenvalue $\lambda_j \in \sigma(\Lambda)$ is generically 1. That is, pairs of matrices (Λ, X) of which a complex-valued eigenvalue has a linear elementary divisor forms a set of measure zero.

Proof Real-valued eigenvalues correspond to those $\lambda_j \in \sigma(\Lambda)$ with $j = 2\ell + 1, \ldots, k$. We have already seen in Theorem 5.12 that $Q(\lambda_j)\mathbf{x}_j = 0$, where $\mathbf{x}_j$ is the j-th column of X. Suppose that the $\mathcal{N}(Q(\lambda_j))$ has dimension greater than 1. From (5.61), it must be that

$$\operatorname{rank}\left(\lambda_j V^\top + U^\top\right) \leq n - 2. \tag{5.72}$$

Rewrite (5.56) as

$$\begin{bmatrix} X^\top & \Lambda^\top X^\top \end{bmatrix} \begin{bmatrix} I & -\lambda_j I \\ 0 & I \end{bmatrix} \begin{bmatrix} I & \lambda_j I \\ 0 & I \end{bmatrix} \begin{bmatrix} U^\top \\ V^\top \end{bmatrix} = 0, \tag{5.73}$$

which is equivalent to

$$\begin{bmatrix} X^\top & (\Lambda^\top - \lambda_j I)X^\top \end{bmatrix} \begin{bmatrix} \lambda_j V^\top + U^\top \\ V^\top \end{bmatrix} = 0. \tag{5.74}$$

Note that, since Λ has a distinct spectrum and $X^\top$ has full row-rank,

$$\operatorname{rank}\left((\Lambda^\top - \lambda_j I)X^\top\right) = k - 1,$$

or equivalently,

$$\dim\left(\mathcal{N}\left((\Lambda^\top - \lambda_j I)X^\top\right)\right) = n - k + 1. \tag{5.75}$$

On the other hand, there exists an orthogonal $G_j \in \mathbb{R}^{m\times m}$ such that

$$\begin{bmatrix} \lambda_j V^\top + U^\top \\ V^\top \end{bmatrix} G_j = \begin{bmatrix} U_{j1}^\top & 0 \\ V_{j1}^\top & V_{j2}^\top \end{bmatrix}, \tag{5.76}$$

where, due to (5.72), $V_{j2}^\top$ has at least $m-(n-2) = n-k+2$ linearly independent columns. We then see from (5.74) that

$$(\Lambda^\top - \lambda_j I)X^\top V_{j2}^\top = 0,$$

implying that $\dim\left(\mathcal{N}\left((\Lambda^\top - \lambda_j I)X^\top\right)\right) \geq n - k + 2$. This contradicts (5.75).

To examine the complex-valued case, notice that (5.52) can be rewritten as

$$M(XR)(R^H\Lambda^2 R) + C(XR)(R^H\Lambda R) + KXR = 0,$$

where R is defined in (5.49). In particular from (5.50) and (5.51), for $1 \leq j \leq 2\ell$, we have $Q(\lambda_j)\mathbf{x}_j = 0$.

We first consider the case $k = n$. We observe two facts. First, the matrix V in the basis $\begin{bmatrix} U^\top \\ V^\top \end{bmatrix}$ for the null space $\mathcal{N}\left(\left[X^\top \ \Lambda^\top X^\top\right]\right)$ can be an arbitrary nonsingular matrix. Secondly, if there exists another vector $\mathbf{z} \in \mathbb{C}^n$ independent of $\mathbf{x}_j$ such that $Q(\lambda_j)\mathbf{z} = 0$, we claim that for this kind of eigenvalue the matrix $(V^\top V)^{-1}$ must satisfy some kind of algebraic varieties in $\mathbb{R}^{n\times n}$. Putting these two facts together, we conclude that any complex-valued eigenvalue having a linear elementary divisor must come from a set of measure zero.

To see the claim concerning the algebraic varieties for $(V^\top V)^{-1}$, we use (5.62) and (5.50) to rewrite $\lambda_j V + U$ as

$$\lambda_j V + U = VXR(\lambda_j I - \Lambda)R^H X^{-1},$$

and thus factorize $Q(\lambda_j)$ as

$$\begin{aligned} Q(\lambda_j) &= (\lambda_j V^\top + U^\top)(\lambda_j V + U) \\ &= X^{-\top}\bar{R}(\lambda_j I - \Lambda)R^\top X^\top V^\top VXR(\lambda_j I - \Lambda)R^H X^{-1}. \end{aligned} \tag{5.77}$$

If $Q(\lambda_j)\mathbf{z} = 0$, from (5.77) we have

$$R^\top X^\top V^\top VXR(\lambda_j I - \Lambda)R^H X^{-1}\mathbf{z} = \tau \mathbf{e}_j, \tag{5.78}$$

where $\mathbf{e}_j$ is the j-th standard unit vector and τ is some scalar. Rewrite (5.78) as

$$(\lambda_j I - \Lambda)R^H X^{-1}\mathbf{z} = \tau R^H X^{-1}(V^\top V)^{-1}X^{-\top}\bar{R}\mathbf{e}_j.$$

Hence, a necessary condition for the existence of $\mathbf{z}$ is that $(V^\top V)^{-1}$ must satisfy the algebraic equation

$$\mathbf{e}_j^\top R^H X^{-1}(V^\top V)^{-1}X^{-\top}\bar{R}\mathbf{e}_j = 0. \tag{5.79}$$

We note in passing that the condition (5.79) for $(V^\top V)^{-1}$ is also sufficient since the above argument can be reversed to show the existence of a vector $\mathbf{z}$ in the null space of $Q(\lambda_j)$.

For the case $k < n$, a similar argument holds. Indeed, using the decompositions (5.67) and (5.69) given in Theorem 5.15, a sufficient and necessary condition for the existence of $\mathbf{z}$ is exactly the same as (5.79) where X and V are replaced by X_{11} and V_{11}, respectively.

In either case, we have proved that outside the algebraic variety the elementary divisor of a generically prescribed complex eigenvalue therefore is of degree 2. □

To further demonstrate the subtlety of the second statement in Theorem 5.16, we make an interesting observation as follows.

Example 5.2. Suppose in the given (Λ, X) that X has full column rank and that Λ has a distinct spectrum. Assume further that $\{\pm i\} \subset \sigma(\Lambda)$. Construct the quadratic pencil $Q(\lambda)$ by taking an orthogonal basis $[U, V]^\top$ for the null space

$\mathcal{N}(\Omega)$. Then the dimension of $\mathcal{N}(Q(\pm i))$ is 2. In other words, in this nongeneric case, both eigenvalues $\pm i$ have two linear elementary divisors.

To see that the sufficient condition (5.79) is satisfied in this case, we observe from (5.70) that $W(X^\top X + \Lambda^\top X^\top X\Lambda)W^\top = I_n$. It follows that

$$W^\top W = (X^\top X + \Lambda^\top X^\top X\Lambda)^{-1}.$$

The last equation in (5.71) gives rise to

$$(V^\top V)^{-1} = (I - X\Lambda W^\top W\Lambda^\top X^\top)^{-1}.$$

Upon substitution, it holds that

$$\begin{aligned}
&X^{-1}(V^\top V)^{-1}X^{-\top}\\
&\quad = X^{-1}(I - X\Lambda W^\top W\Lambda^\top X^\top)^{-1}X^{-\top}\\
&\quad = X^{-1}(I - X\Lambda(X^\top X + \Lambda^\top X^\top X\Lambda)^{-1}\Lambda^\top X^\top)^{-1}X^{-\top}\\
&\quad = X^{-1}(I - X\Lambda X^{-1}(I + X^{-\top}\Lambda^\top X^\top X\Lambda X^{-1})^{-1}X^\top\Lambda^\top X^\top)^{-1}X^{-\top}\\
&\quad = X^{-1}(I + X\Lambda X^{-1}X^{-\top}\Lambda^\top X^\top)X^{-\top}\\
&\quad = X^{-1}X^{-\top} + \Lambda X^{-1}X^{-\top}\Lambda^\top,
\end{aligned} \tag{5.80}$$

where the fourth equality is obtained by using the Sherman–Morrison–Woodburg formula. Substituting (5.80) into (5.79) and assuming that j is the index that defines $\lambda_j = \pm i$, we find that

$$\begin{aligned}
&\mathbf{e}_j^\top R^H X^{-1}(V^\top V)^{-1}X^{-\top}\bar{R}\mathbf{e}_j\\
&\quad = \mathbf{e}_j^\top(R^H X^{-1}X^{-\top}\bar{R} + R^H\Lambda R R^H X^{-1}X^{-\top}\bar{R}R^\top\Lambda^\top\bar{R})\mathbf{e}_j\\
&\quad = \mathbf{e}_j^\top(\tilde{X}^{-1}\tilde{X}^{-\top} + \tilde{\Lambda}\tilde{X}^{-1}\tilde{X}^{-\top}\tilde{\Lambda}^\top)\mathbf{e}_j\\
&\quad = 0.
\end{aligned}$$

5.3.3 *Numerical experiment*

In this section we intend to highlight two main points by numerical examples. All calculation are done by using MATLAB in its default (double) precision. For ease of running text, however, we shall report all numerals to five digits only.

We first demonstrate the eigenstructure of a solution to a typical SQIEP. From the discussion in the preceding sections, we already have a pretty good idea about how the eigenstructure should look. Our point here is to illustrate numerically how the selection of U and V might affect the geometric multiplicity of the double eigenvalue.

To fix ideas, we randomly generate $(\Lambda,\ X) \in \mathbb{R}^{5\times 5} \times \mathbb{R}^{5\times 5}$ as the partial eigenstructure for an SQIEP. Assume

$$X = \begin{bmatrix} -0.4132 & 5.2801 & 2.9437 & -6.6098 & -9.6715 \\ -4.3518 & 3.2758 & -5.1656 & 9.1024 & -9.1357 \\ -0.1336 & -4.0588 & 2.5321 & 3.3049 & -4.4715 \\ -5.1414 & 4.4003 & -2.2721 & 5.2872 & 6.9659 \\ 8.6146 & -4.0112 & -6.9380 & 1.4345 & -4.4708 \end{bmatrix},$$

and

$$\Lambda = \begin{bmatrix} -0.2168 & -4.3159 & 0 & 0 & 0 \\ 4.3159 & -0.2168 & 0 & 0 & 0 \\ 0 & 0 & 2.0675 & -0.9597 & 0 \\ 0 & 0 & 0.9597 & 2.0675 & 0 \\ 0 & 0 & 0 & 0 & -0.3064 \end{bmatrix}.$$

Example 5.3. Choose a basis $\begin{bmatrix} U_1^\top \\ V_1^\top \end{bmatrix}$ for the null space $\mathcal{N}\left(\left[X^\top\ \Lambda^\top X^\top\right]\right)$, say,

$$U_1^\top = \begin{bmatrix} 0.26861 & 0.56448 & -0.08687 & 0.39491 & -0.24252 \\ 0.32690 & -0.24385 & 0.00804 & -0.32844 & 0.42471 \\ -0.33739 & 0.27725 & -0.15949 & -0.05883 & 0.58406 \\ -0.13374 & 0.43824 & 0.09638 & 0.28605 & 0.46936 \\ -0.42433 & 0.17867 & 0.69977 & -0.12829 & -0.16140 \end{bmatrix},$$

$$V_1^\top = \begin{bmatrix} 0.51817 & 0.09467 & 0.20341 & -0.04075 & 0.32693 \\ 0.25575 & 0.38674 & -0.09339 & -0.32830 & -0.22850 \\ 0.31749 & -0.02297 & 0.63841 & 0.01156 & 0.05987 \\ -0.02434 & -0.40196 & 0.09987 & 0.65755 & 0.09646 \\ 0.27184 & 0.02061 & -0.01859 & 0.30413 & -0.03669 \end{bmatrix},$$

and construct

$$Q_1(\lambda) = \lambda^2(V_1^\top V_1) + \lambda(V_1^\top U_1 + U_1^\top V_1) + (U_1^\top U_1).$$

It can be checked that this quadratic pencil has double eigenvalue λ_j for each $\lambda_j \in \sigma(\Lambda)$, according to Theorem 5.15. Furthermore, we compute the singular values of each pencil $Q(\lambda_j)$ and find that

$$\varpi(Q_1(-0.21683 \pm 4.3159i)) = \{17.394, 15.039, 4.3974, 2.6136, 1.2483 \times 10^{-15}\},$$
$$\varpi(Q_1(2.0675 \pm 0.95974i)) = \{5.9380, 4.9789, 1.1788, 0.45926, 4.6449 \times 10^{-16}\},$$
$$\varpi(Q_1(-0.30635)) = \{1.0937, 1.0346, 0.89436, 0.18528, 3.8467 \times 10^{-17}\}.$$

The number of near machine-zero singular values indicates the degree of rank deficiency. The above results imply that the dimension of the null space $Q(\lambda_j)$ is precisely 1 for each $\lambda_j \in \sigma(\Lambda)$, as is guaranteed by Theorem 5.16.

There are infinitely many ways to select the basis. Indeed, we can change from one basis to the other via a right multiplication by a nonsingular matrix.

Example 5.4. Suppose we choose a special basis for $\mathcal{N}\left(\begin{bmatrix} X^\top & \Lambda^\top X^\top \end{bmatrix}\right)$ by

$$\begin{bmatrix} U_2^\top \\ V_2^\top \end{bmatrix} = \begin{bmatrix} U_1^\top V_1^{-\top} X^{-1} \\ X^{-1} \end{bmatrix},$$

and construct

$$Q_2(\lambda) = \lambda^2 (V_2^\top V_2) + \lambda (V_2^\top U_2 + U_2^\top V_2) + (U_2^\top U_2).$$

This basis is so special that we find that each of the four complex-valued eigenvalues of $\sigma(\Lambda)$ has linear elementary divisors. This is evidenced by the corresponding singular values of $Q(\lambda_j)$ which are computed to be

$$\begin{aligned}
&\varpi(Q_2(-0.21683 \pm 4.3159i)) \\
&\quad = \{15.517,\ 0.12145,\ 0.07626,\ 3.4880 \times 10^{-15},\ 7.9629 \times 10^{-16}\}, \\
&\varpi(Q_2(2.0675 \pm 0.95974i)) \\
&\quad = \{21.064,\ 0.16325,\ 0.02540,\ 3.2321 \times 10^{-15},\ 5.2233 \times 10^{-16}\}, \\
&\varpi(Q_2(-0.30635)) \\
&\quad = \{20.995,\ 0.19733,\ 0.08264,\ 0.02977,\ 1.6927 \times 10^{-15}\}.
\end{aligned}$$

We next demonstrate that by exploiting the freedom in the selection of basis for the null space $\mathcal{N}(\Omega)$ some additional optimization constraints can be incorporated into the construction of a solution to SQIEP. These additional constraints are imposed for some logistic reasons with the hope to better approximate a real physical system. We also experiment with the effect of feeding various amounts of information on eigenvalues and eigenvectors into the construction. In particular, we compare the discrepancy between a given (analytic) quadratic pencil and the resulting SQIEP approximation by varying the values of k and the optimal constraints.

To fix ideas, we first generate randomly a 10×10 symmetric quadratic pencil $\hat{Q}(\lambda) = \lambda^2 \hat{M} + \lambda \hat{C} + \hat{K}$, where $\hat{M}$ and $\hat{K}$ are also positive definite, as an analytic model. We then compare the effect of k on its SQIEP approximations for $k = 1, \ldots, 10$. To save space, we shall not report the data of these test matrices $\hat{M}$, $\hat{C}$, and $\hat{K}$ in this presentation, but will make them available upon request. We merely report that the spectrum of $\hat{Q}(\lambda)$ turns out to be the following 10 pairs of complex-conjugate values,

$$\{-0.27589 \pm 1.8585i,\ -0.19201 \pm 1.5026i,\ -0.15147 \pm 1.0972i,\\ -0.11832 \pm 0.54054i, -0.07890 \pm 1.3399i,\ -0.07785 \pm 0.76383i,\\ -0.07716 \pm 0.86045i,\ -0.07254 \pm 1.1576i,\\ -0.06276 \pm 0.97722i,\ -0.05868 \pm 0.18925i\}.$$

These eigenvalues are not arranged in any specific order. Without loss of generality, we shall *pretend* that the first five pairs in the above list are the partially described eigenvalues and wish to reconstruct the quadratic pencil. For $\ell = 1, \ldots, 5$ (and hence $k = 2\ell$), denote these eigenvalues as $\alpha_\ell \pm i\beta_\ell$. Also, define partial eigenpairs $(\Lambda_{2\ell}, X_{2\ell})$ of $\hat{Q}(\lambda)$ according to (5.46) and (5.48), that is,

$$\Lambda_{2l} = \operatorname{diag}\left\{\begin{bmatrix} \alpha_1 & \beta_1 \\ -\beta_1 & \alpha_1 \end{bmatrix}, \ldots, \begin{bmatrix} \alpha_\ell & \beta_\ell \\ -\beta_\ell & \alpha_\ell \end{bmatrix}\right\}, \tag{5.81}$$

$$X_{2\ell} = [x_{1R}, x_{1I}, \ldots, x_{\ell R}, x_{\ell I}], \tag{5.82}$$

where $x_{\ell R} \pm i x_{\ell I}$ is the eigenvector of $\hat{Q}(\lambda)$ corresponding to $\alpha_\ell \pm i\beta_\ell$.

Let $\begin{bmatrix} U_\ell^\top \\ V_\ell^\top \end{bmatrix} \in \mathbb{R}^{2n\times(2n-2\ell)}$ be an orthogonal basis for $\mathcal{N}([X_{2\ell}^\top \ \Lambda_{2\ell}^\top X_{2\ell}^\top])$. We now illustrate three ways to select a *new* basis for $\mathcal{N}([X_{2\ell}^\top \ \Lambda_{2\ell}^\top X_{2\ell}^\top])$, each of which is done for a different optimization purpose.

Example 5.5. Suppose $\hat{K} = L_{\hat{K}} L_{\hat{K}}^\top$ and $\hat{M} = L_{\hat{M}} L_{\hat{M}}^\top$ are the Cholesky factorizations of $\hat{K}$ and $\hat{M}$ in the model pencil, respectively. Find a matrix $G_{\ell 1}^\top \in \mathbb{R}^{(2n-2\ell)\times(2n-2\ell)}$ by solving the sequence of least-square problems

$$\min \left\| \begin{bmatrix} U_\ell^\top \\ V_\ell^\top \end{bmatrix} G_{\ell 1}^\top(:, j) - \begin{bmatrix} L_{\hat{K}} & 0_{n-2\ell} \\ 0_{n-2\ell} & L_{\hat{M}} \end{bmatrix} (:, j) \right\|_2, \tag{5.83}$$

for each of its columns $G_{\ell 1}^\top(:, j)$, $j = 1, \ldots, 2n - 2\ell$, where, for convenience, we have adopted the MATLAB notation $(:, j)$ to denote the j-th column of a matrix.

The solution of (5.83) is intended to not only solve the SQIEP, but also best approximate the original $\hat{K}$ and $\hat{M}$ in the sense that the quantity

$$\|U_\ell^\top G_{\ell 1}^\top G_{\ell 1} U_\ell - \hat{K}\|_F^2 + \|V_\ell^\top G_{\ell 1}^\top G_{\ell 1} V_\ell - \hat{M}\|_F^2 \tag{5.84}$$

is minimized among all possible $G_{\ell 1}^\top \in \mathbb{R}^{(2n-2\ell)\times(2n-2\ell)}$. Once such a matrix $G_{\ell 1}^\top$ is found, we compute the coefficient matrices according to our recipe, that is,

$$\begin{aligned} M_{\ell 1} &= V_\ell^\top G_{\ell 1}^\top G_{\ell 1} V_\ell, \\ K_{\ell 1} &= U_\ell^\top G_{\ell 1}^\top G_{\ell 1} U_\ell, \\ C_{\ell 1} &= U_\ell^\top G_{\ell 1}^\top G_{\ell 1} V_\ell + V_\ell^\top G_{\ell 1}^\top G_{\ell 1} U_\ell, \end{aligned} \tag{5.85}$$

and define the quadratic pencil

$$Q_{\ell 1}(\lambda) = \lambda^2 M_{\ell 1} + \lambda C_{\ell 1} + K_{\ell 1}, \tag{5.86}$$

according to $\ell = 1, \ldots, 5$.

Example 5.6. We first transform $V_\ell^\top$ to $[V_{\ell 0}^\top, 0]$ by an orthogonal transformation. Then we find a matrix $G_{\ell 2}^\top \in \mathbb{R}^{(2n-2\ell)\times(2n-2\ell)}$ in the form

$$G_{\ell 2}^\top = \begin{bmatrix} E_{\ell 2}^\top & 0 \\ 0 & F_{\ell 2}^\top \end{bmatrix}, \tag{5.87}$$

where $E_{\ell 2}^\top = V_{\ell 0}^{-\top} L_{\hat{M}}$ and $F_{\ell 2}^\top$ is an arbitrary $(n-2\ell) \times (n-2\ell)$ orthogonal matrix.

Example 5.7. We transform $U_\ell^\top$ to $[U_{\ell 0}^\top, 0]$ by an orthogonal transformation. Then we find a matrix $G_{\ell 3}^\top \in \mathbb{R}^{(2n-2\ell)\times(2n-2\ell)}$ in the form

$$G_{\ell 3}^\top = \begin{bmatrix} E_{\ell 3}^\top & 0 \\ 0 & F_{\ell 3}^\top \end{bmatrix}, \tag{5.88}$$

where $E_{\ell 3}^\top = U_{\ell 0}^{-\top} L_{\hat{K}}$ and $F_{\ell 3}^\top$ is an arbitrary $(n-2\ell) \times (n-2\ell)$ orthogonal matrix.

The purpose of finding $G_{\ell 2}^\top$ and $G_{\ell 3}^\top$ in the form of (5.87) and (5.88) is to not only solve the SQIEP, but also best approximate the original $\hat{M}$ and $\hat{K}$ in the sense that

$$\|V_\ell^\top G_{\ell 2}^\top G_{\ell 2} V_\ell - \hat{M}\|_F \tag{5.89}$$

and

$$\|U_\ell^\top G_{\ell 3}^\top G_{\ell 3} U_\ell - \hat{K}\|_F \tag{5.90}$$

are minimized by $G_{\ell 2}^\top$ and $G_{\ell 3}^\top$, respectively. Once these matrices are found, we define quadratic pencils $Q_{\ell 2}(\lambda)$ and $Q_{\ell 3}(\lambda)$ in exactly the same way as we define $Q_{\ell 1}(\lambda)$.

It would be interesting to see how the reconstructed quadratic pencils for the SQIEP, with the above-mentioned optimization in mind, approximate the original pencil. Toward that end, we measure the total difference

$$d_{\ell j} = \|M_{\ell j} - \hat{M}\|_F + \|C_{\ell j} - \hat{C}\|_F + \|K_{\ell j} - \hat{K}\|_F, \tag{5.91}$$

between the original pencil and the reconstructed pencil for each $j = 1, 2, 3$ and $\ell = 1, \ldots, 5$.

Example 5.8. In Figure 5.1 we plot the *error* $d_{\ell j}$ between $\hat{Q}(\lambda)$ and $Q_{\ell j}(\lambda)$ for the various cases. Not surprisingly, we notice that the quadratic pencil $Q_{\ell 1}(\lambda)$ constructed from $G_{\ell 1}^\top$ is superior to the other two. What might be interesting to note is that in the formulation of Example 5.5 the amount of eigeninformation available to the SQIEP does not seem to make any significant difference in the

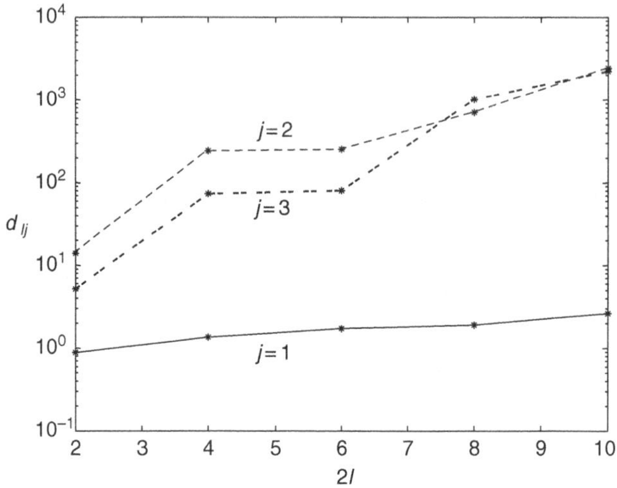

Figure 5.1. *Errors of SQIEP approximations*

measurement of $d_{\ell 1}$. That is, all $d_{\ell 1}$ seems to be of the same order regardless of the value of ℓ. We think an explanation for this is because $G_{\ell 1}^{\top}$ has somewhat more freedom to choose so that $M_{\ell 1}$ and $K_{\ell 1}$ better approximate $\hat{M}$ and $\hat{K}$, respectively.

In applications, the stiffness matrix $\hat{K}$ and the mass matrix $\hat{M}$ of a model usually can be obtained by finite element or finite difference methods. It is the damping matrix $\hat{C}$ in such a system that is generally not known. If some partial eigenstructure can be measured by experiment, then the construction proposed in Example 5.5 might be a good way to recover the original system by best approximating the stiffness matrix and the mass matrix in the sense of minimizing (5.84).

5.4 Monic quadratic inverse eigenvalue problem

With the existence theory established in the preceding section for the SQIEP where $k = n$ plays a vital role in deciding whether the resulting quadratic pencil is singular, it is interesting to study in this section yet another scenario of the QIEP.

In the MQIEP, the leading matrix coefficient M is known and fixed and only symmetric C and K are to be determined. We have already suggested earlier by counting the cardinality of unknowns and equations that the number of prescribed eigenpairs could go up to $k = n+1$. Since the prescribed eigenvectors now form a matrix X of size $n \times (n+1)$, by assuming that X is of full rank, there is at least one column in the given $n \times (n+1)$ matrix X which depends linearly on the other columns. The following analysis is contingent on whether this linearly dependent column is real-valued or complex-valued. We separate the discussion into two cases. Either case shows a way to solve the MQIEP.

5.4.1 *Real linearly dependent eigenvectors*

Assume that the linearly dependent column vector is real-valued. By rearranging the columns if necessary, we may assume without loss of generality that this vector is $\mathbf{x}_{n+1}$. It follows that the $n \times n$ submatrix

$$X_1 := [\mathbf{x}_1, \bar{\mathbf{x}}_1, \ldots, \mathbf{x}_{2\ell-1}, \bar{\mathbf{x}}_{2\ell-1}, \mathbf{x}_{2\ell+1}, \ldots, \mathbf{x}_n], \tag{5.92}$$

of $\tilde{X}$ defined in (5.51) is nonsingular. Let

$$\Lambda_1 := \operatorname{diag}\{\lambda_1, \bar{\lambda}_1, \ldots, \lambda_{2\ell-1}, \bar{\lambda}_{2\ell-1}, \lambda_{2\ell+1}, \ldots, \lambda_n\}, \tag{5.93}$$

be the corresponding submatrix of $\tilde{\Lambda}$ defined in (5.50). Both matrices are closed under complex conjugation in the sense defined before.

Define

$$S := X_1 \Lambda_1 X_1^{-1}. \tag{5.94}$$

Note that, due to complex conjugation, S is a real-valued $n \times n$ matrix. Define a quadratic pencil $Q(\lambda)$ via the factorization

$$Q(\lambda) := (\lambda I_n + S + C)(\lambda I_n - S), \tag{5.95}$$

where C is yet to be determined. Upon comparing the expression of (5.95) with (5.55), we see that

$$K = -(S + C)S. \tag{5.96}$$

The first criterion for solving the MQIEP is that both matrices C and K be real-valued and symmetric. Thus the undetermined real-valued matrix C must first satisfy the following two equations simultaneously:

$$\begin{cases} C^\top = C, \\ S^\top C - CS = S^2 - (S^\top)^2. \end{cases} \tag{5.97}$$

The following result provides a partial characterization of the matrix C we are looking for.

Theorem 5.17. The general solution to (5.97) is given by the formula

$$C = -(S + S^\top) + \sum_{j=1}^{n} \gamma_j \mathbf{y}_j \mathbf{y}_j^\top, \tag{5.98}$$

where the vectors $\mathbf{y}_j$, $j = 1, \ldots, n$, are the columns of the matrix

$$Y_1 := X_1^{-\top} = [\mathbf{y}_1, \ldots, \mathbf{y}_{2\ell-1}, \mathbf{y}_{2\ell}, \mathbf{y}_{2\ell+1}, \ldots, \mathbf{y}_n], \tag{5.99}$$

and the scalars γ_j, $j = 1, \ldots, n$, are arbitrary complex numbers.

Proof It is easy to see that $-(S+S^\top)$ is a particular solution of (5.97). The formula thus follows from an established result (Lancaster and Tismenetsky, 1985, Theorem 1, Section 12.5). □

It might be worth mentioning that the columns of Y_1 are also closed under complex conjugation and, hence, C is real-valued if and only if the corresponding coefficients γ_j are complex conjugate. It only remains to determine these combination coefficients in (5.98) so that the MQIEP is solved. Toward that end, observe first that

$$X_1\Lambda_1^2 + CX_1\Lambda_1 + KX_1 = 0, \tag{5.100}$$

regardless of how the scalars γ_j, $j=1,\ldots,n$, are chosen. In other words, n pairs of the given data have already satisfied the spectral constraint in the MQIEP. We use the fact that the last pair $(\lambda_{n+1}, \mathbf{x}_{n+1}) \in \mathbb{R}\times\mathbb{R}^n$ in the given data must also be an eigenpair of $Q(\lambda)$ in (5.95) to determine the parameters γ_j, $j=1,\ldots,n$.

Define

$$\mathbf{z} := (\lambda_{n+1}I - S)\mathbf{x}_{n+1} \in \mathbb{R}^n. \tag{5.101}$$

Plugging the eigenpair $(\lambda_{n+1}, \mathbf{x}_{n+1})$ into (5.95) and using (5.98), we obtain the equation

$$\lambda_{n+1}\mathbf{z} = S^\top\mathbf{z} - \sum_{j=1}^{n}\gamma_j\mathbf{y}_j\mathbf{y}_j^\top\mathbf{z}$$

which can be written as

$$-X_1^\top(\lambda_{n+1}\mathbf{z} - S^\top\mathbf{z}) = \operatorname{diag}\{\mathbf{y}_1^\top\mathbf{z},\ldots,\mathbf{y}_n^\top\mathbf{z}\}\begin{bmatrix}\gamma_1\\ \vdots\\ \gamma_n\end{bmatrix}. \tag{5.102}$$

Obviously, values of $\gamma_j\mathbf{y}_j^\top\mathbf{z}$, $j=1,\ldots,n$, are uniquely determined. However, the value of γ_j is unique only if

$$\mathbf{y}_j^\top\mathbf{z} = \mathbf{e}_j^\top X_1^{-1}\mathbf{z} \neq 0, \tag{5.103}$$

where $\mathbf{e}_j$ denotes the j-th standard unit vector. In terms of the original data, the condition can be written equivalently as

$$\mathbf{e}_j^\top(\lambda_{n+1}I - \Lambda_1)X_1^{-1}\mathbf{x}_{n+1} \neq 0. \tag{5.104}$$

If we assume that the condition (5.103) holds for all $j=1,\ldots,n$, then the last step in solving the MQIEP is to show that elements in the solution $\{\gamma_1,\ldots,\gamma_n\}$ to (5.102) are closed under complex conjugation in exactly the same way as columns of Y_1.

For convenience, we shall denote

$$\mathbf{r} := \lambda_{n+1}\mathbf{z} - S^\top\mathbf{z} \in \mathbb{R}^n. \tag{5.105}$$

The first 2ℓ elements in equation (5.102) are

$$-\mathbf{x}_{2j-1}^{\top}\mathbf{r} = \mathbf{y}_{2j-1}^{\top}\mathbf{z}\gamma_{2j-1}, \tag{5.106}$$

$$-\mathbf{x}_{2j}^{\top}\mathbf{r} = \mathbf{y}_{2j}^{\top}\mathbf{z}\gamma_{2j}, \quad \text{for } j = 1, \ldots, \ell. \tag{5.107}$$

Recall $\mathbf{x}_{2j-1} = \bar{\mathbf{x}}_{2j}$ and $\mathbf{y}_{2j-1} = \bar{\mathbf{y}}_{2j}$ for $j = 1, \ldots, \ell$. Upon taking the conjugation of (5.106) and comparing with (5.107), we conclude that

$$\gamma_{2j} = \bar{\gamma}_{2j-1}, \quad \text{for } j = 1, \ldots, \ell. \tag{5.108}$$

Similarly, $\gamma_k \in \mathbb{R}$, for $k = 2\ell + 1, \ldots, n$. It is now finally proved that both C and K are indeed real-valued and symmetric. We summarize our first major result as follows.

Theorem 5.18. Let $(\tilde{\Lambda}, \tilde{X}) \in \mathbb{C}^{(n+1)\times(n+1)} \times \mathbb{C}^{n\times(n+1)}$ be given as in (5.50) and (5.51). Assume that one eigenvector, say $\mathbf{x}_{n+1} \in \mathbb{R}^n$, depends linearly on the remaining eigenvectors $\mathbf{x}_1, \ldots, \mathbf{x}_n$ which are linearly independent. If the condition (5.104) is satisfied for all $j = 1, \ldots, n$, then the MQIEP has a unique solution.

We point out quickly that the equation (5.102) is not necessarily consistent. In particular, a possible scenario is as follows.

Corollary 5.19. Under the same assumptions as in Theorem 5.18, if $\mathbf{e}_j^{\top} X_1^{-1}\mathbf{z} = 0$ and $\mathbf{e}_j^{\top}(\lambda_{n+1} I - \Lambda_1^{\top})X_1^{\top}\mathbf{z} \neq 0$ for some j, then the MQIEP has no solution.

5.4.2 *Complex linearly dependent eigenvectors*

Assume that the linearly dependent column vector is complex-valued. By rearranging the columns if necessary, we may assume without loss of generality that this vector is $\mathbf{x}_{2\ell}$. It follows that the $n \times n$ matrix

$$X_1 = [\underbrace{\mathbf{x}_1, \bar{\mathbf{x}}_1, \ldots, \mathbf{x}_{2\ell-3}, \bar{\mathbf{x}}_{2\ell-3}}_{\text{complex-conjugated}}, \underbrace{\mathbf{x}_{2\ell+1}, \ldots, \mathbf{x}_{n+1}}_{\text{real-valued}}, \underbrace{\mathbf{x}_{2\ell-1}}_{\text{complex-valued}}]$$

is nonsingular. For convenience, we shall re-index the sequence of the above column vectors by successive integers. Without causing ambiguity, we shall use the same notation for the renumbered vectors. Specifically, we rewrite the above X_1 as

$$X_1 = [\underbrace{\mathbf{x}_1, \bar{\mathbf{x}}_1, \ldots, \mathbf{x}_{2m-1}, \bar{\mathbf{x}}_{2m-1}}_{\text{complex-conjugated}}, \underbrace{\mathbf{x}_{2m+1}, \ldots, \mathbf{x}_{n-1}}_{\text{real-valued}}, \underbrace{\mathbf{x}_n}_{\text{complex-valued}}], \tag{5.109}$$

column by column but only rename the indices, and define the corresponding

$$\Lambda_1 = \operatorname{diag}\{\lambda_1, \bar{\lambda}_1, \ldots, \lambda_{2m-1}, \bar{\lambda}_{2m-1}, \lambda_{2m+1}, \ldots, \lambda_{n-1}, \lambda_n\}. \tag{5.110}$$

We could further assume in (5.109) that

$$\bar{\mathbf{x}}_n \in \operatorname{span}\{\mathbf{x}_1, \bar{\mathbf{x}}_1, \dots, \mathbf{x}_{2m-1}, \bar{\mathbf{x}}_{2m-1}, \mathbf{x}_n\}, \tag{5.111}$$

since otherwise one of the real-valued eigenvectors (and this is possible only if $2m+1 < n$) must be linearly dependent and we would go back to the case in Section 5.4.1. The following argument is analogous to that of Section 5.4.1, but additional details need to be filled in.

Following (5.94) through (5.96) except that S is now complex-valued, we want to determine the matrix C in the factorization (5.95) and the corresponding K via several steps. We first require both C and K to be Hermitian. That is, the matrix C must satisfy the following equations:

$$\begin{cases} C^H = C \in \mathbb{C}^{n\times n}, \\ S^H C - CS = S^2 - (S^H)^2 \in \mathbb{C}^{n\times n}. \end{cases} \tag{5.112}$$

In contrast to Theorem 5.17, the characterization of C is a little bit more complicated.

Theorem 5.20. The general solution to (5.112) is given by the formula

$$\begin{aligned} C = -(S+S^H) &+ \gamma_1 \mathbf{y}_1 \mathbf{y}_2^H + \gamma_2 \mathbf{y}_2 \mathbf{y}_1^H + \cdots + \gamma_{2m-1} \mathbf{y}_{2m-1} \mathbf{y}_{2m}^H + \gamma_{2m} \mathbf{y}_{2m} \mathbf{y}_{2m-1}^H \\ &+ \gamma_{2m+1} \mathbf{y}_{2m+1} \mathbf{y}_{2m+1}^H + \cdots + \gamma_{n-1} \mathbf{y}_{n-1} \mathbf{y}_{n-1}^H \end{aligned} \tag{5.113}$$

where the vectors $\mathbf{y}_i$, $i = 1, \dots, n$ are the columns of the matrix

$$Y_1 := X_1^{-H} = [\mathbf{y}_1, \mathbf{y}_2, \dots, \mathbf{y}_{2m}, \mathbf{y}_{2m+1}, \dots, \mathbf{y}_n]. \tag{5.114}$$

Proof Again, the formula is similar to that in Theorem 5.17 using exactly the same established result (Lancaster and Tismenetsky, 1985, Theorem 1, Section 12.5). The slight complication is due to the fact that the first $2m$ eigenvalues of S and S^H coincide in a conjugated way and the last eigenvalues of S and S^H are distinct. □

Note that for $j = 1, \dots, m$, $\mathbf{y}_{2j-1}$ and $\mathbf{y}_{2j} = \bar{\mathbf{y}}_{2j-1}$ are the eigenvectors of S^H with corresponding eigenvalues $\bar{\lambda}_{2j-1}$ and λ_{2j-1}, respectively. Likewise, for $k = 2m+1, \dots, n-1$, $\mathbf{y}_k \in \mathbb{R}^n$ is the eigenvector of S^H corresponding to $\lambda_j \in \mathbb{R}$. Finally, $\mathbf{y}_n \in \mathbb{C}^n$ is the eigenvector of S^H corresponding to $\bar{\lambda}_n \in \mathbb{C}$.

By construction, we already know that (5.100) is satisfied with X_1 defined by (5.109) and C defined by (5.113). It remains to determine the coefficients $\gamma_1, \dots, \gamma_{n-1}$ so that the deleted linearly dependent vector $\bar{\mathbf{x}}_n$ (the original $\mathbf{x}_{2\ell}$ before the re-indexing) is also an eigenvector with eigenvalue $\bar{\lambda}_n$. Of course, we also need to make sure that the resulting C and K are ultimately real-valued.

Let

$$\mathbf{z} = (\bar{\lambda}_n I_n - S)\bar{\mathbf{x}}_n. \tag{5.115}$$

Substituting the eigenpair $(\bar{\lambda}_n, \bar{\mathbf{x}}_n)$ into (5.95) and using (5.113), we obtain

$$(\bar{\lambda}_n I_n - S^H)\mathbf{z} = -[\mathbf{y}_1, \ldots, \mathbf{y}_{2m}, \mathbf{y}_{2m+1}, \ldots, \mathbf{y}_{n-1}] \begin{bmatrix} \gamma_1 \mathbf{y}_2^H \mathbf{z} \\ \gamma_2 \mathbf{y}_1^H \mathbf{z} \\ \vdots \\ \gamma_{2m-1} \mathbf{y}_{2m}^H \mathbf{z} \\ \gamma_{2m} \mathbf{y}_{2m-1}^H \mathbf{z} \\ \gamma_{2m+1} \mathbf{y}_{2m+1}^H \mathbf{z} \\ \vdots \\ \gamma_{n-1} \mathbf{y}_{n-1}^H \mathbf{z} \end{bmatrix}. \tag{5.116}$$

With the assumption of (5.111), it is not difficult to see that

$$\mathbf{y}_j^H \mathbf{z} = 0, \quad \text{for } j = 2m+1, \ldots, n-1. \tag{5.117}$$

Equation (5.116) is equivalent to the equation

$$(\bar{\lambda}_n I_n - \Lambda_1^H) X_1^H \mathbf{z} = - \begin{bmatrix} \gamma_1 \mathbf{y}_2^H \mathbf{z} \\ \gamma_2 \mathbf{y}_1^H \mathbf{z} \\ \vdots \\ \gamma_{2m-1} \mathbf{y}_{2m}^H \mathbf{z} \\ \gamma_{2m} \mathbf{y}_{2m-1}^H \mathbf{z} \\ 0 \\ \vdots \\ 0 \end{bmatrix}. \tag{5.118}$$

The left-hand side of (5.118) is completely known. It is now clear that the coefficients $\gamma_1, \ldots, \gamma_{2m}$ are uniquely determined if

$$\mathbf{y}_j^H \mathbf{z} \neq 0, \quad \text{for } j = 1, \ldots, 2m, \tag{5.119}$$

whereas the coefficients $\gamma_{2m+1}, \ldots, \gamma_{n-1}$ in (5.116) (and hence in (5.113)) can be arbitrary real numbers so long as the last $n - 2m$ equations in (5.118) are consistent, that is,

$$\mathbf{x}_j^H \mathbf{z} = 0, \quad \text{for } j = 2m+1, \ldots, n-1. \tag{5.120}$$

Assuming (5.119) and (5.120), we now show that the resulting matrix C in (5.113) is Hermitian. Toward that end, it suffices to show that $\gamma_{2j-1} = \bar{\gamma}_{2j}$, for $j = 1, \ldots, m$. Based on (5.117) and (5.120), we introduce the following two vectors for convenience.

$$\mathbf{p} := X_1^H \mathbf{z} = [p_1, \ldots, p_{2m}, 0, \ldots, 0, p_n]^\top \in \mathbb{C}^n, \tag{5.121}$$

$$\mathbf{q} := X_1^{-1} \bar{\mathbf{x}}_n = [q_1, \ldots, q_{2m}, 0, \ldots, 0, q_n]^\top \in \mathbb{C}^n. \tag{5.122}$$

For $j = 1, \ldots, m$, the $(2j-1)$-th and the $(2j)$-th components of (5.118) are, respectively,

$$(\bar{\lambda}_n - \bar{\lambda}_{2j-1})p_{2j-1} = -\gamma_{2j-1}\mathbf{y}_{2j}^H\mathbf{z} = -\gamma_{2j-1}(\bar{\lambda}_n - \lambda_{2j})q_{2j},$$
$$(\bar{\lambda}_n - \bar{\lambda}_{2j})p_{2j} = -\gamma_{2j}\mathbf{y}_{2j-1}^H\mathbf{z} = -\gamma_{2j}(\bar{\lambda}_n - \lambda_{2j-1})q_{2j-1}.$$

Since $\lambda_{2j-1} = \bar{\lambda}_{2j}$, it follows that

$$p_{2j-1} = -\gamma_{2j-1}q_{2j}, \tag{5.123}$$
$$p_{2j} = -\gamma_{2j}q_{2j-1}. \tag{5.124}$$

On the other hand, observe that

$$\mathbf{z} = (\bar{\lambda}_n I_n - S)\bar{\mathbf{x}}_n = (\bar{\lambda}_n I_n - \bar{S} + \bar{S} - S)\bar{\mathbf{x}}_n = (\bar{S} - S)\bar{\mathbf{x}}_n, \tag{5.125}$$

since $\bar{\mathbf{x}}_n$ is an eigenvector of $\bar{S}$. Observe also that

$$(\bar{S} - S)\mathbf{x}_j = 0, \quad \text{for } j = 1, \ldots, n-1, \tag{5.126}$$

because of the complex conjugation. It follows that

$$(\bar{S} - S)\bar{\mathbf{x}}_n = q_n(\bar{S} - S)\mathbf{x}_n m, \tag{5.127}$$
$$\mathbf{x}_{2j-1}^H(\bar{S} - S)\bar{\mathbf{x}}_n = -\gamma_{2j-1}q_{2j}, \tag{5.128}$$
$$\mathbf{x}_{2j}^H(\bar{S} - S)\bar{\mathbf{x}}_n = -\gamma_{2j}q_{2j-1}. \tag{5.129}$$

Taking conjugation of (5.128) and using (5.127), we obtain

$$-\bar{q}_n\bar{\mathbf{x}}_{2j-1}^H(\bar{S} - S)\bar{\mathbf{x}}_n = -\bar{\gamma}_{2j-1}\bar{q}_{2j}. \tag{5.130}$$

Comparing (5.129) and (5.130), since $\bar{\mathbf{x}}_{2j-1} = \mathbf{x}_{2j}$ for all $j = 1, \ldots, m$, we obtain the critical relationship that

$$\bar{q}_n\gamma_{2j}q_{2j-1} = -\bar{\gamma}_{2j-1}\bar{q}_{2j}, \quad \text{for} \quad j = 1, \ldots, m. \tag{5.131}$$

Now we are ready to show that $\gamma_{2j} = \bar{\gamma}_{2j-1}$, $j = 1, \ldots, m$. We rewrite $\mathbf{x}_n = \bar{X}_1\bar{\mathbf{q}}$ from (5.122) as

$$\begin{aligned}\mathbf{x}_n &= \bar{q}_1\bar{\mathbf{x}}_1 + \bar{q}_2\bar{\mathbf{x}}_2 + \cdots + \bar{q}_{2m-1}\bar{\mathbf{x}}_{2m-1} + \bar{q}_{2m}\bar{\mathbf{x}}_{2m} + \bar{q}_n\bar{\mathbf{x}}_n \\ &= \bar{q}_1\mathbf{x}_2 + \bar{q}_2\mathbf{x}_1 + \cdots + \bar{q}_{2m-1}\mathbf{x}_{2m} + \bar{q}_{2m}\mathbf{x}_{2m-1} + \bar{q}_n\bar{\mathbf{x}}_n.\end{aligned} \tag{5.132}$$

Replacing the last term by

$$\bar{q}_n\bar{\mathbf{x}}_n = \bar{q}_n q_1\mathbf{x}_1 + \bar{q}_n q_2\mathbf{x}_2 + \cdots + \bar{q}_n q_{2m}\mathbf{x}_{2m} + |q_n|^2\mathbf{x}_n,$$

we obtain the equality

$$\begin{aligned}\mathbf{x}_n = {} & (\bar{q}_n q_1 + \bar{q}_2)\mathbf{x}_1 + (\bar{q}_n q_2 + \bar{q}_1)\mathbf{x}_2 + \cdots + (\bar{q}_n q_{2m-1} + \bar{q}_{2m})\mathbf{x}_{2m-1} \\ & + (\bar{q}_n q_{2m} + \bar{q}_{2m-1})\mathbf{x}_{2m} + |q_n|^2\mathbf{x}_n.\end{aligned}$$

Since $\{\mathbf{x}_1, \ldots, \mathbf{x}_{2m}, \mathbf{x}_n\}$ are linearly independent, then

$$\bar{q}_n q_{2j-1} + \bar{q}_{2j} = 0, \quad \text{for } j = 1, \ldots, m. \tag{5.133}$$

Substituting (5.133) into (5.131), we have proved that $\gamma_{2j} = \bar{\gamma}_{2j-1}$, for $j = 1, \ldots, m$.

By now, we have completed the proof that the matrix C constructed using (5.118) is Hermitian. We are ready to state our second major result.

Theorem 5.21. Let $(\tilde{\Lambda}, \tilde{X}) \in \mathbb{C}^{(n+1)\times(n+1)} \times \mathbb{C}^{n\times(n+1)}$ be given as in (5.50) and (5.51). Assume that one eigenvector, say $\mathbf{x}_{2\ell} \in \mathbb{C}^n$, depends linearly on the remaining eigenvectors which are linearly independent. Then

1. Suppose $\ell = \frac{n+1}{2}$, that is, suppose that there is no real-valued vector at all in X. If the condition (5.119) is satisfied for $j = 1, \ldots, n-1$, then the MQIEP has a unique solution.
2. Suppose $\ell < \frac{n+1}{2}$ and that (5.111) holds. If the condition (5.119) is satisfied for $j = 1, \ldots, 2\ell - 2$ and the condition (5.120) is satisfied for $j = 2\ell + 1, \ldots, n+1$, then the MQIEP has infinite many solutions; otherwise it has no solution.

Proof Thus far, we have already shown that both matrices C and K can be constructed uniquely and are Hermitian. It only remains to show that C and K are real symmetric. It suffices to prove that $C = \bar{C}$ and $K = \bar{K}$.

Consider the MQIEP associated with the spectral data $(\bar{\tilde{\Lambda}}, \bar{\tilde{X}})$, the complex conjugate of the original data $(\tilde{\Lambda}, \tilde{X})$. Then the sufficient condition (5.119) for the problem associated with $(\tilde{\Lambda}, \tilde{X})$ applies equally well to the new problem associated with $(\bar{\tilde{\Lambda}}, \bar{\tilde{X}})$. A quadratic pencil therefore can be constructed to solve the new MQIEP. Indeed, by repeating the procedure of construction described above, it is not difficult to see that the constructed pencil for $(\bar{\tilde{\Lambda}}, \bar{\tilde{X}})$ is of the form

$$\widetilde{Q}(\lambda) = \lambda^2 I_n + \lambda \bar{C} + \bar{K}.$$

Since Λ and X are closed under complex conjugation, the spectral information $(\bar{\tilde{\Lambda}}, \bar{\tilde{X}})$ is actually a reshuffle of (Λ, X). As a matter of fact, these two MQIEPs are the same problem. In the first case where $\ell = \frac{n+1}{2}$, the solution is already unique. In the second case where $\ell < \frac{n+1}{2}$ and (5.111) holds, so long as the arbitrarily selected real coefficients $\gamma_{2m+1}, \ldots, \gamma_{n-1}$ remain fix, the complex-conjugated coefficients $\gamma_1, \ldots, \gamma_{2m}$ are also uniquely determined. In either case, we must have that $C = \bar{C} = C^H$ and $K = \bar{K} = K^H$. □

5.4.3 *Numerical examples*

The argument presented in the preceding section offers a constructive way to solve the MQIEP. In this section we use numerical examples to illustrate the two cases discussed above. For ease of running text, we report all numbers to five significant digits only, though all calculations are carried out in full precision.

To generate test data, we first randomly generate a 5×5 real symmetric quadratic pencil $Q(\lambda) = \lambda^2 I + \lambda C + K$ and compute its "exact" eigenpairs (Λ_e, X_e) numerically. We obtain that $\Lambda_e = \operatorname{diag}\{\lambda_1, \ldots, \lambda_{10}\}$, $X_e = [\mathbf{x}_1, \ldots, \mathbf{x}_{10}]$ with eigenvalues

$$\lambda_1 = \bar{\lambda}_2 = -0.31828 + 0.86754i, \quad \lambda_3 = \bar{\lambda}_4 = -0.95669 + 0.17379i,$$
$$\lambda_5 = -4.4955, \quad \lambda_6 = 1.5135, \quad \lambda_7 = \bar{\lambda}_8 = -0.24119 + 0.029864i,$$
$$\lambda_9 = 0.91800, \quad \lambda_{10} = -1.7359,$$

and the corresponding eigenvectors

$$\mathbf{x}_1 = \bar{\mathbf{x}}_2 = \begin{bmatrix} 15.159 - 11.123i \\ -77.470 - 14.809i \\ 2.1930 - 10.275i \\ 0.38210 + 16.329i \\ 57.042 + 18.419i \end{bmatrix}, \quad \mathbf{x}_3 = \bar{\mathbf{x}}_4 = \begin{bmatrix} 65.621 + 34.379i \\ 22.625 + 24.189i \\ -37.062 + 15.825i \\ -9.6496 + 14.401i \\ -0.61893 + 25.609i \end{bmatrix},$$

$$\mathbf{x}_5 = \begin{bmatrix} 2.2245 \\ 1.5893 \\ 2.1455 \\ 2.1752 \\ 1.6586 \end{bmatrix}, \quad \mathbf{x}_6 = \begin{bmatrix} 34.676 \\ -5.8995 \\ 37.801 \\ -66.071 \\ -6.6174 \end{bmatrix}, \quad \mathbf{x}_7 = \bar{\mathbf{x}}_8 = \begin{bmatrix} 35.257 - 0.31888i \\ -25.619 - 4.2156i \\ 98.914 - 1.0863i \\ -21.348 + 5.8290i \\ -97.711 - 1.0693i \end{bmatrix},$$

$$\mathbf{x}_9 = \begin{bmatrix} -97.828 \\ 10.879 \\ 100.00 \\ -4.3638 \\ 22.282 \end{bmatrix}, \quad \mathbf{x}_{10} = \begin{bmatrix} -1.3832 \\ 4.4564 \\ -1.1960 \\ -4.0934 \\ 5.7607 \end{bmatrix}.$$

The above spectral data are not arranged in any specific order. According to our theory, any $n+1$ eigenpairs satisfying the specification of (5.92) and (5.93) and the sufficient condition (5.104) or (5.119), depending upon whether assumptions in Section 5.4.1 or Section 5.4.2 with $\ell = \frac{n+1}{2}$ are applicable, should ensure the full recovery of the original pencil.

Example 5.9. Suppose the prescribed partial eigeninformation is given by

$$(\tilde{\Lambda}, \tilde{X}) = (\operatorname{diag}\{\lambda_1, \lambda_2, \lambda_3, \lambda_4, \lambda_5, \lambda_6\}, \ [\mathbf{x}_1, \mathbf{x}_2, \mathbf{x}_3, \mathbf{x}_4, \mathbf{x}_5, \mathbf{x}_6]).$$

It is easy to check that the real-valued eigenvector $\mathbf{x}_6$ depends linearly on the first five eigenvectors which are linearly independent. This fits the situation discussed in Section 5.4.1 where we choose to work with

$$(\hat{\Lambda}_1, \hat{X}_1) = \left(\operatorname{diag}\{\lambda_1, \bar{\lambda}_1, \lambda_3, \bar{\lambda}_3, \lambda_5\}, \ [\mathbf{x}_1, \bar{\mathbf{x}}_1, \mathbf{x}_3, \bar{\mathbf{x}}_3, \mathbf{x}_5]\right).$$

We construct the unique real symmetric quadratic pencil

$$\hat{Q}(\lambda) = \lambda^2 I_5 + \lambda \hat{C} + \hat{K},$$

Table 5.2. *Residual* $\|\hat{Q}_1(\lambda_j)\mathbf{x}_j\|_2$ *and coefficient errors for Example 5.9*

Eigenpairs	Residual $\|\hat{Q}_1(\lambda_j)\mathbf{x}_j\|_2$
$(\lambda_1, \mathbf{x}_1)$	$2.2612e{-}015$
$(\lambda_2, \mathbf{x}_2)$	$2.2612e{-}015$
$(\lambda_3, \mathbf{x}_3)$	$2.9827e{-}015$
$(\lambda_4, \mathbf{x}_4)$	$2.9827e{-}015$
$(\lambda_5, \mathbf{x}_5)$	$2.0381e{-}015$
$(\lambda_6, \mathbf{x}_6)$	$1.8494e{-}014$
$(\lambda_7, \mathbf{x}_7)$	$7.9955e{-}014$
$(\lambda_8, \mathbf{x}_8)$	$7.9955e{-}014$
$(\lambda_9, \mathbf{x}_9)$	$4.4264e{-}014$
$(\lambda_{10}, \mathbf{x}_{10})$	$4.5495e{-}014$
$\|\hat{C} - C\|_2$	$1.8977e{-}014$
$\|\hat{K} - K\|_2$	$7.3897e{-}014$

by the method described in the proof of Theorem 5.18. In Table 5.2, we show the residual $\|\hat{Q}(\lambda_j)\mathbf{x}_j\|_2$, where $(\lambda_j, \mathbf{x}_j)$ are the computed eigenpairs of $Q(\lambda)$, for $j = 1, \ldots, 10$, as well as the difference $\|\hat{C} - C\|_2$ and $\|\hat{K} - K\|_2$, respectively.

Example 5.10. Suppose the prescribed spectral information is given by

$$(\tilde{\Lambda}, \tilde{X}) = (\mathrm{diag}\{\lambda_1, \lambda_2, \lambda_3, \lambda_4, \lambda_7, \lambda_8\},\ [\mathbf{x}_1, \mathbf{x}_2, \mathbf{x}_3, \mathbf{x}_4, \mathbf{x}_7, \mathbf{x}_8]).$$

Note that all eigenvectors are complex-valued. This fits the situation discussed in Section 5.4.2 with $\ell = \frac{n+1}{2}$ where we choose to work with

$$(\check{\Lambda}_1, \check{X}_1) = \left(\mathrm{diag}\{\lambda_1, \bar{\lambda}_1, \lambda_3, \bar{\lambda}_3, \lambda_7\},\ [\mathbf{x}_1, \bar{\mathbf{x}}_1, \mathbf{x}_3, \bar{\mathbf{x}}_3, \mathbf{x}_7]\right).$$

We construct the unique real symmetric quadratic pencil

$$\check{Q}(\lambda) = \lambda^2 I_5 + \lambda \check{C} + \check{K},$$

by the method described in the proof of Theorem 5.21. In Table 5.3, we show the residual $\|\check{Q}(\lambda_j)\mathbf{x}_j\|_2$, for $j = 1, \ldots, 10$, as well as the difference $\|\check{C} - C\|_2$ and $\|\check{K} - K\|_2$, respectively.

It can be checked that the two examples above satisfy the sufficient conditions (5.104) and (5.119), respectively. The solution is unique and the errors shown in the tables seem to be quite satisfactory.

Now we demonstrate the second situation in Theorem 5.21 when both $\ell < \frac{n+1}{2}$ and (5.111) take place. Our theory asserts that there will be either infinitely many solutions to the MQIEP or no solution at all.

Table 5.3. *Residual* $\|\hat{Q}_1(\lambda_j)\mathbf{x}_j\|_2$ *and coefficient errors for Example 5.10*

Eigenpairs	Residual $\|\check{Q}(\lambda_j)\mathbf{x}_j\|_2$
$(\lambda_1, \mathbf{x}_1)$	4.5422e−016
$(\lambda_2, \mathbf{x}_2)$	4.5422e−016
$(\lambda_3, \mathbf{x}_3)$	7.8025e−016
$(\lambda_4, \mathbf{x}_4)$	7.8025e−016
$(\lambda_5, \mathbf{x}_5)$	3.7137e−014
$(\lambda_6, \mathbf{x}_6)$	2.9549e−014
$(\lambda_7, \mathbf{x}_7)$	9.4143e−016
$(\lambda_8, \mathbf{x}_8)$	9.4143e−016
$(\lambda_9, \mathbf{x}_9)$	6.0018e−014
$(\lambda_{10}, \mathbf{x}_{10})$	4.6464e−014
$\|\check{C} - C\|_2$	1.9222e−014
$\|\check{K} - K\|_2$	1.7951e−014

Example 5.11. Consider the case where $n = 4$ and the prescribed eigenvalues are given by

$$\lambda_1 = \bar{\lambda}_2 = 3.3068 + 8.1301i, \quad \lambda_3 = \bar{\lambda}_4 = 1.8702 + 2.7268i, \quad \lambda_5 = 5.4385,$$

with corresponding eigenvectors

$$\mathbf{x}_1 = \bar{\mathbf{x}}_2 = \begin{bmatrix} 0 \\ 9.2963 + 1.5007i \\ 2.3695 + 1.9623i \\ 3.8789 + 1.0480i \end{bmatrix}, \quad \mathbf{x}_3 = \bar{\mathbf{x}}_4 = \begin{bmatrix} 0 \\ 6.5809 + 8.3476i \\ 4.9742 + 8.0904i \\ 1.1356 + 5.5542i \end{bmatrix}, \quad \mathbf{x}_5 = \begin{bmatrix} 1 \\ 0 \\ 0 \\ 0 \end{bmatrix},$$

respectively. It is obvious upon inspection that the linearly dependent vector in the above X must be a complex-valued vector. Let this linearly dependent vector be $\mathbf{x}_4$. Then the real symmetric quadratic pencil

$$Q(\lambda) = \lambda^2 I + \lambda C + K,$$

where $C = -(S + S^H) + \gamma_1 \mathbf{y}_1 \mathbf{y}_2^H + \gamma_2 \mathbf{y}_2 \mathbf{y}_1^H + \gamma_3 \mathbf{y}_3 \mathbf{y}_3^H$ and $K = -(S + C)S$, can be constructed with arbitrary $\gamma_3 \in \mathbb{R}$. In Table 5.4 we show the residual $\|Q(\lambda_j)\mathbf{x}_j\|_2$, for $j = 1, \ldots, 5$ with various values of γ_3.

Table 5.4. *Residual* $\|Q(\lambda_j)\mathbf{x}_j\|_2$ *for Example 5.11*

	Residual	Residual	Residual
Eigenpairs	$\gamma_3 = 2.56$	$\gamma_3 = 40.6$	$\gamma_3 = 506$
$(\lambda_1, \mathbf{x}_1)$	$2.9334e{-}011$	$2.9334e{-}011$	$2.9334e{-}011$
$(\lambda_2, \mathbf{x}_2)$	$2.9334e{-}011$	$2.9334e{-}011$	$2.9334e{-}011$
$(\lambda_3, \mathbf{x}_3)$	$7.8802e{-}011$	$7.8802e{-}011$	$7.8802e{-}011$
$(\lambda_4, \mathbf{x}_4)$	$7.8802e{-}011$	$7.8802e{-}011$	$7.8802e{-}011$
$(\lambda_5, \mathbf{x}_5)$	$1.7764e{-}015$	$2.8422e{-}014$	$4.5475e{-}013$

Example 5.12. Suppose we modify the first entries of the complex eigenvectors to

$$\mathbf{x}_1 = \bar{\mathbf{x}}_2 = \begin{bmatrix} 9.2963 + 1.5007i \\ 9.2963 + 1.5007i \\ 2.3695 + 1.9623i \\ 3.8789 + 1.0480i \end{bmatrix}, \quad \mathbf{x}_3 = \bar{\mathbf{x}}_4 = \begin{bmatrix} 6.5809 + 8.3476i \\ 6.5809 + 8.3476i \\ 4.9742 + 8.0904i \\ 1.1356 + 5.5542i \end{bmatrix}, \quad \mathbf{x}_5 = \begin{bmatrix} 1 \\ 0 \\ 0 \\ 0 \end{bmatrix}.$$

Still, we see that the linearly dependent vector in the corresponding X must be a complex-valued vector, say $\mathbf{x}_4$. However, we find that the condition (5.120) is not satisfied because

$$\mathbf{x}_3^H \mathbf{z} = \mathbf{x}_3^H (\bar{\lambda}_4 I_4 - S)\bar{\mathbf{x}}_4 = -115.54 + 600.67i \neq 0.$$

The system (5.118) being inconsistent, the real coefficient γ_3 in (5.113) is not solvable. We conclude that the prescribed vectors and the corresponding scalars λ_i, $i = 1, \ldots, 5$, indicated above cannot be part of the spectrum of any 4×4 real-valued, symmetric, and monic quadratic pencil.

5.5 Summary

Partially described inverse eigenvalue problems arise in many important applications. Because eigenpair information sometimes provides additional intrinsic relationships, especially for structured matrices, one of the emphases in this chapter has been to answer how many pairs of information are needed to determine such a matrix.

We have shown by symbolic computation that the dimension of the subspace $\mathcal{S}(\mathbf{u}) \cap \mathcal{S}(\mathbf{v})$ of Toeplitz matrices with two generically prescribed eigenvectors $\mathbf{u}$ and $\mathbf{v}$ is independent of the size of the problem. We have further shown that the dimension is either two, three, or four, depending upon whether the eigenvectors are symmetric or skew-symmetric. All the cases are justified to the extent that the transformation matrices that result in the desired elimination are fully described in terms of the components of u and $\mathbf{v}$. Only one proof (Lemma 5.4) is detailed, but the rest can be done in a very similar way. This result extends that in Cybenko (1984) where only one eigenvector is prescribed. It is an interesting and remarkable observation that the dimension of the set $S(\mathbf{u}) \cap S(\mathbf{v})$ is independent of the size of the problem.

We have also studied the inverse problem of constructing a Toeplitz matrix from two prescribed eigenpairs. We have shown that in almost every direction of $\ker(\tilde{M})$, there is one and only one Toeplitz matrix with the prescribed eigenpairs. In particular, it is shown that if n is odd and if at least one of the given eigenvectors is symmetric, or if n is even and one eigenvector is symmetric and the other is skew-symmetric, then the Toeplitz matrix is unique.

We have further studied the quadratic inverse eigenvalue problems with a great many details. This is a very practical problem in that in a large or complicated system, often it is the case that only partial eigeninformation is available. To understand how a physical system modelled by a quadratic pencil should be modified with only partial eigeninformation at hand, it will be very helpful to first understand how the QIEP should be solved. Some general theory toward that end has been presented in this chapter.

We found that the SQIEP is solvable, provided that the number of given eigenpairs is less than or equal to the size of matrices and that the given vectors are linearly independent. A simple recipe for constructing such a matrix was described, which can serve as the basis for numerical computation. We also found that the unspecified eigenstructure of the reconstructed quadratic pencil is quite limited in the sense discussed in Section 5.3.2. We demonstrated three different ways for the construction that not only satisfied the spectral constraints but also best approximated the original analytical model in some least squares sense. The existence theory for the MQIEP when the leading matrix coefficient M is known and fixed is entirely different from that for the SQIEP. In both cases, the procedure used in the proof can also provide a basis for numerical computation.

We note that the stiffness matrix K is normally more complicated than the mass matrix M. The requirement of maintaining physical feasibility also imposes constraints on the stiffness matrix, making it less flexible and more difficult to construct. Thus, one usual way of formulating an inverse eigenvalue problem is to have the stiffness matrix K determined and fixed from the existing structure, known as the static constraints, and then to find the mass matrix M so that some desired natural frequencies are achieved. This is sometimes so desired even without the damping term C. By exchanging the roles of M and K, the discussion in this paper could be applied equally well to the inverse quadratic eigenvalue problem formed with the aforementioned static constraints in mind.

The study made in this chapter should have shed light on the long standing question of how much a quadratic pencil could be updated, modified, or tuned if some of its eigenvalues and eigenvectors are to be kept invariant. Finally, we should point out that there are unfinished tasks in this study. Among these, sensitivity analysis in the case of a unique solution, robustness in the case of multiple solutions, and existence theory where M or K are specially structured, are just a few interesting topics that have yet to be further investigated.

6

LEAST SQUARES INVERSE EIGENVALUE PROBLEMS

6.1 Overview

Thus far, we have insisted that the solution of an IEP must satisfy both the spectral constraint and the structural constraint. Under these stringent conditions, the IEP generally has no real-valued solution. We know from Theorem 3.13, for example, that the MIEP is always solvable over the complex field and that the number of solutions is at most $n!$, provided all the principal minors of A are distinct from zero. When restricted to the real field, however, only a few sufficient conditions can guarantee the existence of a solution for the MIEP (Theorem 3.14). This nonexistence of a real solution can easily be seen by examples in other types of IEPs. An approximate solution in the sense of least squares is sometimes desirable.

We have indicated earlier in Figure 1.1 that every IEP has a natural generalization to a least squares formulation. More than just being of theoretical interest, this generalization sometimes carries significant purposes in application. In the event that the spectral constraint and the structural constraint cannot be satisfied simultaneously, a decision on which constraint could be compromised has to be made. To effect the physical realizability, for example, it may occur that one of the two constraints in an IEP has to be enforced more critically than the other. Without this, the physical system simply cannot be built. On the other hand, inaccurate measurement of the spectrum implies uncertain information. Under the physical uncertainty, one constraint could be more relaxed than the other.

It is important to note that the least squares approximation can be applied to either constraint in an IEP. That is, we can either enforce the structural constraint and seek to minimize the discrepancy between the inherent spectrum of the structured matrix and the prescribed spectrum in the least squares sense, or enforce the spectral constraint and seek to best fit the desirable structure in the least squares sense. Depending upon which constraint is being enforced, two types of least squares problems can be formulated.

Though it remains interesting to recast every type of IEP we have discussed so far in the least squares setting and to investigate the many associated open questions, such a task would become too immense and trite to be covered in one single chapter. Instead, we shall highlight only some of the main notions when considering a least squares IEP in general and leave details for specific types of IEPs to curious readers.

We shall explore both formulations for the case of LiPIEP2 in some more detail in this chapter. Our main purpose is to demonstrate how a least squares problem should be formulated and how that formulation might affect the answer. Our approach furnishes an example of general least squares IEPs. Perhaps the most exciting result is that, in the context of LiPIEP2, both types of least squares formulations are ultimately equivalent. We are not sure whether such a result can be generalized to least squares formulations of other types of IEPs.

6.2 An example of MIEP

We have already seen in Example 3.5 the range of solvable $\{\lambda_1, \lambda_2\}$ for a two-dimensional MIEP2. It might be interesting to broaden the discussion by dropping the requirement of nonnegativity of $X \in \mathcal{D}_{\mathcal{R}}(n)$. Recall that an MIEP can be converted into a LiPIEP2 via the basis matrices (3.61). The following example also demonstrates an interesting connection between a least squares MIEP and an optimal preconditioner discussed in Greenbaum and Rodrigue (1989) and Greenbaum (1992).

Example 6.1. Given $A = \begin{bmatrix} a & b \\ b & c \end{bmatrix}$ and $\lambda := \{\lambda_1, \lambda_2\}$, the characteristic polynomial of $A(\mathbf{x}) = XA$ for $X = \operatorname{diag}(x_1, x_2) \in \mathbb{R}^2$ is

$$c_A(\lambda) = \lambda^2 - (ax_1 + cx_2)\lambda + (ac - b^2)x_1x_2. \tag{6.1}$$

The MIEP is equivalent to selecting x_1 and x_2 so that λ_1 and λ_2 are the roots of (6.1). Toward this end, we solve x_1 and x_2 in terms of λ_1 and λ_2. It is not difficult to see that a necessary and sufficient condition for the 2×2 MIEP to have a real solution is

$$(a^2c^2 - 2acb^2 + b^4)\lambda_1^2 - 2(a^2c^2 - b^4)\lambda_1\lambda_2 + (a^2c^2 - 2acb^2 + b^4)\lambda_2^2 \geq 0. \tag{6.2}$$

After rotating the (λ_1, λ_2)-axes counterclockwise by an angle of $\pi/4$, the condition (6.2) is reduced to

$$(b^2 - ac)(b^2\hat{\lambda}_1^2 - ac\,\hat{\lambda}_2^2) \geq 0. \tag{6.3}$$

Given any values of a, b, and c, the inequality (6.3) provides an easy way to check whether the MIEP with a given pair of eigenvalues λ_1, λ_2 is solvable.

When A is symmetric and positive definite, a typical domain of feasible λ_1 and λ_2 occupies a shaded region like that in Figure 6.1. If we assume that $\lambda_1 \leq \lambda_2$, as in the discussion, then only the shaded area above the main diagonal line is needed. It is clear from Figure 6.1 that there is a large region in which a given pair of eigenvalues (λ_1, λ_2) fails to be feasible for the MIEP.

Suppose now we want to consider a least squares problem of MIEP in the sense of finding $\mathbf{x}^* \in \mathbb{R}^2$ so that $\|\sigma(\operatorname{diag}(\mathbf{x}^*)A) - \boldsymbol{\lambda}\|_2 = \min_{\mathbf{x}\in\mathbb{R}^2} \|\sigma(\operatorname{diag}(\mathbf{x})A) - \boldsymbol{\lambda}\|_2$. This is the situation where the structural constraint,

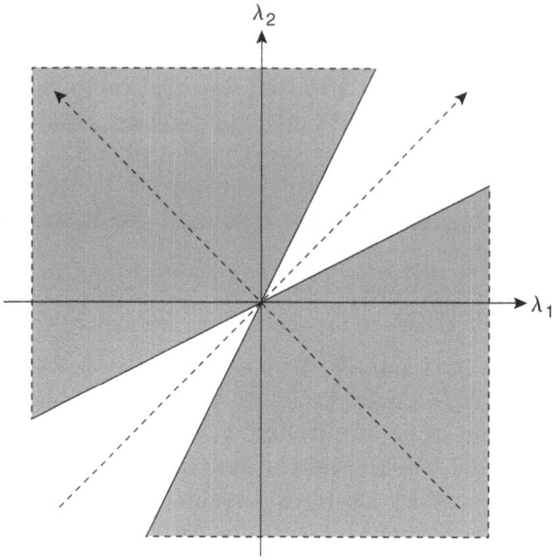

Figure 6.1. *Region of (λ_1, λ_2) on which the MIEP is solvable*

namely, the pre-multiplication by a diagonal matrix, is enforced and the spectral constraint is least squares approximated. Consider the case when the prescribed eigenpair $\lambda = (\lambda_1, \lambda_2)$ is outside the shaded region. The drawing in Figure 6.1 clearly suggests that the corresponding $\sigma(\mathrm{diag}(\mathbf{x}^*)A)$ of the optimal $\mathbf{x}^*$ should be exactly the point on the borderline that is closest to the point $\boldsymbol{\lambda}$.

Observer further that if the diagonal matrix X is required to be positive definite, then the condition number $\kappa(XA)$ referred to in Greenbaum and Rodrigue (1989) is precisely the ratio $\lambda_2(\mathbf{x})/\lambda_1(\mathbf{x}) > 0$. The optimal positive definite diagonal preconditioner in the sense defined in Forsythe and Straus (1955), Greenbaum and Rodrigue (1989), Greenbaum (1992), and van der Sluis (1969/1970b) is given by $X := \mathrm{diag}(1/a, 1/c)$. In this case, the eigenvalues of the corresponding XA are precisely those lying on the upper borderline in the first quadrant of Figure 6.1.

6.3 Least squares LiPIEP2

Without specific *a priori* knowledge about whether an IEP is solvable or not, we might as well begin solving an IEP by casting the problem under the framework of least squares with the hope that a least squares solution will coincide with the exact solution, if the latter exists. We have mentioned earlier that the least squares approximation can be imposed on either constraint while the other constraint is observed.

To solidify the idea and to fix the notation, we restate the definition of a LiPIEP2, that is, Problem 3.4, as follows:

Problem 6.1 *(LiPIEP2)*
Given matrices $A_0, A_1, \ldots, A_n$ in $\mathcal{S}(n)$ and real numbers $\lambda_1 \leq \cdots \leq \lambda_n$, find $\mathbf{d} \in \mathbb{R}^n$ such that eigenvalues $\lambda_1(\mathbf{d}) \leq \cdots \leq \lambda_n(\mathbf{d})$ of

$$A(\mathbf{d}) := A_0 + \sum_{k=1}^{n} d_k A_k, \tag{6.4}$$

satisfy $\lambda_i(\mathbf{d}) = \lambda_i$ for $i = 1, 2, \ldots, n$.

Be aware in the definition that the eigenvalues are arranged in ascending order. This is possible because the underlying matrices are symmetric. The symmetry in the description of LiPIEP2 above and in the discussion of least squares approximations hereafter is not essential. The work can be generalized to more general matrices where complex-valued eigenvalues may be involved. In that case, an extra burden of pairing off the eigenvalues is necessary. This task will become clear in the sequel.

6.3.1 *Formulation*

Our first formulation is simply to generalize the conventional LiPIEP2. We shall assume that the structure imposed by the parameters is fixed. However, we relax the problem to the extent that the set of eigenvalues to be matched is not necessarily the entire spectrum and the number of free parameters in $\mathbf{d}$ is not necessarily the same as the dimension of the underlying matrices. This is more in line with the spirit of the generic PIEP that we originally described in Problem 3.1. It will become clear that in order to solve such a least squares problem, a combinatorics problem naturally arises. This formulation also inherits the troublesome non-smoothness of eigenvalues.

Problem 6.2 *(LSIEP1)*
Given matrices $A_0, A_1, \ldots, A_\ell$ in $\mathcal{S}(n)$ and real numbers $\lambda_1 \leq \cdots \leq \lambda_m$ $(m \leq n)$, find $\mathbf{d} \in \mathbb{R}^\ell$ and a permutation $\rho = \{\rho_1, \ldots, \rho_m\}$ with $1 \leq \rho_1 < \cdots < \rho_m \leq n$ such that the function

$$F(\mathbf{d}, \rho) := \frac{1}{2} \sum_{i=1}^{m} (\lambda_{\rho_i}(\mathbf{d}) - \lambda_i)^2, \tag{6.5}$$

where $\lambda_i(\mathbf{d})$, $i = 1, \ldots, n$, are eigenvalues of the matrix

$$A(\mathbf{d}) = A_0 + \sum_{i=1}^{\ell} d_i A_i, \tag{6.6}$$

is minimized.

It is worth noting that the set of prescribed eigenvalues has cardinality m which might be less than n and that the set of parameters $\mathbf{d}$ has cardinality ℓ which might be different from m. Consequently, associated with the LSIEP1 is a combinatorics problem

$$\min_{1 \le \rho_1 < \cdots < \rho_m \le n} \sum_{i=1}^{m} (\lambda_{\rho_i}(\mathbf{d}) - \lambda_i)^2, \tag{6.7}$$

that looks for the closest match between a subset of the spectrum of $A(\mathbf{d})$ and the prescribed eigenvalues. Note also that the number of available parameters for adjusting the matrix $A(\mathbf{d})$ is ℓ which could be greater than or less than n.

Clearly, the LSIEP1 is a non-linear least squares problem. It is important to observe that the function $F(\mathbf{d}, \rho)$ is not differentiable in $\mathbf{d}$ when the permutation ρ is changed. Still, many standard optimization techniques might be applicable. In particular, the Newton method for solving $\nabla F = 0$ is possible. At a substantial cost, the Hessian matrix of F can be explicitly formulated (see (6.27)). We shall see that the Hessian matrix still exists even if multiple eigenvalues of $A(\mathbf{d})$ are present. (The unbounded terms in (6.27) will eventually be cancelled out at coalescent eigenvalues.) As the separation of eigenvalues decreases, however, the Hessian matrix becomes increasingly ill-conditioned.

Our second formulation is to cast the least squares approximation problem as a problem of finding the distance between two interesting geometric objects. In this case, the spectral constraint is enforced. More specifically, for a given $\Lambda_m := \operatorname{diag}\{\lambda_1, \ldots, \lambda_m\}$, consider the subset

$$\Gamma := \left\{ Q \operatorname{diag}(\Lambda_m, \Lambda) Q^\top \,|\, Q \in \mathcal{O}(n), \Lambda \in \mathcal{D}_{\mathcal{R}}(n-m) \right\}, \tag{6.8}$$

and the affine subspace

$$\mathcal{A} := \{A(\mathbf{d}) \,|\, \mathbf{d} \in \mathbb{R}^\ell\}, \tag{6.9}$$

where $A(\mathbf{d})$ is defined by (6.6). Note that Γ contains all symmetric matrices in $\mathbb{R}^{n \times n}$ with $\lambda_1, \ldots, \lambda_m$ as part of the spectrum. We adjust $Q \in \mathcal{O}(n)$, $\mathbf{d} \in \mathbb{R}^\ell$, in searching for the shortest distance between $\mathcal{A}$ and Γ. This idea constitutes another meaningful least squares approximation.

Problem 6.3 *(LSIEP2)*
Find $\mathbf{d} \in \mathbb{R}^\ell$, $Q \in \mathcal{O}(n)$, and $\Lambda \in \mathcal{D}_{\mathcal{R}}(n-m)$ such that the function

$$G(\mathbf{d}, Q, \Lambda) := \frac{1}{2}\|A(\mathbf{d}) - Q\mathrm{diag}\,(\Lambda_m, \Lambda)\,Q^\top\|^2, \tag{6.10}$$

where $\|\cdot\|$ denotes the Frobenius matrix norm, is minimized.

Note that in the formulation of LSIEP2, no differentiation of eigenvalues is required. This approach has the advantages that a linearly convergent iterative method can be introduced to generate good starting values for other faster but more expensive and locally convergent methods. The idea can be applied to multiplicative inverse eigenvalue problems for the purpose of preconditioning. We shall illustrate our ideas by numerical examples.

6.3.2 *Equivalence*

The most surprising result we are about to discover is that, in spite of their different appearance, these two formulations (for LiPIEP2) are in fact equivalent. At first glance, LSIEP1 and LSIEP2 appear to be very different. In particular, it appears that no permutation of eigenvalues is involved in LSIEP2. However, a process of implicit sorting is indeed happening inside LSIEP2. That implicitly defined sorting leads to the fact that LSIEP1 and LSIEP2 are indeed equivalent. We hasten to point out that it is not clear whether the results developed for LiPIEP2 remain true for other types of IEPs. Lots of open questions deserve further investigation.

We first observe the following result for an arbitrarily given vector $\mathbf{d} \in \mathbb{R}^\ell$ and permutation $\rho = \{\rho_1, \ldots, \rho_m\}$ with $1 \le \rho_1 < \cdots < \rho_m \le n$. For convenience, define $\Lambda_\rho(\mathbf{d}) := \mathrm{diag}\{\lambda_{\rho_1}(\mathbf{d}), \ldots, \lambda_{\rho_m}(\mathbf{d})\}$.

Theorem 6.1. Let $\overline{\rho}$ denote the complement of ρ over the set $\{1, \ldots, n\}$. Define $\Lambda_\sharp := \Lambda_{\overline{\rho}}(\mathbf{d})$. Let $Q_\sharp$ denote the matrix whose columns are orthonormal eigenvectors of $A(\mathbf{d})$ arranged in such a way that

$$Q_\sharp^\top A(\mathbf{d}) Q_\sharp = \mathrm{diag}(\Lambda_\rho(\mathbf{d}), \Lambda_{\overline{\rho}}(\mathbf{d})). \tag{6.11}$$

Then $G(\mathbf{d}, Q_\sharp, \Lambda_\sharp) = F(\mathbf{d}, \rho)$.

Proof From (6.11), we have

$$\begin{aligned} G(\mathbf{d}, Q_\sharp, \Lambda_\sharp) &= \tfrac{1}{2}\|A(\mathbf{d}) - Q_\sharp \mathrm{diag}(\Lambda_m, \Lambda_\sharp) Q_\sharp^\top\|^2 = \tfrac{1}{2}\|Q_\sharp^\top A(\mathbf{d}) Q_\sharp - \mathrm{diag}(\Lambda_m, \Lambda_\sharp)\|^2 \\ &= \tfrac{1}{2}\|\mathrm{diag}(\Lambda_\rho(\mathbf{d}) - \Lambda_m, \Lambda_{\overline{\rho}}(\mathbf{d}) - \Lambda_\sharp)\|^2 = \tfrac{1}{2}\|\Lambda_\rho(\mathbf{d}) - \Lambda_m\|^2. \end{aligned}$$

The assertion therefore follows. □

In the above theorem, we call readers' attention to the fact that a rearrangement of the eigenvectors has been made. We now establish the relationship between LSIEP1 and LSIEP2.

Theorem 6.2. Suppose $(\mathbf{d}^*, \rho^*)$ and $(\mathbf{d}_+, Q_+, \Lambda_+)$ are the global minimizers of LSIEP1 and LSIEP2, respectively. Then:

(i) With $\mathbf{d} = \mathbf{d}^*$, the permutation ρ^* solves (6.7).
(ii) $\mathbf{d}^* = \mathbf{d}_+$.
(iii) The columns of Q_+ are orthonormal eigenvectors of $A(\mathbf{d}^*)$ arranged in such a way that $Q_+^\top A(\mathbf{d}^*)Q_+ = \operatorname{diag}\left(\Lambda_{\rho^*}(\mathbf{d}^*), \Lambda_{\overline{\rho^*}}(\mathbf{d}^*)\right)$.
(iv) $\Lambda_+ = \Lambda_{\overline{\rho^*}}(\mathbf{d}^*)$.
(v) $F(\mathbf{d}^*, \rho^*) = G(\mathbf{d}_+, Q_+, \Lambda_+)$.

Proof Since

$$F(\mathbf{d}^*, \rho^*) = \min_{\rho} \min_{\mathbf{d}} F(\mathbf{d}, \rho) = \min_{\rho} F(\mathbf{d}^*, \rho),$$

it is obvious that ρ^* must solve (6.7) with $\mathbf{d} = \mathbf{d}^*$.

Let $Q(\mathbf{d}^*)$ denote the orthogonal matrix of which columns are eigenvectors of $A(\mathbf{d}^*)$ arranged in such a way that $Q(\mathbf{d}^*)^\top A(\mathbf{d}^*)Q(\mathbf{d}^*) = \operatorname{diag}\left(\Lambda_{\rho^*}(\mathbf{d}^*), \Lambda_{\overline{\rho^*}}(\mathbf{d}^*)\right)$. It follows from Theorem 6.1 that

$$G(\mathbf{d}_+, Q_+, \Lambda_+) \leq G(\mathbf{d}^*, Q(\mathbf{d}^*), \Lambda_{\overline{\rho^*}}(\mathbf{d}^*)) = F(\mathbf{d}^*, \rho^*). \tag{6.12}$$

On the other hand, we have

$$\begin{aligned} G(\mathbf{d}_+, Q_+, \Lambda_+) &= \min_{\mathbf{d},\Lambda} \min_{Q} \tfrac{1}{2}\|A(\mathbf{d}) - Q\operatorname{diag}(\Lambda_m^*, \Lambda)Q^\top\|_F^2 \\ &= \min_{\mathbf{d},\Lambda} \tfrac{1}{2}\|A(\mathbf{d}) - Q(\mathbf{d})\operatorname{diag}\left(\Lambda_m^*, \Lambda\right) Q(\mathbf{d})^\top\|_F^2 &(6.13) \\ &= \min_{\mathbf{d},\Lambda} \tfrac{1}{2}\|\operatorname{diag}\left(\Lambda_\rho(\mathbf{d}) - \Lambda_m^*, \Lambda_{\overline{\rho}}(\mathbf{d}) - \Lambda\right)\|_F^2 &(6.14) \\ &\geq \min_{\mathbf{d}} \tfrac{1}{2}\|\Lambda_\rho(\mathbf{d}) - \Lambda_m^*\|_F^2. &(6.15) \end{aligned}$$

The equality (6.13) follows from the Wielandt–Hoffman theorem (Horn and Johnson, 1991, Theorem 6.3.5) where columns of the orthogonal matrix $Q(\mathbf{d})$ are eigenvectors of $A(\mathbf{d})$ arranged in such a way that elements in the diagonal matrix $Q(\mathbf{d})^\top A(\mathbf{d})Q(\mathbf{d})$ are in the same ordering as those of $\operatorname{diag}(\Lambda_m^*, \Lambda)$. The permutation ρ in (6.14) simply reflects such an rearrangement of eigenvalues of $A(\mathbf{d})$. (This is the implicit sorting referred to above.)

Together with (6.12), we find that the equality in (6.15) holds if and only if $\mathbf{d} = \mathbf{d}^* = \mathbf{d}_+$, $\rho = \rho^*$ and $\Lambda_+ = \Lambda_{\overline{\rho^*}}(\mathbf{d}^*)$. □

It is important to note that Theorem 6.2 holds only at the *global* minimizer. Be aware that LSIEP1 and LSIEP2 are nonlinear least squares problems and, hence, each might have many local minimizers. The equivalence at the local solutions is not guaranteed. At the global solution, Theorem 6.2 warrants that LSIEP1 can be solved by dealing with LSIEP2 only. LSIEP2 also provides a geometric interpretation of LSIEP1. We think such a connection is quite intriguing.

6.3.3 *Lift and projection*

Taking advantage of the equivalence between LSIEP1 and LSIEP2, we now propose an iterative method that alternates points between Γ and $\mathcal{A}$. The idea is essentially the same as the so-called alternating projection method for convex sets (Cheney and Goldstein, 1959; Deutsch, 2001; Han, 1988), except that one of our sets, namely Γ, is not convex. Our main point here is to show that the proximity maps can still be well defined. We call our algorithm a *lift-and-projection* (**LP**) method. A similar idea of lift and projection has been used in Chu and Watterson (1993). For LSIEP2, an extra combinatorics problem is involved. The cost of computation is one spectral decomposition plus one sorting per lift and two triangular linear system solving per projection.

Before introducing the LP method, we stress that no differentiation of eigenvalues is involved in this algorithm. We also note that the LP method converges slowly but globally. We may, therefore, take advantage of a hybrid method by first applying the LP method to attain a low order of accuracy and then switching to a faster but locally convergent method for a high order of accuracy. The LP method may also be used at a step where $A(\mathbf{d})$ appears to have multiple or nearly coalescent eigenvalues. The method is outlined below.

Algorithm 6.1 (The LP method)
Given $\mathbf{d}^{(0)} \in \mathbb{R}^{\ell}$, repeat the following two steps for $k = 0, 1, \ldots$ until convergence:

1. **(Lift)** Find the point $Z^{(k)} \in \Gamma$ such that $\operatorname{dist}(A(\mathbf{d}^{(k)}), Z^{(k)}) = \operatorname{dist}(A(\mathbf{d}^{(k)}), \Gamma)$. We call $Z^{(k)}$ a *lift* of $A(\mathbf{d}^{(k)})$ onto Γ.
2. **(Projection)** Find the point $\mathbf{d}^{(k+1)} \in R^{\ell}$ such that $\operatorname{dist}(A(\mathbf{d}^{(k+1)}), Z^{(k)}) = \operatorname{dist}(Z^{(k)}, \mathcal{A})$. The point $A(\mathbf{d}^{(k+1)}) \in \mathcal{A}$ is called a *projection* of $Z^{(k)}$ onto $\mathcal{A}$.

A schematic diagram of the iteration is depicted in Figure 6.2 although the topology of Γ is much more complicated.

The projection of $Z^{(k)} \in \mathbb{R}^{n \times n}$ onto $\mathcal{A}$ is easy to do. The vector $\mathbf{d}^{(k+1)}$ is the solution of the linear system

$$\sum_{i=1}^{\ell} \langle A_i, A_j \rangle d_i^{(k+1)} = \langle Z^{(k)} - A_0, A_j \rangle, \quad j = 1, \ldots, \ell, \tag{6.16}$$

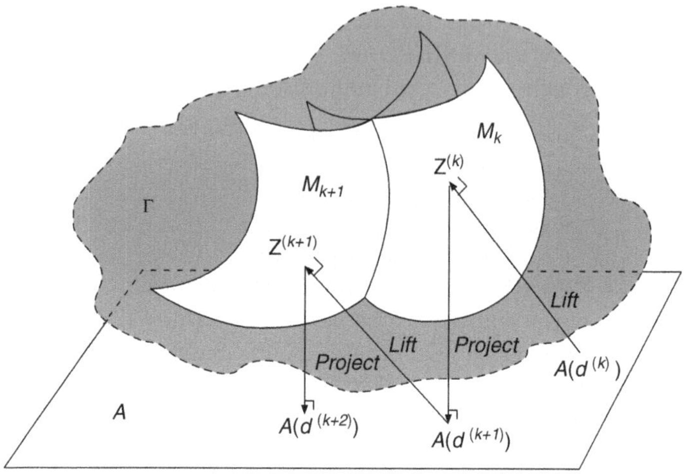

Figure 6.2. *Geometric sketch of lift and projection*

where $\langle A, B\rangle := \text{trace}(A^\top B)$ is the Frobenius inner product for matrix A and B. Note that the coefficient matrix in (6.16) is independent of k. So the left-hand side of (6.16) needs to be factorized only once.

The lift step is not as easy because elements in Γ involve $n - m$ undetermined eigenvalues. Motivated by the proof of Theorem 6.2, however, the step can proceed as follows. Suppose

$$A(\mathbf{d}^{(k)}) = Q(\mathbf{d}^{(k)})\text{diag}\left(\Lambda_{\rho^{(k)}}(\mathbf{d}^{(k)}), \Lambda_{\overline{\rho^{(k)}}}(\mathbf{d}^{(k)})\right) Q(\mathbf{d}^{(k)})^\top,$$

is the spectral decomposition of $A(\mathbf{d}^{(k)})$ where $\rho^{(k)} = \rho^{(k)}(\mathbf{d}^{(k)})$ is the permutation that solves the combinatorics problem (6.7) with $\mathbf{d} = \mathbf{d}^{(k)}$ and $Q(\mathbf{d}^{(k)})$ is the corresponding orthogonal matrix of eigenvectors. Then the shortest distance between $A(\mathbf{d}^{(k)})$ and Γ is attained at the point (Brockett, 1991; Chu, 1992a)

$$Z^{(k)} := Q(\mathbf{d}^{(k)})\text{diag}\left(\Lambda_m^*, \Lambda_{\overline{\rho^{(k)}}}(\mathbf{d}^{(k)})\right) Q(\mathbf{d}^{(k)})^\top. \tag{6.17}$$

In other words, in order to find the shortest distance from $A(\mathbf{d}^{(k)})$ to Γ, it suffices to find the shortest distance from $A(\mathbf{d}^{(k)})$ to a *substructure* $\mathcal{M}_k$ of Γ, where the substructure

$$\mathcal{M}_k := \left\{Q\text{diag}\left(\Lambda_M^*, \Lambda_{\overline{\rho^{(k)}}}(\mathbf{d}^{(k)})\right) Q^\top | Q \in \mathcal{O}(n)\right\}, \tag{6.18}$$

has a much simpler topology than Γ because the diagonal elements are fixed. (See Figure 6.2.) The price we pay for this roundabout is to solve (6.7) per step. It is worth noting that when the iterates are reaching convergence the permutations $\rho^{(k)}$ should become stablized.

Theorem 6.3. The LP method is a descent method in the sense that

$$\|A(\mathbf{d}^{(k+1)}) - Z^{(k+1)}\|^2 \leq \|A(\mathbf{d}^{(k+1)}) - Z^{(k)}\|^2 \leq \|A(\mathbf{d}^{(k)}) - Z^{(k)}\|^2. \quad (6.19)$$

Thus the LP method generates a sequence of matrix pairs $\{(Z^{(k)}, A(\mathbf{d}^{(k)}))\}$ that converges to a (local) stationary point for the problem of minimizing (6.10).

Proof The second inequality is obvious because $A(\mathbf{d}^{(k+1)})$ is the projection point of $Z^{(k)}$ onto $\mathcal{A}$. The first inequality follows from the Wielandt–Hoffman theorem. More precisely, we observe that

$$\begin{aligned}
&\|A(\mathbf{d}^{(k+1)}) - Z^{(k)}\|^2 \\
&= \left\| A(\mathbf{d}^{(k+1)}) - Q(\mathbf{d}^{(k)})\mathrm{diag}\left(\Lambda_m^*, \Lambda_{\overline{\rho^{(k)}}}(\mathbf{d}^{(k)})\right) Q(\mathbf{d}^{(k)})^\top \right\|^2 \\
&\geq \left\| A(\mathbf{d}^{(k+1)}) - Q(\mathbf{d}^{(k+1)})\mathrm{diag}\left(\Lambda_m^*, \Lambda_{\overline{\rho^{(k+1)}}}(\mathbf{d}^{(k+1)})\right) Q(\mathbf{d}^{(k+1)})^\top \right\|^2 \\
&= \|A(\mathbf{d}^{(k+1)}) - Z^{(k+1)}\|^2,
\end{aligned}$$

where the inequality follows from the definition of $Q(\mathbf{d}^{(k+1)})$. □

The alternating projection method (Cheney and Goldstein, 1959; Deutsch, 2001; Han, 1988) has been used to tackle the IEP by first reformulating the problem so as to create convex constraints. In contrast, our application to the LSIEP2 is quite direct. Our approach is interesting and remarkable in two respects. One is that, even though the set Γ is complicated, we can simply work with one of its substructures. The other is that, even though the substructure $\mathcal{M}_k$ is not convex, the so-called proximity map can still be formulated by using the Wielandt–Hoffman theorem.

6.3.4 *The Newton method*

We have indicated earlier that the function $F(\mathbf{d}, \rho)$ is not smooth in ρ. Since the permutation ρ is only a discrete variable, such a non-smoothness does not necessarily preclude a proper application of classical least squares techniques to LSIEP1. In this section we briefly describe how this can be done by the Newton method.

For any $A(\mathbf{d})$ defined by (6.6), let $(\mathbf{q}_i(\mathbf{d}), \lambda_i(\mathbf{d}))$ denote an eigenpair of $A(\mathbf{d})$, that is,

$$A(\mathbf{d})\mathbf{q}_i(\mathbf{d}) = \lambda_i(\mathbf{d})\mathbf{q}_i(\mathbf{d}), \quad i = 1, 2, \ldots, n. \quad (6.20)$$

Having found an optimal permutation $\rho = \{\rho_1, \ldots, \rho_m\}$ for problem (6.7), we categorize (6.20) into two groups:

$$\mathbf{q}_{\rho_i}(\mathbf{d})^\top A(\mathbf{d})\mathbf{q}_{\rho_i}(\mathbf{d}) = \lambda_{\rho_i}(\mathbf{d}), \quad i = 1, 2, \ldots, m; \quad (6.21)$$

$$\mathbf{q}_{\overline{\rho}_j}(\mathbf{d})^\top A(\mathbf{d})\mathbf{q}_{\overline{\rho}_j}(\mathbf{d}) = \lambda_{\overline{\rho}_j}(\mathbf{d}), \quad j = 1, 2, \ldots, n-m. \quad (6.22)$$

The notion here is to assume that ρ is fixed and to iterate the variable $\mathbf{d}$ based on the relationship (6.21) since only elements in $\lambda_\rho(\mathbf{d}) = [\lambda_{\rho_1}, \ldots, \lambda_{\rho_m}]^\top$ are involved in (6.6). It is important to mention that in our numerical experiments, we have seen that the permutation ρ gets changed at the initial stage of iteration.

Upon differentiating both sides of (6.21), we obtain the Jacobian matrix $J(\mathbf{d}) = [J_{ik}(\mathbf{d})]$ of $\lambda_\rho(\mathbf{d})$ where

$$J_{ik}(\mathbf{d}) := \frac{\partial \lambda_{\rho_i}(\mathbf{d})}{\partial d_k} = \mathbf{q}_{\rho_i}(\mathbf{d})^\top A_k \mathbf{q}_{\rho_i}(\mathbf{d}), \quad i = 1, \ldots, m; \; k = 1, \ldots, \ell. \tag{6.23}$$

It is not difficult to see that the first and second derivatives of $F(\mathbf{d}, \rho)$ are given, respectively, by

$$\nabla F(\mathbf{d}, \rho) = J(\mathbf{d})^\top (\lambda_\rho(\mathbf{d}) - \lambda), \tag{6.24}$$

and

$$\nabla^2 F(\mathbf{d}, \rho) = J(\mathbf{d})^\top J(\mathbf{d}) + S(\mathbf{d}), \tag{6.25}$$

where

$$S(\mathbf{d}) := \sum_{i=1}^{m} (\lambda_{\rho_i}(\mathbf{d}) - \lambda_i) \nabla^2 (\lambda_{\rho_i}(\mathbf{d}) - \lambda_i). \tag{6.26}$$

The Hessian matrix involved in (6.26) can be calculated, for example, by using the formula (Lancaster, 1964)

$$\frac{\partial^2 (\lambda_i(\mathbf{d}))}{\partial d_k \partial d_j} = 2 \sum_{\substack{t=1 \\ \lambda_t \neq \lambda_i}}^{n} \frac{[\mathbf{q}_t(\mathbf{d})^\top A_k \mathbf{q}_i(\mathbf{d})][\mathbf{q}_t(\mathbf{d})^\top A_j \mathbf{q}_i(\mathbf{d})]}{\lambda_i(\mathbf{d}) - \lambda_t(\mathbf{d})}. \tag{6.27}$$

Note that the summation in (6.27) is over those t for which $\lambda_t \neq \lambda_i$, so the formula is valid even if λ_i is repeated. One step of the conventional Newton method applied to LSIEP1 amounts to solving the linear system

$$(J(\mathbf{d}^{(k)})^\top J(\mathbf{d}^{(k)}) + S(\mathbf{d}^{(k)}))\Delta \mathbf{d}^{(k)} = -J(\mathbf{d}^{(k)})^\top (\lambda_\rho(\mathbf{d}^{(k)}) - \lambda^*), \tag{6.28}$$

and then advancing to $\mathbf{d}^{(k+1)} := \mathbf{d}^{(k)} + \Delta \mathbf{d}^{(k)}$.

The Newton method, especially in the forming of $S(\mathbf{d})$, is very expensive. A possible strategy for remedying this situation is to employ some kind of hybrid method. For example, we could use the LP method in the initial stage to reach convergence at a relatively low order of accuracy. We then switch to the Newton method for achieving high order of accuracy. The approach also has the advantage that the permutation ρ might get stabilized before the Newton method is called. More precisely, we propose the following hybrid method:

Algorithm 6.2 (LP–Newton method)

1. Choose an arbitrary starting vector $\mathbf{d}^{(0)} \in R^\ell$.
2. For $k = 0, 1, 2, \ldots$, do the LP iteration as follows:
   (a) Compute the spectrum decomposition $Q(\mathbf{d}^{(k)})$ and $\Lambda(\mathbf{d}^{(k)})$.
   (b) Find $\rho^{(k)}$ that solves (6.7) with $\mathbf{d} = \mathbf{d}^{(k)}$.

   (c) Form $Z^{(k)}$ according to (6.17).
   (d) Compute $\mathbf{d}^{(k+1)}$ from (6.16).
   (e) Stop if $\|\mathbf{d}^{(k+1)} - \mathbf{d}^{(k)}\| < \epsilon_1$.
3. Set $\mathbf{d}^{(0)} :=$ the limit point of the LP iteration.
4. For $k = 0, 1, 2, \ldots$, do the Newton iteration as follows:
   (a) Generate the Jacobian matrix (6.23).
   (b) Solve the linear equation (6.28) for $\Delta\mathbf{d}^{(k)}$.
   (c) $\mathbf{d}^{(k+1)} := \mathbf{d}^{(k)} + \Delta\mathbf{d}^{(k)}$.
   (d) Stop if $\|\mathbf{d}^{(k+1)} - \mathbf{d}^{(k)}\| < \epsilon_2$.

Choosing ϵ_1 small enough will ensure the global convergence of this method.

6.3.5 *Numerical experiment*

In this section we present some test results of our methods. We understand that there are many other algorithms for solving nonlinear least squares problems. Some locally convergent methods with cost reduction in mind include, for example, the Gauss–Newton method that does not compute $S(\mathbf{d})$ and the Shamanskii method that does not evaluate the Hessian so frequently. However, we choose to implement the three methods discussed in this section – the LP method, the LP–Newton method, and the Newton method, partially to fulfill our curiosity of comparing their performance.

The experiment was carried out by MATLAB. Initial values $\mathbf{d}^{(0)}$ were generated randomly. To assess the efficiency, we carefully measured the CPU time for each test. Numerical results indicate that the LP method usually approached a stationary point quickly in the first few steps. The improvement then slowed down. This is a common phenomenon of linear convergence. In contrast, the Newton method converged slowly at the initial stage, but eventually the rates were picked up and became quadratic. We have also observed cases where the Newton method failed to converge. All of these observations seem to suggest that a hybrid method should be more suitable for least squares inverse eigenvalue problems.

It must be cautioned that the programming is harder than it looks because of the additional combinatorics problem (6.7) involved. This associated combinatorics problem was converted into a linear sum assignment problem. That is, for each given $\mathbf{d}$, we first create the cost matrix $C = [c_{ij}]$ where

$$c_{ij} := \begin{cases} |\lambda_i(\mathbf{d}) - \lambda_j|, & \text{if } 1 \le j \le m, \\ 0, & \text{otherwise.} \end{cases}$$

We then apply an existing algorithm LSAPR (Burkard and Derigs, 1980) to find a permutation ϕ^* that solves

$$\min_{\phi \in S_n} \sum_{i=1}^{n} c_{i,\phi(i)}, \tag{6.29}$$

where S_n is the symmetric group of all permutations of $\{1, 2, \ldots, n\}$. The core of LSAPR is the so-called shortest augmenting path technique which we shall not elaborate here. Once such an optimal ϕ^* is found, the solution ρ to (6.7) is given by

$$\rho = \{i | \phi^*(i) \leq m\}. \tag{6.30}$$

In the case when $m = n$, that is, when the entire spectrum of A is to be matched, it is not necessary to solve (6.29) since the optimal ρ is simply the permutation that arranges eigenvalues $\lambda_i(d)$ in the same ordering as λ_i^*.

After each iteration we measure $e_k = \|\mathbf{d}^{(k)} - \mathbf{d}^{(k-1)}\|$. The iteration is terminated whenever e_k is small enough. For our experiment, the threshold is set at 10^{-8} for all examples. For the LP–Newton method, the LP iteration is terminated when $e_k < \epsilon_1$. It remains an open question as to how small ϵ_1 should be so that the overall cost of computation would be optimized.

Example 6.2. In this example we test the performance of each individual method. In particular, we compare the performance of the LP–Newton method by specifying the number of LP iterations allowed in the algorithm. The same test data:

$$A_0 = \begin{bmatrix} 0 & -1 & 0 & 0 & 0 \\ -1 & 0 & -1 & 0 & 0 \\ 0 & -1 & 0 & -1 & 0 \\ 0 & 0 & -1 & 0 & -1 \\ 0 & 0 & 0 & -1 & 0 \end{bmatrix},$$

$$A_k = 4e_k e_k^\top, \quad k = 1, 2, \ldots, 5,$$

$$\mathbf{d}^{(0)} = [0.63160, 0.23780, 0.90920, 0.98660, 0.50070]^\top,$$

$$\lambda = [1, 1, 2, 3, 4]^\top,$$

are used in all experiments.

Test results, measured in terms of the CPU time in seconds versus the number n of LP iterations allowed in the LP step, are recorded in Figure 6.3. Incidentally, it happens in this case that all three methods converge to the same least squares solution:

$$\mathbf{d}^* = [0.44230, 0.60440, 0.65660, 0.60440, 0.44230]^\top,$$

$$\lambda(\mathbf{d}^*) = [0.58884, 1.0422, 2.07421, 3.1446, 4.1501]^\top.$$

So it offers an ideal comparison.

Under the same stopping criterion $\|e_k\| \leq 10^{-8}$, the result at $n = 0$ represents the performance of the pure Newton method, whereas the result at $n = 250$ represents the performance of the pure LP method. A careful interpretation of

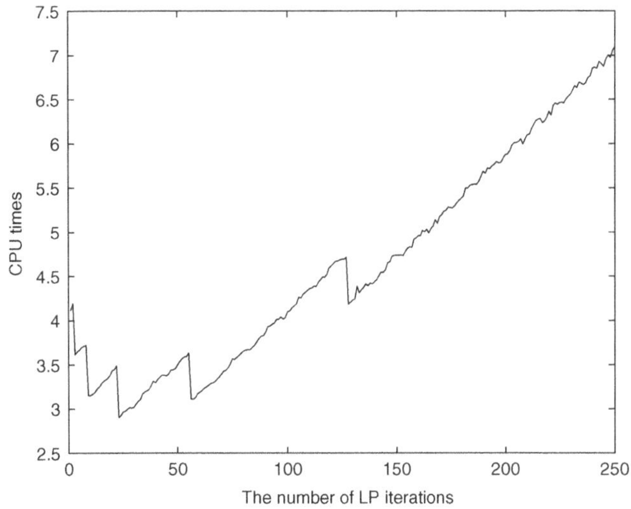

Figure 6.3. *The numbers of LP iterations vs the CPU times for Example 6.2*

the drawing in Figure 6.3 provides several useful insights on the performance of these methods.

The sudden drop of the CPU time in Figure 6.3 is interesting and important. Since we know the cost of the LP iteration is linear in n, the drop of the CPU should be attributed solely to *one less* Newton iteration required to reach the accuracy of convergence. Note that the magnitude of each drop is approximately the same, supporting the conjecture that the drop is caused by one single and resembling event.

The staircase in Figure 6.3 is also interesting. It indicates that the extra LP iterations between two consecutive drops are futile because the number of Newton iterations required to reach the same accuracy has not been reduced.

Although the size of our test problem is small so that the Newton method is not unbearably expensive, Figure 6.3 does suggest that switching to the Newton method after a certain number (23 in this case) of LP iterations will minimize the overall CPU times.

Example 6.3. In this example we want to construct a symmetric Toeplitz matrix with partially prescribed eigenvalues. This is a special case of the notorious ToIEP since only a portion of the eigenvalues are specified. The Toeplitz structure can be written as an LiPIEP2 with the basis matrices:

$$A_0 = 0,$$

$$A_k = [A_{ij}^{(k)}] \in R^{20\times 20} \quad \text{with } A_{ij}^{(k)} := \begin{cases} 1, & \text{if } |i-j| = k-1, \\ 0, & \text{otherwise.} \end{cases}$$

We formulate the problem as a least squares problem with test data:

$$\begin{aligned}\mathbf{d}^{(0)} = [&1.1650, 0.6268, 0.0751, 0.3516, -0.6965, 1.6961, 0.0591,\\ &1.7971, 0.2641, 0.8717, -1.4462, -0.7012, 1.2460, -0.6390,\\ &0.5773, -0.3600, -0.1356, -1.3493, -1.2704, 0.9845]^\top,\\ \boldsymbol{\lambda} = [&-5, -4, -3, -2, -1, 0, 1, 2, 3, 4, 5]^\top.\end{aligned}$$

Note that the spectrum is not complete.

This time we terminate the LP step in the LP–Newton method according to Algorithm 5.1. We have observed two interesting results.

First, although both methods start from the same point $\mathbf{d}^{(0)}$, the LP–Newton and the Newton method converge to distinct solutions. They are, respectively,

$$\begin{aligned}\mathbf{d}^*_{\mathrm{LP-Newton}} = [&0.8486, 0.8424, -0.0050, 0.3076, -0.5089, 1.6325, -0.0659,\\ &1.72764, -0.00038, 1.1018, -1.5155, -0.8286, 1.1952, -0.7433,\\ &0.0336, -0.0737, 0.0356, -1.5870, -0.1220, -0.2275]^\top,\end{aligned}$$

and

$$\begin{aligned}\mathbf{d}^*_{\mathrm{Newton}} = [&7.3464, 1.5410, -6.9636, -1.5020, 5.6980, 5.0777, -4.4890,\\ &-6.4796, 3.3743, 8.2720, -0.6983, -9.4107, -1.0345, 6.8718,\\ &3.1606, -8.8348, -4.3902, 2.4603, 6.7335, -4.9145]^\top.\end{aligned}$$

Secondly, although the resulting matrices $A(\mathbf{d}^*)$ are different, both solutions produce eigenvalues that agree with the given λ almost perfectly, that is, $F(\mathbf{d}^*, \rho^*) \approx 10^{-8}$. Had we known that a perfect match was about to happen, we could have used the cheaper Gauss–Newton method instead of the Newton method to achieve the ultimate quadratic rate of convergence. In general, however, the Gauss–Newton method is less in favor because the LP method does not need the differentiation of eigenvalues.

The computational cost is summarized in Table 6.1. As far as the CPU time is concerned, it is seen that the LP–Newton method converges almost three times faster than the Newton method. It also indicates that improving the accuracy of the LP step from $\epsilon_1 = 0.01$ to $\epsilon_1 = 0.001$ is not necessarily advantageous even though the number of Newton iterations is reduced by 2.

The history of errors e_k is plotted in Figure 6.4. Observe that the Newton method wanders for quite a few steps in the initial stage before it eventually converges to a solution. The LP method, with its descent property, helps to move into a better region for starting the Newton iteration.

Table 6.1. *Computational cost for Example 6.3*

Algorithm	Number of iterations	CPU time (seconds)
1. Newton	24	917.215
2. LP–Newton ($\epsilon_1 = 0.01$)	(LP) 57 (Newton) 7	305.156
3. LP–Newton ($\epsilon_1 = 0.001$)	(LP) 434 (Newton) 5	349.195

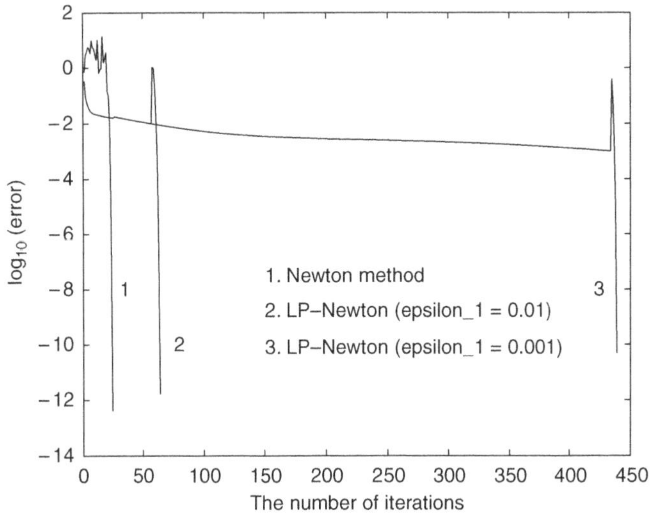

Figure 6.4. *The plot of* $\log_{10}|e_k|$ *for Example 6.3*

Example 6.4. In this example we test the MIEP with partially prescribed eigenvalues. That is, we want to precondition a positive definite matrix:

$$A = \begin{bmatrix} T & E & 0 & 0 \\ E & T & E & 0 \\ 0 & E & T & E \\ 0 & 0 & E & T \end{bmatrix},$$

where

$$T = \begin{bmatrix} 4 & -1 & 0 & 0 \\ -1 & 4 & -1 & 0 \\ 0 & -1 & 4 & -1 \\ 0 & 0 & -1 & 4 \end{bmatrix},$$

$$E = -I_4,$$

by a diagonal premultiplication D so as to attain certain specified eigenvalues. We first convert the MIEP into a LiPIEP2 by means of the Cholesky factorization $A = LL^\top$. In other words, we are now considering a LiPIEP2 with the following test data:

$$\begin{aligned}
A_0 &= 0,\\
A_k &= L^\top E_k L,\\
d^{(0)} &= [1.5578, -2.4443, -1.0982, 1.1226, 0.5817,\\
&\quad -0.2714, 0.4142, -0.9778, -1.0215, 0.3177,\\
&\quad 1.5161, 0.7494, -0.5077, 0.8853, -0.2481, -0.7262]^\top,\\
\boldsymbol{\lambda} &= [1, 5, 10, 15, 20, 25, 30, 35, 40, 45, 50]^\top.
\end{aligned}$$

The LP–Newton method converges to a least squares solution

$$\begin{aligned}
\mathbf{d}^* = [&10.2309, -3.0078, -1.6975, 10.1958, 7.2102,\\
&2.4626, 5.8098, -1.9979, -1.53203.7608,\\
&10.0604, 8.5959, 0.2992, 8.0485, 3.4645, -1.0845]^\top,
\end{aligned}$$

at which again $\boldsymbol{\lambda}$ is perfectly matched with 11 eigenvalues of $A(\mathbf{d}^*) = \operatorname{diag}(\mathbf{d}^*)A$. In contrast, it seems reasonable to conclude from Figure 6.5 that the Newton method diverges after 40 iterates. On the other hand, with a good start point

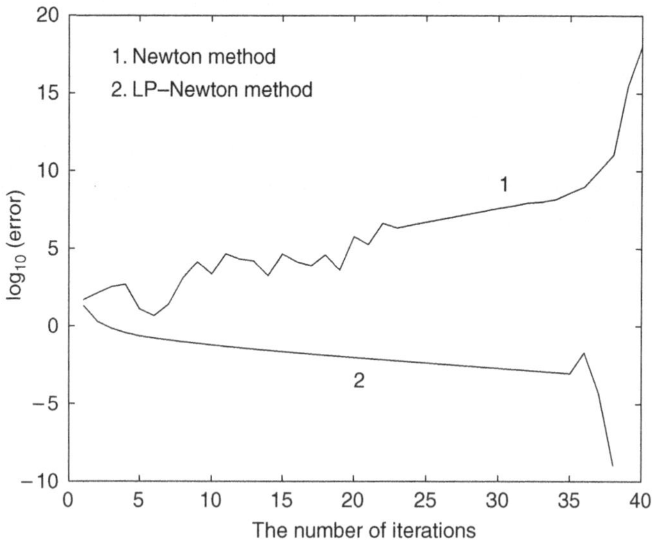

Figure 6.5. *The plot of* $\log_{10}|e_k|$ *for Example 6.4*

Table 6.2. *Computational cost for Example 6.4*

Algorithm	Number of iterations	CPU time (seconds)
1. Newton	≥40	≥923.123
2. LP–Newton	(LP) 35	92.644
	(Newton) 3	

provided by the LP method (with $\epsilon_1 = 0.001$), the Newton method converges to a solution with desired accuracy $\epsilon = 10^{-8}$ within three iterates.

The computational cost is summarized in Table 6.2. None of the methods is cheap, but the advantage of the LP–Newton method is obvious in this case.

6.4 Least squares PDIEP

The discussion thus far, though only limited to the LiPIEP2, points to a common structure shared by spectrally constrained approximations. We shall elaborate upon this general theme in Chapter 7. Before moving into that investigation, we outline in this section yet another kind of least squares problem which is constrained by the partially described eigeninformation.

For simplicity, we shall limit the field to $\mathbb{R}$. Let columns of $V = [\mathbf{v}^{(1)}, \ldots, \mathbf{v}^{(k)}] \in \mathbb{R}^{n\times k}$ and diagonal elements of $\Lambda = \operatorname{diag}\{\lambda_1, \ldots, \lambda_k\} \in \mathbb{R}^{k\times k}$ denote a partial list of eigeninformation. Consider the set

$$\mathcal{M}(V,\Lambda) := \left\{X \in \mathbb{R}^{n\times n} | XV = V\Lambda\right\}, \tag{6.31}$$

of matrices X maintaining the eigeninformation. Assume, without the loss of generality, that columns of V are linearly independent. Then it is easy to see that $X \in \mathcal{M}(V,\Lambda)$ if and only if

$$X = V\Lambda V^{\dagger} + Z(I - VV^{\dagger}), \tag{6.32}$$

for some $Z \in \mathbb{R}^{n\times n}$, where $V^{\dagger}$ denotes the Moore–Penrose generalized inverse of V. In Problem 5.1, we seek the intersection of $\mathcal{M}(V,\Lambda)$ and the set $\mathcal{N}$ of certain structured matrices, which could be empty. Similar to the least squares LiPIEP2, we thus ask for the best least squares approximation. In particular, the following question is of interest.

Problem 6.4 *(Least squares PDIEP of a single matrix)*
Given a matrix $\Phi \in \mathbb{R}^{n\times n}$, find $X \in \mathcal{M}(V,\Lambda)$ such that $\|X - \Phi\|_F$ is minimized.

It is easy to see that $\mathcal{M}(V,\Lambda)$ is a closed convex set. It follows that Problem 6.4 has a unique solution.

Theorem 6.4. The unique solution X_0 to Problem 6.4 is given by

$$X_0 = V\Lambda V^{\dagger} + \Phi(I - VV^{\dagger}). \tag{6.33}$$

Proof We shall use the well-known fact that a necessary and sufficient condition for $\mathbf{k}_0$ being the nearest point on a closed convex set K to a given point ϕ is that

$$\langle \phi - \mathbf{k}_0, \mathbf{k} - \mathbf{k}_0 \rangle \leq 0, \tag{6.34}$$

for all $\mathbf{k} \in K$ (Luenberger, 1969). In our situation, $\mathcal{M}(V, \Lambda)$ is an affine subspace, so the inequality is actually equality due to the orthogonality. Replacing ϕ by Φ, $\mathbf{k}$ by (6.32), and $\mathbf{k}_0$ by (6.33), we see that

$$\begin{aligned}
\langle \Phi - X_0, X - X_0 \rangle &= \langle \Phi VV^{\dagger} - V\Lambda V^{\dagger}, (Z - \Phi)(I - VV^{\dagger}) \rangle \\
&= \langle (\Phi VV^{\dagger} - V\Lambda V^{\dagger})(I - VV^{\dagger}), Z - \Phi \rangle \\
&= 0,
\end{aligned}$$

because $V^{\dagger}VV^{\dagger} = V^{\dagger}$. □

In a more general setting, let $P(X)$ denote the projection of a matrix $X \in \mathbb{R}^{n \times n}$ onto $\mathcal{N}$ with respect to the Frobenius inner product. Then an analogue of LSIEP2 is the following approximation problem.

Problem 6.5 *(LSPDIEP2)*
Find $X \in \mathcal{M}(V, \Lambda)$ that minimizes the function

$$F(X) := \frac{1}{2}||X - P(X)||^2 \tag{6.35}$$

with respect to the Frobenius matrix norm.

If the set $\mathcal{N}$ of structured matrices forms an affine subspace, it is convex. The solution to the LSPDIEP2 is again unique. Indeed, with (6.33) in mind, the above LSPDIEP2 can be solved via a lift-and-projection algorithm. The projection is accomplished by the mapping $P(X)$ whereas the formula (6.33) provides the lift mechanism.

We name the above problem LSPDIEP2 because there should be a problem LSPDIEP1 analogous to LSIEP1 where the eigeninformation is least squares approximated. We are not sure how the analogy should be set up, but it will be an interesting question to investigate whether the equivalence between LSIEP1 and LSIEP2 continue to hold between LSPDIEP1 and LSPDIEP2.

6.5 Summary

An inverse eigenvalue problem, where a matrix is to be constructed from some or all of its eigenvalues, may not have a real-valued solution at all. More specifically, we often observe that the spectral constraint and the structural constraint in a real-valued IEP cannot be satisfied simultaneously. In this case we make a decision on which constraint could be compromised, and solve the IEP in a least squares setting.

Every IEP has a natural generalization to a least squares problem. This setting is particularly convenient when the cardinality of available eigeninformation and the degree of free parameters are not the same.

Using LiPIEP2 as a model, we studied and formulated two types of least squares problems in this chapter. In spite of their different appearance, we have discovered a rather pleasant surprise that the two formulations for LiPIEP2 are equivalent at the global solution. We thus proposed a lift-and-projection scheme, modified from the conventional alternating projection method, as a reasonable numerical method. The method converges linearly and globally, and can be used to generate good starting values for other faster but more expensive and locally convergent methods.

7

SPECTRALLY CONSTRAINED APPROXIMATION

7.1 Overview

We have explored a wide range of IEPs. Starting from scrupulous attention to the exact satisfaction of both the structural constraint and the spectral constraint, we gradually evolve into a more relaxed realm of least squares approximation. This development reverberates what (Gladwell, 1996) has called a distinction between determination and estimation, or between an essentially mathematical problem and an essentially engineering problem. In the preceding chapter, we began to see how a least squares IEP should be formulated. In this chapter, we want to bring forth another prospective showing that the problems of computing least squares approximations for various types of real and symmetric matrices subject to spectral constraints share a common structure. We want to describe a general framework by using the projected gradient method. In the end, we want to demonstrate that the spectrally constrained problem has a even broader generalization than the original context of IEPs.

The key to our success is that the projected gradient of the objective function on the manifold of constraints usually can be formulated explicitly. This gives rise to the construction of a descent flow that can be followed numerically. The explicit form also facilitates the computation of the second-order optimality conditions. Examples of applications are discussed. With slight modifications, the procedure can be extended to solve least squares problems for general matrices subject to singular value constraints.

7.1.1 *Spectral constraint*

To set forth the discussion, we shall first limit our discussion to the subspace $\mathcal{S}(n)$ of all symmetric matrices in $\mathbb{R}^{n\times n}$. Once we get used to the concept, we shall see that such a restriction is entirely unnecessary. Some of the following notions have been utilized in earlier sections, notably in Section 3.2.5 for LiPIEP2. For completeness, we reintroduce the isospectral surface $\mathcal{M}(\Lambda)$ of a given matrix $\Lambda \in \mathcal{S}(n)$ by

$$\mathcal{M}(\Lambda) := \{X \in \mathbb{R}^{n\times n} | X = Q^\top \Lambda Q, Q \in \mathcal{O}(n)\}, \tag{7.1}$$

where $\mathcal{O}(n)$ is the collection of all orthogonal matrices in $\mathbb{R}^{n\times n}$. Let Φ represent either a single matrix or a subspace in $\mathcal{S}(n)$. For every $X \in \mathcal{S}(n)$, the projection of X into Φ is denoted as $P(X)$. If Φ is a single matrix, then $P(X) \equiv \Phi$; otherwise, the projection is taken with respect to the Frobenius inner product.

We consider the following least squares spectrally constrained approximation problem (**SCAP**).

> **Problem 7.1** *(Generic SCAP)*
> Find $X \in \mathcal{M}(\Lambda)$ that minimizes the function
> $$F(X) := \frac{1}{2}||X - P(X)||^2 \tag{7.2}$$
> with respect to the Frobenius matrix norm.

There is a considerable similarity between Probelm 7.1 and Problem 6.5. The main difference, however, is that $\mathcal{M}(V, \Lambda)$ is a linear subspace whereas $\mathcal{M}(\Lambda)$ is a much more complicated nonlinear manifold. Problem 7.1, seemingly less constrained, is harder than Problem 6.5. Depending on how the matrix Λ and the subset Φ are defined, the SCAP represents a wide range of variations. To be more concrete, we mention below a partial list of problems that can be formulated in the above setting.

Consider the case $\Phi \equiv \hat{A}$ and $\Lambda = \operatorname{diag}\{\lambda_1, \dots, \lambda_n\}$. Then we are looking for a least squares approximation to a fixed matrix $\hat{A}$.

> **Problem 7.2** *(SCAP1)*
> Given a real symmetric matrix $\hat{A}$, find a least squares approximation of $\hat{A}$ that is still symmetric but has a prescribed set of eigenvalues $\{\lambda_1, \dots, \lambda_n\}$.

Consider the case where Φ is the subspace of all symmetric Toeplitz matrices and $\Lambda = \operatorname{diag}\{\lambda_1, \dots, \lambda_n\}$. Then this is precisely the Toeplitz inverse eigenvalue problem that we have considered earlier in Section 4.3.

If we choose Φ to be the subspace of all diagonal matrices and $\Lambda = X_0$, then the SCAP becomes a continuous Jacobi method by minimizing the off-diagonal entries.

> **Problem 7.3** *(SCAP2)*
> Given X_0, find its eigenvalues.

Although some of the above problems may be resolved by other means such as those discussed in Friedland et al. (1987), Gill et al. (1981), Golub and Van Loan (1996), Laurie (1988), Lawson and Hanson (1995), the general setting as in Problem 7.1 carries intrinsically some very interesting geometric properties. By exploring this geometry, we show in this chapter that the projected gradient of the objective function $F(X)$ onto the manifold $\mathcal{M}(\Lambda)$ can be calculated explicitly. Consequently, a vector field on the manifold $\mathcal{M}(\Lambda)$ that flows in a descent direction of $F(X)$ can be constructed. As another consequence, the explicit form of the projected gradient facilitates the computation of the second-order optimality conditions. We shall see that this information, in turn, offers some new insights into the classification of the stationary points.

Computational efficiency "temporarily" will not be a major concern in this presentation, although numerical construction of a solution remains one of our ultimate goals. There are several reasons that we refrain from computational details in the current discussion. First of all, our approach does offer a globally convergent numerical method in its own setting. The vector field defined by the projected gradient can readily be integrated by any available software for initial value problems. The recently developed techniques of geometric integration with respect to the underlying geometric structure are particularly appealing. See, for example (Dieci et al., 1994; Hairer et al., 2002; Iserles et al., 2000). Geometric integration is fast developing and ongoing research. It remains to be investigated which numerical integrator is most suitable for our continuation approach. Secondly, we caution that this continuation approach may well be as slow as the usual steepest descent methods regardless of whichever integrator is used. Since most of the time all we really want is the asymptotically stable limit point of the vector field, we are not interested in error for the transitional stage, just in a fast convergence to the limit point. Some cheap but suitable large step methods together with some refining strategies might be just as satisfactory as the more sophisticated integrators. Furthermore, since we also know the projected Hessian of the objective function $F(X)$, convergence certainly can be improved by employing other standard optimization techniques (Gill et al., 1981).

Our emphasis here is that the approach by optimization as proposed in Problem 7.1 is so versatile that one may use the subspace Φ to specify any desired (linear) structure on the optimal solution. LiPIEP2 is just one such example. If the subspace Φ does intersect the surface $\mathcal{M}(\Lambda)$, then of course both the structural constraint and the spectral constraint are met. Otherwise, our approach still finds a point on the isospectral surface $\mathcal{M}(\Lambda)$ that is structurally a least squares approximation to Φ.

Example 7.1. One may wish to find an isospectral solution that carries a certain specific zero pattern. We are not aware of very many discrete numerical methods capable of solving this kind of problem. In Chu and Norris (1988) it is shown (see Chapter 9) that a symmetric matrix with any kind of prescribed off-diagonal zero pattern, such as the one shown in Figure 7.1, was always reachable by following a specially formulated isospectral flow.

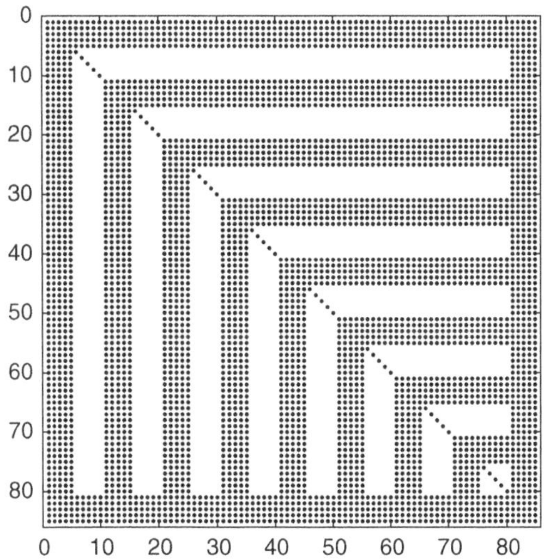

Figure 7.1. *Can such a* 85×85 *matrix exist with arbitrary spectrum?*

7.1.2 *Singular value constraint*

Suppose now Σ is a general matrix in $\mathbb{R}^{m\times n}$ and Ψ is either a single matrix or a subspace of $\mathbb{R}^{m\times n}$. Then analogous to the above notions we may consider the surface $\mathcal{W}(\Sigma)$ defined by

$$\mathcal{W}(\Sigma) := \{X \in \mathbb{R}^{m\times n} | X = U^\top \Sigma V, U \in O(m), V \in \mathcal{O}(n)\}, \tag{7.3}$$

which contains all matrices X that have the same singular values as Σ, and the singular value constrained approximation problem (SVCAP):

Problem 7.4 *(Generic SVCAP)*
Find $X \in \mathcal{W}(\Sigma)$ that minimizes the function

$$F(X) := \frac{1}{2}||X - P(X)||^2, \tag{7.4}$$

where $P(X)$ denotes the projection of X onto Ψ with respect to the Frobenius inner product.

Again, different choices of Σ and Ψ allow the SVCAP to represent different types of problems. We mention a few variations of practical interest.

Choose $\Psi \equiv \hat{A}$ and $\Sigma = \operatorname{diag}\{\sigma_1, \ldots, \sigma_n\}$ where diag is understood to mean an $m \times n$ matrix with extra rows filled by zeros.

Problem 7.5 *(SVCAP1)*
Given a matrix $\hat{A} \in \mathbb{R}^{m\times n}$ (say, $m \geq n$), find a least squares approximation of $\hat{A}$ that has a prescribed set of singular values $\{\sigma_1, \ldots, \sigma_n\}$.

Let $\Sigma = \hat{A}$ be a fixed matrix. If we choose Ψ to be the subspace of all diagonal matrices in $\mathbb{R}^{m\times n}$, then the SVCAP becomes a singular value computation of $\hat{A}$.

Problem 7.6 *(SVCAP2)*
Given a matrix $\hat{A} \in \mathbb{R}^{m\times n}$ (say, $m \geq n$), find the singular values of $\hat{A}$.

The SVCAP is almost parallel to the SCAP, except that the manifold $\mathcal{W}(\Sigma)$ is parameterized by two parameters in $\mathcal{O}(m) \times \mathcal{O}(n)$ and $\mathcal{M}(\Lambda)$ is parameterized by one parameter in $\mathcal{O}(n)$. Such a distinction is immaterial, as we will see in Chapter 9 that both $\mathcal{M}(\Lambda)$ and $\mathcal{W}(\Sigma)$ are *orbits* of some group actions. Therefore, many properties understood for the SCAP can be generalized for the SVCAP. In particular, we shall see that with slight modifications, the procedures developed for SCAP can easily be extended to SVCAP.

7.1.3 *Constrained optimization*

Before we move into matrix calculus, we first review some basic facts from classical optimization theory. In particular, we want to justify a shortcut way of computing the projected Hessian.

Consider the following basic equality constrained optimization problem:

$$\begin{aligned} \text{Minimize} \quad & F(\mathbf{x}), \\ \text{subject to} \quad & C(\mathbf{x}) = 0, \end{aligned} \tag{7.5}$$

where $x \in \mathbb{R}^n, F : \mathbb{R}^n \longrightarrow R$, and $C : \mathbb{R}^n \longrightarrow \mathbb{R}^k$ with $k < n$ are sufficiently smooth functions. Let

$$\mathcal{M} := \{\mathbf{x} \in \mathbb{R}^n | C(\mathbf{x}) = 0\} \tag{7.6}$$

denote the feasible set. We shall assume that $\mathcal{M}$ is a regular surface – that is, for all $\mathbf{x} \in \mathcal{M}$, the set $\{\nabla C_i(\mathbf{x})\}_{i=1}^k$ of gradient vectors of $C(\mathbf{x})$ is linearly independent. It follows that $\mathcal{M}$ is a smooth $(n-k)$-dimensional manifold (Guillemin and Pollack, 1974). Furthermore, for any $\mathbf{x} \in \mathcal{M}$, the space tangent to $\mathcal{M}$ at a $\mathbf{x}$ is given by

$$T_{\mathbf{x}}\mathcal{M} = \{\mathbf{y} \in \mathbb{R}^n | C'(\mathbf{x})\mathbf{y} = 0\}. \tag{7.7}$$

It is a fundamental fact (Gill et al., 1981) that in order for $\hat{\mathbf{x}}$ to be optimal, it is necessary that the gradient vector $\nabla F(\hat{\mathbf{x}})$ is perpendicular to the manifold $\mathcal{M}$.

Let $Z(\mathbf{x}) \in \mathbb{R}^{n\times(n-k)}$ denote a matrix whose columns form an orthonormal basis for $T_{\mathbf{x}}\mathcal{M}$. Then the projection $\mathbf{g}(\mathbf{x})$ of $\nabla F(\mathbf{x})$ onto the tangent space $T_x\mathcal{M}$ is given by $\mathbf{g}(\mathbf{x}) := Z(\mathbf{x})Z(\mathbf{x})^\top \nabla F(\mathbf{x})$. Note that $-\mathbf{g}(\mathbf{x})$ also represents the "steepest descent" direction of F on the manifold $\mathcal{M}$. Obviously a necessary condition for $\hat{\mathbf{x}}$ to be optimal is that

$$Z(\hat{\mathbf{x}})^\top \nabla F(\hat{\mathbf{x}}) = 0. \tag{7.8}$$

For each $\mathbf{x} \in \mathcal{M}$, we may rewrite

$$\mathbf{g}(\mathbf{x}) = \nabla F(\mathbf{x}) - \sum_{i=1}^{k} \lambda_i(\mathbf{x})\nabla C_i(\mathbf{x}) \tag{7.9}$$

for some appropriate scalar functions $\lambda_i(\mathbf{x})$, since the second term on the right-hand side of (7.9) represents the component of $\nabla F(\mathbf{x})$ normal to $T_{\mathbf{x}}\mathcal{M}$. We now suggest a rather simple way of deriving the quadratic form of the projected Hessian. This shortcut may not work for general nonlinear optimization problems, but it proves convenient and is valid for our consideration.

Suppose that the function $\mathbf{g}$ can be smoothly extended to the entire space $\mathbb{R}^n$, that is, suppose the function

$$G(\mathbf{x}) := \nabla F(\mathbf{x}) - \sum_{i=1}^{k} \lambda_i(\mathbf{x})\nabla C_i(\mathbf{x}) \tag{7.10}$$

is defined for every $\mathbf{x} \in \mathbb{R}^n$ and is smooth. Then for every vector $\mathbf{x}, \mathbf{v} \in \mathbb{R}^n$, we can calculate that

$$\mathbf{v}^\top G'(\mathbf{x})\mathbf{v} = \mathbf{v}^\top \left(\nabla^2 F(\mathbf{x}) - \sum_{i=1}^{k}(\mathbf{x})\nabla^2 C_i(\mathbf{x})\right)\mathbf{v} - \mathbf{v}^\top \left(\sum_{i=1}^{k} \nabla C_i(\mathbf{x})(\nabla\lambda_i(\mathbf{x}))^\top\right)\mathbf{v}. \tag{7.11}$$

In particular, if $\mathbf{x} \in \mathcal{M}$ and $\mathbf{v} \in T_{\mathbf{x}}\mathcal{M}$, then (7.11) is reduced to

$$\mathbf{v}^\top G'(\mathbf{x})\mathbf{v} = \mathbf{v}^\top(\nabla^2 F(\mathbf{x}) - \sum_{i=1}^{k} \lambda_i(\mathbf{x})\nabla^2 C_i(\mathbf{x}))\mathbf{v}, \tag{7.12}$$

because $\mathbf{v} \perp \nabla C_i(\mathbf{x})$. We note from (7.12) that the condition

$$\mathbf{v}^\top G'(\mathbf{x})\mathbf{v} \geq 0, \quad \text{for every } \mathbf{v} \in T_{\mathbf{x}}\mathcal{M}, \tag{7.13}$$

is precisely the well-known second-order necessary optimality condition for the problem (7.5). In the sequel, we shall use (7.13) with $G(\mathbf{x})$ defined by (7.10) as a means to calculated the projected Hessian.

7.2 Central framework

Equipped with the above device, we now show how to calculate the projected gradient and to construct a continuous steepest descent flow on the manifold

$\mathcal{M}(\Lambda)$ for Problem 7.1. Because $X = Q^\top \Lambda Q$, we can rewrite the minimization of Problem 7.1 for $X \in \mathcal{M}(\Lambda)$ as the following optimization problem for $Q \in \mathcal{O}(n)$:

$$\begin{aligned} \text{Minimize} \quad & F(Q) := \tfrac{1}{2}\langle Q^\top \Lambda Q - P(Q^\top \Lambda Q), Q^\top \Lambda Q - P(Q^\top \Lambda Q)\rangle \\ \text{subject to} \quad & Q^\top Q = I. \end{aligned} \tag{7.14}$$

In the above, $\langle A, B\rangle$ denotes the Frobenius inner product of matrices A and B. Also, without causing ambiguity, we have used the same notation F for the objective function.

It is well known (and easy to prove) that under the Frobenius inner product the orthogonal complement of $\mathcal{S}(n)$ is given by

$$\mathcal{S}(n)^\perp = \{\text{all skew-symmetric matrices}\}. \tag{7.15}$$

The feasible set $\mathcal{O}(n)$ can be regarded as the zero set of the function $C(X) := \frac{1}{2}(X^\top X - I)$. We obtain from (7.7) that the tangent space of $\mathcal{O}(n)$ at any orthogonal matrix Q is given by

$$\begin{aligned} \mathcal{T}_Q\mathcal{O}(n) &= \{H | H^\top Q + Q^\top H = 0\} = \{H | Q^\top H \text{ is skew-symmetric}\} \\ &= Q\mathcal{S}(n)^\perp \quad (\text{since } Q^\top = Q^{-1}). \end{aligned} \tag{7.16}$$

It follows that the orthogonal complement of $\mathcal{T}_Q\mathcal{O}(n)$ in $\mathbb{R}^{n\times n}$ is given by

$$N_Q\mathcal{O}(n) = Q\mathcal{S}(n). \tag{7.17}$$

This is the space normal to $\mathcal{O}(n)$ at Q.

It is obvious that the objective function F in (7.14) is differentiable for every general matrix $A \in \mathbb{R}^{n\times n}$. For convenience, let $\alpha(A) := \frac{1}{2}\langle A, A\rangle$ and $\beta(A) := A^\top \Lambda A - P(A^\top \Lambda A)$. By the chain rule and the product rule, we can compute the the Fréchet derivative of F at $A \in \mathbb{R}^{n\times n}$ acting on $B \in \mathbb{R}^{n\times n}$ as follows:

$$\begin{aligned} F'(A)B &= \alpha'(\beta(A))(\beta'(A)B) = \langle \beta(A), \beta'(A)B\rangle \\ &= \langle \beta(A), A^\top \Lambda B - P'(A^\top \Lambda A)A^\top \Lambda B + B^\top \Lambda A - P'(A^\top \Lambda A)B^\top \Lambda A\rangle \\ &= 2\langle \beta(A), A^\top \Lambda B - P'(A^\top \Lambda A)A^\top \Lambda B\rangle \quad (\text{since } \beta(A) \text{ is symmetric}) \\ &= 2\langle \beta(A), A^\top \Lambda B\rangle \quad (\text{since either } P' \equiv 0, \text{ or } P' = P; \text{ and } \beta(A) \perp \Phi) \\ &= 2\langle \Lambda A \beta(A), B\rangle. \end{aligned} \tag{7.18}$$

By the Riesz representation theorem, it follows from (7.18) that with respect to the Frobenius inner product the gradient of F for a general matrix A can be interpreted as the matrix

$$\nabla F(A) = 2\Lambda A\beta(A). \tag{7.19}$$

With (7.19) on hand, we can identify a stationary point via the first-order optimality condition.

Theorem 7.1. A necessary condition for $Q \in \mathcal{O}(n)$ to be a stationary point for Problem 7.1 is that the matrix $X := Q^\top \Lambda Q$ commutes with its own projection $P(X)$.

Proof From (7.8) we know that Q is a stationary point for F only if $\nabla F(Q)$ is perpendicular to $\mathcal{T}_Q\mathcal{O}(n)$. By (7.17) and (7.19), this condition is equivalent to $\Lambda Q\beta(Q) \in \mathcal{N}_Q\mathcal{O}(n) = Q\mathcal{S}(n)$. Since $Q^{-1} = Q^\top$, it follows that $X\beta(Q) = X(X - P(X)) \in \mathcal{S}(n)$. Thus it must be that $XP(X) = P(X)X$. □

7.2.1 *Projected gradient*

We can further identify the projected gradient of $F(Q)$ on the manifold $\mathcal{O}(n)$. We have seen that

$$\mathbb{R}^{n\times n} = \mathcal{T}_Q\mathcal{O}(n) \oplus \mathcal{N}_Q\mathcal{O}(n) = Q\mathcal{S}(n)^\perp \oplus Q\mathcal{S}(n). \tag{7.20}$$

Therefore any matrix $X \in \mathbb{R}^{n\times n}$ has a unique orthogonal splitting,

$$X = Q\left\{\tfrac{1}{2}(Q^\top X - X^\top Q)\right\} + Q\left\{\tfrac{1}{2}(Q^\top X + X^\top Q)\right\}, \tag{7.21}$$

as the direct sum of elements from $\mathcal{T}_Q\mathcal{O}(n)$ and $\mathcal{N}_Q\mathcal{O}(n)$. Accordingly, the projection $\mathbf{g}(Q)$ of $\nabla F(Q)$ into the tangent space $\mathcal{T}_Q\mathcal{O}(n)$ can be calculated explicitly as follows:

$$\begin{aligned}\mathbf{g}(Q) &= \tfrac{1}{2}Q\left\{(Q^\top 2\Lambda Q\beta(Q) - 2\beta(Q)Q^\top \Lambda Q)\right\}\\ &= Q\left\{X\beta(Q) - \beta(Q)X\right\} = Q[X, \beta(Q)]\\ &= Q[P(X), X],\end{aligned} \tag{7.22}$$

where $[A, B] := AB - BA$ denotes the Lie bracket operation. Note that Theorem 7.1 follows easily from the right-hand side of (7.22), that is,

$$\mathbf{g}(Q) = 0 \quad \text{if and only if} \quad [P(X), X] = 0. \tag{7.23}$$

From (7.22) it is clear that the vector field defined by the system:

$$\frac{dQ(t)}{dt} = Q(t)[Q(t)^\top \Lambda Q(t), P(Q(t)^\top \Lambda Q(t))] \tag{7.24}$$

defines a (steepest) descent flow $Q(t)$, if $Q(0) \in \mathcal{O}(n)$, on the manifold $\mathcal{O}(n)$ for the objective function $F(Q)$. At the same time, let $X(t) := Q(t)^\top \Lambda Q(t)$. Then $X(t)$ is governed by the ordinary differential equation:

$$\begin{aligned}\frac{dX(t)}{dt} &= \frac{dQ(t)^\top}{dt}\Lambda Q(t) + Q(t)^\top \Lambda \frac{dQ(t)}{dt}\\ &= -[X(t), P(X(t))]X(t) + X(t)[X(t), P(X(t))]\\ &= [[P(X(t)), X(t)], X(t)].\end{aligned} \tag{7.25}$$

By definition, the flow $X(t)$ defined by (7.25) stays on the isospectral surface $\mathcal{M}(\Lambda)$ with any initial value $X(0) = X_0 \in \mathcal{M}(\Lambda)$. Furthermore, the value of the objective function $F(X)$ in (7.1) is guaranteed to be nonincreasing along the forward flow $X(t)$. (Indeed, it decreases in the steepest direction for most of the time.) Problem 7.1, therefore, may be solved simply by integrating the initial value problem

$$\begin{cases} \dot{X} = [[P(X), X], X], \\ X(0) = X_0 \in \mathcal{M}(\Lambda), \end{cases} \tag{7.26}$$

which is known as Brockett's double bracket flow (Bloch, 1990; Brockett, 1991).

7.2.2 *Projected Hessian*

The explicit formula of the projected gradient $\mathbf{g}(Q)$ in (7.22) may be used to calculate the second-order derivative condition for the objective function in the same way as we mentioned in the preceding section. We first extend the function g to the function $G : \mathbb{R}^{n\times n} \longrightarrow \mathbb{R}^{n\times n}$ by defining

$$G(Z) := Z[P(Z^\top \Lambda Z), Z^\top \Lambda Z]. \tag{7.27}$$

By the product rule, it is easy to see that for any $Z, H \in \mathbb{R}^{n\times n}$,

$$\begin{aligned} G'(Z)H = {} & H[P(Z^\top \Lambda Z), Z^\top \Lambda Z] + Z[P(Z^\top \Lambda Z), Z^\top \Lambda H + H^\top \Lambda Z] \\ & + Z[P'(Z^\top \Lambda Z)(Z^\top \Lambda H + H^\top \Lambda Z), Z^\top \Lambda Z]. \end{aligned} \tag{7.28}$$

Consider the case when $Z = Q \in \mathcal{O}(n)$ and $H \in \mathcal{T}_Q\mathcal{O}(n)$. Then $H = QK$ for some $K \in \mathcal{S}(n)^\perp$. Let $X := Q^\top \Lambda Q$. Upon substitution, we have

$$\begin{aligned} \langle G'(Q)QK, QK\rangle &= \langle QK[P(X), X] + Q[P(X), [X, K]] + Q[P'(X)[X, K], X], QK\rangle \\ &= \langle K[P(X), X], K\rangle + \langle [P(X), [X, K]], K\rangle \\ &\quad + \langle [P'(X)[X, K], X], K\rangle. \end{aligned} \tag{7.29}$$

At a stationary point, $[P(X), X] = 0$. This first-order optimality condition greatly simplifies (7.29) to become

$$\langle G'(Q)(QK), QK\rangle = \langle [P(X), K] - P'(X)[X, K].[X, K]\rangle. \tag{7.30}$$

We want to see that $\langle G'(Q)(QK), QK\rangle \geq 0$ for all $K \in \mathcal{S}(n)^\perp$. Note that $P'(X)$ is either P itself or identically zero. So (7.30) can be further simplified and thus provides additional information about the stationary points. We shall demonstrate how these formulas can be used in the next section.

7.3 Applications

In this section we illustrate how the framework established can be applied to answer some of the questions raised earlier.

7.3.1 *Approximation with fixed spectrum*

To answer the SCAP1, the projection mapping is simply the constant $P(X) \equiv \hat{A}$. Let $\Lambda = \operatorname{diag}\{\lambda_1, \ldots, \lambda_n\}$. According to (7.22), the projected gradient is given by

$$\mathbf{g}(Q) = Q[\hat{A}, Q^\top \Lambda Q]. \tag{7.31}$$

The solution $X(t)$ to the initial value problem:

$$\begin{cases} \dot{X} = [[\hat{A}, X], X], \\ X(0) = \Lambda, \end{cases} \tag{7.32}$$

determines an isospectral flow that converges to a stationary solution of the least squares problem.

There is something very interesting about the limit point of the system (7.32). The information can be retrieved from the second-order condition. For simplicity, assume that

$$\lambda_1 > \lambda_2 > \cdots > \lambda_n, \tag{7.33}$$

and that the eigenvalues of $\hat{A}$ are ordered as

$$\mu_1 > \mu_2 > \cdots > \mu_n. \tag{7.34}$$

Let Q be a stationary point of F on $\mathcal{O}(n)$. We define $X := Q^\top \Lambda Q$ and

$$E := Q\hat{A}Q^\top. \tag{7.35}$$

By Lemma 7.1 we should have $[\hat{A}, X] = 0$. It follows that E must be a diagonal matrix since E commutes with the diagonal matrix Λ (Lancaster and Tismenetsky, 1985) (the assumption (7.33) is used here). Since E and $\hat{A}$ are similar, the diagonal elements $\{e_1, \ldots, e_n\}$ of E must be a permutation of $\{\mu_1, \ldots, \mu_n\}$. The equation (7.30) at the stationary point Q becomes:

$$\begin{aligned} \langle G'(Q)(QK), QK \rangle &= \langle [\hat{A}, K], [X, K] \rangle \\ &= \langle Q^\top EQK - KQ^\top EQ, Q^\top \Lambda QK - KQ^\top \Lambda Q \rangle \\ &= \langle E\hat{K} - \hat{K}E, \Lambda\hat{K} - \hat{K}\Lambda \rangle, \end{aligned} \tag{7.36}$$

where the matrix $\hat{K} = QKQ^\top$ is still skew-symmetric. Let $\hat{k}_{ij}$ denote the (i, j)-component of the matrix $\hat{K}$. It is easy to see that equation (7.36) can be expressed as

$$\langle G'(Q)(QK), QK \rangle = 2 \sum_{i<j} (\lambda_i - \lambda_j)(e_i - e_j)\hat{k}_{ij}^2. \tag{7.37}$$

From equation (7.37) and assumption (7.33) we see that the second-order optimality condition has the following equivalent statements:

$$\begin{aligned}&\langle G'(Q)QK, QK\rangle > 0, \quad \text{for every } K \in S(n)^{\perp}\\ &\Leftrightarrow (\lambda_i - \lambda_j)(e_i - e_j) \geq 0, \quad \text{for all } i < j\\ &\Leftrightarrow e_1 > e_2 > \cdots > e_n\\ &\Leftrightarrow e_i = \mu_i, \quad \text{for every } i.\end{aligned} \tag{7.38}$$

Putting together (7.35) and (7.38), the solution to the SCAP1 is now completely characterized as follows.

Theorem 7.2. Under the assumptions (7.33) and (7.34), a stationary point Q is a local minimizer of F on $\mathcal{O}(n)$ if and only if the columns $\mathbf{q}_1, \ldots, \mathbf{q}_n$ of the matrix $Q^{\top}$ are the normalized eigenvectors of $\hat{A}$ corresponding respectively to $\mu_1, \ldots, \mu_n$. In this case, the solution to the SCAP1 is the unique global minimizer and is given by

$$X = \lambda_1 \mathbf{q}_1 \mathbf{q}_1^{\top} + \cdots + \lambda_n \mathbf{q}_n \mathbf{q}_n^{\top}. \tag{7.39}$$

The above result can be generalized with slight modifications in case multiple eigenvalue are present. The only difference is that the least squares solution X is no longer unique if the matrix $\hat{A}$ has multiple eigenvalues, but the objective values are the same.

Theorem 7.2 may also be regarded as a re-proof of the well-known Wielandt–Hoffman theorem (Hoffman and Wielandt, 1953; Horn and Johnson, 1991; Wilkinson, 1965). The interpretation is interesting. Suppose that matrices $A, A + E$ and $E \in \mathcal{S}(n)$ have eigenvalues $\mu_1 > \cdots > \mu_n, \lambda_1 > \cdots > \lambda_n$ and $\tau_1 > \cdots > \tau_n$, respectively. Then the Wielandt–Hoffman theorem asserts that

$$\sum_{i=1}^{n} (\lambda_i - \mu_i)^2 \leq \sum_{i=1}^{n} \tau_i^2. \tag{7.40}$$

Theorem 7.2 asserts that the equality in (7.40) holds when the matrix $X = A+E$ is given by (7.39) where the Frobenius norm of the perturbation matrix E is minimized. We think the proof, being different from both the original proof of Hoffman and Wielandt (1953) and the one given in Wilkinson (1965), is of interest in its own right.

The dynamics of (7.32) enjoys a special *sorting property* as we have seen in proving Theorem 7.2. The continuous realization idea can therefore be applied to solve the data matching problem (Brockett, 1989, 1991) and a variety of generic combinatorial optimizations, including the link between the Toda lattice and gradient flow (Bloch et al., 1990) and hence the sorting of eigenvalues observed in the QR algorithm, the link with the total least squares problem (Bloch, 1990), and applications in linear programming and, in particular, to interior point methods (Bloch, 1990; Bayer and Lagarias, 1989; Faybusovich,

1991a, b). A glimpse of the progress in this fascinating area can be found in Bloch (1994).

Example 7.2. A typical linear programming problem concerns

$$\begin{aligned} &\text{Maximize} \quad \mathbf{c}^\top \mathbf{x}, \\ &\text{Subject to} \quad A\mathbf{x} \le \mathbf{b}, \end{aligned} \tag{7.41}$$

where A is a given matrix, and **b**and **c** are given vectors. Assuming that the feasible set $\mathcal{P} := \{\mathbf{x}|\mathbf{x} \in \mathbb{R}^n,\ A\mathbf{x} \le \mathbf{b}\}$ is a convex polytope with vertices at $\mathbf{a}_1, \ldots, \mathbf{a}_p \in \mathbb{R}^n$, it is a well known fact that one of the values $\mu_i := \mathbf{c}^\top \mathbf{a}_i$, $i = 1, \ldots, p$, will be the optimum for problem (7.41).

To sort out this particular vertex, we simply start from the initial value $X_0 := \text{diag}\{1, 0, \ldots, 0\} \in \mathbb{R}^{p\times p}$ and follow the isospectral flow $X(t)$ of (7.32) with $\hat{A} := \text{diag}\{\mu_1, \ldots, \mu_p\}$. By our theory, $X(t)$ converges to a diagonal matrix $\hat{X}$ whose elements must be a permutation of those of X_0 and must be arranged in an order similar to that of $\hat{A}$. By identifying the index corresponding to the value 1 in $\hat{X}$, we locate the optimal vertex.

7.3.2 *Toeplitz inverse eigenvalue problem revisit*

We have already proposed an iterative method for solving the ToIEP in Section 4.3.3. We now justify the system proposed in Example 4.5. For the ToIEP, Φ is taken to be the n-dimensional subspace of all symmetric Toeplitz matrices and $\Lambda := \text{diag}\{\lambda_1, \ldots, \lambda_n\}$. Note that Φ has a natural orthonormal basis $\{E_1, \ldots, E_n\}$ where $E_k := (e_{ij}^{(k)})$ and

$$e_{ij}^{(k)} := \begin{cases} 1/\sqrt{2(n-k+1)}, & \text{if } 1 < k \le n \text{ and } |i-j| = k-1, \\ 1/\sqrt{n}, & \text{if } k = 1 \text{ and } i = j, \\ 0, & \text{otherwise.} \end{cases} \tag{7.42}$$

Thus the projection P is easy to compute and is given by

$$P(X) = \sum_{k=1}^{n} \langle X, E_k \rangle E_k. \tag{7.43}$$

In contrast to the iterative method discussed in Section 4.3.3, the descent flow approach offers a globally convergent method of computation. By using the Adams–Bashforth–Moulton based subroutine ODE in Shampine and Gordon (1975) as the integrator, for example, we have never failed to get convergence in our numerical experimentation.

Unfortunately, such an approach for the ToIEP suffers from two shortcomings. First, any diagonal matrix is necessarily a stationary point. So the initial value $X(0)$ of the differential equation (7.26) cannot be chosen to be just Λ. This restriction is not serious. Second, we do experience cases of convergence to a stable stationary point which is not Toeplitz, although the limit point is always centrosymmetric. By picking up a different initial value, we can change the course of integration and hopefully converge to another stationary point. Our numerical experience indicates that the ToIEPs generally have multiple solutions.

The second-order condition (see (7.30))

$$\langle G'(Q)QK, QK\rangle = \langle [P(X), K] - P[X, K], [X, K]\rangle \tag{7.44}$$

for Problem B becomes more involved now. For the time being we have not tried to use (7.44) to classify the stationary points. It seems plausible that by classifying all stationary points one could answer the theoretical existence question for the inverse Toeplitz eigenvalue problem. This direction certainly deserves further exploration.

Toeplitz annihilator Deviating from the descent flow discussed above, another twist can be made by forming a differential equation of the form

$$\frac{dX}{dt} = [X, k(X)], \tag{7.45}$$

where $k : \mathcal{S}(n) \longrightarrow \mathcal{S}(n)^{\perp}$ is required to be a linear Toeplitz annihilator in the sense that

$$k(X) = 0 \text{ if and only if } X \in \mathcal{T}. \tag{7.46}$$

Chu (1992a) has explored several ways to define such a linear map k with the property (7.46). Perhaps the simplest way is to define the (i, j)-th entry k_{ij} of $k(X)$ by

$$k_{ij} := \begin{cases} x_{i+1,j} - x_{i,j-1}, & \text{if } 1 \le i < j \le n, \\ 0, & \text{if } 1 \le i = j \le n, \\ x_{i-1,j} - x_{i,j+1}, & \text{if } 1 \le j < i \le n. \end{cases} \tag{7.47}$$

The intuitive thinking behind such a definition is as follows: The flow of (7.45) starting with $X(0) = \Lambda := \operatorname{diag}\{\lambda_1, \ldots, \lambda_n\}$ stays on the isospectral manifold

$$\mathcal{M}(\Lambda) := \{Q^{\top}\Lambda Q | Q \in \mathcal{O}(n)\}. \tag{7.48}$$

Assume the generic case that all prescribed eigenvalues $\lambda_1, \ldots, \lambda_n$ are distinct. Then $[X, k(X)] = 0$ if and only if $k(X)$ is a polynomial of X (Gantmacher, 1959) and, therefore, $k(X) \in \mathcal{S}(n) \cap \mathcal{S}(n)^{\perp} = \{0\}$. If condition (7.46) holds, then all equilibria of the flow $X(t)$ are necessarily in $\mathcal{T}$. Since $\|X(t)\| = \|\Lambda\|$ for all $t \in R$, $X(t)$ must have a non-empty invariant ω-limit set. From this, we raise the hope that $X(t)$ might converge to a single point $\hat{X}$ as $t \longrightarrow \infty$ (Aulbach, 1984, Theorem 2.3, p. 23) and, in this way, the ToIEP is solved.

Obviously, the intuition outlined above is inadequate in many aspects. The conjecture that the dynamical system (7.45) does has a simple asymptotic behavior as $t \longrightarrow \infty$ still remains to be proved to this date. On the other hand, extensive numerical experiments with appropriately selected initial values for this Toeplitz annihilator flow on various sets of spectral data strongly suggest the orbit *always* converges to an equilibrium point (Diele and Sgura, 1999). Despite the theoretical difficulty, we suggest that following the solution flow of (7.45) might be a feasible numerical method for solving the ToIEP.

7.3.3 *Jacobi-type eigenvalue computation*

The SCAP2 is a symmetric eigenvalue problem. Numerous algorithms have already be developed for solving the matrix eigenvalue problem. It is of course not our intention to claim that we have a new and effective method. We simply want to point out that, in a way similar to the Toda flow that models the QR algorithm (Symes, 1981/82; Chu and Norris, 1988), the Jacobi method also has a continuous analogue.

Suppose that $\Lambda = X_0$ is the matrix whose eigenvalues are to be found. Recall that the main idea behind the Jacobi method is to systematically reduce the norm of the off-diagonal elements by means of orthogonal similarity transformations. We choose Φ to be the subspace of all diagonal matrices. Since the projection $P(X) = \text{diag}(X)$ is exactly the diagonal matrix of X, we see that the objective of Problem 7.1 is the same as that of the Jacobi method. The gradient flow (7.25) defined by the initial value problem:

$$\begin{cases} \dot{X} = [[\text{diag}X, X], X], \\ X(0) = X_0, \end{cases} \tag{7.49}$$

therefore, may be regarded as a continuous analogue of the iterates generated by the Jacobi method. It is fairly easy to characterize the dynamical behavior of (7.49).

The necessary condition for X to be a stationary point, according to Theorem 7.1, is

$$[\text{diag}X, X] = 0. \tag{7.50}$$

The second-order sufficient condition for optimality at a stationary point, according to (7.30), is

$$\langle G'(Q)(QK), QK\rangle = \langle[\text{diag}X, K] - \text{diag}[X, K], [X, K]\rangle > 0 \tag{7.51}$$

for every skew-symmetric matrix K. By using (7.50) and (7.51) we are able to analyze all stationary points as follows (Driessel, 1987a, pp. 33–36).

Theorem 7.3. Let X be a stationary point for Problem 7.1 with Λ and Φ defined as those for Problem 7.3.

(i) If X is a diagonal matrix, then X is an isolated global minimizer.

(ii) If X is not a diagonal matrix but $\text{diag}(X)$ is a scalar matrix (i.e., $\text{diag}(X) = cI$ for some scalar c), then X is a global maximizer.

(iii) If X is not a diagonal matrix and $\text{diag}(X)$ is not a scalar matrix, then X is a saddle point.

We finally remark that the gradient flow (7.49) is moving by it own nature in a descent direction of the function $F(X)$. So the existence of the latter two cases in Theorem 7.3 should not cause any annoyance in the computation for eigenvalues.

7.4 Extensions

The framework discussed in the preceding section can be extended to the SVCAP without too much trouble. The key to our approach is to recognize an inner product over the product space $\mathbb{R}^{m\times m} \times \mathbb{R}^{n\times n}$ through the induced Frobenius inner product:

$$\langle (A_1, A_2), (B_1, B_2) \rangle := \langle A_1, B_1 \rangle + \langle A_2, B_2 \rangle. \tag{7.52}$$

Regarding the feasible set $\mathcal{O}(m) \times \mathcal{O}(n)$ as the zero set of the function $C(A_1, A_2) := (\frac{1}{2}(A_1^\top A_1 - I), \frac{1}{2}(A_2^\top A_2 - I))$, we can show that the tangent space and the normal space of $\mathcal{O}(m) \times \mathcal{O}(n)$ at a point (Q_1, Q_2) are given by, respectively,

$$\mathcal{T}_{(Q_1,Q_2)}\mathcal{O}(m) \times \mathcal{O}(n) = Q_1\mathcal{S}(m)^\perp \times Q_2\mathcal{S}(n)^\perp \tag{7.53}$$

and

$$\mathcal{N}_{(Q_1,Q_2)}\mathcal{O}(m) \times \mathcal{O}(n) = Q_1\mathcal{S}(m) \times Q_2\mathcal{S}(n). \tag{7.54}$$

Without reiterating every detailed step, we now demonstrate how Problems 7.5 and 7.6 can be resolved.

7.4.1 *Approximation with fixed singular values*

It is easy to see that Problem 7.5 is equivalent to the following formulation:

$$\begin{array}{ll} \text{Minimize} & F(Q_1, Q_2) := \frac{1}{2}\|\Sigma - Q_1^\top \hat{A} Q_2\|^2 \\ \text{subject to} & Q_1^\top Q_1 = I, \quad Q_2^\top Q_2 = I. \end{array} \tag{7.55}$$

Analogous to (7.18), we find that the Fréchet derivative of F at a general point $(A_1, A_2) \in \mathbb{R}^{m\times m} \times \mathbb{R}^{n\times n}$ acting on $(B_1, B_2) \in \mathbb{R}^{m\times m} \times \mathbb{R}^{n\times n}$ is given by

$$\begin{aligned} F'(A_1, A_2)(B_1, B_2) &= \langle \Sigma - A_1^\top \hat{A} A_2, -B_1^\top \hat{A} A_2 - A_1^\top \hat{A} B_2 \rangle \\ &= \langle \Sigma - A_1^\top \hat{A} A_2, -B_1^\top \hat{A} A_2 \rangle + \langle \Sigma - A_1^\top \hat{A} A_2, -A_1^\top \hat{A} B_2 \rangle \\ &= \langle -(\Sigma - A_1^\top \hat{A} A_2) A_2^\top \hat{A}^\top, B_1^\top \rangle + \langle -\hat{A}^\top A_1 (\Sigma - A_1^\top \hat{A} A_2), B_2 \rangle. \end{aligned} \tag{7.56}$$

Therefore, with respect to the inner product (7.52), we may interpret that the gradient of F at (A_1, A_2) is given by the pair:

$$\nabla F(A_1, A_2) = \Big(-\hat{A} A_2 (\Sigma - A_1^\top \hat{A} A_2)^\top, -\hat{A}^\top A_1 (\Sigma - A_1^\top \hat{A} A_2) \Big). \tag{7.57}$$

A necessary condition for $(Q_1, Q_2) \in \mathcal{O}(m) \times \mathcal{O}(n)$ to be a stationary point of the objective function F is $\nabla F(Q_1, Q_2) \perp \mathcal{T}_{(Q_1,Q_2)}\mathcal{O}(m) \times \mathcal{O}(n)$. Using (7.53), this is equivalent to $\hat{A}Q_2(\Sigma - Q_1^\top \hat{A} Q_2) \in Q_1\mathcal{S}(m)$ and $\hat{A}^\top Q_1(\Sigma - Q_1^\top \hat{A} Q_2) \in Q_2\mathcal{S}(n)$. Let

$$X := Q_1^\top \hat{A} Q_2. \tag{7.58}$$

It is not difficult to see that the above necessary condition is equivalent to

$$X\Sigma^\top = \Sigma X^\top \quad \text{and} \quad X^\top \Sigma = \Sigma^\top X. \tag{7.59}$$

For simplicity, assume that

$$\sigma_1 > \sigma_2 > \cdots > \sigma_n > 0, \tag{7.60}$$

and that the singular values of $\hat{A}$ are ordered (in the generic case) as

$$\mu_1 > \mu_2 > \cdots > \mu_n > 0. \tag{7.61}$$

Then it follows from the two equations in (7.59) that the $m \times n$ matrix X must be a diagonal matrix where the extra rows are filled with zeros. Because of the relationship (7.58), we know therefore that the diagonal elements, say, $e_1, \ldots, e_n$ of X must be a permutation of singular values of $\hat{A}$.

The projection of $\nabla F(Q_1, Q_2)$ into the tangent space $\mathcal{T}_{(Q_1,Q_2)}\mathcal{O}(m) \times \mathcal{O}(n)$ can be calculated according to the same principle as in (7.22). We claim that the projection of $\nabla F(Q_1, Q_2)$ is given by

$$\begin{aligned} \mathbf{g}(Q_1, Q_2) &= \left(\tfrac{1}{2}\{Q_1 \Sigma Q_2^2 \hat{A}^\top Q_1 - \hat{A} Q_2 \Sigma^\top\}, \tfrac{1}{2}\{Q_2 \Sigma^\top Q_1^\top \hat{A} Q_2 - \hat{A}^\top Q_1 \Sigma\}\right) \\ &= \left(\tfrac{1}{2} Q_1(\Sigma X^\top - X\Sigma^\top), \tfrac{1}{2} Q_2(\Sigma^\top X - X^\top \Sigma)\right). \end{aligned} \tag{7.62}$$

Readers are invited to furnish the proof by themselves. As is suggested, we may define a flow $X(t)$ by

$$\begin{cases} \frac{dX}{dt} = \frac{1}{2}(\Sigma X^\top X - X\Sigma^\top X + XX^\top \Sigma - X\Sigma^\top X), \\ X(0) = \hat{A}, \end{cases} \tag{7.63}$$

which will move in a descent direction of F. Furthermore, we may extend the function $\mathbf{g}$ to

$$G(Z_1, Z_2) := \left(\tfrac{1}{2}\{Z_1 \Sigma Z_2^\top \hat{A}^\top Z_1 - \hat{A} Z_2 \Sigma^\top\}, \tfrac{1}{2}\{Z_2 \Sigma^\top Z_1^\top \hat{A} Z_2 - \hat{A}^\top Z_1 \Sigma\}\right) \tag{7.64}$$

for general matrices $(Z_1, Z_2) \in \mathbb{R}^{m \times m} \times \mathbb{R}^{n \times n}$ and take its derivative. In particular, we claim that at a stationary point (Q_1, Q_2) the projected Hessian of F

acting on a tangent vector (Q_1K_1, Q_2K_2) where $K_1 \in \mathcal{S}(m)^\perp$ and $K_2 \in \mathcal{S}(n)^\perp$ is given by the bilinear form

$$\langle (Q_1K_1, Q_2K_2), G'(Q_1, Q_2)(Q_1K_1, Q_2K_2)\rangle = \langle K_1\Sigma - \Sigma K_2, K_1X - XK_2\rangle. \tag{7.65}$$

Again, readers are invited to fill in the details. We note here the similarity between (7.65) and (7.37).

Let $k_{ij,1}$ and $k_{ij,2}$ denote the (i,j)-components of the skew matrices K_1 and K_2, respectively. Then (7.65) can be expressed as

$$\begin{aligned}\langle (Q_1K_1, Q_2K_2), G'(Q_1, Q_2)(Q_1K_1, Q_2K_2)\rangle = &\sum_{i=1}^{n}\sum_{p=n+1}^{n} e_i\sigma_i k_{pi,1}^2\\ &+\sum_{i\neq k}\{(e_i\sigma_i + e_k\sigma_k)k_{ik,1}^2\\ &+(e_i\sigma_i + e_k\sigma_k)k_{ik,2}^2\\ &-2(e_i\sigma_k + e_k\sigma_i)k_{ik,1}k_{ik,2}\}\end{aligned}$$

since $k_{ii,j} = 0$ and $k_{ik,j} = -k_{ki,j}$ for all $1 \leq i,k \leq n$ and $j = 1$ or 2. The second-order optimality condition has the following equivalent statements:

$$\begin{aligned}&\langle (Q_1K_1, Q_2K_2), G'(Q_1, Q_2)(Q_1K_1, Q_2K_2)\rangle > 0,\\ &\text{ for every } K_1 \in \mathcal{S}(m)^\perp, K_2 \in \mathcal{S}(n)^\perp,\\ \Leftrightarrow\ &(e_i\sigma_i + e_k\sigma_k)k_{ik,1}^2 + (e_i\sigma_i + e_k\sigma_k)k_{ik,2}^2 - 2(e_i\sigma_k + e_k\sigma_i)k_{ik,1}k_{ik,2} > 0,\\ &\text{for every } k_{ik,1}, k_{ik,2} \in \mathbb{R},\\ \Leftrightarrow\ &\text{the discriminant } (e_i + e_k)(\sigma_i + \sigma_k)(e_i - e_k)(\sigma_i - \sigma_k) > 0,\\ &\text{for every } i \text{ and } k,\\ \Leftrightarrow\ &e_1 > e_2 > \cdots > e_n\\ \Leftrightarrow\ &e_i = \mu_i \text{ for every } i.\end{aligned} \tag{7.67}$$

In summary, we have proved the following theorem.

Theorem 7.4. Under the assumptions (7.60) and (7.61), a pair of matrices (Q_1, Q_2) is a local minimizer of F on $\mathcal{O}(m) \times \mathcal{O}(n)$ if and only if the columns of Q_1 and the columns of Q_2 are, respectively, the left and right singular vectors of $\hat{A}$. In this case, the unique least squares approximation to $\hat{A}$ subject to the singular values constraints is the global minimizer and is given by

$$X = Q_1\Sigma Q_2^\top. \tag{7.68}$$

Again, using the above theorem, we now have in hand a re-proof of the Wielandt–Hoffman theorem for singular values (Horn and Johnson, 1991; Golub and Van Loan, 1996) in the same way as that for eigenvalues. In other words, suppose that A and $A + E$ have singular values $\mu_1 > \mu_2 > \cdots > \mu_n > 0$ and $\sigma_1 > \sigma_2 > \cdots > \sigma_n > 0$, respectively. Then

$$\sum_{i=1}^{n} (\sigma_i - \mu_i)^2 \leq ||E||^2. \tag{7.69}$$

We remark that the initial value problem (7.63) defines a descent flow $X(t)$, regardless of the assumptions (7.60) and (7.61). It is always the case in our approach that the flow is descending without any knowledge of the second-order derivative. As a descent flow on a compact surface, we know that the flow must have a limit point which determines a local minimum.

Finally, we make a point that the flow (7.63) also enjoys a sorting property that realigns singular values.

7.4.2 *Jacobi-type singular value computation*

The SVCAP2 is a singular value computation. This problem can be handled in a similar way to Problem 7.3. We choose Ψ to be the the subspace of all diagonal matrices and consider a Jacobi-type flow from the following optimization problem:

$$\begin{aligned} \text{Minimize} \quad & F(Q_1, Q_2) := \tfrac{1}{2}||Q_1^\top \hat{A} Q_2 - \text{diag}(Q_1^\top \hat{A} Q_2)||^2 \\ \text{subject to} \quad & Q_1^\top Q_1 = I, \quad Q_2^\top Q_2 = I. \end{aligned} \tag{7.70}$$

We can formulate the projected gradient of F explicitly. In particular, we claim without given details that the initial value problem

$$\begin{cases} \dot{X} = \frac{1}{2} \left\{ \left((\text{diag}X)X^\top - X(\text{diag}X)^\top \right) X - X \left((\text{diag}X)^\top X - X^\top (\text{diag}X) \right) \right\}, \\ X(0) = \hat{A}, \end{cases} \tag{7.71}$$

defines a descent flow on the manifold $W(\hat{A})$ for the function F. As before, we can further classify the equilibrium points of (7.71) by means of the projected Hessian form. The details can be found in Driessel (1987b) and we shall mention only the major result without proof:

Theorem 7.5. Let $\hat{A} \in \mathbb{R}^{m \times n}$ have distinct, nonzero singular values. Let X be an equilibrium point of the differential equation (7.71). Then X is stable if and only if X is a diagonal matrix.

7.5 Simultaneous reduction

Up to this point, we have considered minimizing $\|X - P(X)\|$ of only a single matrix X. It is certainly plausible to extend this idea beyond one single matrix and this leads to the problem of simultaneous reduction of multiple matrices. For illustration purposes, we shall employ only orthogonal similarity or orthogonal equivalence as the means of transformations, but this limitation is not necessary. Based on the Jacobi idea, given the desired form to which matrices are reduced, the sum of squares of the complementary part is to be mininimized. The projected gradient of the objective function gives rise to a system of ordinary differential equations that can be readily solved by numerical software.

The advantages of this approach are that the desired form to which matrices are reduced can be almost arbitrary, and that if a desired form is not attainable, then the limit point of the corresponding differential equation gives a way of measuring the distance from the best reduced matrices to the nearest matrices that have the desired form. We outline the general procedure for deriving these differential equations in this section. Again, we shall be interested mainly in real-valued matrices although the discussion in the sequel can be generalized to the complex-valued case. We shall illustrate one such generalization when dealing with the nearest normal matrix problem in Section 7.6.

7.5.1 *Background review*

Simultaneous reduction has been of interest both in theory and in application. It might be appropriate to furnish some historic background to unify the general settings before we move into computational details. The following is a classical problem in the field of algebra:

Problem 7.7 *(Orbit of general matrices by similarity actions)*
Given k arbitrary matrices $A_1, \ldots, A_k \in \mathbb{R}^{n\times n}$, identify the similarity class

$$\{(B_1, \ldots, B_k) | B_i = T^{-1}A_iT, \quad i = 1, \ldots, k; \ T \in \mathcal{G}l(n)\}. \tag{7.72}$$

It is known that the classification of similarity classes of k-tuples of matrices can be reduced to the classification of simultaneous similarity of commuting pairs of matrices (Gel′fand and Ponomarev, 1969). Only recently have the complex-valued versions of Problem 7.7 has been theoretically solved in the work by Friedland (1983) and Dias da Silva and Laffey (1999). The technique used is highly algebraic in nature. Roughly speaking, the orbit is determined by the values of certain rational functions in the entries of $A_1, \ldots, A_k$.

Various applications in which the classification of orbits is needed and various partial results can be found in Friedland (1983), Dias da Silva and Laffey (1999) and the references contained therein. But no numerical procedure has ever been attempted when $k > 2$.

An associated problem when limited to symmetric matrices is:

> **Problem 7.8** *(Orbit of symmetric matrices by orthogonal similarity actions)*
> Given k arbitrary matrices $A_1, \ldots, A_k \in \mathcal{S}(n)$, identify the similarity class
> $$\{(B_1, \ldots, B_k) | B_i = Q^\top A_i Q, \quad i = 1, \ldots, k; \ Q \in \mathcal{O}(n)\}. \tag{7.73}$$

Because of concerns about numerical stability, numerical analysts usually prefer orthogonal transformations to general invertible transformations. Therefore, it is of practical interest to consider the following problem.

> **Problem 7.9** *(Orbit of general matrices by orthogonal similarity actions)*
> Given k arbitrary matrices $A_1, \ldots, A_k \in \mathbb{R}^{n \times n}$, identify the similarity class
> $$\{(B_1, \ldots, B_k) | B_i = Q^\top A_i Q, \quad i = 1, \ldots, k; \ Q \in \mathcal{O}(n)\}. \tag{7.74}$$

The type of transformation $Q^\top AQ$ with $Q \in \mathcal{O}(n)$ will be referred to, henceforth, as the *(real) orthogonal similarity transformation.* The only difference between Problems 7.8 and 7.9 is that we have replaced symmetric matrices by general matrices. We have reasons to believe that this replacement makes Problem 7.9 harder to analyze. We mention, for example, the well-known real Schur decomposition theorem (Golub and Van Loan, 1996, p. 362) that offers a glimpse into similarity classes inquired by Problem 7.9 for the case $k = 1$.

Theorem 7.6 (RSD). If $A \in \mathbb{R}^{n \times n}$, then there exists an orthogonal matrix $Q \in \mathcal{O}(n)$ such that $Q^\top AQ$ is upper quasi-triangular, that is, $Q^\top AQ$ is block upper-triangular where each diagonal block is either a 1×1 matrix or a 2×2 matrix having complex conjugate eigenvalues.

As it is, the RSD theorem falls short of fully identifying the orthogonal similarity orbits of a general matrix $A \in \mathbb{R}^{n \times n}$ because the off-diagonal blocks are not characterized. Only when A is symmetric, then the similarity orbits of A in $\mathcal{S}(n)$, being diagonalizable through the transformation $Q^\top AQ$, are perfectly classified.

We usually are interested in identifying a matrix by its *canonical form.* In application, quite often a canonical form is meant to be of a special *matrix structure.* The structure could be, for instance, a diagonal matrix, a bidiagonal matrix, an upper triangular matrix, and so on. The Jordan canonical form is a more complicated structure. A different view of Problem 7.9, therefore, is to consider the following problem:

> **Problem 7.10** *(Simultaneous reduction by orthogonal similarity)*
> Given k specified (but possibly the same) canonical forms for matrices in $\mathbb{R}^{n\times n}$, determine if the orbit of k matrices $A_1, \ldots, A_k \in \mathbb{R}^{n\times n}$ under the action of $\mathcal{O}(n)$ contains an element such that each $Q^\top A_i Q$ has the specified structure.

We mention a related but slightly different problem to illustrate an application of Problem 7.10. For decades, the problem of simultaneous diagonalization of two symmetric matrices has received much attention. See, for example, Berman and Ben-Israel (1971), Golub and Van Loan (1996), Majindar (1979), Uhlig (1973, 1976), and the historical survey (Uhlig, 1979). A classical result in this direction is stated as follows.

Theorem 7.7. If A is symmetric and B is symmetric and positive definite, then there exists a nonsingular X such that both $X^\top AX$ and $X^\top BX$ are diagonal matrices.

Be aware that the diagonalization processes, even for symmetric matrices, involve nonsingular *congruence* transformations which usually are not orthogonal. This is partly due to the reason that orthogonal transformations are too limited to result in the diagonal form. But then it is curious to know how much reduction orthogonal transformations can accomplish.

In addition to the orthogonal similarity transformation, there is another type of transformation, $Q^\top AZ$ with $Q, Z \in \mathcal{O}(n)$, referred to as *(real) orthogonal equivalence transformation,* used in numerical analysis. The importance of the real orthogonal equivalence transformation is evidenced by the singular value decomposition theorem (Golub and Van Loan, 1996, p. 71).

Theorem 7.8 (SVD). If $A \in \mathbb{R}^{m\times n}$, then there exist orthogonal matrices $Q \in \mathcal{O}(m)$ and $Z \in \mathcal{O}(n)$ such that $Q^\top AZ$ is a diagonal matrix.

Analogous to (7.74), the *equivalence orbit* of any given k arbitrary matrices $A_1, \ldots, A_k \in \mathbb{R}^{m\times n}$ (under the action of $\mathcal{O}(m)$ and $\mathcal{O}(n)$) is defined to be the set

$$\{(B_1, \ldots, B_k) | B_i = Q^\top A_i Z, i = 1, \ldots, k; \quad Q \in \mathcal{O}(m),\ Z \in \mathcal{O}(n)\}. \tag{7.75}$$

Motivated by Problem 7.10, we ask the following question.

Problem 7.11 *(Simultaneous reduction by orthogonal equivalence)* Given k specified (but possibly the same) canonical forms for matrices in $\mathbb{R}^{m\times n}$, determine if the equivalence orbit of k matrices $A_1, \ldots, A_k \in \mathbb{R}^{m\times n}$ contains an element such that each $Q^\top A_i Z$ has the specified structure.

The SVD theorem settles the special case $k = 1$ in Problem 7.11. When $k = 2$, then Problem 7.11 is partially answered by the so-called generalized real Schur decomposition theorem (Golub and Van Loan, 1996, p. 396):

Theorem 7.9 (GRSD). If $A, B \in \mathbb{R}^{n\times n}$, then there exist orthogonal matrices Q and Z such that $Q^\top AZ$ is upper quasi-triangular and $Q^\top BZ$ is upper triangular.

We should distinguish GRSD from another analogous but different application known as the generalized singular value decomposition theorem (Paige and Saunders, 1981; Van Loan, 1976):

Theorem 7.10 (GSVD). If $A \in \mathbb{R}^{m\times n}$ and $B \in \mathbb{R}^{p\times n}$, then there exist orthogonal $U \in \mathcal{O}(m)$, $V \in \mathcal{O}(p)$ and invertible $X \in \mathbb{R}^{n\times n}$ such that $U^\top AX$ and $V^\top BX$ are diagonal matrices.

We note that besides the generality of dimensions of A and B, the GSVD is fundamentally different from GRSD in that the orthogonal matrices are not the same for A as for B and that the right transformation matrix X is only required to be nonsingular.

Every one of the aforementioned special cases of either Problem 7.10 or Problem 7.11 has significant applications in numerical analysis. Enormous amounts of effort have been devoted to the study of these special matrix decompositions. See the book by Golub and Van Loan (1996) and the references cited therein. On the other hand, for the general cases of either Problems 7.10 or 7.11, little is known in the literature. Conceivably, when more matrices are involved, the simultaneous reduction problem becomes more difficult both theoretically and numerically.

Example 7.3. Recall that a numerically stable method, called the QZ algorithm (Moler and Stewart, 1973), has been developed to solve the important generalized eigenvalue problem $A\mathbf{x} = \lambda B\mathbf{x}$. This algorithm is based on Theorem 7.9.

In this section we recast the simultaneous reduction problem as an equality-constrained optimization problem and apply the projected gradient method. We develop a differential equation approach that can be used as a numerical method for answering both Problems 7.10 and 7.11. Our approach is flexible in at least two respects. One is that the differential equations for various types of *canonical forms* can easily be derived within a uniform framework for a given k. The other is that the framework can easily be modified if k is changed. In view of these advantages, the tool we are about to develop allows one to experiment with combinations of many different canonical forms with only slight modifications in the computer program. Furthermore, if the desired form is not attainable, then the limit point of the corresponding differential equation gives a way of measuring the distance from the best reduced matrices to the *nearest* matrices that have the desired form. This information is sometimes useful in applications.

The QR algorithm, the SVD algorithm and the QZ algorithm are a few of the iterative methods that play very prominent roles in matrix computations. Differential equations that model these iterative processes has been reviewed in the paper by Chu (1988). Most of the ideas there have been based on the fact that a finite, nonperiodic Toda lattice is a continuous analogue of the QR algorithm (Symes, 1981/82). The differential equation approach developed in this section more or less is an extension of the continuous analogue of the Jacobi method for symmetric eigenvalue problems that we have described earlier.

A collection of variations and references of the Jacob method can be found in Golub and Van Loan (1996, pp. 444–459). In the past, attempts have been made to extend the Jacobi iteration to other classes of matrices and to push through corresponding convergence results. But success has been reported only for normal matrices (Goldstine and Horwitz, 1959) which, then, was employed to solve the closest normal matrix problem (Ruhe, 1987). For nonnormal matrices, the situation is considerably more difficult. Simultaneous reduction of more than one general matrices is thus an even harder problem. It turns out that our differential equation approach offers a fairly easy but systematic reduction procedure. In fact, the approach is so versatile that one can examine the (similarity or equivalence) orbit of k given matrices for many different combinations of reduced forms.

7.5.2 *Orthogonal similarity transformation*

In the next few sections, we first derive a general framework of constructing differential equations for Problem 7.10. The development is parallel to that done in Section 7.2. In particular, we show how the projected gradient can be calculated. We then demonstrate a special application to the simultaneous diagonalization of two symmetric matrices. Differential equations for Problem 7.11 can similarly be derived. All these techniques can be combined and generalized to the complex-valued case. We shall be brief since most of the calculus has been done before.

We first define some notation. Let $A_i \in \mathbb{R}^{n\times n}$, $i = 1,\dots,k$, denote k given matrices. For each i, let $V_i \subset \mathbb{R}^{n\times n}$ denote the set of all matrices having the specified form to which A_i is supposed to be reduced. For simplicity, we shall assume that each V_i is linear. Given any $X \in \mathbb{R}^{n\times n}$, its projection onto the subspace V_i is denoted as $P_i(X)$. For any matrix $X \in \mathbb{R}^{n\times n}$, we define the residual operator,

$$\alpha_i(X) := X^\top A_i X - P_i(X^\top A_i X), \quad i = 1, \ldots, k. \tag{7.76}$$

We remark here that the choice of the subspace V_i can be quite arbitrary and need not be the same for every i. For example, V_1 may be taken to be the subspace of all diagonal matrices, V_2 the subspace of all upper Hessenberg matrices, and so on. Our idea in approaching Problem 7.10 is to consider the following optimization problem.

Problem 7.12 *(Minimization on orbits of $\mathcal{O}(n)$)*
Given matrices $A_i \in \mathbb{R}^{n\times n}$ and desirable forms V_i, $i = 1, \ldots, k$,

$$\text{Minimize } F(Q) := \frac{1}{2}\sum_{i=1}^{k} \|\alpha_i(Q)\|^2 \tag{7.77}$$

$$\text{subject to } Q^\top Q = I.$$

That is, while moving along the orthogonal similarity orbit of the given matrices $A_1, \ldots, A_k$, we want to minimize the total distance between the point $Q^\top A_i Q$ and the set V_i for all i. Of course, one may regard Problem 7.12 as a standard equality-constrained optimization problem and thus solve the problem by many existing numerical algorithms found in, for example (Gill et al., 1981). In doing so, however, one has to interpret a matrix equation as a collection of n^2 nonlinear equations. The computation of derivatives in the *unpacked form* proves to be very inconvenient.

By now we know the feasible set $\mathcal{O}(n) := \{Q | Q^\top Q = I\}$ very well. The Fréchet derivative of the objective function F in (7.77) at a general $X \in \mathbb{R}^{n\times n}$ acting on a general $Y \in \mathbb{R}^{n\times n}$ can be calculated as follows:

$$\begin{aligned} F'(X)Y &= \sum_{i=1}^{k} \langle \alpha_i(X), \alpha_i'(X)Y \rangle \\ &= \sum_{i=1}^{k} \langle \alpha_i(X), X^\top A_i Y + Y^\top A_i X - P_i(X^\top A_i Y + Y^\top A_i X) \rangle \\ &= \sum_{i=1}^{k} \langle A_i^\top X \alpha_i(X) + A_i X \alpha_i^\top(X), Y \rangle. \end{aligned} \tag{7.78}$$

In the second equation above we have used the fact that the projections P_i are linear. In the third equation above we have used the fact that $\alpha_i(X)$ is perpendicular to V_i. Equation (7.78) suggests that with respect to the Frobenius inner product, we may interpret the *gradient* of F at a general point X as the matrix

$$\nabla F(X) = \sum_{i=1}^{k} (A_i^\top X \alpha_i(X) + A_i X \alpha_i^\top(X)). \tag{7.79}$$

The projection $\mathbf{g}(Q)$ of $\nabla F(Q)$ into the tangent space $T_Q\mathcal{O}(n)$ can be calculated as follows:

$$\begin{aligned}\mathbf{g}(Q) &= \frac{Q}{2}\{Q^\top \nabla F(X) - \nabla F(X)^\top Q\} \\ &= \frac{Q}{2}\sum_{i=1}^{k}([Q^\top A_i^\top Q, \alpha_i(Q)] + [Q^\top A_i Q, \alpha_i^\top(Q)]).\end{aligned} \tag{7.80}$$

Now that $\mathbf{g}(Q)$ is tangent to the manifold $\mathcal{O}(n)$ (note that the big summation in (7.80) ends up with a skew-symmetric matrix), the vector field

$$\frac{dQ}{dt} = -\mathbf{g}(Q) \tag{7.81}$$

defines a flow on the manifold $\mathcal{O}(n)$. By the way we construct it, this flow moves in the *steepest descent* direction for the objective function $F(Q)$. To see the motion in the orbit, we define

$$X_i(t) := Q(t)^\top A_i Q(t) \tag{7.82}$$

for $i = 1, \ldots, k$. Upon differentiation and substitution, we find that each $X_i(t)$ must satisfy the ordinary differential equation:

$$\begin{aligned}\frac{dX_i}{dt} &= \frac{dQ^\top}{dt} A_i Q + Q^\top A_i \frac{dQ}{dt} \\ &= \left[X_i, \frac{1}{2}\sum_{j=1}^{k}([X_j, P_j^\top(X_j)] + [X_j^\top, P_j(X_j)]\right].\end{aligned} \tag{7.83}$$

It is worth noting that the above arguments can be *reversed* (Chu and Norris, 1988). That is, any solution $X(t)$ to (7.83) can be written in the form of (7.82) with $Q(t)$ satisfying (7.81). The big summation in the first bracket of (7.83) is always a skew-symmetric matrix. Therefore, the flow $X_i(t)$ naturally stays on the isospectral surface $\mathcal{M}(A_i) := \{Q^\top A_i Q | Q \in \mathcal{O}(n)\}$ if it starts from an initial value $X_i(0) \in \mathcal{M}(A_i)$. One obvious choice of the initial value will be $X_i(0) = A_i$. The differential system (7.83) may be integrated by many readily available ODE solvers. We do not have general theory about when simultaneous reduction can be achieved or not. Even if the prescribed canonical form is not attainable, the solution flow $X(t)$ still provides a systematic way of simultaneously reducing the norm of the residuals.

We observe from (7.83) that the vector field for each component is, in general, a homogeneous polynomial of degree 3. Such a complicated dynamical system is difficult to analyze theoretically. The initial value problem, however, is easy to solve numerically. By varying the subspaces V_i (and, correspondingly, the projections P_i), therefore, we have established an instrument for testing numerically if a given set of matrices $A_1, \ldots, A_k$ can be simultaneously reduced to certain desired forms through orthogonal similarity transformations. We think the versatility of our approach is quite interesting.

Example 7.4. Consider the case $k = 1$. The differential equation (7.83) becomes

$$\begin{cases} \dot{X} = \left[X, \frac{[X, P_1^\top(X)] - [X, P_1^\top(X)]^\top}{2}\right], \\ X(0) = A_1 \quad \text{(general)}. \end{cases} \tag{7.84}$$

We first choose V_1 to be the subspace of all upper triangular matrices. According to our theory, the solution of (7.84) defines an isospectral flow that moves (for $t \geq 0$) to minimize the norm of the strictly lower triangular elements. This idea clearly generalizes that of the Jacobi method for symmetric eigenvalue problems.

Indeed, if X is symmetric, then so is $\dot{X}$. If the initial value A_1 is symmetric, then so is $X(t)$ for all t. In this case, we may be better off if V_1 is chosen to be the subspace of all diagonal matrices. With this choice, the differential system (7.84) becomes

$$\begin{cases} \dot{X} = [X, [X, diag(X)]], \\ X(0) = A_1 \text{ (symmetric)}, \end{cases} \tag{7.85}$$

which is the system (7.49) we have discussed earlier.

Example 7.5. Starting with the initial matrix

$$A_1 = \begin{bmatrix} 1.0000 & 3.0000 & 5.0000 & 7.0000 \\ -3.0000 & 1.0000 & 2.0000 & 4.0000 \\ 0.0000 & 0.0000 & 3.0000 & 5.0000 \\ 0.0000 & 0.0000 & 0.0000 & 4.0000 \end{bmatrix},$$

we integrated the equation (7.84) and found that the solution flow converged to the limit point

$$\begin{bmatrix} 2.2500 & 3.3497 & 3.1713 & 2.8209 \\ -0.3506 & 2.2500 & 8.0562 & 6.1551 \\ 0.6247 & -0.8432 & 2.2500 & 3.2105 \\ -0.0846 & 0.2727 & -0.3360 & 2.2500 \end{bmatrix}.$$

Although the initial matrix (7.86) is an upper quasi-triangular matrix, the limit point (7.5) is a full matrix. Nonetheless, along the solution flow the norm of the strictly lower triangular elements had been reduced monotonically from 3 to 1.1910. This example confirms that the upper quasi-triangular matrix guaranteed by the RSD theorem is not necessarily a stationary point when minimizing *all* the strictly lower triangular elements (Stewart, 1985).

7.5.3 *A nearest commuting pair problem*

We demonstrate another interesting application of the differential system (7.83). In general, two arbitrarily given matrices A_1 and A_2 do not commute. It might be desirable to determine how far the pair (A_1, A_2) is away from being commutable. This problem can be formulated as follows:

Problem 7.13 *(Nearest commuting pair problem)*
Given two matrices A_1 and A_2, find their nearest commuting pair of matrices E_1 and E_2 in the sense that

$$\begin{aligned} \text{Minimize} \quad & F(E_1, E_2) := \frac{1}{2}\sum_{i=1}^{2} \|E_i - A_i\|^2, \qquad (7.86) \\ \text{subject to} \quad & E_1E_2 - E_2E_1 = 0. \end{aligned}$$

Again, Problem 7.13 is a typical equality constrained optimization problem and can be solved by many available methods. In applying the method of Lagrange multipliers, for example, we need to solve the following system of matrix equations

$$\begin{aligned} E_1 - A_1 + \Lambda E_2^\top - E_2^\top \Lambda &= 0 \\ E_2 - A_2 + E_1^\top \Lambda - \Lambda E_1^\top &= 0 \\ E_1E_2 - E_2E_1 &= 0 \end{aligned} \qquad (7.87)$$

for the variables E_1, E_2 and the multiplier Λ. This approach suffers from some obvious difficulties.

Suppose both A_1 and A_2 are symmetric. A problem slightly less general than Problem 7.13 is to determine how far (A_1, A_2) is away from a symmetric, commuting pair (Bunse-Gerstner et al., 1993). Let E_1 and E_2 be any symmetric, commuting pair. We shall assume further that at least one of these two matrices has distinct eigenvalues (this is the generic case). It is not difficult to show that E_1 and E_2 can be simultaneously diagonalized by an orthogonal similarity transformation (Gantmacher, 1959, p. 222, Corollary 1). Let $D_i = Q^\top E_i Q$, $i = 1, 2$, be the diagonal matrices with orthogonal matrix Q. We observe from the relation

$$\sum_{i=1}^{2} \|E_i - A_i\|^2 = \sum_{i=1}^{2} \|D_i - Q^\top A_i Q\|^2 \qquad (7.88)$$

that the left-hand side of (7.88) will be minimized if one first finds an orthogonal matrix Q such that the matrices $Q^\top A_i Q$ are as close to diagonal matrices as possible, and then sets $D_i = \operatorname{diag}(Q^\top A_i Q)$. In other words, the problem of finding a nearest commuting pair to a given pair of symmetric matrices is boiled down to the problem of simultaneous reduction of off-diagonal elements of the given pair by orthogonal transformations. The latter problem fits in as a special case of our general framework in the preceding section. We simply proceed as follows.

Take both $V_i, i = 1, 2$, to be the subspace of diagonal matrices. According to (7.83), the descent flow is given by the initial value problem:

$$\begin{cases} \dot{X}_i = \left[X_i, \sum_{j=1}^{2}[X_j, \mathrm{diag}(X_j)],\right] \\ X_i(0) = A_i, \quad i = 1, 2, \end{cases} \tag{7.89}$$

since both X_i and $\mathrm{diag}(X_i)$ are symmetric matrices.

In the event that A_1 and A_2 cannot be diagonalized simultaneously, the limit point of the flow gives a way of measuring the distance from (A_1, A_2) to the nearest commuting pair (see (7.88)). Comparing with the system (7.85), one finds immediately that (7.89) is a direct generalization of the Jacobi algorithm. It is known that the straightforward "diagonalize one, then diagonalize the other" approach for simultaneously diagonalizing pairs of symmetric matrices is subject to numerical hazards that may prevent convergence (Bunse-Gerstner et al., 1993). We think our approach gives a new twist to the algorithm.

7.5.4 *Orthogonal equivalence transformation*

Together with the notion of product topology that we have described in Section 7.4, the techniques developed in Section 7.5.2 can be generalized to solve Problem 7.11 which include the GRSD as a special case.

We adopt almost identical notation. Let $A_i \in \mathbb{R}^{m \times n}, i = 1, \ldots, k$, be the given matrices. For each i, let $V_i \subset \mathbb{R}^{m \times n}$ denote the subspace of all matrices having the specified form to which A_i is supposed to be reduced. The projection operator from $\mathbb{R}^{m \times n}$ to V_i is denoted by P_i. For any $X \in \mathbb{R}^{m \times m}$ and $Y \in \mathbb{R}^{n \times n}$, we define

$$\alpha_i(X, Y) := X^\top A_i Y - P_i(X^\top A_i Y). \tag{7.90}$$

We reformulate Problem 7.11 as:

Problem 7.14 *(Minimization on orbits of $\mathcal{O}(m) \times \mathcal{O}(n)$)*
Give matrices $A_i \in \mathbb{R}^{m \times n}$ and desirable form V_i, $i = 1, \ldots, k$,

$$\text{Minimize } F(Q, Z) := \frac{1}{2}\sum_{i=1}^{k} \|\alpha_i(Q, Z)\|^2 \tag{7.91}$$

$$\text{subject to } Q^\top Q = I_m, \; Z^\top Z = I_n.$$

The Fréchet derivative of the objection function in (7.91) at a general $(X, Y) \in \mathbb{R}^{m\times m} \times \mathbb{R}^{n\times n}$ acting on a general $(H, K) \in \mathbb{R}^{m\times m} \times \mathbb{R}^{n\times n}$ is

$$F'(X,Y)(H,K) = \sum_{i=1}^{k} \langle \alpha_i(X,Y), H^\top A_i Y + X^\top A_i K - P_i(H^\top A_i Y + X^\top A_i K)\rangle$$
$$= \sum_{i=1}^{k} \left(\langle A_i Y \alpha_i^\top (X,Y), H\rangle + \langle A_i^\top X \alpha_i(X,Y), K\rangle \right). \tag{7.92}$$

Therefore, with respect to the induced Frobenius inner product, we may interpret the *gradient* of F at (X, Y) as the pair:

$$\nabla F(X,Y) = \left(\sum_{i=1}^{k} A_i Y \alpha_i^\top (X,Y), \sum_{i=1}^{k} A_i^\top X \alpha_i(X,Y) \right). \tag{7.93}$$

We note that there is a considerable similarity between (7.79) and (7.93), except that (7.93) involves a pair of matrices.

Because of the product topology, we may use the same principle as in (7.80) to calculate the projection $\mathbf{g}(Q, Z)$ of $\nabla F(Q, Z)$ into the tangent space $T_{(Q,Z)}\mathcal{O}(m) \times \mathcal{O}(n)$. After simplification, we claim that

$$\mathbf{g}(Q,Z) = \left(\frac{Q}{2} \sum_{i=1}^{k} (Q^\top A_i Z \alpha_i^\top (Q,Z) - \alpha_i(Q,Z) Z^\top A_i^\top Q), \right.$$
$$\left. \frac{Z}{2} \sum_{i=1}^{k} (Z^\top A_i^\top Q \alpha_i(Q,Z) - \alpha_i^\top (Q,Z) Q^\top A_i Z) \right). \tag{7.94}$$

Therefore, the vector field

$$\frac{d(Q,Z)}{dt} = -\mathbf{g}(Q,Z) \tag{7.95}$$

defines a steepest descent flow on the manifold $\mathcal{O}(m) \times \mathcal{O}(n)$ for the objective function $F(Q, Z)$. This flow also defines a motion via

$$X_i(t) := Q(t)^\top A_i Z(t), \quad i = 1, \ldots, k, \tag{7.96}$$

on the manifold of orthogonal equivalence orbit. Upon differentiation and substitution, it is not difficult to see that each $X_i(t)$ satisfies the equation

$$\frac{dX_i}{dt} = \sum_{j=1}^{k} \left\{ X_i \frac{X_j^\top P_j(X_j) - P_j^\top (X_j) X_j}{2} + \frac{P_j(X_j) X_j^\top - X_j P_j^\top (X_j)}{2} X_i \right\}. \tag{7.97}$$

By specifying the initial values, say $X_i(0) = A_i$, and the subspaces V_i, we now have an instrument in hand to explore various simultaneous reduction problems numerically simply by integrating equation (7.97).

Example 7.6. The Jacobi SVD flow defined by system (7.71) is a special case of (7.97) by taking $k = 1$ and V_1 to be the subspace of all diagonal matrices in $\mathbb{R}^{m \times n}$.

7.6 Closest normal matrix problem

All the techniques discussed in Sections 7.5.2 and 7.5.4 can be generalized to the complex-valued case. We conclude this chapter with a demonstration of how this generalization should be done by working on the closest normal matrix problem.

Nearest matrix approximation has received considerable attention in the literature. See, for example, the review article by Higham (1989) and the references contained therein. The determination of a closest normal matrix to a given square complex matrix in the Frobenius norm has been completely solved in (Ruhe, 1987) with considerable algebraic manipulations. In contrast, we shall cast this problem into our framework from which we obtain a new and clear geometric characterization of the first and the second order optimality condition. Additionally, we will see that the optimality conditions are much easier to derive than the conventional way employed in the literature.

Our first task is to identify a normal matrix. In the paper by Grone et al. (1987), a list of 70 conditions each of which is equivalent to the normality of a given matrix is given. Another 20 or so conditions were added to the list by Elsner and Ikramov (1998) one decade later. Among these, we find the following well-known fact most useful to us. In the space $\mathbb{C}^{n \times n}$ of $n \times n$ complex-valued matrices, let $\mathcal{U}(n)$ denote the group of all unitary matrices and $\mathcal{D}_{\mathbb{C}}(n)$ denote the subspace of all diagonal matrices.

Theorem 7.6.1. A matrix $Z \in \mathbb{C}^{n \times n}$ is normal if and only if there exists a unitary $U \in \mathcal{U}(n)$ such that $U^*ZU \in \mathcal{D}_{\mathbb{C}}(n)$.

Therefore, the closest normal matrix problem (**NNMP**) can be formulated as follows.

Problem 7.15 *(Nearest normal matrix problem)*
Given an arbitrary matrix $A \in \mathbb{C}^{n \times n}$, find $U \in \mathcal{U}(n)$ and $D \in \mathcal{D}_{\mathbb{C}}(n)$ that minimize

$$F(U, D) := \tfrac{1}{2}\|A - UDU^*\|^2. \tag{7.98}$$

At first glance, we note that the two matrix variables U and D in the minimization of (7.98) are considered to be independent of each other. It appears that the NNMP is a two-variable minimization problem. Letting $Z := UDU^*$, however, we observe that the relationship

$$\|A - Z\|^2 = \|U^*AU - D\|^2 \tag{7.99}$$

holds. Obviously, for any given $U \in \mathcal{U}(n)$, the best $D \in \mathcal{D}_{\mathbb{C}}(n)$ that will minimize the right-hand side of (7.99) is $\text{diag}(U^*AU)$. Therefore, at *global* extrema, Problem 7.15 is equivalent to a simplified approximation problem where we only need to work with one single variable $U \in \mathcal{U}(n)$.

Problem 7.16 *(NNMP1)*
Given $A \in \mathbb{C}^{n\times n}$, find $U \in \mathcal{U}(n)$ that minimizes

$$F(U) := \frac{1}{2}\|U^*AU - \text{diag}(U^*AU)\|^2. \tag{7.100}$$

Since unitary transformations do not alter the Frobenius norm of a matrix, minimizing the sum of squares of off-diagonal elements of a matrix is equivalent to maximizing the sum of squares of diagonal elements. From (7.100), we immediately conclude that the closest normal matrix is characterized by the following classical result.

Theorem 7.11. (Causey, 1964; Gabriel, 1979) Given $A \in \mathbb{C}^{n\times n}$, let $Z = UDU^*$ where $U \in \mathcal{U}(n)$ and $D \in \mathcal{D}_{\mathbb{C}}(n)$. Then Z is a closest normal matrix to A in the Frobenius norm if and only if

(i) the unitary matrix U maximizes $\|\text{diag}(V^*AV)\|$ among all $V \in \mathcal{U}(n)$;
(ii) the diagonal matrix D is such that $D = \text{diag}(U^*AU)$.

We see from Problem 7.16 and Theorem 7.11 that except being complex-valued, the situation in the NNMP is just like that discussed in Section 7.5.2. We want to minimize the norm of the off-diagonal elements by unitary similarity transformations on A.

7.6.1 *First-order optimality condition*

The ideas discussed in Section 7.5.2 can be applied almost without change to the complex-valued case. We briefly describe our procedure as follows: We shall regard $\mathbb{C}^{n\times n}$ as the vector space $\mathbb{R}^{n\times n}\times\mathbb{R}^{n\times n}$ over the field of real numbers. That is, we shall identify the complex matrix Z as a pair of real matrices $(\Re Z, \Im Z)$, where $\Re Z$ and $\Im Z$ represent the real and the imaginary part of Z, respectively. The inner product on $\mathbb{C}^{n\times n}$ is defined by

$$\langle X, Y\rangle_C := \langle \Re X, \Re Y\rangle + \langle \Im X, \Im Y\rangle. \tag{7.101}$$

We note that $\langle Z, Z\rangle_C = \|Z\|^2$. The topology imposed on $\mathbb{C}^{n\times n}$ by (7.101) resembles that on $\mathbb{R}^{m\times m} \times \mathbb{R}^{n\times n}$ given by (7.52). We may thus also take advantage of the techniques developed in Section 7.5.4. In this context, the analogue to (7.16) is that the tangent space to $\mathcal{U}(n)$ at any unitary matrix U is given by

$$T_U\mathcal{U}(n) = UH(n)^{\perp} \tag{7.102}$$

where $H(n)$ is the collection of all Hermitian matrices in $\mathbb{C}^{n\times n}$. Furthermore, identifying $Z = (\Re Z, \Im Z)$, one can calculate the Fréchet derivative and the gradient for the objective function F in (7.100). It is not difficult to prove that all the calculation can be carried out formally just as in the real-valued case. In particular, one can show that the projected gradient $\mathbf{g}(U)$ of F onto the manifold $\mathcal{U}(n)$ is given by

$$\mathbf{g}(U) = \frac{U}{2}\{[\text{diag}(U^*AU), U^*A^*U] - [\text{diag}(U^*AU), U^*A^*U]^*\}. \tag{7.103}$$

From (7.103), we obtain the following first-order optimality condition.

Theorem 7.6.2. Let $W := U^*AU$. Then for U to be a stationary point of Problem 7.16, it is necessary that

$$[\text{diag}(W), W^*] = [\text{diag}(W), W^*]^*. \tag{7.104}$$

Write $W = [w_{ij}]_{i,j=1}^n$. It is easy to see that the condition (7.104) is equivalent to

$$\overline{w}_{ji}(w_{ii} - w_{jj}) = w_{ij}(\overline{w}_{jj} - \overline{w}_{ii}). \tag{7.105}$$

If we define a matrix $H = [h_{ij}]_{i,j=1}^n$ by

$$h_{ij} = \begin{cases} \frac{w_{ij}}{w_{ii}-w_{jj}}, & \text{if } w_{ii} \neq w_{jj}, \\ 0, & \text{if } w_{ii} = w_{jj}, \end{cases} \tag{7.106}$$

then the condition (7.105) is equivalent to the condition that H be Hermitian. This observation is in concordance with the notion of ΔH*-matrix* introduced in Gabriel (1979, 1987), Higham (1989), Ruhe (1987). We think our derivation, being different from those done in the literature, is of interest in its own right.

7.6.2 *Second-order optimality condition*

We can take one step further. The explicit form of the projected gradient (7.103) can significantly facilitate the computation of the projected Hessian on the tangent space of $\mathcal{U}(n)$.

We first extend the projected gradient function $\mathbf{g}$ formally to the entire space $\mathbb{C}^{n\times n}$, that is, we assume equation (7.103) is defined for general complex matrices. Since the extended $\mathbf{g}$ is smooth, we may formally take its Fréchet derivative. In Section 7.1.3 we have observed that the quadratic form of the extended Fréchet derivative applied to tangent vectors corresponds exactly to the projected Hessian of the Lagrangian function. We recall that the tangent space of our feasible $\mathcal{U}(n)$ is given by $UH(n)^\perp$. Therefore, we are able to calculate the quadratic form

$$\langle UK, \mathbf{g}'(U)UK\rangle = \langle[\text{diag}(W), K] - \text{diag}[W, K]), [W, K]\rangle_C \tag{7.107}$$

with unitary U and skew-Hermitian K. In this way, the second-order optimality condition for Problem 7.16 is hereby established.

Theorem 7.13. Let $W := U^*AU$. Then necessary (sufficient) conditions for $U \in \mathcal{U}(n)$ to be a local minimizer of Problem 7.16 are that:

(i) the matrix $[\mathrm{diag}(W), W^*]$ is Hermitian;

(ii) the quadratic form $\langle [\mathrm{diag}(W), K] - \mathrm{diag}[W, K]), [W, K]\rangle_C$ is nonnegative (positive) for every skew-Hermitian matrix K.

We note that the approach in Ruhe (1987) utilized the Lagrangian function with a Hermitian matrix as the Lagrange multipliers. The second-order condition (either formula (12) or formula (15) in Ruhe (1987)) also involved the Lagrangian multipliers. Our description in the above theorem does not need any information about the Lagrangian multipliers. We think that the result in Theorem 7.13 is more convenient.

7.6.3 *Numerical methods*

There are at least two ways to solve the NNMP numerically. The iterative method proposed in Ruhe (1987) employs a special Jacobi algorithm for normal matrices (Goldstine and Horwitz, 1959). The idea is to transform the matrix A by rotations into U^*AU which is a $\Delta - H$ matrix. Since this satisfies the first-order optimality condition, it is suggested that $Z := U\mathrm{diag}(U^*AU)U^*$ is a putative nearest normal matrix. This approach requires careful calculation of certain shifts, phases, and rotation angles in the process.

Based on our previous experience, we can propose a continuous analogue by simply integrating the differential system

$$\begin{aligned} \frac{dU}{dt} &= U\frac{[W, \mathrm{diag}(W^*)] - [W, \mathrm{diag}(W^*)]^*}{2} \\ \frac{dW}{dt} &= \left[W, \frac{[W, \mathrm{diag}(W^*)] - [W, \mathrm{diag}(W^*)]^*}{2}\right] \end{aligned} \tag{7.108}$$

for the unitary matrix $U(t)$ and the variable $W(t) := U(t)^*AU(t)$ until convergence occurs. By then, the matrix $Z := \tilde{U}\mathrm{diag}(\tilde{W})\tilde{U}^*$, where $\tilde{}$ denotes a limit point of (7.108), will be a putative nearest normal matrix.

Example 7.7. Integrate the system (7.108) with initial values $U(0) = I$ and

$$W(0) = \begin{bmatrix} 0.7616 + 1.2296i & -1.4740 - 0.4577i \\ -1.6290 - 2.6378i & 0.1885 - 0.8575i \end{bmatrix}.$$

The approximated limit point of (7.108) is given by

$$\tilde{W} \approx \begin{bmatrix} 2.2671167250 + 1.9152270486i & 0.4052706333 + 0.8956586233i \\ -0.9095591045 - 0.3730293488i & -1.3170167250 - 1.5431270486i \end{bmatrix}$$

and

$$\tilde{U} \approx \begin{bmatrix} 0.8285289301 - 0.0206962995i & 0.5350877833 - 0.1636842669i \\ -0.5350877833 - 0.1636842669i & 0.8285289301 + 0.0206962995i \end{bmatrix}.$$

The matrix $\tilde{W}$ agrees with the one given in (Ruhe, 1987) only in its diagonal elements. However, after substitution, the matrix $Z = \tilde{U}\text{diag}(\tilde{W})\tilde{U}^*$ is the same as that given in Ruhe (1987).

Other than the difference that complex-valued matrices are involved, it is interesting to note the obvious similarity between (7.84) and (7.108).

7.7 Summary

This chapter represents a very important milestone in our discussion of IEPs. The techniques we have developed have allowed us to explore territory that is far beyond merely IEPs. The nearest normal matrix approximation, for instance, would have nothing to do with IEPs since eigeninformation has never come into play. However, the projected gradient method indeed originated from solving (least squares) IEPs. In a word, the IEPs or the ISVPs are but a special case of the much larger problems of computing least squares approximations subject to spectral or singular value constraints. These problems share a common structure that allows us to introduce a general procedure by using the projected gradient method in this chapter.

The notion of either isospectrum or iso-singular value are further extended to include two types of transformations, orthogonal similarity and orthogonal equivalence, as constrained optimization problems. We then find applications such as simultaneous reductions or nearest normal matrix approximation. We hereby declare that even the notion of orthogonality can be replaced by any kind of matrix groups and, hence, open up many more possibilities and open questions. This will be the main topic in Chapter 9.

Throughout this chapter, we have seen the theme repeatedly many times, namely, the objective functions are formed by following the spirit of the Jacobi method. The framework in deriving these equations is quite general in that the number of the given matrices and the desired forms to which the given matrices are supposed to be reduced can be almost arbitrary. By integrating the corresponding differential equation, we have thus established a general numerical tool that can be used to tackle, for example, Problems 7.10 and 7.11 for various reduced forms. In the event that a specified form is not attainable, the limit point of the corresponding differential equation still gives a way of measuring how far the matrices can be reduced.

8

STRUCTURED LOW RANK APPROXIMATION

8.1 Overview

This chapter concerns the construction of a structured low rank matrix that is nearest to a given matrix. Low rank approximation problems can be considered as IEPs because the rank deficiency implies that a specified number of eigenvalues or singular values are preset to zeros. The notion of structured low rank approximation arises when in practice the empirical data collected in a matrix do not maintain either the specified structure or the desirable rank as is expected in the original system. On another front, low rank approximation often has the advantages of reducing the dimensionality or retrieving principal factors of otherwise a very large, complex and redundant system. The task to retrieve useful information while maintaining the underlying physical feasibility often necessitates the search for a good structured lower rank approximation of the data matrix. We shall address some of the theoretical and numerical issues involved in this kind of problem.

Finding a structured low rank approximation of a general data matrix is a critical task in many disciplines. The list of applications includes image compression, signal enhancement, model reduction, protein folding, computer algebra, noise reduction, seismic inversion, latent semantic indexing, principal component analysis, regularization for ill-posed problems, and so on. However, just like IEPs, there does not seem to be a unified approach to each of these applications.

A practical means to tackle the low rank approximation problem, if the 2-norm or the Frobenius norm is used in the measurement of closeness, is the truncated singular value decomposition (TSVD) method. The theory of TSVD and its applications are well understood (Cadzow, 1988; Cadzow and Wilkes, 1990; Golub and Van Loan, 1996; Hansen, 1987). When the desired rank is relatively low and the data matrix is large and sparse, a complete SVD becomes too expensive. Some less expensive alternatives for numerical computation, for example, the Lanczos bidiagonalization process (Millhauser et al., 1989; Simon and Zha, 2000) and the Monte Carlo algorithm (Frieze et al., 1998), are available. None of these methods, however, can address the underlying matrix structure that is also part of the constraint. Recall that in Section 4.8.3 we have pointed out that finding a rank deficient matrix, even in the case of affine structure, is very difficult.

The purpose of this chapter is to provide some preliminary investigations into this structured low rank approximation problem. We shall treat some mathematical properties, point out some interesting applications, and report some numerical experiments. Our discussion by no means is complete. Readers will soon realize that we are dealing with these applications case by case and that we have raised more questions than can be answered. There is plenty of unfinished work in both theoretical and numerical aspects. There appears to be little information available in the literature. We thus consider the presentation in this chapter as merely a beginning step toward fully understanding the problem.

To grasp the ideas, we first describe the structured low rank approximation problem (**SLRAP**) in a fairly general setting as follows.

> **Problem 8.1** *(Generic SLRAP)*
> Given a matrix $A \in \mathbb{R}^{m\times n}$, an integer k, $1 \leq k < \operatorname{rank}(A)$, a class of matrices Ω, and a fixed matrix norm $\|\cdot\|$, find a matrix $\hat{B} \in \Omega$ of rank k such that
> $$\|A - \hat{B}\| = \min_{B\in\Omega,\ \operatorname{rank}(B)=k} \|A - B\|. \tag{8.1}$$

Note in the above that any feasible approximation B must satisfy both a structural constraint specified by Ω and the rank constraint specified by k. For this reason, such a problem is sometimes referred to as a *structure preserving rank reduction problem.* The structural constraint is immaterial if Ω is simply the entire space $\mathbb{R}^{m\times n}$. On the other hand, the low rank approximation by specially structured matrices becomes a much harder problem. We should also point out that the measurement used in Problem (8.1) need not be the usual 2-norm or the Frobenius norm. If other norms are used, solving (8.1) presents another degree of difficulty for TSVD-based methods. We shall see subsequently that some of our approaches are capable of handling this general case.

We point out that the notion of low rank approximation can be viewed from different standpoints. Suppose a given matrix A is known *a priori* to have k singular values larger than ϵ. An idea in (van der Veen, 1996), for instance, is to find all rank-k approximates $\hat{A}$ such that $\|A - \hat{A}\|_2 < \epsilon$. The objective in van der Veen (1996) is not to compute an approximate $\hat{A}$ of rank k that minimizes $\|A - \hat{A}\|_2$, but rather to compute the one in which the approximation error is limited. In contrast, there are no *a priori* knowledge or restrictions on the singular values of A in our formulation. Furthermore, our methods compute a best approximate matrix $\hat{A}$ belonging to a specified affine subspace Ω.

A serious challenge associated with the SLRAP is that generally there is no easy way to characterize, either algebraically or analytically, a given class of structured lower rank matrices. This lack of explicit description of the feasible set makes it difficult to apply classical optimization techniques. We thus wish to stress two main points. First, we provide some theoretical insight into the structure preserving low rank approximation problem. Secondly, we propose some numerical procedures to tackle this structure preserving rank reduction problem.

8.2 Low rank Toeplitz approximation

We begin our discussion with the case where the structured matrices form an affine subspace Ω. This class of constraint arises naturally in applications due to the interrelation of matrix elements in some prescribed fashion. The so-called *linear structure* in Cadzow (1988), for instance, is an affine constraint. Some examples of linear structure include symmetric Toeplitz, block Toeplitz, Hankel, upper Hessenberg, or banded matrices with fixed bandwidth. On the other hand, the low rank condition is often predestined and is inherent to the system behind the physical setup. In addition to being of theoretical interest in its own right, the structure preserving rank reduction problem deserves consideration also for practical reasons. Some applications have already been outlined in Section 2.7.

The special case of symmetric Toeplitz structure using the Frobenius matrix norm is used to exemplify the ideas throughout the discussion (Cybenko, 1982). We stress that the notion we are about to derive is so general that our procedures can be applied to problems of any rank, any linear structure, and any matrix norm.

Be cautious that the solvability of Problem 8.1 even after knowing that the feasible set is not empty, is not clear at all. The greatest difficulty is that of tracking down systematically a structured low rank matrix. The lack of complete understanding of the theory does not necessarily preclude the problem from being solved numerically. One possible approach is to employ the singular value decomposition to characterize low rank matrices. The desired matrix structure is then formulated as a set of equalities among the singular values and singular vectors. The resulting formulation becomes an equality constrained optimization problem. Another possibility is to introduce a lift-and-project method that alternates iterations between low rank matrices and structured matrices. This iterative scheme gives rise to a point-to-point map $\mathcal{P}_k$ from any given initial matrix T to its limit point $\mathcal{P}_k(T)$ that is expected to satisfy both the structural and the rank conditions. Using this point-to-point map $\mathcal{P}_k(T)$ as a handle to capture structured low rank matrices, the resulting formulation becomes an unconstrained optimization problem.

8.2.1 *Theoretical considerations*

This section provides some primary notions related to the approximation by structured low rank matrices. Two basic questions are raised in this section with only partial answers. The first question concerns the feasible set. Can structured matrices have arbitrary rank? The second question concerns the solvability.

Can a given matrix always be approximated by a matrix with a specific structure and a specific rank?

Substance of feasible set Before any numerical method is attempted, a fundamental question associated with Problem (8.1) is whether low rank matrices with specified structure actually exist. Toward that end, we first observe the following result.

Theorem 8.1. Given a matrix $A \in R^{m\times n}$, an integer k, $1 \leq k < \text{rank}(A)$, a class of matrices Ω, and a fixed matrix norm $\|\cdot\|$, the matrix approximation problem

$$\min_{B\in\Omega,\ \text{rank}(B)\leq k} \|A - B\| \tag{8.2}$$

is *always* solvable, as long as the feasible set is non-empty.

Proof The collection of *all* rank deficient square matrices, being the pre-image set $\det^{-1}(\{0\})$ under the determinant map, forms a *closed* set. Since the rank of a matrix is equal to the order of its largest nonzero minor, a similar determinant argument implies the set of all $m \times n$ matrices with rank $\leq k$ is a closed set. The minimization in (8.2) therefore can be considered as taking place over a compact neighborhood of A. If this neighborhood is not empty, the minimum exists. □

We quickly point out that Problem (8.2) is different from Problem (8.1). In (8.2) the rank condition is less than or equal to k while in (8.1) the rank condition is required to be exactly equal to k. In (8.2) the feasible matrices form a closed set while in (8.1) the feasible set might be open. Pathologically it is possible that a given target matrix A does not have a nearest structured rank-k matrix approximation, but does have a nearest structured matrix approximation that is of rank $k - 1$ or lower. That is, it is pathologically possible that Problem (8.2) has a solution while (8.1) does not.

For special classes of matrices, it is sometime possible to prove that low rank matrices with specified structures do exist. We mention two specific examples below.

Theorem 8.2. Symmetric Toeplitz matrices can have arbitrary rank.

Proof We shall give a constructive proof. Let n be the dimension of symmetric Toeplitz matrices being considered. Given any integer $1 \leq k < n$, first construct a $k \times k$ symmetric Toeplitz matrix

$$S = T([a_1, a_2, \ldots, a_k])$$

where $\{a_i\}_{i=1}^k$, are arbitrary real numbers satisfying the relationship

$$a_j = a_{k+2-j}$$

for $j = 2, \ldots, k$. Note that the choices of $\{a_i\}_{i=1}^k$ such that $\det(S) = 0$ form a closed, low dimensional manifold. Thus, with numbers $\{a_i\}_{i=1}^k$ selected out of an open and dense set, that is, with probability one if $\{a_i\}_{i=1}^k$ are selected randomly, the matrix S is of rank k.

Now expand the matrix S to an $n \times n$ matrix T by *augmenting* and *stacking* S to itself. Repeatedly use copies of S if necessary. Fill in the right and the lower borders of T with copies of the first $k' := n(\text{mod})k$ columns and rows of S. Fill in the lower right corner of T with the leading $k' \times k'$ principal submatrix of S. For example, if $n = 2k + 3$, then T would be

$$\begin{bmatrix} S & S & S(:, 1:3) \\ S & S & S(:, 1:3) \\ S(1:3, :) & S(1:3, :) & S(1:3, 1:3) \end{bmatrix}.$$

Since the matrix has repeated columns, it is readily seen that T is symmetric Toeplitz and that $\text{rank}(T) = k$. □

Theorem 8.3. There are $n \times n$ Hankel matrices with any given rank.

Proof The matrix ΞT is a Hankel matrix for any Toeplitz matrix T and vice versa. So the result follows from the preceding theorem. However, we give a different proof below.

Let n be the dimension of the Hankel matrices being considered. Let k be the desired rank. Choose arbitrary nonzero numbers $\beta_1, \ldots, \beta_k$ and $z_1, \ldots, z_k$ where $z_i \neq z_j$ whenever $i \neq j$. Define

$$h_j := \sum_{i=1}^{k} \beta_i z_i^j, \quad j = 1, 2, \ldots, 2n - 1.$$

Then the matrix

$$H := \begin{bmatrix} h_1 & h_2 & \ldots & h_n \\ h_2 & h_3 & \ldots & h_{n+1} \\ \vdots & & & \vdots \\ h_n & h_{n+1} & \ldots & h_{2n-1} \end{bmatrix},$$

is a Hankel matrix of rank k. □

The construction in Theorem 8.3 in fact is a known correspondence between low rank Hankel matrices and noiseless time-domain signals comprising k components of exponentially decaying sinusoids (de Beer, 1995; Cadzow and Wilkes, 1990; Mittelmann and Cadzow, 1987; Park et al., 1999). When noise is added to H, the rank k is lost. In this instance, one of the prevailing reasons for considering (8.1) where A stands for the noisy data matrix is to gain insight into the original signal by removing the noise, maintaining the Hankel structure, and reducing the rank. A specific and detailed application of the structure preserving rank reduction problem to medical science can be found in the lecture notes by de Beer (1995).

For engineering applications, the existence question might not be as important as deriving a reliable method for achieving a closest possible approximation matrix. The existence question itself, however, is an important step before such an achievement can be obtained. For other types of linear structures, the existence question generally is a challenging algebraic problem that is of interest in its own right. We are not aware of any definitive studies on this aspect in the literature.

Solvability of nearest approximation Once it is established that the feasible set is not empty, what remains is to find from within the feasible set a nearest approximation to the given target (noisy) matrix. We must be cautious, however, that the nonemptiness of a feasible set does not necessarily settle the structure preserving rank reduction Problem 8.1. The existence of lower rank matrices of specified structure does not guarantee that one of such matrices will be *closest* to a given target matrix. Minimization over an open set, such as the trivial example of minimizing $f(x) = 1/x$ over $x > 0$, does not necessarily have a solution.

It appears that very little work has been done about the solvability of Problem 8.1, even for Toeplitz matrices. Similar comments were echoed in the discussion (Park et al., 1999). While Theorem 8.1 guarantees that a lower rank approximation is always possible, being able to determine that Problem 8.1 is solvable for a specific structure Ω and for a specific rank k would be a significant accomplishment.

We speculate that one of the difficulties in proving the solvability of Problem 8.1 could be due to its *finite dimensionality.* For the infinite dimensional case, the solvability issue sometimes can be settled. For example, the following result asserts that the low rank Hankel approximation to an infinite dimensional Hankel matrix always exists. It does not seem possible that the proof can be extended to finite-dimensional matrices.

Theorem 8.4. (Adamjan et al., 1971) Suppose the underlying matrices are of infinite dimension. Then the closest approximation to a Hankel matrix by a low rank Hankel matrix always exists and is unique.

The difference between finding a structured low rank matrix and finding the closest structured low rank approximation to a given target matrix needs to be carefully discerned. Indeed, in many practices somehow this distinction has been overlooked. We shall point out later that the popular Cadzow algorithm (de Beer, 1995; Cadzow and Wilkes, 1990), for example, only finds a structured low rank matrix that is *near* a given target matrix, but the algorithm alone does not produce the nearest structured low rank approximation.

Symmetric Toeplitz matrices To illustrate the algebraic complexity of Problem 8.1, we shall study the structure of symmetric Toeplitz matrices in a little more detail. Recall that a symmetric Toeplitz matrix T can be identified by its first row $T([t_1, \ldots, t_n])$. Let $\mathcal{T}_n$ denote the affine subspace of all $n \times n$ symmetric Toeplitz matrices. We examine the problem of characterizing symmetric Toeplitz matrices with prescribed rank. A related discussion on representing Hankel matrices by Vandermonde factorization can be found in (Boley et al., 1997).

Any $n \times n$ symmetric matrix M of rank k can be decomposed into the form

$$M = \sum_{i=1}^{k} \alpha_i \mathbf{y}^{(i)} \mathbf{y}^{(i)\top} \tag{8.3}$$

with $\alpha_i \in \mathbb{R}$ and $\mathbf{y}^{(i)} \in \mathbb{R}^n$. For the matrix M in (8.3) to be Toeplitz, say, $M = T([t_1, \ldots, t_n])$, the following system of $n(n-1)/2$ equations must be satisfied by the α_i's and $\mathbf{y}^{(i)}$'s:

$$\sum_{i=1}^{k} \alpha_i y_j^{(i)} y_{j+s}^{(i)} = t_{s+1}, \quad s = 0, 1, \ldots, n-2, \ 1 \leq j \leq n-s. \tag{8.4}$$

We stress that (8.4) is necessary but not sufficient for ensuring that the matrix M is of rank k. Note that the rank of M is exactly k if and only $\alpha_i \neq 0$ for all i.

Example 8.1. The case $k = 1$ is easy and shows that the collection of rank one Toeplitz matrices forms an open set.

Any vector $\mathbf{y}^{(1)} \in \mathbb{R}^n$ such that $\mathbf{y}^{(1)} \mathbf{y}^{(1)\top}$ is Toeplitz must be of the form

$$\mathbf{y}^{(1)} = c\mathbf{e}, \tag{8.5}$$

where

$$\mathbf{e} = [1, \ldots, 1]^\top \quad \text{or} \quad \mathbf{e} = [1, -1, 1, -1, \ldots, (-1)^{n-1}]^\top,$$

and c is an arbitrary scalar. Thus rank one Toeplitz matrices form two simple one-parameter families:

$$\mathcal{S} = \left\{\alpha_1 T([1, \ldots, 1]) \text{ or } \alpha_1 T([1, -1, 1, \ldots, (-1)^{n-1}]) | \alpha_1 \neq 0\right\}.$$

Even for $k = 2$, the characterization of scalars $\alpha_i \in \mathbb{R}$ and vectors $\mathbf{y}^{(i)} \in \mathbb{R}^n$, $i = 1, 2$, so that the matrix $\alpha_1 \mathbf{y}^{(1)} \mathbf{y}^{(1)\top} + \alpha_2 \mathbf{y}^{(2)} \mathbf{y}^{(2)\top}$ is Toeplitz, becomes considerably more complicated.

Example 8.2. All rank two 4×4 Toeplitz matrices form a five-dimensional algebraic variety.

There are a total of two scalars $\alpha_i \in \mathbb{R}$ and eight components in $\mathbf{y}^{(i)} \in \mathbb{R}^4$ involved. The necessary conditions on $y_3^{(1)}, y_4^{(1)}, y_3^{(2)}, y_4^{(2)}$ and α_1 in terms of the five parameters $y_1^{(1)}, y_2^{(1)}, y_1^{(2)}, y_2^{(2)}$ and α_2 in order that the summation

$\alpha_1 \mathbf{y}^{(1)} \mathbf{y}^{(1)^\top} + \alpha_2 \mathbf{y}^{(2)} \mathbf{y}^{(2)^\top}$ forms a 4×4 Toeplitz matrix of rank two are as follows:

$$\begin{cases} \alpha_1 := \dfrac{\alpha_2 \left({y_1^{(2)}}^2 - {y_2^{(2)}}^2 \right)}{-{y_1^{(1)}}^2 + {y_2^{(1)}}^2}, \\[2ex] y_3^{(1)} := \dfrac{y_2^{(1)} y_1^{(2)} y_1^{(1)} + 2\, y_2^{(2)} {y_2^{(1)}}^2 - y_2^{(2)} {y_1^{(1)}}^2}{y_2^{(1)} y_1^{(2)} + y_1^{(1)} y_2^{(2)}}, \\[2ex] y_4^{(1)} := -\dfrac{\left[\begin{array}{c} {y_2^{(1)}}^3 {y_1^{(2)}}^2 - 4\, {y_2^{(1)}}^3 {y_2^{(2)}}^2 - 4\, y_1^{(1)} y_1^{(2)} y_2^{(2)} {y_2^{(1)}}^2 \\ -2\, y_2^{(1)} {y_1^{(1)}}^2 {y_1^{(2)}}^2 + 3\, y_2^{(1)} {y_2^{(2)}}^2 {y_1^{(1)}}^2 + 2\, y_1^{(2)} y_2^{(2)} {y_1^{(1)}}^3 \end{array} \right]}{{y_2^{(1)}}^2 {y_1^{(2)}}^2 + 2\, y_2^{(1)} y_1^{(2)} y_1^{(1)} y_2^{(2)} + {y_1^{(1)}}^2 {y_2^{(2)}}^2}, \\[2ex] y_3^{(2)} := -\dfrac{y_2^{(1)} {y_1^{(2)}}^2 - 2\, y_2^{(1)} {y_2^{(2)}}^2 - y_1^{(2)} y_2^{(2)} y_1^{(1)}}{y_2^{(1)} y_1^{(2)} + y_1^{(1)} y_2^{(2)}}, \\[2ex] y_4^{(2)} := -\dfrac{\left[\begin{array}{c} 3\, {y_2^{(1)}}^2 {y_1^{(2)}}^2 y_2^{(2)} - 4\, {y_2^{(1)}}^2 {y_2^{(2)}}^3 + 2\, y_2^{(1)} y_1^{(1)} {y_1^{(2)}}^3 \\ -4\, y_2^{(1)} y_1^{(1)} {y_2^{(2)}}^2 y_1^{(2)} - 2\, y_2^{(2)} {y_1^{(1)}}^2 {y_1^{(2)}}^2 + {y_1^{(1)}}^2 {y_2^{(2)}}^3 \end{array} \right]}{{y_2^{(1)}}^2 {y_1^{(2)}}^2 + 2\, y_2^{(1)} y_1^{(2)} y_1^{(1)} y_2^{(2)} + {y_1^{(1)}}^2 {y_2^{(2)}}^2}. \end{cases}$$

It is conceivable from this simple illustration that an explicit description of algebraic equations for low rank symmetric Toeplitz matrices will become more complex for higher dimensional cases.

Example 8.3. For a 3×3 Toeplitz matrix $T = T([t_1, t_2, t_3])$ to be rank deficient, its determinant $\det(T) = (t_1 - t_3)(t_1^2 + t_1 t_3 - 2t_2^2)$ must be zero. The set

$$\{(t_1, t_2, t_3) \in \mathbb{R}^3 | \det(T([t_1, t_2, t_3])) = 0\}$$

is a union of two algebraic varieties as is shown in Figure 8.1.

The rank one matrices correspond to the conditions $t_1 = t_2 = t_3$ or $t_1 = -t_2 = t_3$. These matrices are identified as the two diagonal lines, excluding the origin, in Figure 8.1. The union of the remaining nontrivial two-dimensional surfaces in Figure 8.1 represents all 3×3 symmetric Toeplitz matrices of rank two. This simple illustration in Figure 8.1 points out an important fact: the solution to the structured lower rank approximation problem is not unique.

Example 8.4. According to the symmetric Wedderburn rank reduction formula (Chu, 1998; Chu et al., 1995), B is a rank-one reduced matrix from A if and only if

$$B = A - \frac{A\mathbf{x}\mathbf{x}^\top A}{\mathbf{x}^\top A \mathbf{x}}, \tag{8.6}$$

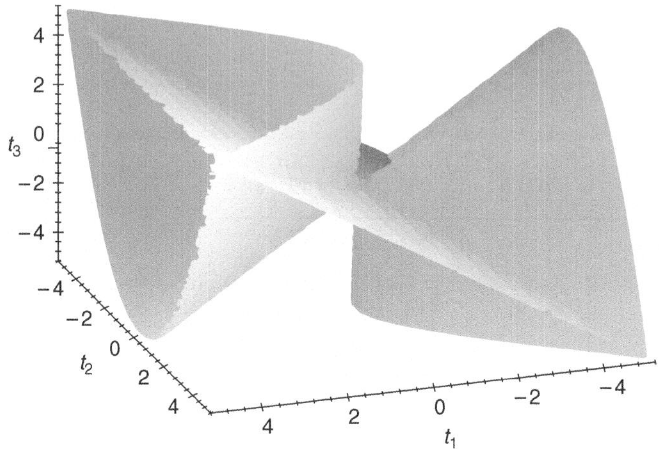

Figure 8.1. *Lower rank, symmetric, 3×3 Toeplitz matrices identified in $\mathbb{R}^3$*

for some $\mathbf{x} \in \mathbb{R}^n$ satisfying $x^\top Ax \neq 0$. For such a rank reduced matrix B to be Toeplitz it is necessary that $\mathbf{x}$ satisfies the equation

$$A\mathbf{x} = c\mathbf{e} \tag{8.7}$$

for some $c \in \mathbb{R}$. The system is solvable only when $\mathbf{e}$ is in the range space of A and the resulting lower rank matrix B

$$B = A - \frac{\mathbf{e}\mathbf{e}^\top}{\mathbf{e}^\top A^\dagger \mathbf{e}}, \tag{8.8}$$

where $A^\dagger$ is the Perron–Frobenius generalized inverse, is independent of c. This choice of B is very restrictive. In other words, the least squares low rank approximation to a given matrix A cannot be accomplished by merely reducing the rank of A by the rank modification formula.

8.2.2 *Tracking structured low rank matrices*

If low rank matrices with a specified structure cannot be characterized analytically, we must devise other means to accomplish this construction. In this section we introduce a mechanism that is capable of constructing low rank matrices of any linear structure iteratively, if such a matrix exists. The idea is equivalent to a special application of Cadzow's composite property mapping algorithm (Cadzow, 1988) that, in turn, is a variation of von Neumann's alternating projection scheme (Cheney and Goldstein, 1959; Deutsch, 2001; Han, 1988).

We shall use the Toeplitz structure to illustrate the procedure. Similar approaches have been used in (Glunt et al., 1990) for Euclidean distance matrix structure and in (Higham, 2002) for correlation matrix structure. We stress once again that the scheme can be used to construct low rank matrices of *any* linear

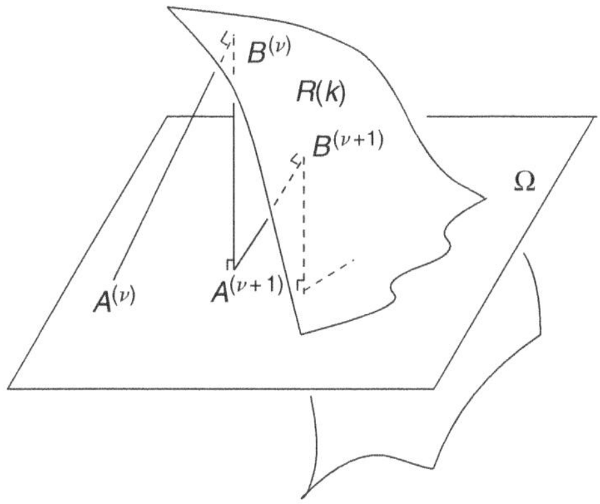

Figure 8.2. *Geometry of lift and projection between $\mathcal{R}(k)$ and Ω*

structure, if such a matrix exist. For structures other than Toeplitz, only minor modification in the projection part is needed.

Alternating projection method Imagine that the set of all rank-k matrices forms a surface $\mathcal{R}(k)$ and the set of matrices with the specified structure forms another surface Ω. Then the desired set of structured rank-k matrices can be regarded as the intersection of these two geometric entities. A linearly convergent method whose spirit is parallel to the lift-and-projection method described in Section 6.3.3 can be formulated to find points at this intersection. The basic idea is to alternate projections between these two sets so that the rank constraint and the structural constraint are satisfied alternatively while the distance in between is being reduced. The algorithm is outlined below. The geometry of lift and projection is depicted in Figure 8.2.

Algorithm 8.1 (The lift-and-projection method)
Let Ω denote the set of matrices with the specified structure. Starting with an arbitrary $A^{(0)} = A \in \Omega$, iterate the following two steps for $\nu = 0, 1, \ldots$ until convergence.

1. **Lift**. Compute the rank-k matrix $B^{(\nu)}$ in $R(k)$ that is nearest to $A^{(\nu)}$.
2. **Projection**. Compute the projection $A^{(\nu+1)}$ of $B^{(\nu)}$ onto the subspace Ω.

To carry out these steps in action, we remark that the lift usually can be done by the TSVD. That is, from $A^{(\nu)} \in \Omega$, first compute its SVD

$$A^{(\nu)} = U^{(\nu)} \Sigma^{(\nu)} {V^{(\nu)}}^{\top}.$$

Let $s_1^{(\nu)}, s_2^{(\nu)}, \ldots$ denote the singular values along the diagonal of $\Sigma^{(\nu)}$. Replace the matrix $\Sigma^{(\nu)}$ by $\operatorname{diag}\{s_1^{(\nu)}, \ldots, s_k^{(\nu)}, 0, \ldots, 0\}$ and define

$$B^{(\nu)} := U^{(\nu)}\Sigma^{(\nu)}V^{(\nu)^{\top}}.$$

In turn, the projection usually involves solving a few simple (linear) equations whose setup depends upon the structure in Ω.

Example 8.5. If $\Omega = \mathcal{T}_n$, then the projection in Algorithm 8.1 is particularly simple. The diagonals of $A^{(\nu+1)}$ are simply the averages of diagonals of $B^{(\nu)}$, respectively.

In the process of lift and projection, the sequence $\{A^{(\nu)}\}$ of matrices will not necessarily have the desirable rank k. However, it is clear that

$$\|A^{(\nu+1)} - B^{(\nu+1)}\|_F \leq \|A^{(\nu+1)} - B^{(\nu)}\|_F \leq \|A^{(\nu)} - B^{(\nu)}\|_F. \tag{8.9}$$

Thus, Algorithm 8.1 is a descent method.

In the above, we should clarify that this descent property (8.9) is measured only with respect to the Frobenius norm which is not necessarily the same norm used in Problem 8.1. The descent property (8.9) of Algorithm 8.1 guarantees that if all $A^{(\nu)}$ are distinct then the iteration converges to a structured matrix of rank k for all matrix norms. In principle, it is possible that the iteration could be trapped in an impasse where $A^{(\nu)}$ and $B^{(\nu)}$ will not improve any more. This will be the case when, for example, $\mathcal{R}(k) \bigcap \Omega = \emptyset$. In our experience in dealing with symmetric Toeplitz structure, we have not encountered such a difficulty thus far.

A point-to-point map The above lift-and-projection approach provides a means to calculate a point $P_k(T) \in \Omega \bigcap \mathcal{R}(k)$ for each given $T \in \Omega$. The map

$$P_k : \Omega \longrightarrow \Omega \bigcap \mathcal{R}(k) \tag{8.10}$$

is defined to be the limit point $P_k(T)$ if the above lift-and-projection iteration procedure starting with T converges.

Despite the descent property (8.9), one must not be mistaken in thinking that $P_k(T)$ is the *closest* rank-k matrix in Ω to T. Given a target matrix A, the resulting $P_k(A)$ is *not* the solution to the structure preserving rank reduction approximation Problem (8.1). Unfortunately, we have seen many references citing that $P_k(A)$ is the solution.

Example 8.6. It is worth noting that, even in the case $n = 2$, the iteration procedure in Algorithm 8.1 with $k = 1$ applied to Toeplitz matrices $T([t_1, 0])$ or $T([0, t_2])$ converges to the zero matrix, instead of a rank one matrix. This observation suggests that the map P_1 can be at most piecewise continuous. This observation also reiterates what we have discussed in Theorem 8.1, that the nearest point is not necessarily of the desired rank-k matrix. The limit point of the iteration could degenerate into lower rank.

8.2.3 *Numerical methods*

In this section we begin to touch upon numerical methods for solving the structure preserving rank reduction Problem (8.1). We understand that, especially for the Toeplitz structure and the Hankel structure, there is a considerable volume of published research on this topic alone in the engineering literature. For instance, Park et al. (1999) have proposed using what they called total least norm with penalty techniques to tackle (8.1). Thus, our purpose is not to evaluate or compare existing methods. Rather, we are proposing a general computational framework that can accommodate any kind of structure, any kind of norm, and any lower rank for the approximation.

In the following, we offer two existing optimization packages to solve the problem. But, again, our purpose is to demonstrate how the framework can be adapted to available software. It is the framework, not the implementation details, that we want to emphasize in this section.

We shall rewrite the problem in two different formulations that can be solved by readily available optimization routines. The difference between these two formulations lies in the way we parameterize the structured low rank matrices. Along with the discussion, we shall also report some preliminary experiences in applying software packages to our methods.

Explicit optimization One convenient way to parameterize low rank matrices is via the singular value decomposition. That is, any rank k square matrix M can be identified by the triplet (U, Σ, V) if $M = U\Sigma V^\top$, where U and V are orthogonal matrices and $\Sigma = \operatorname{diag}\{s_1, \ldots, s_k, 0, \ldots, 0\}$ with $s_1 \geq \ldots \geq s_k > 0$. We use the singular values $s_1, \ldots, s_k$ as well as entries in U and V as parameters to specify a low rank matrix. Any structural constraints can then be qualified via a set of algebraic equalities among these variables. A rewriting of (8.1) in this way is called an *explicit formulation*, inferring that every constraint is explicitly represented in the description of the problem.

If symmetry is part of the structural constraint, then nonzero eigenvalues and the corresponding eigenvectors in the spectral decomposition could be used as the parameterization variables instead. This would effectively reduce the number of unknowns. Even so, it is clear that this explicit formulation involves lots of variables. Often, the underlying structural constraint implies that many of these variables are not independent of each other. The earlier Example 8.2 clearly indicates this kind of redundancy.

Example 8.7. Suppose that the symmetric Toeplitz structure, that is, $\Omega = \mathcal{T}_n$, is being considered in the SLRAP. Using the parameterization (8.3) and writing $M(\alpha_1, \ldots, \alpha_k, \mathbf{y}^{(1)}, \ldots, \mathbf{y}^{(k)}) = [m_{ij}]_{i,j=1}^n$, the explicit formulation gives rise to an equality constrained optimization problem.

Problem 8.2 *(Toeplitz low rank approximation problem)*
Given $A \in \mathbb{R}^{n \times n}$,

$$\text{Minimize} \quad \|A - M(\alpha_1, \ldots, \alpha_k, \mathbf{y}^{(1)}, \ldots, \mathbf{y}^{(k)})\|, \tag{8.11}$$

$$\text{Subject to} \quad m_{j,j+s-1} = m_{1,s},\ s = 1, \ldots, n-1,\ j = 2, \ldots, n-s+1. \tag{8.12}$$

In the above, the objective function in (8.11) is described in terms of the nonzero eigenvalues $\alpha_1, \ldots, \alpha_k$ and the corresponding eigenvectors $\mathbf{y}^{(1)}, \ldots, \mathbf{y}^{(k)}$ of M. The constraints (8.12) that involve algebraic interrelationship among variables $\alpha_1, \ldots, \alpha_k, y^{(1)}, \ldots, y^{(k)}$ are used to ensure that M is symmetric and Toeplitz.

The idea in forming an equality constrained optimization problem such as (8.11) and (8.12) can be extended to the general case. For nonsymmetric or rectangular matrices, singular values and singular vectors are used as variables. To reflect various types of structures, we only need to modify the constraint statements accordingly. Note again that the norm used in (8.11) can be arbitrary.

The notion of explicit formulation outlined above is fairly robust. Almost all kinds of structure-preserving low rank approximation problems can be rewritten in this way. The drawback, however, is that it induces considerable redundancy in the description of the problem. As an example, recall that symmetric Toeplitz matrices have special spectral properties, that is, $\lceil n/2 \rceil$ of the eigenvectors are even and $\lfloor n/2 \rfloor$ are odd (Cantoni and Bulter, 1976). However, the formulation in (8.12) does not take this additional structure into account in the eigenvectors $\mathbf{y}^{(j)}$. Even if we want to include this structure in the constraint statements, we face another dilemma because we are not sure which eigenvalue should be associated with, say, an even eigenvector. The formulation has introduced more variables than needed. The formulation has overlooked any additional internal relationship among the current $n(n-1)/2$ equality constraints. This could have caused, inadvertently, additional computational complexity and difficulties as we report below.

In our first experiment, we take advantage of the matrix manipulation capacity of MATLAB and use its existing routines in the Optimization Toolbox for our application. The algorithm FMINCON in MATLAB employs a sequential quadratic programming technique to focus on the solution of the Kuhn–Tucker equations which, in turn, are solved by a quasi-Newton updating procedure. By default, the code estimates any required derivative information by finite difference approximations. The routine seems appropriate for solving the above problem.

Example 8.8. Consider the symmetric Toeplitz matrix $A = T([1, 2, 3, 4, 5, 6])$ as the given target matrix. Suppose that we want to find least squares approximations of A with rank $k = 5, 4, 3, 2, 1$, respectively.

We set the termination tolerance required on the solution as well as on the objective function at 10^{-6}. For each specified rank k, we call the routine FMINCON to solve Problem 8.11 subject to 8.12, using $T^{(0)} = P_k(A)$ as the starting value. The exit condition of FMINCON reports that the optimization is terminated successfully. Denote, for each k, the calculated optimal solution by T_k^*.

We find that T_5^* (whose values are listed later in Table 8.2) does give a smaller objective value than $P_5(A)$. For all other $k < 5$, however, we we find that the difference $\|T_k^* - P_k(A)\|$ is in the range of 10^{-6}. That is, the optimal solution claimed is considered to be the same as the initial point.

Seeing the above computation, one might mistakenly conclude (and many have done so) that the lift-and-projection method (or Cadzow's algorithm) gives precisely $T_k^* = P_k(A)$ for $k = 4, 3, 2, 1$. However, by using other software or algorithms under the same circumstances, we find that this conclusion is simply not true. We speculate that there are three possible reasons that cause this "failure" in our low rank approximation: (1) the constraints in (8.12) might be "too" linearly dependent; (2) the termination criteria in FMINCON might not be adequate; or (3) the feasible set forms so complicated a geometric entity that the constraints are not easy to satisfy. In any event, this simple toy problem raises some concerns. We wonder whether concerns (1) and (3) also cause problems to other optimization software. In the next experiment, we resort to some more sophisticated optimization packages to solve (8.12).

We believe that many other software packages, such as those described by Czyzyk et al. (1996) and Moré and Wright (1993), can also be used. We choose the package LANCELOT as the next optimization solver. This code is a standard Fortran 77 package for solving large-scale nonlinearly constrained optimization problems. The details of LANCELOT are too complicated to describe in just a few words here. We refer readers to the book by Conn et al. (1992) for more information. Fortunately, the package LANCELOT is available on the NEOS Server (Czyzyk et al., 1996; Gropp and Moré, 1997). In addition, the NEOS Server for LANCELOT uses the ADIFOR automatic differentiation tool (Bischof et al., 1995).

Example 8.9. LANCELOT is able to find optimal low rank solutions for all values of k without any difficulties. Furthermore, the solutions obtained by using LANCELOT agree, up to the required accuracy 10^{-6}, with those obtained by using our second method (see Table 8.2).

The overhead cost of using LANCELOT for our test problem is reported in Table 8.1. The "# of variables" used by LANCELOT is $(n + 1)k$ for a problem of size n and desired rank k. The "# of f/c calls" refers to the number of function/gradient evaluations as well the constraint/Jacobian evaluations. The total time in seconds given in Table 8.1 seems to suggest that the choice of the rank affects the computational cost nonlinearly.

Table 8.1. *Cost overhead in using* LANCELOT *for* $n = 6$

Rank k	5	4	3	2	1
# of variables	35	28	21	14	7
# of f/c calls	108	56	47	43	19
Total time	12.99	4.850	3.120	1.280	0.4300

The experience that FMINCON in MATLAB fails and LANCELOT in NEOS succeeds in solving the very same Problem (8.12) is a clear signal that any numerical solutions obtained by one algorithm need to be compared carefully with results from other algorithms. We should stress that it is not clear whether the proposed formulation would run into the same difficulties even with LANCELOT as those we have experienced with FMINCON, when the problem size becomes larger. We also should point out that there are many other algorithms available from NEOS (Czyzyk et al., 1996; Gropp and Moré, 1997), but we have not tried these for our problem. The explicit formulation makes it possible to tackle the SLRAP by conventional optimization techniques.

Implicit optimization By regarding the mapping $P_k(T)$ in (8.10) as a parameterization of low rank matrices, the rank reduction problem SLRAP can now be formulated as an "unconstrained" optimization problem:

$$\text{minimize}_{T \in \Omega} \|A - P_k(T)\|, \tag{8.13}$$

where A is the given target matrix whose nearest low rank matrix is being sought inside Ω. Despite the fact that $P_k(T)$ may not be defined for all T, the above formulation at least provides us with a handle to manipulate the objective function

$$f(T) := \|A - P_k(T)\|.$$

We remind readers that the norm used in (8.13) can be any matrix norm, but the projections in computing $P_k(T)$ via Algorithm 8.1 use the Frobenius norm.

It is important to note that $P_k(T)$ is *not* necessarily the closest rank k Toeplitz matrix to T. Nowhere in Algorithm 8.1 is it suggested that $P_k(A)$ is a solution to Problem 8.1. Unfortunately, in the literature and in many applications, $P_k(A)$ has been mistaken to be the nearest approximation to A. For instance, a quote by Cadzow (Cadzow, 1988; Cadzow and Wilkes, 1990) claims that Algorithm 8.1 alone, namely, the limit point $P_k(A)$ of the iteration, would serve "as a cleansing process whereby any corrupting noise, measurement distortion or theoretical mismatch present in the given data set (namely, A) is removed." A similar misconception seems to prevail in many other applications using Cadzow's algorithm (de Beer, 1995; Dendrinos et al., 1991; Li, 1997a) and in the discussion in Park et al. (1999). We emphasize that more needs to be done in order to find the *closest* structured low rank approximation to the given A, since $P_k(A)$ is known to be merely one candidate in the feasible set. The fact that more has to be done to obtain the *closest* structured low rank matrix has somehow been overlooked. An interesting discussion in (De Moor, 1994) suggests that this situation does have some impact on real applications.

With the formulation (8.13) in hand, the structure preserving rank reduction problems become more tractable. The constraint that $T \in \Omega$ can easily be handled and built into the problem directly.

Example 8.10. If $\Omega = \mathcal{T}_n$, then only n independent variables $t_1, \ldots, t_n$ are needed to specify $T = T([t_1, \ldots, t_n])$. In this case, the Toeplitz structure can be built into the problem directly by writing the function $P_k(T)$ as $P_k(t_1, \ldots, t_n)$. The function $P_k(t_1, \ldots, t_n)$ can be evaluated point by point as we have described earlier.

We call (8.13) an *implicit formulation* of the structure preserving rank reduction problem, referring to the fact that the constraints, particularly the rank condition, are hidden inside the point-to-point map $P_k(T)$. We stress again that this implicit formulation is in fact an *unconstrained* optimization problem because the structural constraint $T \in \Omega$ can be replaced by independent variables. This is in contrast to the explicit formulation discussed in the preceding section.

Knowing how to compute $P_k(T)$ is sufficient for the application of a wide spectrum of unconstrained optimization methods to Problem (8.13). There are at least two techniques available. The Nelder–Mead simplex search method, for example, is an *ad hoc* optimization technique that does not use the gradient information. The simplex search method requires only function evaluations, a feature that our point-to-point map $P_k(T)$ can satisfactorily provide. For more sophisticated methods, the gradient information can be calculated by using either numerical or automatic differentiation techniques. Using exactly the same example as in the previous section for the explicit formulation, we illustrate the idea of the implicit formulation next.

We shall first employ the MATLAB routine FMINUNC as our optimization solver for Problem (8.13). The routine FMINUNC employs a BFGS quasi-Newton scheme equipped with a mixed quadratic and cubic line search procedure. Obviously, this choice of using FMINUNC is only for convenience and illustration. Our point is to illustrate that the point-to-point mapping $\mathcal{P}_k(T)$ is workable by standard optimization skills. Any other package, LANCELOT included, could have been used to solve (8.13).

Example 8.11. The termination tolerance for both parameters $TolFun$ and $TolX$ in FMINUNC is set at 10^{-6}. We start with $T^{(0)} = A$ as the initial value. Since no gradient information is analytically given, a finite difference approximation is used in FMINUNC by default.

Table 8.2. *Test results for a 6×6 symmetric Toeplitz structure using* FMINUNC

Rank k	5	4	3	2	1
Iterations	59	49	39	29	18
SVDs	1013	2914	2200	1860	589
T_k^*	$\begin{bmatrix}1.1046\\1.8880\\3.1045\\3.9106\\5.0635\\5.9697\end{bmatrix}$	$\begin{bmatrix}1.2408\\1.8030\\3.0352\\4.1132\\4.8553\\6.0759\end{bmatrix}$	$\begin{bmatrix}1.4128\\1.7980\\2.8171\\4.1089\\5.2156\\5.7450\end{bmatrix}$	$\begin{bmatrix}1.9591\\2.1059\\2.5683\\3.4157\\4.7749\\6.8497\end{bmatrix}$	$\begin{bmatrix}2.9444\\2.9444\\2.9444\\2.9444\\2.9444\\2.9444\end{bmatrix}$
$\|A - T_k^*\|$	0.5868	0.9581	1.4440	3.2890	8.5959
$\varpi(T_k^*)$	$\begin{bmatrix}17.9851\\7.4557\\2.2866\\0.9989\\0.6164\\3.4638e{-15}\end{bmatrix}$	$\begin{bmatrix}17.9980\\7.4321\\2.2836\\0.8376\\2.2454e{-14}\\2.0130e{-14}\end{bmatrix}$	$\begin{bmatrix}18.0125\\7.4135\\2.1222\\1.9865e{-14}\\9.0753e{-15}\\6.5255e{-15}\end{bmatrix}$	$\begin{bmatrix}18.2486\\6.4939\\2.0884e{-14}\\7.5607e{-15}\\3.8479e{-15}\\2.5896e{-15}\end{bmatrix}$	$\begin{bmatrix}17.6667\\2.0828e{-14}\\9.8954e{-15}\\6.0286e{-15}\\2.6494e{-15}\\2.1171e{-15}\end{bmatrix}$

Table 8.2 summarizes the test statistics for symmetric Toeplitz low rank approximations T_k^* with $k = 5, 4, 3, 2, 1$. The second row "Iterations" refers to the number of function evaluations of the map P called by the search in FMINUNC. Each call of P requires the computation of SVD several times. The third row "SVDs" is the number of lifts done in Algorithm 8.1 for the entire calculation. The large number of SVD calls even for this small size toy problem seems to indicate the degree of difficulty for this structure preserving rank reduction problem. The last row $\varpi(T_k^*)$ denotes the computed singular values of T_k^*. The nearly machine-zero values indicate the degrees of rank deficiency.

Three important observations come about from this experiment. First, we are able to calculate *all* low rank matrices while maintaining the symmetric Toeplitz structure. This is somewhat surprising. We know from Theorem 8.2 that symmetric Toeplitz matrices can have arbitrary lower rank, but there is no general theory that guarantees the "nearest" symmetric Toeplitz approximation of any lower rank to a given matrix. Is this observation true in general? Does it extend to other structures? This existence question of a solution to Problem 8.1 deserves further investigation. Secondly, we note in Table 8.2 that the distance between A and T_k^* is quite significant, strongly indicating that the noisy data could be substantially "filtered" using our numerical procedures. Thirdly, a runtime history of intermediate results for the case $k = 5$ is reported in Table 8.3. In this case, the optimal Toeplitz T_5^* given in Table 8.2 has a calculated singular value 3.4638×10^{-15}, suggesting that T_5^* is computationally of rank 5. We remark that T_5^* can only be perceived as a local minimizer, as is generally expected for nonlinear optimization problems. Nevertheless, we note that $\|T_5^* - A\| \approx 0.586797 < 0.589229 \approx \|P_5(A) - A\|$. Although this difference only represents

Table 8.3. *Output of intermediate results from* FMINUNC

Iteration	Func-count	f(x)	Step-size	Directional derivative
1	6	0.589229	1	−0.0587
2	16	0.587051	0.0747333	0.000147
3	27	0.586845	0.104418	1.83e−06
4	37	0.586807	0.125006	−7.37e−07
5	47	0.586797	0.194334	−1.77e−07
6	57	0.586797	0.0900986	1.18e−08

Table 8.4. *Test results of using* FMINSEARCH *for symmetric Toeplitz structure*

Rank k	5	4	3	2	1
Iterations	318	258	248	264	255
SVDs	5167	15127	13684	16432	7946

a slight improvement on the objective value, it is evident that the Cadzow initial iteration does not give rise to an optimal approximation to the noisy data.

To solve the SLRAP, we could also use the Nelder–Mead simplex search method FMINSEARCH in MATLAB as another optimization solver. In this case, only functional calls to $P_k(T)$ are needed, and there is no need of derivative information.

Example 8.12. In our experiments, the algorithm FMINSEARCH returns the same optimal solutions as FMINUNC. However, test statistics in Table 8.4 show that the search method costs more function evaluations due to its slow convergence. In turn, the method requires more SVD computations.

8.2.4 *Summary*

The structure preserving rank reduction problem concerns the construction of the nearest approximation to a given matrix by matrices with a specific rank and a specific structure. Elements in the feasible set, if not empty, are themselves solutions to certain IEPs. This structured low rank approximation is needed in many important applications.

In this section we have merely begun to investigate some theoretical and numerical issues associated with structure preserving rank reduction problems. At present, we have mainly concentrated on linear structure constraints. Even so, many questions remain to be answered. We have proposed two general frameworks to reformulate the problem. The explicit formulation is robust and versatile, but it is difficult to take into account any special spectral properties of structured matrices. The implicit formulation is easy to use, but the computation is slow in convergence. In either case, we have illustrated how the resulting optimization problems can then be tackled numerically by utilizing existing software packages. We have reported some preliminary numerical experiments.

The idea of reformulation has been illustrated by the symmetric Toeplitz structure with Frobenius norm throughout this discussion. Our approach can readily be generalized to rank reduction problems of any given linear structure and of any given matrix norm.

8.3 Low rank circulant approximation

Circulant matrices arise in many important applications. Approximation by low rank circulant matrices, in particular, has important applications in signal and image processing. Circulant structure is a special type of linear structure, so the technique discussed in the preceding section is applicable. However, circulant structure enjoys a distinct property, that is, it is possible to devise a procedure using the fast Fourier transform (**FFT**) to compute the nearest *real* circulant approximation C, with a specific rank, to a given real $n \times n$ matrix A. Besides the speed gained by using the FFT method, the low rank circulant real approximation problem is not as straightforward as the usual truncated singular value decomposition since a conjugate-even set of eigenvalues must be maintained to guarantee that C is real. Extensions of this work to the block circulant structure with circulant blocks C is straightforward, leading to an efficient way of preconditioning image post-processing computations.

8.3.1 *Preliminaries*

An $n \times n$ matrix C of the form

$$C = \begin{bmatrix} c_0 & c_1 & & \cdots & c_{n-1} \\ c_{n-1} & c_0 & c_1 & \cdots & c_{n-2} \\ c_{n-2} & c_{n-1} & c_0 & \cdots & c_{n-3} \\ \vdots & \vdots & \ddots & \ddots & \vdots \\ c_1 & c_2 & & c_{n-1} & c_0 \end{bmatrix}$$

is called a *circulant matrix*. As each row of a circulant matrix is just the previous row cycled forward one step, a circulant matrix is uniquely determined by the entries of its first row. We shall denote a circulant matrix by $Circul(\mathbf{c})$ if its first row is $\mathbf{c}$. In this section, we are mainly concerned with the case when $\mathbf{c} \in \mathbb{R}^n$.

Let Π ($= \Pi_n$) denote the specific permutation matrix of order n,

$$\Pi := \begin{bmatrix} 0 & 1 & 0 & \dots & 0 \\ 0 & 0 & 1 & & 0 \\ \vdots & & \ddots & \ddots & \vdots \\ 0 & & & & 1 \\ 1 & 0 & & \dots & 0 \end{bmatrix}, \tag{8.14}$$

which is call the *forward shift matrix*. It is easy to see that

$$C = \sum_{k=0}^{n-1} c_k \Pi^k \tag{8.15}$$

if and only if $C = Circul(\mathbf{c})$ with $\mathbf{c} := [c_0, \dots, c_{n-1}]$. It is convenient to represent this relationship as

$$Circul(\mathbf{c}) = P_{\mathbf{c}}(\Pi) \tag{8.16}$$

where

$$P_{\mathbf{c}}(x) = \sum_{k=0}^{n-1} c_k x^k \tag{8.17}$$

is called the characteristic polynomial of $Circul(\mathbf{c})$. Because of this representation, it follows that circulant matrices are closed under multiplication. It is also clear that circulant matrices commute under multiplication.

Many important properties of circulant matrices can be traced back mainly to those of the matrix Π. The circulant structure often makes it possible to resolve many matrix-theoretic questions by "closed form" answers. See, for example, Chan (1988). The book by Davis (1979) is generally considered to be the most complete reference on circulant matrices. It is also known that circulant matrices are closely related to the Fourier analysis (Van Loan, 1992) that will be used to develop a fast algorithm in this section.

Circulant matrices have important applications to diverse areas of disciplines including acoustics, electrodynamics, image processing, mathematical statistics, number theory, numerical analysis, and stationary time series. For several important applications in linear estimation, see the recent book by T. Kailath et al. (2000). For other diverse applications in adaptive optics and image restoration, see, for example, Chan et al., 1994. Our goal in this section is to retrieve as much information as possible from a given real matrix A while enforcing a circulant structure by approximating A as a real circulant matrix C with a certain desired rank. Extensions of this work to block circulant with circulant blocks C is straightforward. This extension might lead to an efficient computation of an optimal block circulant with circulant blocks regularized preconditioner, as used in Hanke et al. (1993) and Nagy et al. (1997) for image reconstruction.

8.3.2 *Basic spectral properties*

For completeness, we briefly review some of the basic spectral properties relevant to our study. Most of the proofs can be found in (Davis, 1979; Van Loan, 1992).

Let $i := \sqrt{-1}$. For a fixed integer $n \geq 1$, let ω $(= \omega_n)$ denote the primitive n-th root of unity

$$\omega := \exp\left(\frac{2\pi i}{n}\right) \tag{8.18}$$

and consider the vector

$$\boldsymbol{\vartheta} = [1, \omega, \omega^2, \dots, \omega^{n-1}]^\top \in \mathbb{C}^n.$$

Let Ω $(= \Omega_n)$ denote the diagonal matrix

$$\Omega := \operatorname{diag}(\boldsymbol{\vartheta}), \tag{8.19}$$

and let F $(= F_n)$ denote the so-called Fourier matrix whose Hermitian adjoint F^* is defined by

$$F^* := \frac{1}{\sqrt{n}} \begin{bmatrix} 1 & 1 & 1 & \dots & 1 \\ 1 & \omega & \omega^2 & \dots & \omega^{n-1} \\ 1 & \omega^2 & \omega^4 & \dots & \omega^{2n-2} \\ \vdots & & & & \vdots \\ 1 & \omega^{n-1} & \omega^{n-2} & \dots & \omega \end{bmatrix}, \tag{8.20}$$

that is, $\sqrt{n}F^*$ is the Vandermonde matrix generated by the vector $\boldsymbol{\vartheta}$. Observe that F is a unitary matrix. The following spectral decomposition is a key to our discussion.

Theorem 8.5. The forward shift matrix Π is unitarily diagonalizable. Indeed,

$$\Pi = F^* \Omega F. \tag{8.21}$$

The circulant matrix $Circul(\mathbf{c})$ with any given row vector c has a spectral decomposition

$$Circul(\mathbf{c}) = F^* P_{\mathbf{c}}(\Omega) F. \tag{8.22}$$

Note the vector of eigenvalues $\boldsymbol{\lambda} = [P_{\mathbf{c}}(1), \dots, P_{\mathbf{c}}(\omega^{n-1})]$ of a circulant matrix $Circul(\mathbf{c})$ can quickly be calculated from

$$\boldsymbol{\lambda}^\top = \sqrt{n} F^* \mathbf{c}^\top. \tag{8.23}$$

Now, from (8.23), the inverse eigenvalue problem of finding a circulant matrix with a prescribed spectrum can easily be answered. Given any vector $\boldsymbol{\lambda} := [\lambda_0, \dots, \lambda_{n-1}]$, the circulant matrix $Circul(\mathbf{c})$ with $\mathbf{c}$ defined by

$$\mathbf{c}^\top = \frac{1}{\sqrt{n}} F \boldsymbol{\lambda}^\top \tag{8.24}$$

will have eigenvalues $\{\lambda_1, \ldots, \lambda_n\}$. It is important to note that the matrix–vector multiplication involved in either (8.23) or (8.24) is precisely that involved in the fast Fourier transform (FFT). Thus both the eigenvalue problem and the inverse problem for circulant matrices can be answered in $O(n \log_2 n)$ floating point operations. Observe also that if all the eigenvalues are distinct then there are precisely $n!$ distinct circulant matrices with the prescribed spectrum.

For real circulant matrices, every complex-valued eigenvalue has the corresponding complex conjugate as another eigenvalue. Indeed, the spectrum of any real circulant matrix necessarily appears in a more special order, called *conjugate-even* in (Van Loan, 1992). Thus to obtain a *real-valued* circulant matrix using the FFT in (8.24) for the inverse eigenvalue problem, the vector $\boldsymbol{\lambda}$ of the prescribed eigenvalues must also be arranged in a conjugate-even order. More precisely, the following arrangement of eigenvalues allows efficient FFT calculation for real data.

Theorem 8.6. (Van Loan, 1992) If the eigenvalues are arranged in the order

(i) $\boldsymbol{\lambda} := [\lambda_0, \lambda_1, \ldots, \lambda_{m-1}, \lambda_m, \overline{\lambda_{m-1}}, \ldots, \overline{\lambda_1}]$, where $\lambda_0, \lambda_m \in \mathbb{R}$ and $n = 2m$; or

(ii) $\boldsymbol{\lambda} := [\lambda_0, \lambda_1, \ldots, \lambda_m, \overline{\lambda_m}, \ldots, \overline{\lambda_1}]$, where $\lambda_0 \in \mathbb{R}$ and $n = 2m + 1$,

then the circulant matrix $Circul(\mathbf{c})$ with c obtained from (8.24) is real-valued and has the prescribed $\boldsymbol{\lambda}$ as its spectrum.

For later reference, we shall refer to λ_0 and λ_m, if $n = 2m$, and λ_0, if $n = 2m + 1$, in the above theorem as the *absolutely real* elements in $\boldsymbol{\lambda}$.

The singular value decomposition of $Circul(\mathbf{c})$ follows by rewriting the expression (8.22) as

$$Circul(\mathbf{c}) = (F^* P_{\mathbf{c}}(\Omega)|P_{\mathbf{c}}(\Omega)|^{-1})|P_{\mathbf{c}}(\Omega)|F \tag{8.25}$$

where $|X|$ means the absolute value of the elements of X. The singular values of $Circul(\mathbf{c})$ are $|P_{\mathbf{c}}(\omega^k)|$, $k = 0, 1, \ldots, n-1$. It follows that any $n \times n$ real circulant matrix can have at most $\lceil (n+1)/2 \rceil$ distinct singular values. More precisely, the singular values must appear as either $\sigma_{n_0}, \sigma_{n_1}, \sigma_{n_1}, \ldots, \sigma_{n_{m-1}}, \sigma_{n_{m-1}}, \sigma_{n_m}$, if $n = 2m$; or $\sigma_{n_0}, \sigma_{n_1}, \sigma_{n_1}, \ldots, \sigma_{n_m}, \sigma_{n_m}$, if $n = 2m + 1$. Here, the indices do not necessarily reflect the magnitudes of the singular values.

8.3.3 *Conjugate-even approximation*

Given a general matrix $A \in \mathbb{R}^{n \times n}$, its nearest circulant matrix approximation measured in the Frobenius norm is simply the matrix $Circul(\mathbf{c})$ obtained by averaging over diagonals of A, as shown in (Chan, 1988). If $\mathbf{c} = [c_0, \ldots, c_{n-1}]$, then c_k is simply the projection

$$c_k := \frac{1}{n}\langle A, \Pi^k \rangle, \quad k = 0, \dots, n-1, \tag{8.26}$$

of A onto Π^k with respect to the Frobenius inner product. This projection $Circul(\mathbf{c})$ is generally of full rank even if A has lower rank to begin with.

Best complex approximation It is known that the truncated singular value decomposition gives rise to the nearest low rank approximation in Frobenius norm. Observe further that the low rank approximation $Circul(\hat{\mathbf{c}})$ of a circulant matrix $Circul(\mathbf{c})$ by the truncated singular value decomposition is automatically circulant. We thus have the following algorithm for low rank circulant approximation.

Algorithm 8.2 (Circulant low rank approximation – complex)
Given a general $n \times n$ matrix A, the matrix $Circul(\hat{\mathbf{c}})$ computed below is a nearest circulant matrix to A with rank no higher than $\ell \leq n$.

1. Use the projection (8.26) to find the nearest circulant matrix approximation $Circul(\mathbf{c})$ of A.
2. Use the inverse FFT (8.23) to calculate the spectrum $\boldsymbol{\lambda}$ of the matrix $Circul(\mathbf{c})$.
3. Arrange all elements of $|\boldsymbol{\lambda}|$ in descending order, including those with equal modulus. Let $\hat{\boldsymbol{\lambda}}$ be the vector consisting of elements of $\boldsymbol{\lambda}$, but those corresponding to the last $n - \ell$ singular values in the descending order are set to zero.
4. Apply the FFT (8.24) to $\hat{\boldsymbol{\lambda}}$ to compute a nearest circulant matrix $Circul(\hat{\mathbf{c}})$ of rank ℓ to A.

The above algorithm is fast due to the employment of efficient FFT calculations. The resulting matrix $Circul(\hat{\mathbf{c}})$, however, is complex-valued in general. To construct the real-valued low rank approximation, the truncated singular values must be specifically selected so that the resulting vector $\hat{\boldsymbol{\lambda}}$ of *truncated* eigenvalues is conjugate-even. Recall that many of the singular values are paired. Thus to preserve the conjugate-even property, the deletion of one complex eigenvalue often necessitates the deletion of its complex conjugate as well. To achieve the desired rank, the criteria for truncation must be modified, as we discuss next.

Best conjugate-even approximation It is clear from Theorem 8.5 that all circulant matrices of the same size have the same set of unitary eigenvectors. The low rank real circulant approximation problem, therefore, can be described as the following data matching Problem (**DMP**).

Problem 8.3 *(DMP)*
Given a conjugate-even vector $\boldsymbol{\lambda} \in \mathbb{C}^n$, find its nearest conjugate-even approximation $\hat{\boldsymbol{\lambda}} \in \mathbb{C}^n$ subject to the constraint that $\hat{\boldsymbol{\lambda}}$ has exactly $n - \ell$ zeros.

If there were no conjugate-even constraint, the DMP could easily be answered. See, for example (Brockett, 1989; Chu, 1990). Without loss of generality, we may write $\hat{\boldsymbol{\lambda}} = [\hat{\boldsymbol{\lambda}}_1, \mathbf{0}] \in \mathbb{C}^n$ with $\hat{\boldsymbol{\lambda}}_1 \in \mathbb{C}^\ell$ being arbitrary and consider the problem of minimizing

$$F(P, \hat{\boldsymbol{\lambda}}) = \|P\hat{\boldsymbol{\lambda}}^\top - \lambda^\top\|^2$$

where the permutation matrix P is used to search the match. Write $P = [P_1, P_2]$ with $P_1 \in \mathbb{R}^{n\times\ell}$. We then have the least squares problem

$$F(P, \hat{\boldsymbol{\lambda}}) = \|P_1\hat{\boldsymbol{\lambda}}_1^\top - \boldsymbol{\lambda}^\top\|^2$$

that obviously has its optimal solution with

$$\hat{\boldsymbol{\lambda}}_1 = \boldsymbol{\lambda}P_1.$$

This shows that the entries of $\hat{\boldsymbol{\lambda}}_1$ must come from the rearrangement of a portion of $\boldsymbol{\lambda}$. Indeed, the objective function becomes

$$F(P, \hat{\boldsymbol{\lambda}}) = \|(P_1P_1^\top - I)\boldsymbol{\lambda}\|^2$$

where $P_1P_1^\top - I$ represents a projection. Thus the optimal permutation P should be such that $P_1P_1^\top$ projects $\boldsymbol{\lambda}$ to its first ℓ components with largest modulus. In other words, without the conjugate-even constraints, the answer to the data matching problem corresponds precisely to the usual selection criterion mentioned in Algorithm 8.2, that is, $\hat{\boldsymbol{\lambda}}$ is obtained by setting to zero $n - \ell$ elements of $\boldsymbol{\lambda}$ with smallest modulus. With the conjugate-even constraint, the above criterion remains effective but the truncation also depends on the conjugate-even structure inside $\boldsymbol{\lambda}$ as we shall now explain.

Conjugate-even structure

Example 8.13. Consider the case $n = 6$. We shall first assume that neither λ_1 nor λ_2 is a real number. There are six possible conjugate-even structures. For convenience, we shall denote each structure by a tree graph. Each node in the tree represents an element of $\boldsymbol{\lambda}$. Arrange the nodes from top to bottom according to the descending order of their moduli. In case of a tie, arrange the complex conjugate nodes at the same level and place the real node below the complex nodes. Thus the conjugate-even structure $\lambda_1, \overline{\lambda_1}, \lambda_0, \lambda_2, \overline{\lambda_2}, \lambda_3$, arranged in the descending order of their modulus, will be denoted by the tree in Figure 8.3.

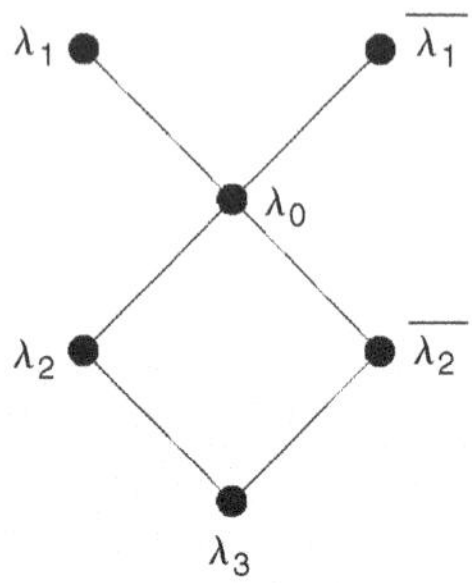

Figure 8.3. *Tree graph of* $\lambda_1, \overline{\lambda_1}, \lambda_0, \lambda_2, \overline{\lambda_2}, \lambda_3$ *with* $|\lambda_1| \geq |\lambda_0| > |\lambda_2| \geq |\lambda_3|$

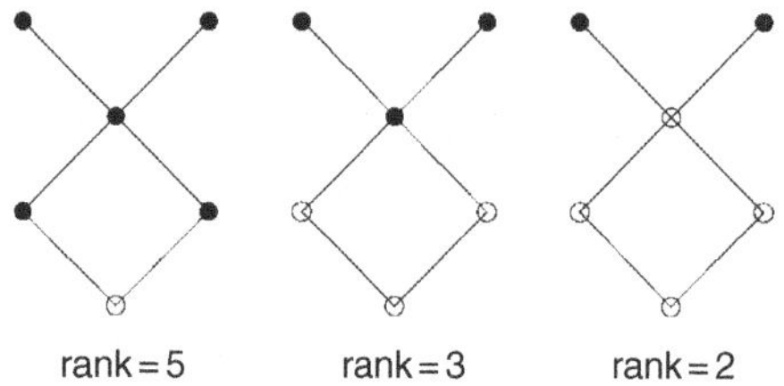

Figure 8.4. *Tree graphs of* $\hat{\boldsymbol{\lambda}}$ *with rank 5, 3, and 2*

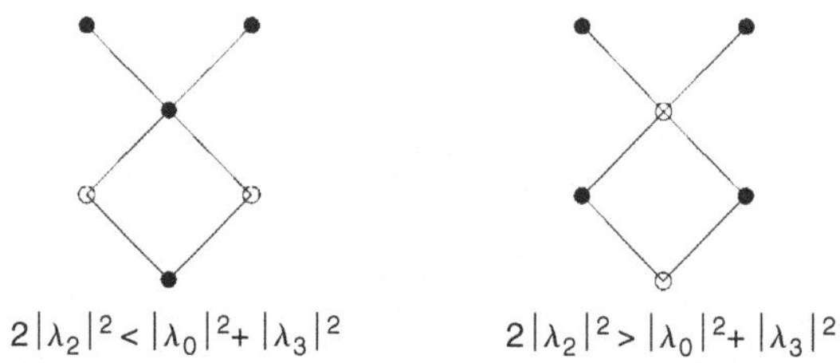

Figure 8.5. *Tree graphs of* $\hat{\boldsymbol{\lambda}}$ *with rank 4*

The nearest conjugate-even vectors to $\boldsymbol{\lambda}$ of rank 5, 3, and 2, respectively, are easy to determine. Their trees are given in Figure 8.4 where $\circ$ and $\bullet$ at each node denotes, respectively, that the particular node is being replaced by zero or remains unchanged from the original tree. For these ranks and for this specific structure depicted in Figure 8.3, the conjugate-even requirement has no effect.

However, depending upon whether $2|\lambda_2|^2 > |\lambda_0|^2 + |\lambda_3|^2$, there are two choices for $\hat{\boldsymbol{\lambda}}$ as the nearest conjugate-even approximation of rank 4. See Figure 8.5.

Finally, the nearest rank-1 conjugate-even approximation for the tree of $\boldsymbol{\lambda}$ given by Figure 8.3 is depicted in Figure 8.6.

It should be noted that we have assumed implicitly that if $n = 2m$, then the two absolutely real elements in a conjugate-even $\boldsymbol{\lambda}$ are λ_0 and λ_m and that $|\lambda_0| \geq |\lambda_m|$. We have also assumed that the remaining $2m - 2$ elements

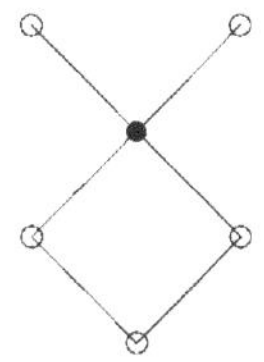

Figure 8.6. *Tree graph of* $\hat{\boldsymbol{\lambda}}$ *with rank 1*

are "potentially" complex-valued (some of which could in fact turn out to be real-valued in special cases), that they are paired up (necessarily), and are arranged in descending order, that is, $|\lambda_1| \geq |\lambda_2| \geq \ldots \geq |\lambda_{m-1}|$. A similar assumption can be made for the case $n = 2m + 1$. See the ordering stated in Theorem 8.6. Nevertheless, we will *never* assume any ordering relationship between the absolutely real element(s) and the potential complex elements. Indeed, it is precisely such an ordering relationship that will determine the truncation criteria as we have demonstrated above for the case $n = 6$. In other words, assuming that there are exactly $m + 1$ distinct absolute values of elements in $\boldsymbol{\lambda}$, then there are exactly $\binom{m+1}{2}$ many possible conjugate-even structures for the case $n = 2m$, depending upon where the moduli of the absolutely real elements fit in the moduli of the potentially complex elements when a *total* ordering is taken.

Example 8.14. Again, under the assumption that both λ_1 and λ_2 are not real numbers, we further illustrate our point of conjugate-even structure by considering other cases for $n = 6$ in Figure 8.7.

The leftmost column in Figure 8.7 represents the six possible conjugate-even structures of $\boldsymbol{\lambda}$ when elements are arranged in descending order of their modulus. For each fixed structure, moving from left to right, Figure 8.7 demonstrates the plan of how the nodes on the original tree should be "pruned" to solve the DMP for various lower rank conditions.

Be aware that there are four cases, A4, A2, B2, and F4, where additional comparisons are needed to further discern which plan should be used. This situation happens when an even number of nodes from a "loop" are to be dropped. We have already discussed the case F4 in Figure 8.5. Other cases can easily be identified.

It is entirely possible that there are real-valued elements other than the two (when n is even) absolutely real elements in a conjugate-even $\boldsymbol{\lambda}$. The eigenvalues of a symmetric circulant matrix, for instance, are conjugate-even and all real. When this happens, these conjugate-even real-valued elements must appear in pairs and the truncation criteria are further complicated.

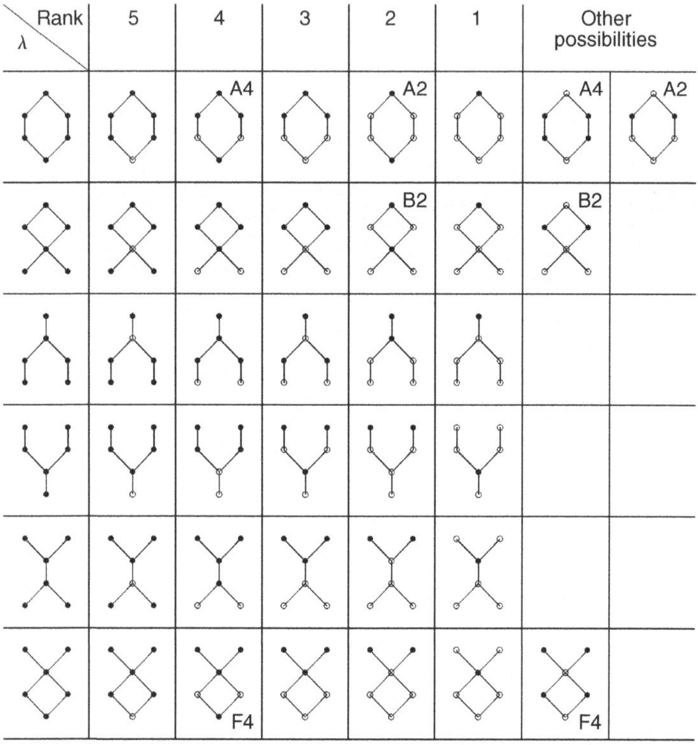

Figure 8.7. *Possible solutions to the DMP when $n = 6$*

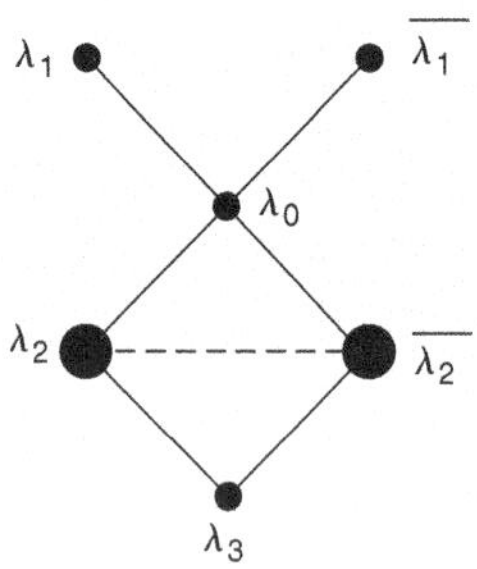

Figure 8.8. *Tree graph of* $\lambda_1, \overline{\lambda_1}, \lambda_0, \lambda_2, \lambda_2, \lambda_3$ *with* $|\lambda_1| \geq |\lambda_0| > |\lambda_2| \geq |\lambda_3|$

Example 8.15. Using the example discussed in Figure 8.3 but assuming further that $\lambda_2 = \overline{\lambda_2}$, we illustrate our point below. First we use a dashed link in Figure 8.8 and larger dots to indicate the occurrence that $\lambda_2 = \overline{\lambda_2}$.

It is important to note that, in contrast to the two drawings in Figure 8.5, the tree graph of the nearest conjugate-even approximation $\hat{\boldsymbol{\lambda}}$ with rank 4 changes its structure in this case. See Figure 8.9.

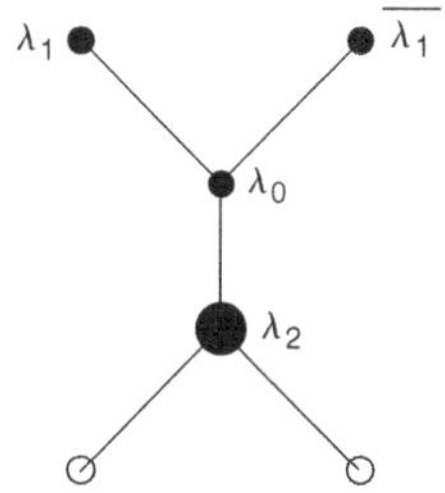

Figure 8.9. *Tree graph of* $\hat{\boldsymbol{\lambda}}$ *with rank 4 when* $\lambda_2 = \overline{\lambda_2}$

8.3.4 *Algorithm*

The above examples give a glimpse of the complexity of a FFT-based algorithm for computing the real-valued low rank circulant approximation. The challenge lies in detecting the conjugate-even structure and building the tree graph automatically. Once this hurdle is passed, the gain in speed is phenomenal.

We will summarize a prototype scheme in this section. In order to highlight the notion on how the singular values of $Circul(\mathbf{c})$ should be truncated and not to be distracted with the programming details, we simplify many computational operations by adopting a pseudo-MATLAB syntax.

Algorithm 8.3 (Circulant low rank approximation – real)
Given $\mathbf{c} \in \mathbb{R}^n$ a positive integer $1 \leq \ell < n$, let $m = \lfloor n/2 \rfloor$. Define $tol = n\epsilon \|J\|$ where ϵ is the machine accuracy as the threshold of system zero. The matrix $Circul(\hat{\mathbf{c}})$ with $\hat{\mathbf{c}}$ computed at the end of the following steps has eigenvalues $\hat{\boldsymbol{\lambda}}$ containing exactly $n - \ell$ zeros and is the nearest approximation to $Circul(\mathbf{c})$.

1. $\boldsymbol{\lambda} = n * \mathbf{ifft}(\mathbf{c});$ (Indices of $\boldsymbol{\lambda}$ start with 1.)
   $\hat{\boldsymbol{\lambda}} = \boldsymbol{\lambda}(1 : m + 1);$
2. if $n = 2m$
   $I_r = \mathbf{find}(\mathbf{abs}(\mathbf{imag}(\boldsymbol{\lambda}(2 : m))) < tol) + 1;$
   $I_c = \mathbf{find}(\tilde{}\,\mathbf{ismember}(2 : m, I_r)) + 1;$
   else
   $I_r = \mathbf{find}(\mathbf{abs}(\mathbf{imag}(\boldsymbol{\lambda}(2 : m + 1))) < tol) + 1;$
   $I_c = \mathbf{find}(\tilde{}\,\mathbf{ismember}(2 : m + 1, I_r)) + 1;$
   end
3. $[t, J] = \mathbf{sort}(\mathbf{abs}(\boldsymbol{\lambda}));$
   $J = \mathbf{fliplr}(J);$ (J is the index set sorting $\boldsymbol{\lambda}$ in descending order.)
   for $i = 1 : m + 1$

$$I(:,:,i) = \begin{cases} [2, 0], & \text{if } \mathbf{ismember}(J(i), I_c); \\ [2, 1], & \text{if } \mathbf{ismember}(J(i), I_r); \\ [1, 1], & \text{otherwise}; \end{cases}$$

   end

4. $\sigma = 0;$
   $s = m + 1;$
   while $\sigma < n - \ell$
   $\quad \sigma = \sigma + I(1, 1, s);$
   $\quad s = s - 1;$
   end
   $idx = s + 1;$ (idx indicates the place where $\boldsymbol{\lambda}$ is to be cut.)
5. if $\sigma = n - \ell$
   $\quad \hat{\boldsymbol{\lambda}}(J(idx : m + 1)) = \mathbf{zeros}(1, m - idx + 2);$
   go to 7
6. $k_\ell = \mathbf{min}(\mathbf{find}(I(1, 1, idx + 1 : m + 1) == 1)) + idx;$
   $k_u = \mathbf{max}(\mathbf{find}(I(1, 1, 1 : idx) == 1));$
   if $I(:, :, idx) == [2, 1]$
   $\quad$ if ~$\mathbf{isempty}(k_\ell)$
   $\quad\quad \hat{\boldsymbol{\lambda}}(J(idx : m + 1)) = \mathbf{zeros}(1, m - idx + 2);$
   $\quad\quad \hat{\boldsymbol{\lambda}}(J(k_\ell)) = \boldsymbol{\lambda}(J(idx));$
   $\quad$ else
   $\quad\quad \hat{\boldsymbol{\lambda}}(J(k_u)) = 0;$
   $\quad\quad \hat{\boldsymbol{\lambda}}(J(idx + 1 : m + 1)) = \mathbf{zeros}(1, m - idx + 1);$
   $\quad$ end
   else
   $\quad$ if ~$\mathbf{isempty}(k_\ell)$
   $\quad\quad$ if $\mathbf{isempty}(k_u)$
   $\quad\quad\quad \hat{\boldsymbol{\lambda}}(J(idx : m + 1)) = \mathbf{zeros}(1, m - idx + 2);$
   $\quad\quad\quad \hat{\boldsymbol{\lambda}}(J(k_\ell)) = \boldsymbol{\lambda}(J(k_\ell));$
   $\quad\quad$ else
   $\quad\quad\quad t_1 = 2 * \mathbf{abs}(\boldsymbol{\lambda}(J(idx)))^2;$
   $\quad\quad\quad t_2 = \mathbf{abs}(\boldsymbol{\lambda}(J(k_u)))^2 + \mathbf{abs}(\boldsymbol{\lambda}(J(k_\ell)))^2;$
   $\quad\quad\quad$ if $t_1 \leq t_2$
   $\quad\quad\quad\quad \hat{\boldsymbol{\lambda}}(J(idx : m + 1)) = \mathbf{zeros}(1, m - idx + 2);$
   $\quad\quad\quad\quad \hat{\boldsymbol{\lambda}}(J(k_\ell)) = \boldsymbol{\lambda}(J(k_\ell));$
   $\quad\quad\quad$ else
   $\quad\quad\quad\quad \hat{\boldsymbol{\lambda}}(J(idx + 1 : m + 1)) = \mathbf{zeros}(1, m - idx + 1);$
   $\quad\quad\quad\quad \hat{\boldsymbol{\lambda}}(J(k_u)) = 0;$
   $\quad\quad\quad$ end
   $\quad\quad$ end

else

$\hat{\boldsymbol{\lambda}}(J(k_u)) = 0;$

$\hat{\boldsymbol{\lambda}}(J(idx+1:m+1)) = \mathbf{zeros}(1, m-idx+1);$

end

end

7. $$\hat{\boldsymbol{\lambda}} = \begin{cases} \left[\hat{\boldsymbol{\lambda}}, \mathbf{fliplr}(\mathbf{conj}(\hat{\boldsymbol{\lambda}}(2:m)))\right], & \text{if } n = 2m; \\ \left[\hat{\boldsymbol{\lambda}}, \mathbf{fliplr}(\mathbf{conj}(\hat{\boldsymbol{\lambda}}(2:m+1)))\right], & \text{if } n = 2m+1; \end{cases}$$

$\hat{\mathbf{c}} = \mathbf{real}(\mathbf{fft}(\hat{\boldsymbol{\lambda}}))/n;$

return

8.3.5 *Numerical experiment*

Applying Algorithm 8.3, we illustrate a variety of interesting facts observed through some numerical experiments. We report all numerics using only four significant digits, although entries of all matrices in consideration and the corresponding eigenvalues are originally in full length double precision.

Consider first the 8×8 symmetric circulant matrix whose first row is given by a randomly generated vector

$$\mathbf{c}_1 = [0.5404, 0.2794, 0.1801, -0.0253, -0.2178, -0.0253, 0.1801, 0.2794].$$

All its eigenvalues are real, and if arranged in descending order of their moduli, are

$$\sigma(Circul(\mathbf{c}_1)) = \{1.1909, 1.1892, 1.1892, 0.3273, 0.3273, 0.1746, -0.0376, -0.0376\}.$$

The singular values clearly are simply the moduli of these eigenvalues. Observe the parity caused by the conjugate-evenness, whereas 1.1909 and 0.1746 are what we call absolutely real eigenvalues.

Example 8.16. By using Algorithm 8.2, the nearest circulant approximation of rank 7 to $Circul(\mathbf{c}_1)$ would simply be the truncated singular value decomposition by setting the last eigenvalue -0.0376 to zero. But such a TSVD approach would result in a complex matrix.

To obtain the nearest real-valued circulant approximation of rank 7, we have to keep the pair of -0.0376 and replace the absolutely real value 0.1746 by zero. Rearranging the resulting eigenvalues in the conjugate-even ordering

$$\hat{\boldsymbol{\lambda}}_1 = [1.1909, 1.1892, -0.0376, 0.3273, 0, 0.3273, -0.0376, 1.1892],$$

we can construct the nearest real-valued rank-7 circulant approximation to $Circul(\mathbf{c}_1)$ via the FFT and obtain the first row vector

$$\hat{\mathbf{c}}_1 = [0.5186, 0.3012, 0.1583, -0.0035, -0.2396, -0.0035, 0.1583, 0.3012].$$

Example 8.17. Using the same data, consider the rank 4 circulant approximation. Four eigenvalues have to be set to zero. Upon examining the conjugate-evenness involved in $\sigma(Circul(\mathbf{c}_1))$, we realize the two smallest eigenvalues -0.0376, the next smallest eigenvalue 0.1746, and one of the double eigenvalues 0.3273 have to be dropped. This results in a topology change in the graph tree in the same way as we have described in Figures 8.8 and 8.9. In other words, the conjugate-even order of eigenvalues becomes

$$\tilde{\boldsymbol{\lambda}}_1 = [1.1909, 1.1892, 0, 0, 0.3273, 0, 0, 1.1892],$$

and the resulting nearest rank 4 circulant approximation is given by the row vector

$$\tilde{\mathbf{c}}_1 = [0.4871, 0.3182, 0.1898, -0.1023, -0.1075, -0.1023, 0.1898, 0.3182].$$

Consider next the 9×9 circulant matrix whose first row is given by

$$\mathbf{c}_2 = [1.6864, 1.7775, 1.9324, 2.9399, 1.9871, 1.7367, 4.0563, 1.2848, 2.5989].$$

The corresponding eigenvalues have conjugate-even structure given by

$$\begin{aligned}\sigma(Circul(\mathbf{c}_2)) = \{&20.0000, -2.8130 \pm 1.9106i, 3.0239 \pm 1.0554i,\\ &-1.3997 \pm 0.7715i, -1.2223 \pm 0.2185i\}.\end{aligned}$$

Note the absolute real eigenvalue $\lambda_1 = 20$ has molulus much larger than any other eigenvalues.

Example 8.18. Based on the above conjugate-even structure, to obtain a real-valued circulant approximation of rank 8 we have no choice but to select the vector (in its ordering)

$$\begin{aligned}\hat{\boldsymbol{\lambda}}_2 = [&0, -2.8130 - 1.9106i, 3.0239 - 1.0554i, -1.3997 - 0.7715i,\\ &-1.2223 + 0.2185i, -1.2223 - 0.2185i, -1.3997 + 0.7715i,\\ &3.0239 + 1.0554i, -2.8130 + 1.9106i]\end{aligned}$$

to produce

$$\hat{\mathbf{c}}_2 = [-0.5358, -0.5872, -1.1736, -0.3212, 1.0198, 1.4013, -0.0761, -0.4115, 0.6844]$$

as the first row of its nearest real-valued circulant approximation. Be aware that the largest singular value has been dropped.

The fact that the *largest* eigenvalue (singular value) of $Circul(\mathbf{c}_2)$ must be set to zero to produce the nearest rank 8 approximation in Example 8.18 is quite counterintuitive to the usual sense of the truncated singular value decomposition approximation. On the other hand, it is worth noting that if we slightly modify our approximation criteria by only requesting a nearest low rank approximation with rank *no greater than 8*, the answer could be completely different. In this particular example, such a nearest matrix turns out to be of rank 7 and is in agreement with the usual TSVD approximation by truncating the *pair* of eigenvalues with the smallest moduli.

In the final experiment, we perturb a given circulant matrix $C_\kappa \in \mathbb{R}^{n\times n}$ of rank κ with random noise and investigate how much information can be retrieved. With probability one any random noise added to C_κ will destroy the circulant structure as well as the rank condition. To establish a comparison, we may assume, without loss of generality, that after the projection step (8.26) mentioned in Algorithm 8.2 the added noise is a circulant matrix. Thus, only the rank condition is destroyed in the perturbation. More precisely, let $E \in \mathbb{R}^{n\times n}$ denote a random but fixed circulant matrix with unit Frobenius norm. Consider the perturbation of C_κ by additive noise of magnitude (in Frobenius norm) 10^{-j}, that is, consider the circulant matrices

$$W_j = C_\kappa + 10^{-j}E, \quad j = 1, \ldots, 12.$$

It is almost certain that under such a random perturbation the matrix W_j will be of full rank. Note that $\|W_j - C_\kappa\| = 10^{-j}$ represents the severity of the perturbation. It will be interesting to see if W_j has any closer circulant matrix approximation of rank κ than C_κ, especially when j is large, and how that compares to C_κ.

Example 8.19. Consider the test case with $n = 100$, $\kappa = 73$, and a predetermined matrix C_{73}. In Figure 8.10 the (continuous) lines depict the distribution of singular values of the perturbed matrices W_j for $j = 1, \ldots, 12$, respectively, whereas the singular values of the original C_{73} are marked by $*$. Observe

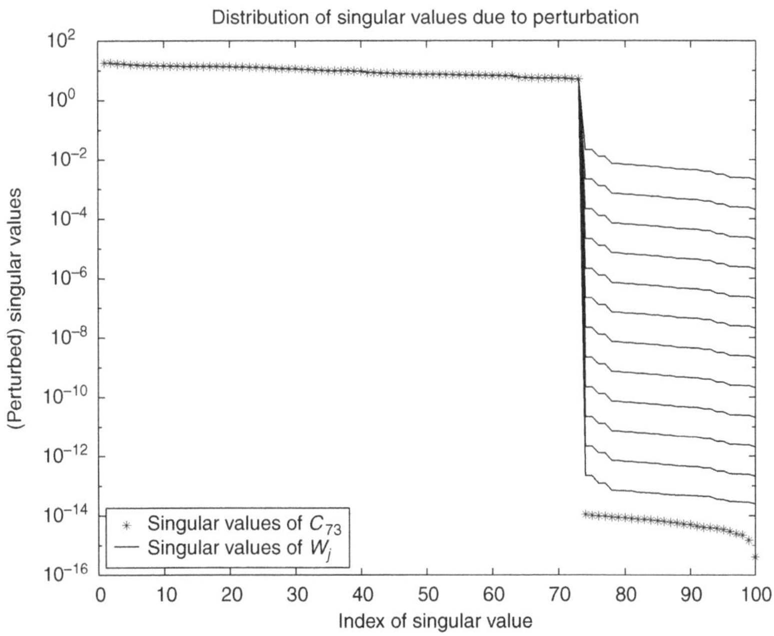

Figure 8.10. *Distribution of singular values*

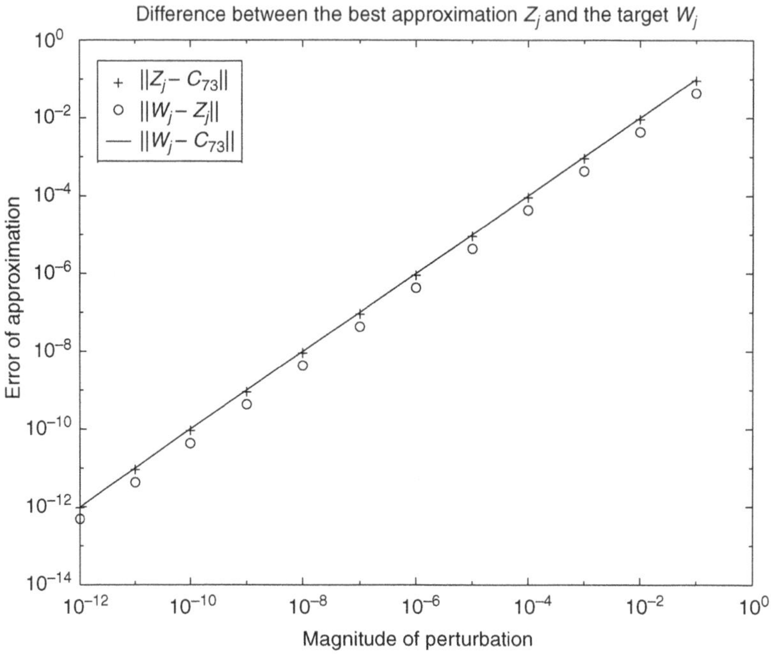

Figure 8.11. *Errors in approximation*

how the perturbation affects the last 27 (machine zero) singular values of C_{73} more significantly than the first 73 (larger) singular values according to the magnitude 10^{-j}.

Using our algorithm to find the best circulant approximation Z_j to W_j, we find that it is always the case that

$$\|W_j - Z_j\| < \|W_j - C_\kappa\|$$

for each j. This is indicated in Figure 8.11 by the fact that the circles $\circ$ are always below the diagonal line. Also marked in Figure 8.11 by + signs is the difference between Z_j and C_κ.

8.3.6 *An application to image reconstruction*

For space invariant image restoration computations, the matrix form H of the point spread function (PSF) is block Toeplitz with Toeplitz blocks whenever Dirichlet boundary conditions are present, as in space object imaging, see for example (Nagy et al., 1997). The restoration process involves solving large scale linear systems with coefficient matrix H, generally by iterative methods such as conjugate gradient type schemes. In (Hanke et al., 1993) it was shown how an effective regularizing block circulant with circulant blocks preconditioner C for equations involving H can be found using the 2-D FFT and replacing the eigenvalues corresponding to the noise subspace by ones. This is an $O(n^2 \log n)$ process if the image has n^2 pixels.

An alternative to that process is to find the nearest low rank BCCB approximation to H, and replace the zero eigenvalues by ones. Extension of the work in this section to the block BCCB case is straightforward. Again, this is an $O(n^2 \log n)$ process for image reconstruction. The extension may possibly lead to a more effective preconditioner.

8.3.7 *Summary*

We have studied the low rank circulant approximation problem. For any given real data matrix, its nearest real circulant approximation can simply be determined from the average of its diagonal entries. Its nearest lower rank approximation can also be determined effectively from the truncated singular value decomposition and the fast Fourier transform. To maintain the circulant structure, to induce a specific lower rank, and to stay real, however, the conjugate-even structure must be taken into account and that can substantially change the truncation criteria. We have proposed a fast algorithm to accomplish all of these objectives, and extensions to the block case with possible applications to image reconstruction are discussed.

8.4 Low rank covariance approximation

An indispensable task in almost every discipline of science is to analyze certain data to search for relationships between a set of exogenous and endogenous variables. The exigencies of such a task become especially strong in this era of information and digital technologies as massive amounts of data are being generated at almost every level of application.

It is generally acknowledged that most of the information gathering devices or methods at present have only finite bandwidth. One thus cannot avoid the fact that the data collected often are not exact. For example, signals received by antenna arrays often are contaminated by instrumental noise; astronomical images acquired by telescopes often are blurred by atmospheric turbulence; databases prepared by document indexing often are biased by subjective judgment; and even empirical data obtained in laboratories often do not satisfy intrinsic physical constraints. Before any deductive sciences can be applied further, it is important to first reconstruct the data matrices so that the inexactness is reduced while certain feasibility conditions are satisfied.

Furthermore, in many situations the data observed from complex phenomena represent the integrated result of several interrelated variables acting together. When these variables are less precisely defined, the actual information contained in the original data matrix might be overlapping, fuzzy, and no longer that clear-cut. A reduced system might provide the same level of fidelity as the original system.

One common factor in the various approaches for noise removal, model reduction, feasibility reconstruction, and so on, is to replace the original data matrix by a lower dimensional representation obtained somehow via subspace approximation or truncation. The truncated singular value decomposition, for example, is one commonly used candidate for replacement. Despite the many reported successes in application and the many seemingly intuitive arguments to support this approach, there appears to be a lack of rigorous mathematics to justify exactly what is really going on behind this low rank approximation. This short section is an attempt to fill that gap from a statistical point of view.

Additionally, needs for efficient generation of samples from random processes and approximations to their covariance arise often in many disciplines. In practice, especially in space–time applications where uncertainty propagates over time, compact covariance approximations are an essential tool for avoiding huge sample collection. Low rank covariance approximation itself is an important subject.

8.4.1 *Low dimensional random variable approximation*

We first consider a general random (column vector) variable $\mathcal{X}$ in $\mathbb{R}^n$ with a certain unspecified distribution. Let $\mathcal{E}[\mathcal{X}]$ denote the expected value of $\mathcal{X}$. Typically, $\mathrm{cov}(\mathcal{X}) := \mathcal{E}[(\mathcal{X} - \mathcal{E}[\mathcal{X}])(\mathcal{X} - \mathcal{E}[\mathcal{X}])^\top] \in \mathbb{R}^{n\times n}$ is defined as the *covariance matrix* of $\mathcal{X}$. Being symmetric and positive semidefinite, the deterministic matrix $\mathrm{cov}(\mathcal{X})$ enjoys a spectral decomposition

$$\mathrm{cov}(\mathcal{X}) = \sum_{j=1}^{n} \lambda_j \boldsymbol{p}_j \boldsymbol{p}_j^\top \tag{8.27}$$

where we also assume that eigenvalues are arranged in the descending order $\lambda_1 \geq \lambda_2 \geq \cdots \geq \lambda_n$. Observe that $\boldsymbol{p}_1, \ldots, \boldsymbol{p}_n$ form an orthonormal basis for $\mathbb{R}^n$. Express the random column variable $\mathcal{X}$ as

$$\mathcal{X} = \sum_{j=1}^{n} (\boldsymbol{p}_j^\top \mathcal{X}) \boldsymbol{p}_j. \tag{8.28}$$

Note that the columns in the matrix $P := [\boldsymbol{p}_1, \ldots, \boldsymbol{p}_n]$ are deterministic vectors themselves. The randomness of $\mathcal{X}$ therefore must come solely from the randomness of each coefficient in (8.28). The following observation gives important insight on the portion of randomness of $\mathcal{X}$ in each of the eigenvector directions.

Theorem 8.7. Let $\boldsymbol{\alpha} := P^\top \mathcal{X}$. Then $\boldsymbol{\alpha}$ is a random variable whose components are mutually stochastically independent. Indeed,

$$\mathcal{E}[\boldsymbol{\alpha}] = P^\top \mathcal{E}[\mathcal{X}], \tag{8.29}$$

$$var(\boldsymbol{\alpha}) = \mathrm{diag}\{\lambda_1, \ldots, \lambda_n\}. \tag{8.30}$$

Proof The expected value (8.29) of $\boldsymbol{\alpha}$ is obvious. The covariance matrix of $\boldsymbol{\alpha}$ is given by

$$\begin{aligned}\operatorname{cov}(\boldsymbol{\alpha}) &= \mathcal{E}[(\boldsymbol{\alpha} - \mathcal{E}[\boldsymbol{\alpha}])(\boldsymbol{\alpha} - \mathcal{E}[\boldsymbol{\alpha}])^{\mathsf{T}}] \\ &= \mathcal{E}[(P^{\mathsf{T}}(\mathcal{X} - \mathcal{E}[\mathcal{X}])(\mathcal{X} - \mathcal{E}[\mathcal{X}])^{\mathsf{T}}P)] \\ &= \operatorname{diag}\{\lambda_1, \dots, \lambda_n\}\end{aligned}$$

where the last equality follows from the definition of (8.27). □

Recall that smaller variance means the values of the random variable are more clustered around the mean. It is fair to say that a random variable with larger variance is harder to *predict* than a random variable with smaller variance.

From Theorem 8.7, we make one important observation. That is, the larger the eigenvalue λ_j of $\operatorname{cov}(\mathcal{X})$ is, the larger the variance of the random (scalar) variable α_j is. Consider the fact from (8.28) that the random variable $\mathcal{X}$ is made of random contributions from each of the n directions $\boldsymbol{p}_j$, $j = 1, \dots, n$. Consider also that the contribution from each direction is governed independently by the distribution of the corresponding random variable α_j. Thus, the less the variance of α_j is, the less unpredictable is the contribution from the direction $\boldsymbol{p}_j$. As far as the random nature of $\mathcal{X}$ is concerned, it is intuitively correct from a statistical point of view that those coefficients α_j with larger variances should represent a more integral part of the stochastic nature of $\mathcal{X}$. It is in this context that we may *rank* the importance of corresponding eigenvectors $\boldsymbol{p}_j$ as *essential* components for the variable $\mathcal{X}$ according to the magnitude of λ_j.

If it becomes desirable to approximate the random variable $\mathcal{X}$ by another unbiased yet *simpler* variable $\hat{\mathcal{X}}$, we see from Theorem 8.7 that $\hat{\mathcal{X}}$ had better capture those components corresponding to larger λ_j in the expression (8.28). We quantify this notion below, which provides the basic idea of truncation.

So that this section is self-contained, we first re-prove a useful result that is classical in estimation theory (Luenberger, 1969; Melsa and Cohn, 1978).

Theorem 8.8. Let $\boldsymbol{x} \in \mathbb{R}^n$ and $\boldsymbol{y} \in \mathbb{R}^r$ denote two random variables with mean zero, respectively. Then the coefficient matrix $K \in \mathbb{R}^{n\times r}$ that gives rise to the best unbiased linear estimation $\hat{\boldsymbol{x}} = K\boldsymbol{y}$ of $\boldsymbol{x}$ in the sense of minimizing $\mathcal{E}[\|\boldsymbol{x} - \hat{\boldsymbol{x}}\|^2]$ is

$$K = \mathcal{E}[\boldsymbol{x}\boldsymbol{y}^{\mathsf{T}}](\mathcal{E}[\boldsymbol{y}\boldsymbol{y}^{\mathsf{T}}])^{-1}. \tag{8.31}$$

In this case, each $\hat{x}_i$ is the minimum-variance estimate of the corresponding x_i, respectively, for $i = 1, \dots, n$.

Proof Let the matrix K be written in rows, i.e., $K = [\boldsymbol{k}_1^{\mathsf{T}}, \dots, \boldsymbol{k}_n^{\mathsf{T}}]^{\mathsf{T}}$. Observe that

$$\mathcal{E}[\|\boldsymbol{x} - \hat{\boldsymbol{x}}\|^2] = \sum_{i=1}^{n} \mathcal{E}[(x_i - \hat{x}_i)^2] = \sum_{i=1}^{n} \mathcal{E}[(\boldsymbol{k}_i^{\mathsf{T}}\boldsymbol{y} - x_i)^2].$$

Thus it suffices to consider minimizing each individual term $g(\boldsymbol{k}_i) := \mathcal{E}[(\boldsymbol{y}^\top \boldsymbol{k}_i - x_i)^2]$, $i = 1, \ldots, n$, in the above summation. (It is in this sense that the term "minimum-variance" unbiased estimate for each component is used.) The first-order optimality condition for g to be minimized at $\boldsymbol{k}_i$ is

$$\nabla g(\boldsymbol{k}_i) = 2\mathcal{E}[(\boldsymbol{y}^\top \boldsymbol{k}_i - x_i)\boldsymbol{y}] = 0. \tag{8.32}$$

We may rewrite the necessary condition for $i = 1, \ldots, n$ collectively as

$$\mathcal{E}[\boldsymbol{y}\boldsymbol{y}^\top]K^\top = \mathcal{E}[\boldsymbol{y}\boldsymbol{x}^\top].$$

It follows that K is given by (8.31). □

Theorem 8.9. Assume that $\hat{\boldsymbol{x}}$ is the minimum-variance estimate of $\boldsymbol{x}$ in the same setting as in the above theorem. Then

$$\operatorname{cov}(\boldsymbol{x} - \hat{\boldsymbol{x}}) = \operatorname{cov}(\boldsymbol{x}) - \operatorname{cov}(\hat{\boldsymbol{x}}), \tag{8.33}$$

$$\mathcal{E}[\|\boldsymbol{x} - \hat{\boldsymbol{x}}\|^2] = \operatorname{trace}(\operatorname{cov}(\boldsymbol{x} - \hat{\boldsymbol{x}})) = \mathcal{E}[\boldsymbol{x}^\top \boldsymbol{x}] - \mathcal{E}[\boldsymbol{x}^\top K \boldsymbol{y}]. \tag{8.34}$$

Proof By definition,

$$\operatorname{cov}(\boldsymbol{x} - \hat{\boldsymbol{x}}) = \operatorname{cov}(\boldsymbol{x}) - \mathcal{E}[\boldsymbol{x}\hat{\boldsymbol{x}}^\top] - \mathcal{E}[\hat{\boldsymbol{x}}\boldsymbol{x}^\top] + \operatorname{cov}(\hat{\boldsymbol{x}}).$$

Observe then by substitution that

$$\mathcal{E}[\boldsymbol{x}\hat{\boldsymbol{x}}^\top] = \mathcal{E}[\hat{\boldsymbol{x}}\boldsymbol{x}^\top] = \mathcal{E}[\hat{\boldsymbol{x}}\hat{\boldsymbol{x}}^\top] = \mathcal{E}[\boldsymbol{x}\boldsymbol{y}^\top](\mathcal{E}[\boldsymbol{y}\boldsymbol{y}^\top])^{-1}\mathcal{E}[\boldsymbol{y}\boldsymbol{x}^\top].$$

The equation (8.33) is proved. The residual (8.34) can be calculated from

$$\mathcal{E}[\|\boldsymbol{x} - \hat{\boldsymbol{x}}\|^2] = \sum_{i=1}^{n} g(\boldsymbol{k}_i) = -\sum_{i=1}^{n} \mathcal{E}[(\boldsymbol{k}_i^\top \boldsymbol{y} - x_i)x_i]$$

by using (8.32). □

Returning to the problem of approximating the random variable $\mathcal{X}$ by an unbiased yet simpler variable $\hat{\mathcal{X}}$, consider the case that by a simpler variable $\tilde{\mathcal{X}}$ we mean a random variable limited to a *lower dimensional* subspace. Our goal then is to find a proper subspace $\mathcal{S}$ and a particular random variable $\tilde{\mathcal{X}}$ on $\mathcal{S}$ such that $\mathcal{E}[\|\mathcal{X} - \tilde{\mathcal{X}}\|^2]$ is minimized.

Observe first that, given any r-dimensional subspace $\mathcal{S}$, there exists a matrix $K \in \mathbb{R}^{n\times r}$ such that columns of the matrix product PK, with P given by (8.27), form a basis for $\mathcal{S}$. Any unbiased random variable $\tilde{\mathcal{X}}$ restricted to $\mathcal{S}$ can then be expressed in the form

$$\tilde{\mathcal{X}} = PK\boldsymbol{\beta}$$

where $\boldsymbol{\beta}$ stands for a certain (column) random variable in $\mathbb{R}^r$. We may further assume that components in $\boldsymbol{\beta}$ are mutually independent because, if otherwise, we may simply do a spectral decomposition of $\boldsymbol{\beta}$ similarly to (8.27) and Theorem 8.7. It follows that $\mathcal{E}[\|\mathcal{X} - \tilde{\mathcal{X}}\|^2] = \mathcal{E}[\|\boldsymbol{\alpha} - K\boldsymbol{\beta}\|^2]$. The minimum-variance problem is now reduced to the problem of finding K and $\boldsymbol{\beta}$ so that $\mathcal{E}[\|\boldsymbol{\alpha} - K\boldsymbol{\beta}\|^2]$ is minimized.

From Theorem 8.8, however, we know that at an optimizer the coefficient matrix K and the variable $\boldsymbol{\beta}$ are not totally unrelated. Indeed, given $\boldsymbol{\beta}$, the optimal matrix K is completely determined and is given by

$$K = \mathcal{E}[\boldsymbol{\alpha}\boldsymbol{\beta}^\top](\mathcal{E}[\boldsymbol{\beta}\boldsymbol{\beta}^\top])^{-1}. \tag{8.35}$$

From (8.34), we further know that to obtain the minimum-variance approximation of $\mathcal{X}$, it only remains to choose $\boldsymbol{\beta}$ so that

$$\mathcal{E}[\boldsymbol{\alpha}^\top K\boldsymbol{\beta}] = \left\langle \mathcal{E}[\boldsymbol{\alpha}\boldsymbol{\beta}^\top](\mathcal{E}[\boldsymbol{\beta}\boldsymbol{\beta}^\top])^{-1}, \mathcal{E}[\boldsymbol{\alpha}\boldsymbol{\beta}^\top] \right\rangle \tag{8.36}$$

is maximized. This nonlinear optimization problem turns out to have a simple solution as we shall see from the proof of the following theorem.

Theorem 8.10. Suppose that $\mathcal{X}$ is a random variable in $\mathbb{R}^n$ with mean zero and that its covariance matrix has a spectral decomposition given by (8.27). Then among all unbiased variables restricted to any r-dimensional subspaces in $\mathbb{R}^n$, the random variable

$$\hat{\mathcal{X}} := \sum_{j=1}^{r} (\boldsymbol{p}_j^\top \mathcal{X})\boldsymbol{p}_j \tag{8.37}$$

is the best linear minimum-variance estimate of $\mathcal{X}$ in the sense that $\mathcal{E}[\|\mathcal{X} - \hat{\mathcal{X}}\|^2]$ is minimized.

Proof We already know that $\mathcal{E}[\|\mathcal{X} - \tilde{\mathcal{X}}\|^2] = \mathcal{E}[\|\boldsymbol{\alpha} - K\boldsymbol{\beta}\|^2]$. From Corollary 8.9, we also know that

$$\mathcal{E}[\|\boldsymbol{\alpha} - K\boldsymbol{\beta}\|^2] = \operatorname{trace}(\operatorname{cov}(\boldsymbol{\alpha})) - \operatorname{trace}(\operatorname{cov}(K\boldsymbol{\beta}))$$

is minimized. In fact, in the proof of Theorem 8.8, we have pointed out that the estimation $\hat{\boldsymbol{\alpha}} = K\boldsymbol{\beta}$ is a component-wise minimum-variance estimation. That is, for each $i = 1, \ldots, n$, the coefficient matrix K has the effect that

$$\mathcal{E}[(\alpha_i - \hat{\alpha}_i)^2] = \mathcal{E}[\alpha_i^2] - \mathcal{E}[\hat{\alpha}_i^2]$$

is minimized. Recall that $\mathcal{E}[\alpha_i^2] = \lambda_i$. Thus, we should somehow select $\boldsymbol{\beta}$ in such a way that the corresponding $\hat{\boldsymbol{\alpha}} = K\boldsymbol{\beta}$ will have $\mathcal{E}[\hat{\alpha}_i^2] = \lambda_i$ for as many i's as possible. More specifically, since the trace is to sum over the differences $\lambda_i - \mathcal{E}[\hat{\alpha}_i^2]$ whereas $\lambda_1 \geq \cdots \geq \lambda_n$ and since we only have r degrees of freedom to determine $\boldsymbol{\beta}$, the best we can hope is to choose $\boldsymbol{\beta}$ so that the first k eigenvalues $\lambda_1, \ldots, \lambda_r$ are matched.

It turns out that if we choose the special case $\boldsymbol{\beta} = [\alpha_1, \ldots, \alpha_r]^\top$, then

$$\hat{\boldsymbol{\alpha}} = [\alpha_1, \ldots, \alpha_r, 0, \ldots, 0]^\top$$

and the corresponding $PK\hat{\boldsymbol{\beta}}$ is precisely given by (8.37). □

It is important to note that in the above linear minimum-variance estimation, the variable $\mathcal{X}$ is *centered* at zero. If $\mathcal{X}$ is not centered at zero, the expression for truncation would be much more complicated. Somehow this centering has been ignored in many practices where low rank approximation is used. Without the centering, we really would like to raise the flag that the resulting truncated data should suffer from the loss of some significant statistical meaning. We shall comment more on this in the next section.

On the other hand, the following theorem shows that the choice of $\hat{\mathcal{X}}$ not only makes the diagonal entries of $\operatorname{cov}(\hat{\mathcal{X}})$ best approximate those of $\operatorname{cov}(\mathcal{X})$, but that the entire matrix $\operatorname{cov}(\hat{\mathcal{X}})$ be reasonably close to $\operatorname{cov}(\mathcal{X})$ as well.

Theorem 8.11. Suppose that $\mathcal{X}$ is a random variable in $\mathbb{R}^n$ with mean zero and that its covariance matrix has a spectral decomposition given by (8.27). Then among all unbiased variables restricted to any r-dimensional subspaces in $\mathbb{R}^n$, the random variable $\hat{\mathcal{X}}$ defined in (8.37) also minimizes $\|\operatorname{cov}(\hat{\mathcal{X}}) - \operatorname{cov}(\mathcal{X})\|$.

Proof The proof is quite straightforward. It is well known that the best rank r approximation to the matrix $\operatorname{cov}(\mathcal{X})$ is given by the truncated summation $\sum_{i=1}^{r} \lambda_i \boldsymbol{p}_i \boldsymbol{p}_i^\top$, which clearly is also the covariance matrix of $\hat{\mathcal{X}}$. □

Example 8.20. Given a random vector $\mathcal{X} \in \mathbb{R}^n$ with mean zero, suppose columns of $P = [\mathbf{p}_1, \ldots, \mathbf{p}_k] \in \mathbb{R}^{n\times k}$ are unit vectors that whiten $\mathcal{X}$. That is, suppose that

$$\operatorname{cov}(\mathcal{Y}) = I_k.$$

where $\mathcal{Y} \in \mathbb{R}^k$ is random vector

$$\mathcal{Y} := P^\top \mathcal{X}. \tag{8.38}$$

By Theorem 8.8, the best linear minimum-variance estimate of $\mathcal{X}$ is given by

$$\hat{\mathcal{X}} = (\mathcal{E}(\mathcal{X}\mathcal{X}^\top)P)(P^\top \mathcal{X}). \tag{8.39}$$

Furthermore, the matrix

$$\operatorname{cov}(\hat{\mathcal{X}}) = (\mathcal{E}(\mathcal{X}\mathcal{X}^\top)P)(\mathcal{E}(\mathcal{X}\mathcal{X}^\top)P)^\top \tag{8.40}$$

is the best rank k covariance approximation to $\operatorname{cov}(\mathcal{X})$. In this case, one can replace $P^\top \mathcal{X}$ in (8.39) by any white random vector $\mathcal{W} \in \mathbb{R}^k$ satisfying $\operatorname{cov}(\mathcal{W}) = I_k$ and obtain

$$\tilde{\mathcal{X}} = (\mathcal{E}(\mathcal{X}\mathcal{X}^\top)P)\mathcal{W}, \tag{8.41}$$

which has the same second-order statistics as $\hat{\mathcal{X}}$.

8.4.2 *Truncated SVD*

The discussion thus far is based on the fact that the random variable $\mathcal{X}$ is completely known. Such an assumption is not realistic in practice since often the probability distribution function of the underlying random variable $\mathcal{X}$ is not known *a priori*. One common practice in application then is to simulate the random variable $\mathcal{X}$ by a collection of ℓ random samples. These samples are recorded in a $n \times \ell$ matrix X. Each column of X represents one random sample of the underlying random (column vector) variable $\mathcal{X} \in \mathbb{R}^n$. It is known that when ℓ is large enough, many of the stochastic properties of $\mathcal{X}$ can be recouped from X.

The question now is how to retrieve a sample data matrix from X to represent the minimum-variance approximation $\hat{\mathcal{X}}$ of $\mathcal{X}$. To begin with, we shall assume that $\mathcal{E}[\mathcal{X}] = 0$ and that the samples X have been centered, that is the mean value of each row is zero. The connection lies in the observations that the covariance matrix of the samples

$$R = \frac{XX^{\top}}{\ell}$$

converges to $\operatorname{cov}(\mathcal{X})$ by the law of large numbers. Analogous to (8.27), let

$$R = \sum_{i=1}^{n} \mu_i \boldsymbol{u}_i \boldsymbol{u}_i^{\top} \tag{8.42}$$

be the spectral decomposition of R with eigenvalues $\mu_1 \geq \cdots \geq \mu_n$ and orthonormal eigenvectors $\boldsymbol{u}_1, \ldots, \boldsymbol{u}_n$. Then it follows from the observation in Theorem 8.10 that the best low dimensional minimum-variance estimate $\hat{\mathcal{X}}$ to $\mathcal{X}$ should be represented by the matrix

$$\hat{X} := \sum_{j=1}^{r} \boldsymbol{u}_j (\boldsymbol{u}_j^{\top} X). \tag{8.43}$$

The *low dimension* estimate $\hat{\mathcal{X}}$ to the (continuous) random variable $\mathcal{X}$ is now comfortably translated into a *low rank* approximation $\hat{X}$ to the (discrete) random sample matrix X.

Indeed, the singular value decomposition of X

$$X = U\Sigma V^{\top} = \sum_{i=1}^{n} \sigma_i \boldsymbol{u}_i \boldsymbol{v}_i^{\top} \tag{8.44}$$

shares the same eigenvectors of R as its left singular vectors, that is $U = [\boldsymbol{u}_1, \ldots, \boldsymbol{u}_n]$ with singular values given by $\sigma_i = \sqrt{\ell \mu_i}$, $i = 1, \ldots, n$, respectively. The notion of the truncated singular value decomposition of X is simply the partial sum $\sum_{i=1}^{r} \sigma_i \boldsymbol{u}_i \boldsymbol{v}_i^{\top}$, which we now see is precisely $\hat{X}$ defined in (8.43).

In this sense, the truncated singular value decomposition of a given data matrix X representing random samples of an unknown random variable $\mathcal{X}$ now has a statistical meaning. That is, the truncated rank r singular value decomposition represents random samples of the best minimum-variance linear estimate $\hat{\mathcal{X}}$ to $\mathcal{X}$ among all possible r-dimensional subspaces.

8.4.3 *Summary*

In many applications, the truncated singular value decomposition of the observed data matrix is used to *filter out the noise*. In this section, we try to convey the notion that the truncation singular value decomposition is in fact the best minimum-variance estimation of the underlying *unknown* random variable, be it contaminated by noises or not. Note that in Theorem 8.8 no relationship between $\boldsymbol{x}$ and $\boldsymbol{y}$ is assumed. Likewise, no relationship between $\hat{\mathcal{X}}$ and $\mathcal{X}$ is assumed prior to the conclusion proved in Theorem 8.10. The notion of *truncation* now is manifested through the notion of best minimum-variance estimation.

Although it is obvious in the context of linear algebra that the truncated rank r singular value decomposition $\hat{X}$ of a given matrix X minimizes the 2-norm or Frobenius norm of the difference $X - Y$ among all possible rank r matrices Y, one must wonder what this low rank approximation means if X is a random matrix (from any kind of unknown distribution). If each column of X represents an *unpredictable* sample of a certain unknown distribution, one must wonder how much fidelity the corresponding column in the truncated matrix $\hat{X}$ really represents and how to measure it. The statistical interpretation of truncated singular value decomposition discussed in this section should fill that gap. The truncated singular value decomposition $\hat{X}$ not only is the best approximation to X in the sense of norm, but also is the closest approximation to X in the sense of statistics. It maintains the most significant stochastic portion of the original data matrix X.

Finally, other than the truncated singular value decomposition, many other types of low rank approximation to the given data matrix X have been proposed. For example, one of the most contentious issues in latent semantic indexing (LSI) for data mining is to find a suitable low rank representation of the original term-document indexing matrix. For this issue alone, many parties are in fierce competition to patent their special techniques. In this section, we have shown the significance of those larger singular values and the corresponding left singular vectors. Generally speaking, any lower rank approximation to an empirical data matrix X should carry properties similar, if not identical, to the truncated singular value decomposition, that is should contain substantial stochastic information about the original random variable $\mathcal{X}$. It perhaps is not too judgemental to say that any low rank approximation (for data mining) without this notion in mind, regardless of how efficient the computation could be, is equivalent to an attempt to *see things through the glass of darkness.*

8.5 Euclidean distance matrix approximation

Crippen and Havel (1988) have referred to the following endeavor as the *fundamental problem in distance geometry.*

> **Problem 8.4** *(Fundamental problem in distance geometry)*
> Given the distance and chirality constraints which define (our state of knowledge of) a mobile molecule, find one or more conformations which satisfy them, or else prove that no such conformations exists.

The notion of distance geometry, initiated by Menger and Schoenbeng in the 1930s, has been an area of active research because of its many important applications, including molecular conformation problems in chemistry (Crippen and Havel, 1988), multidimensional scaling in behavioral sciences (de Leeuw and Heiser, 1982; Mathar, 1985), and multivariate analysis in statistics (Commandeur et al., 1999). The technique, for example, is frequently applied to the generation of three-dimensional structures of macromolecules such as proteins or polynucleotides from the many interatomic distances observed in NMR (nuclear magnetic resonance) spectra (Havel and Wüthrich, 1984). The article by Gower (1982) provides an excellent reference that expounds the framework of Euclidean distance geometry in general. More extensive discussion on the background and applications can be found in the book by Crippen and Havel (1988). This section discusses only some very limited specifics on this subject.

8.5.1 *Preliminaries*

The basic task in distance geometry is to use lists of interpoint distances to generate coordinates of the points that are consistent with the measured interpoint distances. As such, one of the most basic requisites in the study of distance geometry is the information of interpoint relationships. Given n particles at locations $\mathbf{p}_1, \ldots, \mathbf{p}_n$ in the space $\mathbb{R}^m$, the corresponding *Euclidean distance matrix* $Q(\mathbf{p}_1, \ldots, \mathbf{p}_n) = [q_{ij}]$ is the $n \times n$ symmetric and nonnegative matrix whose entry q_{ij} is defined by

$$q_{ij} = \|\mathbf{p}_i - \mathbf{p}_j\|^2, \quad i, j = 1, \ldots, n, \tag{8.45}$$

where $\|\cdot\|$ denotes the Euclidean norm in $\mathbb{R}^m$. In other words, the distance matrix $Q(\mathbf{p}_1, \ldots, \mathbf{p}_n) = [q_{ij}]$ contains an exhaustive record of relative spacing between any two of the n particles in $\mathbb{R}^m$. For most real-world applications, we only need to be concerned about the case $m = 3$. Nonetheless, the theory presented hereinafter can be extended to a general dimension m.

Rank condition Euclidean distance matrices enjoys many interesting properties. The thesis by Dattorro (2004), for example, summarizes several equivalent forms of Euclidean distance matrices. The Euclidean distance matrices are closely connected to positive semidefinite matrices (Johnson and Tarazaga, 1995; Laurent, 1998). Several other theoretical results can be found in articles such as (Bakonyi and Johnson, 1995; Farebrother, 1987; Glunt et al., 1990; Gower, 1985). Among these, we find the rank structure of Euclidean distance matrices most peculiar.

That is, regardless of the size n of the matrix, a Euclidean distance matrix is always rank deficient. For this reason, distance matrix approximation can always be regarded as a structured low rank approximation. The structure is that defined by (8.45). The following result is well established in the literature.

Theorem 8.12. For any $n \geq m+2$, the rank of Q is no greater than $m+2$ and is generically $m+2$.

The Euclidean distance matrix for a geometric entity embedded in $\mathbb{R}^3$ with at least five points, for instance, is generically of rank 5 regardless of the number n of points involved. Once all the interparticle distances are in hand, it is a simple task to construct the (three-dimensional) geometric structure. The fact that a distance matrix is always rank deficient strongly suggests that many entries in the matrix provide redundant information. Indeed, not all the interparticle distances are needed for the construction. To characterize when and what partial information of interparticle distances will be sufficient for the determination of the entire geometric structure is an interesting and very challenging open question by itself. Such a task is referred to as a *completion problem* in the literature.

Completion problems for special cases such as n-vertex polyhedra are not difficult to answer. Theoretical results for more general configurations, however, are known only in terms of abstract graph theory (Bakonyi and Johnson, 1995; Huang et al., 2003) and are NP-hard. Despite the lack of a satisfactory theory, numerical methods for the Euclidean distance matrix completion problem abound (Alfakih et al., 1999; Huang et al., 2003; Trosset, 1997). In the following, we seek to delineate a similar notion to tackle this important completion problem along with a least squares distance matrix approximation formulation. In contrast to existing methods in the literature, we offer a much simpler framework which works for both approximation and completion problems. The procedure can be generalized to arbitrarily higher m. More importantly, our matrix calculus exploits the inherent block structure of the underlying objective function and its derivatives which, in turn, enable us to take advantage of large-scale optimization techniques.

Protein folding Before we formulate our problem and describe a numerical procedure for its solution, it might be fitting to ground the study through an interesting application in biochemistry. A protein molecule is a connected sequence of amino acid molecules. It is known that there are only 20 amino acids in nature. If we represent each amino acid by a letter from a 20-letter alphabet, then a protein is a string of amino acids linked together like a word. Nowadays, most laboratories have the technology to find the ordered sequence of amino acids in a protein. However, merely knowing the long linear chains is not enough. To function properly, the one-dimensional primary amino acid sequence must fold into a particular three-dimensional conformation called its tertiary configuration. This protein then interacts three-dimensionally with other proteins or other groups of molecules called substrates in a manner much like a lock and key arrangement. It is the tertiary structure that mediates how a protein functions. For example, the final folding configuration determines whether a protein acts as an enzyme in chemical reactions or behaves as part of an antibody in the immune system, and so on. In short, how a protein folds determines how the protein works, which ultimately determines how our bodies work. Wrongly folded proteins do not behave normally, and their abnormal function can lead to disease. Understanding how proteins fold into three-dimensional structures given their primary linear sequence thus becomes an incredibly important task.

The literature on the various aspects of protein folding is enormous. A recent survey by Neumaier (Neumaier, 1997) gives a masterly account of the current state of this problem and contains 360 references, many of which also have extensive bibliographies. In this section we intend to address the folding problem under one presumptuous scenario. Biologists try to see the three-dimensional structure by techniques such as X-ray crystallography or nuclear magnetic resonance (NMR) to build molecular dynamics models (Brünger and Nilges, 1993). Yet the noise in the sensing devices or the imperfection of the models often results in indefinite or incomplete images. Thus an imperative task is to retrieve a possible folding structure that is nearest to the observed but possibly inconsistent or imperfect configuration.

More specifically, suppose that $F \in \mathbb{R}^{n \times n}$ denotes an observation of (the squares of) the relative spacing among n particles in $\mathbb{R}^3$. In practice, it is very likely that some of the spacing has not been measured accurately. One prominent sign that something in the data is inconsistent would be that the rank of F is greater than 5. We therefore would like to retrieve whatever feasible information we can out of the matrix F so as to infer a realistic conformation. It is also likely that some of the spacing is unobservable and, consequently, some entries in the matrix F are missing. In this case, we want to complete the matrix F so that it becomes a distance matrix. In either case, the observed matrix F needs to be modified.

Classical approaches Be aware that q_{ij} in (8.45) is defined to be the distance squared. The corresponding problem to the following discussion but *without* the squares has a very different nature and an even longer history. It perhaps dates back to the 17th century when Fermat studied the problem of finding the shortest network interconnection points on a plane. Nowadays the problem without the squares is better known as the Euclidean facilities location problem and the Steiner minimal tree problem, both of which will not be discussed in this book.

One idea of modification is to find a Euclidean distance matrix $Q \in \mathbb{R}^{n \times n}$ such that Q represents the minimal change of F. The criterion used in characterizing the changes affects the way in which the approximation is formulated. In (Chu, 2000), for example, it was suggested that low rank approximation techniques could be applied to find the nearest nonnegative, symmetric matrix of rank 5 to the given F. This approach, however, only partially corrects the problem because a nonnegative and symmetric matrix of rank 5 is not necessarily a distance matrix. The distance matrix has more structure that is yet to be exploited. In (Glunt et al., 1990), efforts have been made to cast the distance matrices as the intersection of two geometric entities. The Dykstra algorithm that alternates projections between two convex sets is used with the intention of minimizing a quadratic functional over the intersection which contains Euclidean distance matrices. This approach, however, suffers from its slow convergence and possible stagnation. We mention that there is a rich discussion for distance matrices without the squares. Gaussian smoothing followed by limited-memory variable-metric continuation techniques (Moré and Wu, 1999) and transformation into standard convex programming in conic form followed by a polynomial time interior point algorithm (Xue and Ye, 1997) are just two of many possible avenues to circumvent the nonsmoothness. Similar approaches by using primal–dual interior point techniques that solves an equivalent semidefinite programming problem have been proposed in (Alfakih et al., 1999; Laurent, 1998; Mathar, 1985; Trosset, 1997). In what follows, we propose to tackle the approximation problem by a least squares formulation where the Euclidean distance matrices are parameterized directly in terms the location vectors $\mathbf{p}_1, \ldots, \mathbf{p}_n$. One immediate advantage of this representation is that the resulting approximation of the imperfect F is guaranteed to be a distance matrix.

While we appreciate the efficiency and elegance of the many existing methods, especially the semidefinite programming techniques, usually there is a large number of specific projections or transformations that must take place before the techniques can be applied. In contrast, our approach is much more straightforward. We work directly with the location vectors. Our idea is not new. Various least squares formulations have been discussed in Crippen and Havel (1988), Glunt et al. (1993), Moré and Wu (1999), Trosset (1997). Our main contribution is to point out a highly organized way of explicitly computing the gradient and the Hessian of the objective function in block forms. Efficient classical optimization techniques therefore can be applied. Furthermore, the derivative information can be characterized block by block in accordance with the location vectors. This capacity gives us the additional flexibility to "assemble" the derivative information and, hence, facilitates the handling of the more complicated situation where some position vectors are either fixed or missing. Our framework works for both approximation and completion problems.

Finally, we point out that some recent work in Burer et al. (1999), Burer and Monteiro (2001), Vanderbei and Benson (1999) actually suggests transforming semidefinite programming problems to standard convex programming problems via Choleskey-type factorizations. These approaches allow the possibility of exploiting rank structures. In some sense, this notion is close to ours in that the Euclidean distance matrix approximation problem is written as a smooth nonlinear programming problem as opposed to a semidefinite programming problem.

8.5.2 *Basic formulation*

For generality, let entries in $F \in \mathbb{R}^{n\times n}$ represent henceforth the squares of observed distances among n particles in $\mathbb{R}^m$. As we begin to lay down the groundwork for discussion, we shall assume initially that *all* entries of F except the diagonal are to be approximated. Later we shall modify the basic formulation to tackle two special cases: the approximation problem where some location vectors of the n particles are known and fixed and the completion problem where partial information of spacing among the n particles, but not necessarily from known location vectors, is known and fixed. Each problem requires different modifications and will be handled separately.

Our basic idea of the Euclidean distance matrix approximation of F is to solve the following least squares problem.

Problem 8.5 *(Least squares euclidean distance matrix approximation)* Given a matrix $F \in \mathbb{R}^{n\times n}$, minimize the objective function

$$f(\mathbf{p}_1,\ldots,\mathbf{p}_n) = \frac{1}{2}\|F - [\langle \mathbf{p}_i - \mathbf{p}_j, \mathbf{p}_i - \mathbf{p}_j\rangle]\|_F^2 \tag{8.46}$$

where $\langle\cdot,\cdot\rangle$ denotes the usual Euclidean inner product and $\|\cdot\|_F$ denotes the Frobenius matrix norm.

Thus far, it is obvious that the solution is not unique because any rigid body motion of a given conformation will produce the same relative spacing. Later we shall show how some of the locations must be fixed as reference points, so not every vector in the entire list of $\mathbf{p}_1,\ldots,\mathbf{p}_n$ is to be determined numerically. Note that the function $f(\mathbf{p}_1,\ldots,\mathbf{p}_n)$ is *not* convex. Unless a global optimization technique is applied, generally we will find local solutions only.

We should point out immediately that in a demonstration of its large-scaled unconstrained nonlinear minimization capability, MATLAB features an M-file called MOLECULAR in the Optimization Toolbox that solves a two-dimensional molecule conformation problem. However, that code appears to be loosely organized when compared to our block form. Our formulation easily extends that in MOLECULAR to any general dimensions. More importantly, we offer in the following a highly organized way of computing the derivative information that not only improves and generalizes the coding in MOLECULAR, but also gives us great flexibility in assembling the derivative information location by location.

8.5.3 *Analytic gradient and Hessian*

For convenience, denote

$$d_{ij} = f_{ij} - \langle \mathbf{p}_i - \mathbf{p}_j, \mathbf{p}_i - \mathbf{p}_j\rangle \tag{8.47}$$

for $i, j = 1, \ldots, n$. We shall consider f as a function defined on the space $\mathbb{R}^m \times \cdots \times \mathbb{R}^m$ equipped with the product topology. In such a normed topological space, the Fréchet derivative of f at the point $(\mathbf{p}_1, \ldots, \mathbf{p}_n)$ acting on an arbitrary *n-fold* vector $(\mathbf{z}_1, \ldots, \mathbf{z}_n) \in \mathbb{R}^m \times \cdots \times \mathbb{R}^m$ can be represented as the multilinear functional

$$f'(\mathbf{p}_1, \ldots, \mathbf{p}_n) \cdot (\mathbf{z}_1, \ldots, \mathbf{z}_n) = \sum_{k=1}^{n} \frac{\partial f}{\partial \mathbf{p_k}}(\mathbf{p}_1, \ldots, \mathbf{p}_n) \cdot \mathbf{z}_k \tag{8.48}$$

where $\cdot$ denotes the operator action and the *partial gradient* $\partial f / \partial \mathbf{p_k}$ for each k can be expressed in the following theorem.

Theorem 8.13. For $k = 1, \ldots, n$, the partial gradient $\partial f / \partial \mathbf{p_k}$ is an m-dimensional vector given by

$$\frac{\partial f}{\partial \mathbf{p}_k} = -4 \sum_{\substack{j=1 \\ j \neq k}}^{n} (\mathbf{p}_k - \mathbf{p}_j) d_{kj}. \tag{8.49}$$

Proof By the Riesz representation theorem, the action of the partial gradient can be considered as the inner product

$$\frac{\partial f}{\partial \mathbf{p_k}}(\mathbf{p}_1, \ldots, \mathbf{p}_n) \cdot \mathbf{z}_k = \left\langle \frac{\partial f}{\partial \mathbf{p_k}}(\mathbf{p}_1, \ldots, \mathbf{p}_n), \mathbf{z}_k \right\rangle.$$

Observe that

$$\left\langle \frac{\partial f}{\partial \mathbf{p_k}}, \mathbf{z}_k \right\rangle = \left\langle \frac{\partial}{\partial \mathbf{p}_k} \left(F - [\langle \mathbf{p}_i - \mathbf{p}_j, \mathbf{p}_i - \mathbf{p}_j \rangle]\right) \cdot \mathbf{z}_k, F - [\langle \mathbf{p}_i - \mathbf{p}_j, \mathbf{p}_i - \mathbf{p}_j \rangle] \right\rangle,$$

whereas it is easy to see that the first action on the right-hand side of the above expression results in an $n \times n$ matrix with zero entries everywhere except the k-th row and column. More precisely,

$$\begin{aligned} &\frac{\partial}{\partial \mathbf{p}_k} \left(F - [\langle \mathbf{p}_i - \mathbf{p}_j, \mathbf{p}_i - \mathbf{p}_j \rangle]\right) \cdot \mathbf{z}_k \\ &= -2 \begin{pmatrix} 0 & \ldots & \langle \mathbf{z}_k, \mathbf{p}_k - \mathbf{p}_1 \rangle & & 0 \\ \vdots & & \vdots & & \vdots \\ \langle \mathbf{z}_k, \mathbf{p}_k - \mathbf{p}_1 \rangle & \ldots & 0 & \ldots & \langle \mathbf{z}_k, \mathbf{p}_k - \mathbf{p}_n \rangle \\ & & \vdots & & \\ 0 & & \langle \mathbf{z}_k, \mathbf{p}_k - \mathbf{p}_n \rangle & & 0 \end{pmatrix}. \end{aligned}$$

It follows that

$$\frac{\partial f}{\partial \mathbf{p_k}}(\mathbf{p}_1, \ldots, \mathbf{p}_n) \cdot \mathbf{z}_k = \langle \mathbf{z}_k, -4 \sum_{\substack{j=1 \\ j \neq k}}^{n} (\mathbf{p}_k - \mathbf{p}_j) d_{kj} \rangle$$

and the assertion is proved. □

Using the product topology again, the gradient ∇f of the objective function f can easily be recognized as the n-fold vector

$$\nabla f(\mathbf{p}_1, \dots, \mathbf{p}_n) = \left(\frac{\partial f}{\partial \mathbf{p}_1}, \dots, \frac{\partial f}{\partial \mathbf{p}_n} \right) \in \mathbb{R}^m \times \cdots \times \mathbb{R}^m. \tag{8.50}$$

With the availability of the gradient, any numerical method utilizing the gradient information can now be employed to solve the approximation problem. For instance, the flow $(\mathbf{p}_1(t), \dots, \mathbf{p}_n(t))$ defined by the differential equation

$$\frac{d\mathbf{p}_k}{dt} = 4 \sum_{\substack{j=1 \\ j \neq k}}^{n} (\mathbf{p}_k - \mathbf{p}_j)(f_{kj} - \langle \mathbf{p}_k - \mathbf{p}_j, \mathbf{p}_k - \mathbf{p}_j \rangle), \quad k = 1, \dots, n \tag{8.51}$$

moves in the steepest descent direction to reduce the values for the objective function f. As a descent flow bounded in a neighborhood of F, its limit point must exist and gives a (local) least squares approximation of the given F.

The critical point of $f(\mathbf{p}_1, \dots, \mathbf{p}_n)$ occurs at wherever the gradient vanishes. It will be more effective for finding such a critical point if the second-derivative information of f is available. By interpreting the gradient of f as an n-fold vector of $m \times 1$ blocks, we can formulate the Hessian of f as an $n \times n$ block matrix with $m \times m$ blocks. Let

$$g : \mathbb{R}^m \times \cdots \times \mathbb{R}^m \to \mathbb{R}^m \times \cdots \times \mathbb{R}^m$$

denote the nonlinear function whose k-th *component* $g_k : \mathbb{R}^m \times \cdots \times \mathbb{R}^m \to \mathbb{R}^m$ is precisely

$$g_k(\mathbf{p}_1, \dots, \mathbf{p}_n) = \frac{\partial f}{\partial \mathbf{p}_k}(\mathbf{p}_1, \dots, \mathbf{p}_n). \tag{8.52}$$

Then the Jacobian matrix of g_k constitutes precisely the k-th row block of the Hessian of f. Indeed, we can perform the block-to-block calculation as follows.

Theorem 8.14. For a fixed k and $i = 1, \dots, n$, the (k, i)-block partial Jacobian $\partial g_k / \partial \mathbf{p}_i$ is given by the $m \times m$ matrix

$$\frac{\partial g_k}{\partial \mathbf{p}_i} = \begin{cases} \sum_{j=1, j \neq k}^{n} \left[-4 d_{kj} I_m + 8(\mathbf{p}_k - \mathbf{p}_j)(\mathbf{p}_k - \mathbf{p}_j)^\top \right], & \text{if } i = k; \\ 4 d_{ki} I_m - 8(\mathbf{p}_k - \mathbf{p}_i)(\mathbf{p}_k - \mathbf{p}_i)^\top, & \text{if } i \neq k, \end{cases} \tag{8.53}$$

where I_m is the $m \times m$ identity matrix.

Proof Under the product topology, the Fréchet derivative $g_k'(\mathbf{p}_1, \dots, \mathbf{p}_n)$ acting at an arbitrary n-fold vector $(\mathbf{w}_1, \dots, \mathbf{w}_n) \in \mathbb{R}^m \times \cdots \times \mathbb{R}^m$ can be considered as the block-to-block operation

$$g_k'(\mathbf{p}_1, \dots, \mathbf{p}_n) \cdot (\mathbf{w}_1, \dots, \mathbf{w}_n) = \sum_{i=1}^{n} \frac{\partial g_k}{\partial \mathbf{p}_i} \cdot \mathbf{w}_i.$$

Based on (8.49), if $i = k$, then

$$\frac{\partial g_k}{\partial \mathbf{p}_k}.\mathbf{w}_k = -4\sum_{\substack{j=1\\ j\neq k}}^{n}\left[\mathbf{w}_k d_{kj} - 2(\mathbf{p}_k - \mathbf{p}_j)\langle \mathbf{w}_k, \mathbf{p}_k - \mathbf{p}_j\rangle\right]$$

$$= \sum_{\substack{j=1\\ j\neq k}}^{n}\left[-4d_{kj}I_m + 8(\mathbf{p}_k - \mathbf{p}_j)(\mathbf{p}_k - \mathbf{p}_j)^{\mathsf{T}}\right]\mathbf{w}_k.$$

If $i \neq k$, then

$$\frac{\partial g_k}{\partial \mathbf{p}_i}.\mathbf{w}_i = -4\left[-\mathbf{w}_i d_{ki} - 2(\mathbf{p}_k - \mathbf{p}_i)\langle -\mathbf{w}_i, \mathbf{p}_k - \mathbf{p}_i\rangle\right]$$

$$= \left[4d_{ki}I_m - 8(\mathbf{p}_k - \mathbf{p}_i)(\mathbf{p}_k - \mathbf{p}_i)^{\mathsf{T}}\right]\mathbf{w}_i.$$

The action of the Fréchet derivative of ${g'}_k$ is precisely the multiplication of the Jacobian matrix with the n-fold vector $[\mathbf{w}_1, \ldots, \mathbf{w}_n]^{\mathsf{T}}$. □

8.5.4 *Modification*

In practice, usually some additional information is available in the geometric conformation. For example, the X-ray crystal structure of each of the 20 amino acids in nature is known. Together with the fact that most of the amino acid sequences of our proteins are also known, it is often the case that, once a certain amino acid is known to be present in the protein, a certain block of the matrix F is already predetermined and fixed. The least squares formulation (8.46) should be modified accordingly to reflect this fact. We stress that the derivative information available in block form as described above is particularly convenient for the overall process of assembling the gradient and the Hessian which are essential for almost all efficient optimization technique to take effect. We briefly outline two situations in this section.

Approximation with partially fixed locations It is clear that any rotations, translations, or reflections of a given conformation will produce the same relative spacing and hence the same distance matrix. To shun rotation, translation, or reflection so as to exactly locate the positions of these particles from a given distance matrix, $m + 1$ positions of these n particles in the embedding space $\mathbb{R}^m$ must be specified and bound as reference points. It is possible that some additional location vectors among $\mathbf{p}_1, \ldots, \mathbf{p}_n$ are also known and fixed beforehand. Let $\mathbf{q}$ denote the indices of known location vectors. Then entries f_{ij} of F where both $i, j \in \mathbf{q}$ correspond to the spacing among these known location vectors. These entries of F should be exact and kept constant.

These known and fixed location vectors should not be altered. As such, derivatives at these points should be zero. The block form of derivatives facilitates the task of keeping these known position vectors invariant by simply nullifying any derivative information at the corresponding blocks. On the other hand, any spacing information between particles that has been missed in the original observation can simply be replaced by zero at the corresponding entry in F. Our optimization scheme should automatically fill in the correct spacing information at the end of its iteration.

Completion with partially fixed F In a completion problem, the matrix F represents a partially specified distance matrix. The task is to find the unspecified elements of F so that the completed F is a full distance matrix. The completion problem differs from the above approximation problem in that the specified entries in F do not necessarily correspond to any known location vectors while still being required to remain the same throughout the whole matrix completion process. Before any attempt at completion is taken, one fundamental issue must be mentioned first. That is, the specified entries in F must be consistent by themselves to begin with. If any of these entries also carries an error, then we are facing a more complicated problem of both completion and approximation. To determine whether the specified entries in F are consistent so that F indeed can be completed as a distance matrix is another challenging open problem (Trosset, 1997).

Let Δ denote the index set of those specified entries of F, that is, let

$$\Delta := \{(i, j) \in \mathbb{Z} \times \mathbb{Z} \,|\, d_{ij} = 0\}.$$

A reasonable least squares formulation of the completion problem would be to minimize the same objective function $f(\mathbf{p}_1, \ldots, \mathbf{p}_n)$ as in (8.46), where $d_{ij} = 0$ whenever $(i, j) \in \Delta$, subject to the additional equality constraints

$$\langle \mathbf{p}_i - \mathbf{p}_j, \mathbf{p}_i - \mathbf{p}_j \rangle = f_{ij} \quad \text{for all } (i, j) \in \Delta. \tag{8.54}$$

In other words, the completion problem is now cast as an equality constrained optimization problem.

8.5.5 *Numerical examples*

The formulas of the gradient (8.49) and Hessian (8.53) are valid for arbitrary dimension m. The discussion stipulated in this section can be applied to the most general formulation (8.46). In order that we can visualize the effect of the least squares approximation, we shall demonstrate some numerical examples by limiting the experiments to the cases where $m = 2$ or 3.

Example 8.21. The parametric equation

$$\begin{cases} x = -10\cos(t) - 2\cos(5t) + 15\sin(2t); \\ y = -15\cos(2t) + 10\sin(t) - 2\sin(5t); \qquad 0 \leq t \leq 2\pi \\ z = 10\cos(3t); \end{cases}$$

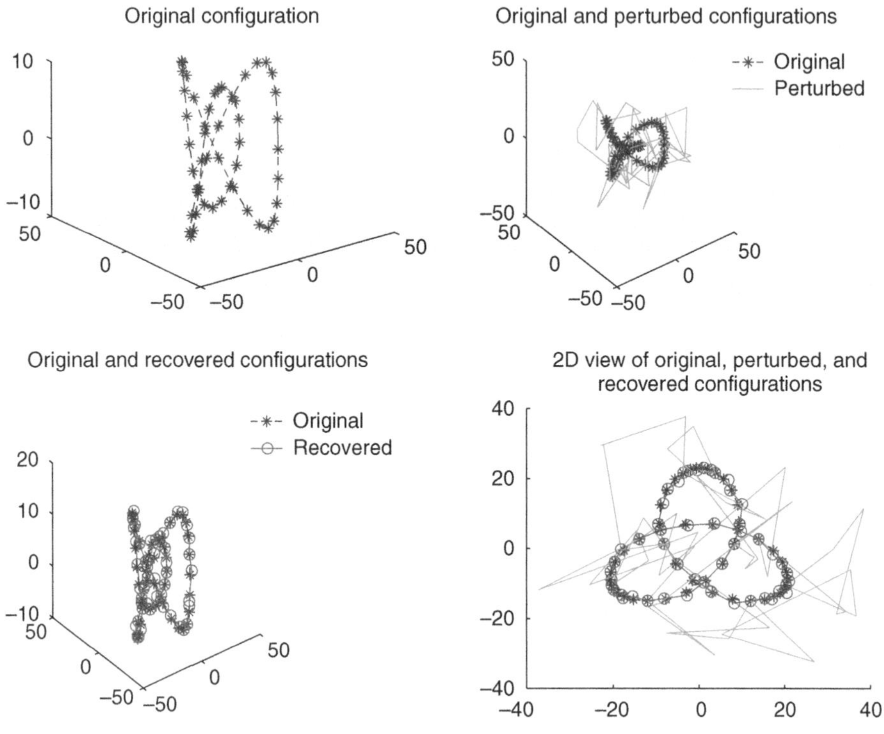

Figure 8.12. *Approximation of a knot*

defines a knot in $\mathbb{R}^3$. See the upper left of Figure 8.12 for a drawing in $\mathbb{R}^3$ and the lower right for its projection onto the xy-plane. To simulate the distance geometry, we represent the knot by finitely many, say n, discrete points marked by "$*$" in Figure 8.12. These n discrete points define a true $n \times n$ distance matrix $Q = [q_{ij}]$. Let $\mathbf{q}$ denote the indices of known location vectors. To avoid rotation, translation, or reflection, the size of $\mathbf{q}$ should be at least four. We perturb the square roots of elements of Q symmetrically at those locations not belonging to $\mathbf{q}$ by random noise with mean zero and variance σ^2, that is, let $F = [f_{ij}]$, where

$$f_{ij} = f_{ji} = \begin{cases} 0, & \text{if } i = j, \\ q_{ij}, & \text{if both } i \text{ and } j \text{ are in } \mathbf{q}, \\ \left(\sqrt{q_{ij}} + \sigma * \mathrm{randn}(1)\right)^2, & \text{if either } i \text{ or } j \text{ is not in } \mathbf{q}, \end{cases}$$

and randn denotes the normally distributed random variable with mean zero and variance one. Then F represents (the squares of) the observed and imperfect spacing among the n location vectors of the knot. The standard deviation σ provides a measurement of how far F is away from a distance matrix. In this example, we take $\sigma = 2$ which suggests that the original distance matrix Q is buried in a significant amount of random noise. We want to find the least squares distance matrix approximation of F and to plot, with the $\mathbf{q}$ location vectors fixed, the resulting configuration.

For illustration purposes, we employ an existing routine FMINUNC in the Optimization Toolbox, Version 2.2, of MATLAB to solve the least squares Problem (8.46), while keeping those location vectors identified by $\mathbf{q}$ fixed throughout the iteration. Obviously, many other software packages can be utilized as well. As the starting value, we perturb each entry of the true location vectors by adding a random variable with uniform distribution over [−20,20]. This would amount to a fairly deviate initial guess. See the upper-right and lower-right drawings in Figure 8.12 for comparison.

Depicted in the lower-left of Figure 8.12 by "○" is the numerical result of the case where $n = 51$ and $\mathbf{q} = 1 : 5 : 51$. Its projection onto the xy-plane is compared with that of the original knot in the lower-right of Figure 8.12. In this particular example, the value of the objective function is reduced from an order of 10^9 to 10^6, indicating that F is far from being a distance matrix. Nevertheless, the drawing in Figure 8.12 shows a remarkable likeness between the recovered and the original configurations.

Example 8.22. Consider the helix defined by

$$\begin{cases} x = 4\cos(3t); \\ y = 4\sin(3t); \quad 0 \leq t \leq 2\pi; \\ z = 2t. \end{cases}$$

We repeat the above experiment except using $n = 101$ and $\mathbf{q} = 1 : 5 : 101$. The original and the recovered configurations are plotted in the upper drawing in Figure 8.13. Although the numerical solution is not as smooth, it captures the essential feature of a helix from afar with a fairly deviate initial guess and a fraudulent F.

To illustrate that the folding of the numerical solution into a helix is not due to the fact that points identified by the current $\mathbf{q}$, i.e., $1 : 5 : 101$, have already roughly outlined a crude helix, we rerun the experiment with $\mathbf{q} = [1 : 3, 99 : 101]$, i.e., with only the first and the last three location vectors fixed. The numerical result is plotted in the lower drawing in Figure 8.13. A close examination shows that "∗" and "○" coincide only at these points and that the computed "helix" deviates further from the true helix than the previous case. However, the feature of spirality is evident.

Example 8.23. The two-dimensional twist defined by

$$\begin{cases} x = t + 3\sin(2t); \\ y = t + 2\sin(3t); \quad -2\pi \leq t \leq 2\pi; \end{cases}$$

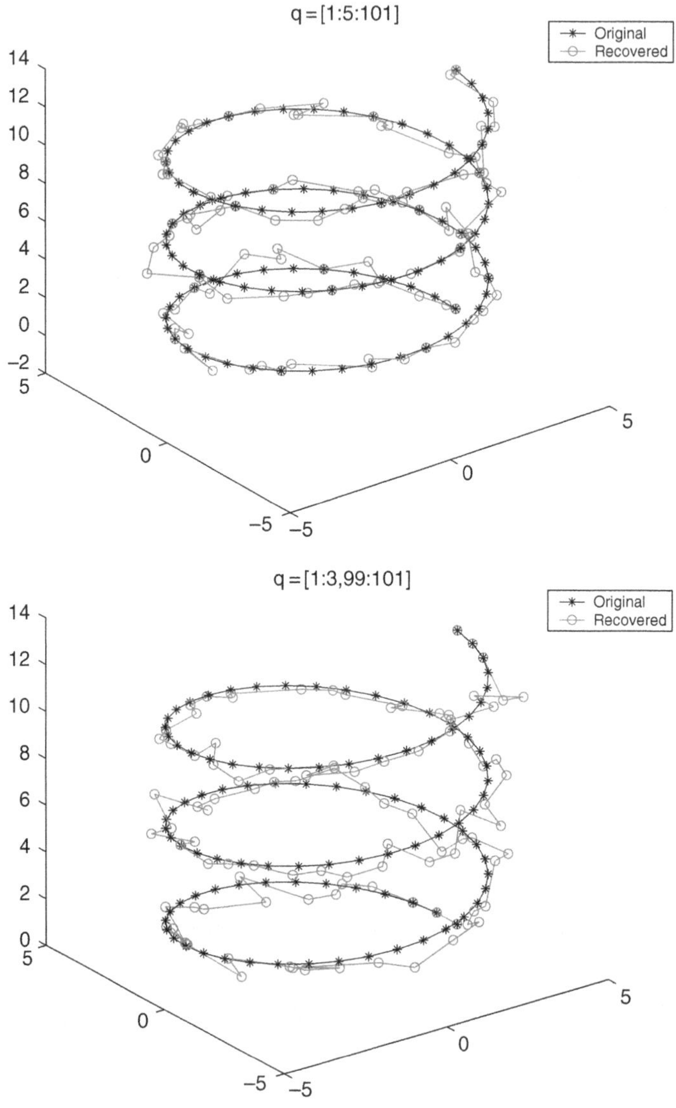

Figure 8.13. *Approximation of a helix*

has many critical turns which are close to each other. A more careful observation of the spacing among the location vectors is necessary. To effect this scenario, we assume a smaller standard deviation $\sigma = 1$ of noise in the simulation. Depicted in Figure 8.14 is the numerical result with $n = 61$ and only the first and last three location vectors are fixed. We find that even with moderately wild initial guesses (not shown but refer to the setting explained in Example 8.21), the numerical

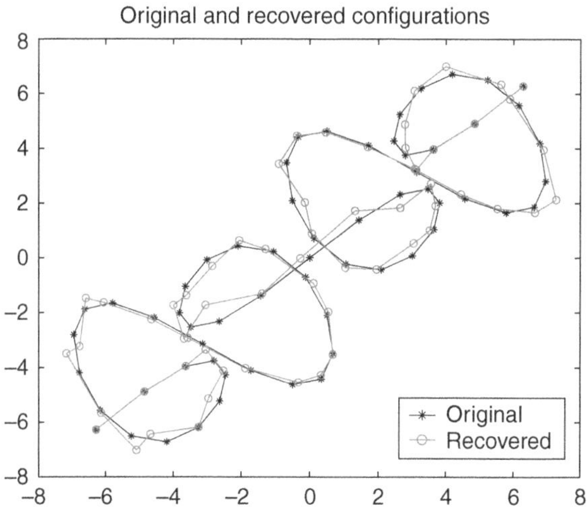

Figure 8.14. *Approximation of a twist*

result is able to pick up the folds of the original curve. Of course, the smaller the σ is and the more known positions are prescribed, the easier and more accurately the least squares approximation can be accomplished.

Example 8.24. Consider the 6×6 partially specified matrix

$$F = \begin{bmatrix} 0 & 3 & 4 & 3 & 4 & 3 \\ 3 & 0 & 1 & x_1 & 5 & x_3 \\ 4 & 1 & 0 & 5 & x_2 & 5 \\ 3 & x_1 & 5 & 0 & 1 & x_4 \\ 4 & 5 & x_2 & 1 & 0 & 5 \\ 3 & x_3 & 5 & x_4 & 5 & 0 \end{bmatrix}$$

where x_i, $i = 1, \ldots, 4$, denotes the unspecified entries. To complete this matrix means to find six location vectors in $\mathbb{R}^m$ whose relative spacing agrees with those already specified in F. It is difficult to tell at a glance at this matrix what the dimension m of the embedding space should be.

It is suggested in Trosset (1997) that no conformation could give rise to this matrix F if $m = 2$. Note that the first 3×3 principal submatrix is completely known, suggesting that three location vectors could have been self-determined. Using $\mathbf{p}_1 = (0,0)$, $\mathbf{p}_2 = (\sqrt{3},0)$, and $\mathbf{p}_3 = (\sqrt{3},1)$ as reference points in $\mathbb{R}^2$, there are eight equality constraints in the form of (8.54) for the remaining three location vectors. If the embedding space is $\mathbb{R}^2$, then the unknown location vectors $\mathbf{p}_j$, $j = 4,5,6$, constitute only six unknowns. There are more constraints than unknowns. It is perhaps for this reason that the overdetermined system has no feasible solution at all in $\mathbb{R}^2$.

Table 8.5. *Examples of entries for completed distance matrix*

x_1	x_2	x_3	x_4
6.6883	3.9512	2.0187	7.3255
1.7434	9.1772	2.2007	2.2006
2.2800	9.4157	2.3913	4.7487
2.7971	5.7203	7.2315	7.2315
2.2723	9.4208	2.3964	4.7398

Table 8.6. *Example of location vectors in* $\mathbb{R}^4$ *for completed distance matrix*

$\mathbf{p}_1$	$\mathbf{p}_2$	$\mathbf{p}_3$	$\mathbf{p}_4$	$\mathbf{p}_5$	$\mathbf{p}_6$
0	1.7321	1.7321	0.9014	0.5774	1.1359
0	0	1.0000	−0.5613	−0.4223	−0.9675
0	0	0	1.3335	1.7980	−0.2711
0	0	0	−0.3067	0.5057	0.8367

Suppose we embed $\mathbf{p}_i$, $i = 1, 2, 3$, in $\mathbb{R}^3$ by adding zeros to their third components and seek to complete the matrix F by location vectors in $\mathbb{R}^3$. We employ an existing routine FMINCON in the Optimization Toolbox of MATLAB by supplying the equality constraints (8.54). Each row in Table 8.5 represents (up to the fifth significant digit) a completed distance matrix of F obtained by a different starting value. It is interesting to note that vectors in Table 8.6 denote a set of location vectors that complete the same matrix F in $\mathbb{R}^4$.

8.5.6 *Summary*

In contrast to the many algorithms developed specifically for the distance matrix approximation problem, we cast the problem under a least squares approximation in terms of the location vectors directly and propose using conventional large-scale optimization techniques instead. We manage the resulting complexity by organizing the gradient and Hessian information in block forms. The matrix calculus makes it particularly easy to assemble the derivatives for existing software packages when some location vectors are known and fixed. Numerical experiments seem to suggest that the conventional methods are efficient and robust in the reconstruction.

8.6 Low rank approximation on unit sphere

This is an era of information and digital technologies. Massive amounts of data are being generated at almost every level of applications in almost every area of disciplines. Quite often the data collected from complex phenomena represent the integrated result of several interrelated variables acting together. When these variables are less precisely defined, it becomes important to distinguish which variable is related to which and how the variables are related before deductive sciences can further be applied. Extracting interesting information from raw data, generally known as data mining, becomes an indispensable task.

Whenever a task of data mining is performed, the practitioners are engaging two notions together consciously or subconsciously. That is, we somehow bear in mind what constitutes "the information" and what information should be regarded as "interesting". Qualifications of information vary from application to application. We could be hunting after, for example, patterns of appearance, association rules between sets of items, clustering of data points, implied concepts or categories, principal components or factors, and so on. The quality that a piece of information is deemed interesting could also vary. Basically, it depends upon, for example, how much confidence and support the retrieved information can furnish, the profundities that the information can contain, the unexpectedness that the information can introduce, and perhaps most important of all, the actionability that the information can offer, that is the ability that the retrieved information can suggest for concrete and profitable decision-making (Berry et al., 1999; Kleinberg et al., 1998).

It is clear that for different information retrievals, different techniques should be used. Some examples in the field include the rank reduction or lower dimension approximation in the context of factor analysis and the k-means method in the context of cluster analysis. What is not clear, and which has led to misuse or inaccurate interpretation in many practices, is that in order to retrieve "credible" information any intentional or unintentional partiality, prejudice, bias, or predisposition should first be removed from the data. The raw data should be preprocessed in some ways, including centering, scaling, and so on before any retrieval technique is applied. The first purpose of this section is to stress this point of necessity by elaborating more upon the preprocessing via the linear model in Section 8.6.1. In particular, we shall use two commonly used data mining techniques, factor retrieval and latent semantic indexing, to exemplify that one kind of standardization process results in rows with unit length.

Generally speaking, data mining involves two major endeavors: to uncover hidden connections and to get the job done fast. Toward that end, it is often necessary to approximate the original (preprocessed) raw data by data that have simpler structure. For various reasons, lower rank is often regarded as a simpler and useful structure. (See, e.g., the discussion in Chu (2000) and the references contained therein for the statistical meaning of the commonly used truncated singular value decomposition.) Obviously, in order not to lose too much authenticity, the reduced data must resemble the algebraic or statistical constituents of the original data in some sense. The second purpose of this section is to consider the computational problem of low rank approximation of data that are originally distributed over the unit sphere. To determine whether a given low rank approximation has retained some of the hereditary characters of the original data, three kinds of cost functions are proposed in this section. The projected gradient method intended to optimize these cost functions is exploited in Section 8.6.2 as a numerical means to find the low rank approximation of data on the unit sphere.

Finally, a modification of the projected gradient method to more compact form is introduced in Section 8.6.3 to reduce the computational cost. Results from some numerical experiments are given in Section 8.6.4 and seem to suggest that the method works reasonably well.

8.6.1 *Linear model*

In this section, we use a fairly general setting of the linear model to demonstrate two important cases where data pre-processing is critical.

Let $Y = [y_{ij}] \in \mathbb{R}^{n \times \ell}$ denote the matrix of "observed" data. We shall be more specific about how the observation is made later. For the time being, let y_{ij} represent, in a broad sense, the *standard score* of variable i by entity j. By a standard score we mean that a raw score per variable has been normalized to have mean 0 and standard deviation 1. Only after this normalization, the matrix

$$R := \frac{1}{\ell} Y Y^{\top}, \tag{8.55}$$

represents the correlation matrix of all n variables. Note that $r_{ii} = 1$ and $|r_{ij}| \leq 1$ for all $i, j = 1, \ldots, n$. In a linear model, it is assumed that the score y_{ij} is a linearly weighted score by entity j based on several factors. We shall temporarily assume that there are m factors, but it is precisely the point that the factors are to be retrieved in the mining process. A linear model, therefore, assumes the relationship

$$Y = AF. \tag{8.56}$$

In the above, $A = [a_{ik}] \in \mathbb{R}^{n \times m}$ is a matrix with a_{ik} denoting loadings of variable i on factor k or, equivalently, the *influence of factor k on variable i*, and $F = [f_{kj}] \in \mathbb{R}^{m \times \ell}$ with f_{kj} denoting the score of factor k by entity j or the *response of entity j to factor k*.

Factor analysis To help better grasp the notion of linear modelling in (8.56), we now demonstrate the first interpretation of the above data. Assume that each of the ℓ columns of the observed matrix Y represents the transcript of a college student (an entity) at his/her freshman year on n fixed subjects (the variables), for example Calculus, English, Chemistry, and so on. It is generally believed that a college freshman's academic performance depends on a number of factors including, for instance, family social status, finance, high school GPA, cultural background, and so on. Upon entering the college, each student could be asked to fill out a questionnaire inquiring these factors of his/her antecedents. In turn, individual responses to those factors are translated into scores and placed in the corresponding column of the scoring matrix F. What is not clear to the educators/administrators is how to choose the factors to compose the questionnaire or how each of the chosen factors would be weighted (the loadings) to reflect their effect on each particular subject. In practice, we usually do not have *a priori* knowledge about the number and character of underlying factors in A. Sometimes we do not even know the factor scores in F. Only the data matrix Y is observable. Explaining the complex phenomena observed in Y with the help of a *minimal number* of factors extracted from the data matrix is the primary and most important goal of factor analysis.

Why is the normalization of raw data Y necessary in this context? Obviously, different professors (even those teaching the same topic) and subjects will have different grading rules and degrees of difficulty. To collectively gather students' academic records together from wide variations for the purpose of analyzing some influential factors in A, it is clear that all these scores should be evaluated on the same basis. Otherwise, the factors cannot be correctly revealed. The same basis of scores means the standardization (of rows) of Y.

It is customary to assume that all sets of factors being considered are uncorrelated with each other. As such, rows of F should be orthogonal to each other. We should further assume that the scores in F for each factor are normalized. Otherwise, one may complain that the factors being considered are not neutral to begin with and the analysis would be biased and misinformed. Under these assumptions, we have

$$\frac{1}{\ell} F F^\top = I_m, \tag{8.57}$$

where I_m stands for the identity matrix in $\mathbb{R}^{m \times m}$. It follows that the correlation matrix R can be expressed directly in terms of the loading matrix A alone, that is,

$$R = AA^\top. \tag{8.58}$$

Factor extraction now becomes a problem of decomposing the correlation matrix R into the product $AA^\top$ using as few factors as possible.

As a whole, the i-th row of A may be interpreted as how the data variable i is weighted across the list of current factors. If the sum of squares of this row, called the *communality* of variable i, is small, it suggests that this specific variable is of little consequence to the current factors. If this happens, either this variable should be dissociated from other variables since it contributes nothing to the analysis or some new factors should be brought in if this variable is required to stay in the pool for analysis. We point out immediately, however, that after the normalization described above, it is always the case that

$$\operatorname{diag}(AA^\top) = I_n. \tag{8.59}$$

For convenience, in the above and henceforth, the notation $\operatorname{diag}(M)$ refers to the diagonal matrix of the matrix M.

On the other hand, the k-th column of A may be interpreted as correlations of the data variables with that particular k-th factor. Those data variables with high factor loadings are considered to be more like the factor in some sense and those with zero or near-zero loadings are treated as being unlike the factor. The quality of this likelihood, which we call the *significance* of the corresponding factor, is measured by the norm of the k-th column of A. One basic idea in factor analysis is to rewrite the loadings of variables over some newly selected factors so as to manifest more clearly the correlation between variables and factors. Suppose the newly selected factors are expressed in terms of columns of the orthogonal matrix

$$V := [\mathbf{v}_1, \ldots, \mathbf{v}_m] \in \mathbb{R}^{m \times m}. \tag{8.60}$$

Then this rewriting of factor loadings with respect to V is mathematically equivalent to a change of basis, that is A is now written as $B := AV$. One of the fundamental questions in the practice of factor analysis is to determine some appropriate new basis for V. Note that because $VV^\top = I_m$, the very same observed data now is decomposed as $Y = AF = (AV)(V^\top F) = BG$ with B and $G = V^\top F$ representing, respectively, the factor loadings and uncorrelated standard factor scores corresponding to the factors in V. From this we also see that the correlation matrix $R = AA^\top = BB^\top \in \mathbb{R}^{n \times n}$ is independent of the factors selected. This is the main reason that in the process of defining new factors it is often desirable to retrieve information directly from the correlation matrix R rather than from any particular loading matrix A.

Latent semantic indexing We now demonstrate the second interpretation of the linear model with the task of finding documents relevant to given queries. The idea in so-called *latent semantic indexing* (LSI) is as follows. The textual documents are usually collected in an *indexing matrix* $H = [h_{ik}]$ in $\mathbb{R}^{n \times m}$. Each document is represented by one row in H. The entry h_{ik} in H represents the *weight* of one particular *term* k in document i whereas each term could be defined by just one single word or a string of phrases. A natural choice of the weight h_{ik} is by counting the number of times that the term k occurs in document i. Many more elaborate weighting schemes that are observed to yield better performance have been discussed in the literature (G. and O'Leary, 1998). For example, to enhance discrimination between various documents and to enhance retrieval effectiveness term-weighting scheme such as

$$h_{ik} = t_{ik} g_k n_i, \tag{8.61}$$

where t_{ik} captures the relative importance of term k in document i, g_k weights the overall importance of term k in the entire set of documents, and

$$n_i = \left(\sum_{k=1}^{m} t_{ik} g_k \right)^{-1/2} \tag{8.62}$$

is the scaling factor for normalization, is commonly used. Each query is represented as a column vector $\mathbf{q}_j^\top = [q_{1j}, \ldots, q_{mj}]$ in $\mathbb{R}^m$ where q_{kj} represents the weight of term k in the query j. Again, the weighting for terms in a query can also use more elaborate schemes. To measure how the query $\mathbf{q}_j$ matches the documents, we calculate the column vector

$$\mathbf{c}_j = H\mathbf{q}_j \tag{8.63}$$

and rank the relevance of documents to $\mathbf{q_j}$ according to the *scores* in $\mathbf{c}_j$.

To put the notation in the context of our discussion in the preceding sections, we observe the follow analogies:

$$\begin{aligned}
\text{indexing matrix } H &\longleftrightarrow \text{loading matrix } A\\
\text{document } i &\longleftrightarrow \text{variable } i\\
\text{term } k &\longleftrightarrow \text{factor } k\\
\text{weight } h_{ik} \text{ of term } k \text{ in document } i &\longleftrightarrow \text{loading of factor } k \text{ on variable } i\\
\text{one query } \mathbf{q}_j &\longleftrightarrow \text{one column in scoring matrix } F\\
\text{weights } q_{kj} \text{ of term } k \text{ in query } \mathbf{q}_j &\longleftrightarrow \text{scores of entity } j \text{ on factor } k\\
\text{scores } \mathbf{c}_j \text{ of the } j\text{-th query} &\longleftrightarrow \text{scores in } j \text{ column of data matrix } Y
\end{aligned}$$

Note that in contrast to the factor retrieval described in the preceding section, the factors are already specified by the predetermined terms. The calculation involved in an LSI application, provided that H is a "reasonable" representation of the relationship between documents and terms, is not so much in computing the factors based on the scores in $\mathbf{c}_j$, $j = 1, \ldots, \ell$, but rather the emphasis is on the vector–matrix multiplication (8.63). Nonetheless, in the search engine application, the matrix H is never exact nor perfect. How to represent the indexing matrix and the queries in a more compact form so as to facilitate the computation of the scores therefore become a major challenge in the field of LSI. The truncated singular value decomposition (TSVD) of H, for example, has been suggested as a replacement of H (Berry et al., 1999). The TSVD is not only the best low rank approximation to H in norm, but more importantly it contains a substantial portion of the stochastic nature of the original H (Chu, 2000). However, we quickly point out an important observation that the normalized TSVD no longer maintains the highest fidelity as does the TSVD, as we shall see from the numerical examples in the sequel.

The idea behind measuring the scores of query $\mathbf{q}_j$ via the product (8.63) warrants some explanation. The i-th entry of $\mathbf{c}_j$ is precisely the inner product of the i-th row of H with $\mathbf{q}_j$. The i-th row of H represents the distribution of all terms over document i. If all rows of H are normalized to the same, say, unit length, then the row that gives rise to the largest value of inner product is the document that is nearest to the query. This matching mechanism is known as the *cosine similarity* in the literature. Other than for the convenience of computing cosine similarities, it is intuitive to see that the normalization is necessary in the context of LSI. If h_{ik} simply represents the frequency of occurrence of term k in document i without normalization, one could artificially inflate the prominence of document i by padding it with repeated pages or volumes. Such trickery could be prevented by normalization. Our point to make is that, similar to (8.59), it should be assumed in any given indexing matrix H that

$$\operatorname{diag}(HH^\top) = I_n. \tag{8.64}$$

Without this normalization, the scores in $\mathbf{c}_j$ is biased and unfair.

One final point is worth mentioning. Ideally, it would be nice to have all terms uncorrelated (orthogonal) to each other, but vocabularies can be ambiguous and indexing is inexact. Indeed, no two authors write the same way and no two people index documents the same way. The matrix H can never be perfect. Considerable research efforts have been devoted to somehow "best" approximating H with some "more simply" structured matrices (Dhillon, 2001; Park et al., 2001). We find it plausible that once we have accumulated enough experiences $C = [\mathbf{c}_1, \ldots, \mathbf{c}_\ell]$ from queries $\mathbf{q}_1, \ldots, \mathbf{q}_\ell$, we should be able to employ the factor retrieval technique described earlier to reconstruct terms (factors). That is, after the standardization of the scores, the information of C may be used as feedback to the process of selecting newer and more suitable terms. Recall from (8.58) that no reference to the queries is needed. This learning process with feedback from C can be iterated to reduce and improve the indexing matrix H. We find the idea of finding the ultimate H, if it exists, via this feedback learning process is interesting and might deserve further study.

8.6.2 *Fidelity of low rank approximation*

The linear model (8.56) can have a much broader interpretation than the two applications we have demonstrated. Its interpretation for factor analysis alone is enough to manifest the critical role that the loading matrix A will play. Data mining in the context of factor analysis means the need to retrieve a few principal factors to represent the original A whose factors might be tangled or not clearly defined. Likewise, the interpretation of (8.56) for the LSI suggests that the indexing matrix H is playing an equally important role as the loading matrix A. Data mining in the context of LSI means the need to derive an approximation of H that allows fast computation. A common problem in data mining under the above framework can be formulated as follows.

Problem 8.6 *(Low rank approximation on unit sphere)*
Given a matrix $A \in \mathbb{R}^{n\times m}$ whose rows are of unit length, find an approximation matrix $Z \in \mathbb{R}^{n\times m}$ of A whose rows are also of unit length, but $\operatorname{rank}(Z) = k$ where $k < \min\{m, n\}$.

In this section, we introduce a numerical procedure that can be employed to solve this algebraically constrained low rank approximation problem.

The quality of the approximation by a lower rank matrix Z to A can be measured in different ways. In the following, we shall describe three kinds of measurements and use them to form our objective functions.

Fidelity test Considering A as the indexing matrix in the LSI, the first way to infer whether Z is a good (but low rank) representation of A is to use documents (rows) in A as queries to inquire about document information from Z. If the document–term relationships embedded in Z adhere to that in A, then the scores of the query $\mathbf{q}_j$ applied to Z, where $\mathbf{q}_j^\top$ is the j-th row of A, should point to the most relevant document j. It should be so for each $j = 1, \ldots, n$. In other words, in the product $ZA^\top$ the highest score per column should occur precisely at the diagonal entry. This notion of self-examination is referred to as the *fidelity test* of Z to A.

Ideally, the fidelity test should product the maximum value 1 along the diagonal. Since the indexing matrix A is never perfect, we might want to consider the collective effect for the sake of easy computation and say that the approximation Z remains in "good" fidelity to the original A if the trace of $ZA^\top$ is sufficiently close to the trace of $AA^\top$, namely, the value n. This low rank approximation problem therefore becomes the problem of maximizing

$$\operatorname{trace}(ZA^\top) = \langle Z, A \rangle, \tag{8.65}$$

where $\langle M, N \rangle$ stands for the Frobenius inner product of matrices M and N in $\mathbb{R}^{n\times m}$. The objective, in short, is to increase the fidelity of the matrix Z to A while maintaining its low rank.

It is difficult to provide an algebraic characterization of the constraint that the matrix Z should have both the specific low rank k and rows of unit length. In this section, we use parameters to carry out the representation. Observe that, given any $X \in \mathbb{R}^{n\times m}$, the rows of the matrix

$$Z = \operatorname{diag}\left(XX^\top\right)^{-1/2} X \tag{8.66}$$

are normalized to unit length. To ensure the low rank condition, we represent X via the parameters (U, S, V) where

$$X = USV^\top \tag{8.67}$$

is the singular value decomposition of X. The task now is equivalent to maximizing the functional

$$F(U, S, V) := \langle \left(\operatorname{diag}\left(USS^\top U^\top\right)\right)^{-1/2} USV^\top, A \rangle, \tag{8.68}$$

subject to the conditions that $U \in \mathcal{O}_n$, $S \in \mathbb{R}^k$, and $V \in \mathcal{O}_m$, where $\mathcal{O}_n$ denotes the group of orthogonal matrices of size $n \times n$. For compatibility, we shall use the same symbol $S \in \mathbb{R}^k$ to denote the "diagonal matrix" in $\mathbb{R}^{n\times m}$ whose first

k diagonal entries are those of S. Observe that by limiting S to $\mathbb{R}^k$, the rank of the matrix $X = USV^\top \in \mathbb{R}^{n\times m}$ is guaranteed to be at most k, and exactly k if S contains no zero entry.

The Fréchet derivative of F at (U, S, V) acting on any $(H, D, K) \in \mathbb{R}^{n\times n} \times \mathbb{R}^k \times \mathbb{R}^{m\times m}$ can be considered as

$$F'(U,S,V)\cdot(H,D,K) = \frac{\partial F}{\partial U}\cdot H + \frac{\partial F}{\partial S}\cdot D + \frac{\partial F}{\partial V}\cdot K, \tag{8.69}$$

where $\Lambda \cdot \eta$ denotes the result of the action by the linear operator Λ on η. For convenience, define

$$\alpha(U,S) := \left(\operatorname{diag}\left(USS^\top U^\top\right)\right)^{-1}. \tag{8.70}$$

We now calculate each action in (8.69) as follows.

First, it is easy to see that

$$\frac{\partial F}{\partial V}\cdot K = \langle \alpha(U,S)^{1/2} USK^\top, A\rangle. \tag{8.71}$$

Using the property that

$$\langle MN, P\rangle = \langle M, PN^\top\rangle = \langle N, M^\top P\rangle \tag{8.72}$$

for any matrices M, N, and P of compatible sizes, it follows from the Riesz representation theorem that, with respect to the Frobenius inner product, the partial gradient can be represented as

$$\frac{\partial F}{\partial V} = A^\top \alpha(U,S)^{1/2} US. \tag{8.73}$$

Observe next that

$$\frac{\partial F}{\partial S}\cdot D = \langle \alpha(U,S)^{1/2} UDV^\top, A\rangle + \left\langle \left(\frac{\partial \alpha(U,S)^{1/2}}{\partial S}\cdot D\right) USV^\top, A\right\rangle, \tag{8.74}$$

whereas by the chain rule we have

$$\begin{aligned}\frac{\partial \alpha(U,S)^{1/2}}{\partial S}\cdot D &= \tfrac{1}{2}\alpha(U,S)^{1/2}(-\alpha(U,S)\operatorname{diag}(UDS^\top U^\top + USD^\top U^\top)\alpha(U,S))\\ &= -\tfrac{1}{2}\alpha(U,S)^{3/2}\operatorname{diag}(UDS^\top U^\top + USD^\top U^\top).\end{aligned} \tag{8.75}$$

In the above, we have used the fact that diagonal matrices are commutative. For convenience, define the diagonal matrix

$$\omega(U,S,V) := \alpha(U,S)^{3/2}\operatorname{diag}(AVS^\top U^\top). \tag{8.76}$$

Taking advantage of the fact that $\langle M, \text{diag}(N)\rangle = \langle \text{diag}(M), \text{diag}(N)\rangle$, we obtain the expression

$$\begin{aligned}\frac{\partial F}{\partial S} \cdot D &= \langle D, U^\top \alpha(U,S)^{1/2} AV\rangle - \tfrac{1}{2}\langle UDS^\top U^\top + USD^\top U^\top, \omega(U,S,V)\rangle \\ &= \langle D, \text{diag}(U^\top \alpha(U,S)^{1/2} AV)\rangle - \langle D, \text{diag}(U^\top \omega(U,S,V)US)\rangle. \qquad (8.77)\end{aligned}$$

We thus conclude that

$$\frac{\partial F}{\partial S} = \text{diag}_k(U^\top \alpha(U,S)^{1/2} AV) - \text{diag}_k(U^\top \omega(U,S,V)US), \qquad (8.78)$$

where, for clarity, we use diag_k to denote the first k diagonal elements and to emphasize that the partial gradient $\partial F/\partial S$ is in fact a vector in $\mathbb{R}^k$.

Finally,

$$\frac{\partial F}{\partial U} \cdot H = \langle \alpha(U,S)^{1/2} HSV^\top, A\rangle + \left\langle \left(\frac{\partial \alpha(U,S)^{1/2}}{\partial U} \cdot H\right) USV^\top, A\right\rangle, \qquad (8.79)$$

whereas

$$\begin{aligned}\frac{\partial \alpha(U,S)^{1/2}}{\partial U} \cdot H &= \tfrac{1}{2}\alpha(U,S)^{1/2}(-\alpha(U,S)\text{diag}(HSS^\top U^\top + USD^\top H^\top)\alpha(U,S)) \\ &= -\tfrac{1}{2}\alpha(U,S)^{3/2}\text{diag}(HSS^\top U^\top + USS^\top H^\top). \qquad (8.80)\end{aligned}$$

Together, we obtain

$$\begin{aligned}\frac{\partial F}{\partial U} \cdot H &= \langle H, \alpha(U,S)^{1/2} AVS^\top\rangle - \tfrac{1}{2}\langle HSS^\top U^\top + USS^\top H^\top, \omega(U,S,V)\rangle \\ &= \langle H, \alpha(U,S)^{1/2} AVS^\top\rangle - \langle H, \omega(U,S,V)USS^\top\rangle, \qquad (8.81)\end{aligned}$$

and hence

$$\frac{\partial F}{\partial U} = \alpha(U,S)^{1/2} AVS^\top - \omega(U,S,V)USS^\top. \qquad (8.82)$$

Formulas (8.73), (8.78), and (8.82) constitute the gradient of the objective function F,

$$\nabla F(U,S,V) = \left\langle \frac{\partial F}{\partial U}, \frac{\partial F}{\partial S}, \frac{\partial F}{\partial V}\right\rangle,$$

in the general ambient space $\mathbb{R}^{n\times n} \times \mathbb{R}^k \times \mathbb{R}^{m\times m}$. For our purpose, it is not enough to merely know the gradient ∇F in the general space, because we are interested in (U,S,V) coming from a special manifold. We have to project this gradient to the manifold $\mathcal{O}_n \times \mathbb{R}^k \times \mathcal{O}_m$.

By taking advantage of the product topology, the tangent space $\mathcal{T}_{(U,S,V)}(\mathcal{O}_n \times \mathbb{R}^k \times \mathcal{O}_m)$ of the product manifold $\mathcal{O}_n \times \mathbb{R}^k \times \mathcal{O}_m$ at $(U, S, V) \in \mathcal{O}_n \times \mathbb{R}^k \times \mathcal{O}_m$ can be decomposed as the product of tangent spaces, that is,

$$\mathcal{T}_{(U,S,V)}(\mathcal{O}_n \times \mathbb{R}^k \times \mathcal{O}_m) = \mathcal{T}_U\mathcal{O}_n \times \mathbb{R}^k \times \mathcal{T}_V\mathcal{O}_m. \tag{8.83}$$

The projection of $\nabla F(U, S, V)$ onto $\mathcal{T}_{(U,S,V)}(\mathcal{O}_n \times \mathbb{R}^k \times \mathcal{O}_m)$, therefore, is the product of the projection of the $\partial F/\partial U$ onto $\mathcal{T}_U\mathcal{O}_n$, the projection of $\partial F/\partial S$ onto $\mathbb{R}^k$, and the projection of $\partial F/\partial V$ onto $\mathcal{T}_V\mathcal{O}_m$. Each of the projections can easily be calculated.

Observe that any matrix $M \in \mathbb{R}^{n\times n}$ has a unique orthogonal splitting,

$$M = U\left\{\tfrac{1}{2}(U^\top M - M^\top U)\right\} + U\left\{\tfrac{1}{2}(U^\top M + M^\top U)\right\}, \tag{8.84}$$

as the sum of elements from the tangent space $\mathcal{T}_U\mathcal{O}_n$ and the normal space $\mathcal{N}_U\mathcal{O}_n$. Thus the projection $\mathcal{P}_{\mathcal{T}_U\mathcal{O}_n}(M)$ of M onto $\mathcal{T}_U\mathcal{O}_n$ is given by the matrix

$$\mathcal{P}_{\mathcal{T}_U\mathcal{O}_n}(M) = U\left\{\tfrac{1}{2}(U^\top M - M^\top U)\right\}. \tag{8.85}$$

Replacing M by $\partial F/\partial U$ in (8.85), we thus obtain the explicit formulation of the projection of $\partial F/\partial U$ onto the tangent space $\mathcal{T}_U\mathcal{O}_n$. Similarly, the projection of $\partial F/\partial V$ onto tangent spaces $\mathcal{T}_V\mathcal{O}_m$ can be calculated. The projection of $\partial F/\partial S$ onto $\mathbb{R}^k$ is just itself. These projections are precisely the so-called projected gradient of the objective function $F(U, S, V)$ on the manifold $\mathcal{O}_n \times \mathbb{R}^k \times \mathcal{O}_m$.

Many iterative methods making use of projected gradient for constrained optimization are available in the literature. What we have derived above leads to an explicit formulation of the projected gradient. This piece of information is valuable and useful to iterative scheme applications. On the other hand, we find it most natural and convenient to use the dynamical system

$$\begin{aligned} \frac{dU}{dt} &= \mathcal{P}_{\mathcal{T}_U\mathcal{O}_n}\left(\frac{\partial F}{\partial U}\right), \\ \frac{dS}{dt} &= \frac{\partial F}{\partial S}, \\ \frac{dV}{dt} &= \mathcal{P}_{\mathcal{T}_V\mathcal{O}_m}\left(\frac{\partial F}{\partial V}\right), \end{aligned} \tag{8.86}$$

where the partial derivative are given by (8.73), (8.78), and (8.82), respectively, to navigate a flow on the manifold $\mathcal{O}_n \times \mathbb{R}^k \times \mathcal{O}_m$. The flow which we shall call the *fidelity flow* for later reference moves along the steepest ascent direction to maximize the objective functional F. Since (8.86) defines an ascent flow by an analytic vector field, it is known by the Łojasiewicz theorem that the flow converges to a single point (U_*, S_*, V_*) of equilibrium at which

$$Z_* = (\operatorname{diag}(U_* S_* S_*^\top U_*^\top))^{-1/2} U_* S_* V_*^\top \tag{8.87}$$

is locally the best rank k approximation to A subject to the constraint of unit row vectors.

Nearness test In the context of LSI, each row of A may be considered as a "distribution" of the corresponding document over the list of terms. In other words, each document is represented by a single point in the term space $\mathbb{R}^m$. It is natural to evaluate the quality of an approximation Z to A by measuring how far Z is from A. The low rank approximation problem therefore becomes the problem of minimizing the functional

$$E(U, S, V) := \tfrac{1}{2}\langle \alpha(U, S)^{1/2} U S V^\top - A, \alpha(U, S)^{1/2} U S V^\top - A \rangle, \tag{8.88}$$

subject to the constraints that $U \in \mathcal{O}_n$, $S \in \mathbb{R}^k$, and $V \in \mathcal{O}_m$. This notion of finding nearest low rank matrices is called the *nearest test*. Nearest low rank approximation problems under different (linear) structures have been considered in (Chu et al., 2003; Dhillon, 2001; Park et al., 2001). By a linear structure, we mean that all matrices of the same structure form an affine subspace. Our structure, defined by the requirement that each row is of unit length, is not a linear structure.

We may carry out a similar calculation of the projected gradient of E as before. However, we quickly point out that

$$\tfrac{1}{2}\langle Z - A, Z - A \rangle = \tfrac{1}{2}(\langle Z, Z \rangle + \langle A, A \rangle) - \langle Z, A \rangle. \tag{8.89}$$

Note that $\langle Z, Z \rangle = \langle A, A \rangle = n$. Therefore, minimizing the functional $E(U, S, V)$ in (8.88) is equivalent to maximizing the functional $F(U, S, V)$ in (8.68). The fidelity test is exactly the same as the nearest test under the Frobenius inner product.

Absolute fidelity test Generally, there is no guarantee that entries of A are nonnegative. In the fidelity test, there is a possibility that cancellations will occur in the trace calculation. To prevent cancellation, we might want to consider maximizing (one half of) the sum of squares of the diagonal elements of $ZA^\top$, that is, the functional

$$G(U, S, V) := \tfrac{1}{2}\langle \operatorname{diag}(\alpha(U, S)^{1/2} U S V^\top A^\top), \operatorname{diag}(\alpha(U, S)^{1/2} U S V^\top A^\top) \rangle, \tag{8.90}$$

subject to the constraints that $U \in \mathcal{O}_n$, $S \in \mathbb{R}^k$, and $V \in \mathcal{O}_m$.

Once again, denote the Fréchet derivative of G at (U, S, V) acting on any $(H, D, K) \in \mathbb{R}^{n\times n} \times \mathbb{R}^k \times \mathbb{R}^{m\times m}$ as

$$G'(U, S, V)\cdot(H, D, K) = \frac{\partial G}{\partial U}\cdot H + \frac{\partial G}{\partial S}\cdot D + \frac{\partial G}{\partial V}\cdot K. \tag{8.91}$$

We can calculate each partial derivative in (8.91) to obtain the gradient ∇G.

For convenience, denote

$$\theta(U, S, V) := \operatorname{diag}(\alpha(U, S)^{1/2} U S V^\top A^\top). \tag{8.92}$$

Observe that $\operatorname{diag}(\cdot)$ is a linear operator, so $\operatorname{diag}' = \operatorname{diag}$. It follows that

$$\frac{\partial G}{\partial V}\cdot K = \langle \alpha(U, S)^{1/2} U S K^\top A^\top, \theta(U, S, V)\rangle. \tag{8.93}$$

Using (8.72) and the fact that diagonal matrices commute, we see that

$$\frac{\partial G}{\partial V} = A^\top \theta(U, S, V)^\top \alpha(U, S)^{1/2} U S = A^\top \alpha(U, S)\operatorname{diag}(U S V^\top A^\top) U S. \tag{8.94}$$

Likewise,

$$\begin{aligned}\frac{\partial G}{\partial S}\cdot D &= \langle \operatorname{diag}(\alpha(U, S)^{1/2} U D V^\top A^\top), \theta(U, S, V)\rangle\\ &\quad + \left\langle \operatorname{diag}\left(\left(\frac{\partial \alpha(U, S)^{1/2}}{\partial S}\cdot D\right) U S V^\top A^\top\right), \theta(U, S, V)\right\rangle,\end{aligned}$$

whereas the action $((\partial\alpha(U, S)^{1/2})/\partial S)\cdot D$ is already calculated in (8.75). Hence,

$$\begin{aligned}\frac{\partial G}{\partial S}\cdot D &= \langle D, \operatorname{diag}(U^\top \alpha(U, S)^{1/2}\theta(U, S, V) A V)\rangle\\ &\quad - \tfrac{1}{2}\langle \operatorname{diag}(U D S^\top U^\top + U S D^\top U^\top),\\ &\quad\ \alpha(U, S)^{3/2}\theta(U, S, V)\operatorname{diag}(A V S^\top U^\top)\rangle\\ &= \langle D, \operatorname{diag}(U^\top \alpha(U, S)^{1/2}\theta(U, S, V) A V)\rangle\\ &\quad - \langle D, \operatorname{diag}(U^\top \alpha(U, S)^{3/2}\theta(U, S, V)\operatorname{diag}(A V S^\top U^\top) U S)\rangle.\end{aligned} \tag{8.95}$$

We thus conclude that

$$\begin{aligned}\frac{\partial G}{\partial S} &= \operatorname{diag}_k(U^\top \alpha(U, S)^{1/2}\theta(U, S, V) A V)\\ &\quad - \operatorname{diag}_k(U^\top \alpha(U, S)^{3/2}\theta(U, S, V)\operatorname{diag}(A V S^\top U^\top) U S).\end{aligned} \tag{8.96}$$

Finally,

$$\frac{\partial G}{\partial U} \cdot H = \langle \mathrm{diag}(\alpha(U,S)^{1/2} H S V^\top A^\top), \theta(U,S,V)\rangle$$
$$+ \left\langle \mathrm{diag}\left(\left(\frac{\partial \alpha(U,S)^{1/2}}{\partial U} \cdot H \right) U S V^\top A^\top \right), \theta(U,S,V) \right\rangle .$$

Substituting the expression of $\partial \alpha(U,S)^{1/2}/\partial U \cdot H$ computed in (8.80) and simplifying, we see that

$$\begin{aligned}\frac{\partial G}{\partial U} \cdot H &= \langle H, \alpha(U,S)^{1/2} \theta(U,S,V) A V S^\top \rangle \\ &\quad - \tfrac{1}{2} \langle \mathrm{diag}\big(H S S^\top U^\top + U S S^\top H^\top\big), \alpha(U,S)^{3/2} \theta(U,S,V) A V S^\top U^\top \rangle \\ &= \langle H, \alpha(U,S)^{1/2} \theta(U,S,V) A V S^\top \rangle \\ &\quad - \langle H, \alpha(U,S)^{3/2} \theta(U,S,V) \mathrm{diag}\big(A V S^\top U^\top\big) U S S^\top \rangle\end{aligned}$$

and obtain

$$\frac{\partial G}{\partial U} = \alpha(U,S)^{1/2} \theta(U,S,V) A V S^\top - \alpha(U,S)^{3/2} \theta(U,S,V) \mathrm{diag}(A V S^\top U^\top) U S S^\top). \tag{8.97}$$

Formulas (8.94), (8.96), and (8.97) constitute the gradient,

$$\nabla G = \left\langle \frac{\partial G}{\partial U}, \frac{\partial G}{\partial S}, \frac{\partial G}{\partial V} \right\rangle ,$$

of the objective function G in the general ambient space $\mathbb{R}^{n \times n} \times \mathbb{R}^k \times \mathbb{R}^{m \times m}$. Similar to the preceding section, we can obtain the projected gradient of G in explicit form by using (8.85). The corresponding projected gradient flow, called the *absolute fidelity flow* for later reference, evolves on the manifold $\mathcal{O}_n \times \mathbb{R}^k \times \mathcal{O}_m$ to maximize $G(U,S,V)$. We shall not bother to prescribe the vector field of the absolute fidelity flow here. Rather, we conclude with a remark about the formulas of the vector fields discussed above. That is, there are many repeated blocks of expressions in the formulas whereas all radicals of matrices acquired involve only diagonal matrices.

8.6.3 *Compact form and Stiefel manifold*

The above discussion provides a basis for computing the low rank approximation. We must notice, however, that a lot of information computed is not needed in defining either X in (8.67) or Z in (8.66). It suffices to know only a portion of columns of $U \in \mathcal{O}_n$ and $V \in \mathcal{O}_m$, particularly when the desirable rank k of X is assumed to be much smaller than m and n. Indeed, we may write X precisely the same as (8.67), but assume that

$$U \in \mathcal{O}(n,k), \quad S \in \mathbb{R}^k, \quad \text{and } V \in \mathcal{O}(m,k), \tag{8.98}$$

where

$$\mathcal{O}(p,q) := \{Q \in \mathbb{R}^{p\times q} | Q^\top Q = I_q\} \tag{8.99}$$

denotes the set of all $p \times q$ real matrices with orthonormal columns. This set, known as the Stiefel manifold, forms a smooth manifold (Stiefel, 1935).

The set $\mathcal{O}(p,q)$ enjoys many properties similar to those of the orthogonal group. In particular, it can be checked that all formulas derived above for the gradients of $F(U,S,V)$ and $G(U,S,V)$ remain valid, even if we restrict U to $\mathcal{O}(n,k)$ and V to $\mathcal{O}(m,k)$. It only remains to compute the projected gradient. We outline in this section some main points for this purpose.

Embedding $\mathcal{O}(p,q)$ in the Euclidean space $\mathbb{R}^{p\times q}$ equipped with the Frobenius inner product, it is easy to see that any vector H in the tangent space $\mathcal{T}_Q\mathcal{O}(p,q)$ is necessarily of the form

$$H = QK + (I_p - QQ^\top)W \tag{8.100}$$

where $K \in \mathbb{R}^{q\times q}$ and $W \in \mathbb{R}^{p\times q}$ are arbitrary, and K is skew-symmetric. Furthermore, the space $\mathbb{R}^{p\times q}$ can be written as the direct sum of three mutually perpendicular subspaces

$$\mathbb{R}^{p\times q} = Q\mathcal{S}(q) \oplus \mathcal{N}(Q^\top) \oplus Q\mathcal{S}(q)^\perp, \tag{8.101}$$

where $\mathcal{S}(q)$ is the subspace of $q \times q$ symmetric matrices, $\mathcal{S}(q)^\perp$ is the subspace of $q \times q$ skew-symmetric matrices, and $\mathcal{N}(Q^\top) := \{X \in \mathbb{R}^{p\times q} | Q^\top X = 0\}$. Any $M \in \mathbb{R}^{p\times q}$ can be uniquely split as

$$M = Q\frac{Q^\top M - M^\top Q}{2} + (I - QQ^\top)M + Q\frac{Q^\top M + M^\top Q}{2}. \tag{8.102}$$

Similar to (8.85), it follows that the projection $\mathcal{P}_{\mathcal{O}(p,q)}(M)$ of $M \in \mathbb{R}^{p\times q}$ onto the tangent space $\mathcal{T}_Q\mathcal{O}(p,q)$ is given by

$$\mathcal{P}_{\mathcal{O}(p,q)}(M) = Q\frac{Q^\top M - M^\top Q}{2} + (I - QQ^\top)M. \tag{8.103}$$

Note that in case $p = q$ so that Q is orthogonal, the second term in (8.103) is identically zero and the above notion is identical to that of (8.85).

Replacing M and $\mathcal{O}(p,q)$ by partial gradients that we have calculated in the preceding sections and corresponding Stiefel subspaces in (8.103), for example, by $\partial F/\partial U$ and $\mathcal{O}(m,k)$, we obtain projected gradients as before except that the dimensionality is much reduced and the computation is much more cost efficient. We remark that $\mathcal{O}(m,k)$ is a subspace of $\mathcal{O}_m$, so the projections of, say, $\partial F/\partial U$ onto $\mathcal{O}(m,k)$ and $\mathcal{O}_m$ are different in general. We also remark that, for a given $A \in \mathbb{R}^{n\times m}$ and a desirable rank k, the gradient flow on the orthogonal group would involve $n^2 + k + m^2$ dependent variables whereas the flow on the Stiefel manifold involves only $(n + m + 1)k$ variables.

8.6.4 *Numerical examples*

In this section we report some experimental results from using the above-mentioned dynamical systems of gradient flows. We have tested projected gradient flows on both the orthogonal group and the Stiefel manifold, but we shall report only the calculation done by the compacted flow on the Stiefel manifold. At the moment, our primary concern is not so much on the efficiency of these methods. Rather, we focus on the *behavior* of the resulting flows from these differential systems. In what follows, we want to study the effect of rank on the fidelity as well the comparison of the two objective functions.

For the purpose of demonstration, we shall employ existing routines in MATLAB as the ODE integrators. It is understood that many other ODE solvers, especially the recently developed geometric integrators, can be used as well. The ODE suite (Shampine and Reichelt, 1997) in MATLAB contains in particular a Klopfenstein–Shampine, quasi-constant step size, stiff system solver **ode15s**. Assuming the original data matrix A is not precise in its own right in practice, high accuracy approximation of A is not needed. We set both local tolerance $AbsTol = RelTol = 10^{-6}$ while maintaining all other parameters at the default values of the MATLAB codes.

The numerical tests have been conducted using matrices A randomly generated and then row-wise normalized. The initial value for each of the dynamical systems is the TSVD of another randomly generated matrix.

Example 8.25. Let A be a 10×7 row-wise normalized random matrix. We measure the fidelity and the absolute fidelity by the ratios

$$r(Z) = \frac{\operatorname{trace}(ZA^\top)}{\operatorname{trace}(AA^\top)} = \frac{\operatorname{trace}(ZA^\top)}{n}, \tag{8.104}$$

$$q(Z) = \frac{\|\operatorname{diag}(ZA^\top)\|_2^2}{\|\operatorname{diag}(AA^\top)\|_2^2} = \frac{\|\operatorname{diag}(ZA^\top)\|_2^2}{n}, \tag{8.105}$$

respectively. The highest attainable value for either $r(Z)$ or $q(Z)$ should be one. To make a distinction between the two gradient flows related to (8.68) and (8.90), we further denote the corresponding solutions by Z_F and Z_G, respectively. Figure 8.15 represents a typical evolution history of the fidelity and the absolute fidelity in t with various rank values. It is seen, for example, that the approximations Z_F and Z_G with rank $k = 4$ can attain about 90% fidelity and 80% absolute fidelity to the original matrix A, respectively. As expected, higher rank approximations in either case yields higher fidelity.

We further compare $r(Z_F)$ with $r(Z_G)$ and $q(Z_F)$ with $q(Z_G)$ for the case $k = 5$ in Figure 8.16. Recall that $Z_F(t)$ and $Z_G(t)$ are gradient flows for different objective functions, (8.68) and (8.90), respectively. It is therefore expected that

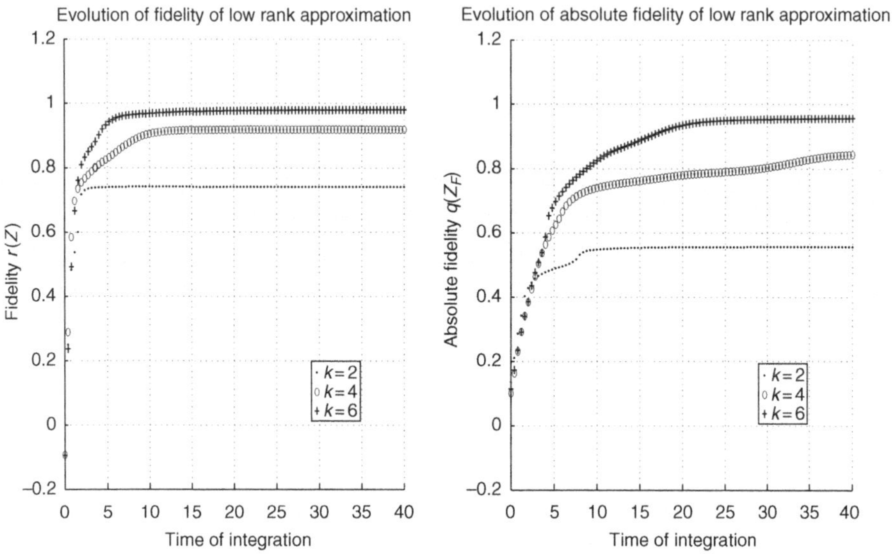

Figure 8.15. *Behavior of fidelity (left) and absolute fidelity (right) with various rank values*

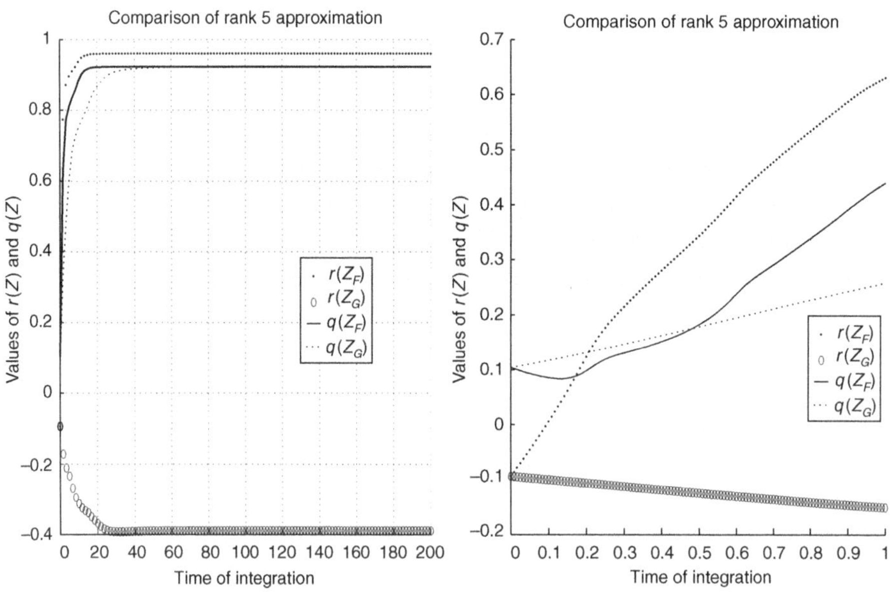

Figure 8.16. *Comparison of cost functions under different dynamical systems when* $k = 5$

$r(Z_F(t))$ and $q(Z_G(t))$ should be monotone increasing. It is interesting to observe that $q(Z_F(t))$ is also monotone increasing in t. Note that each diagonal element in the matrix product $ZA^\top$ has modulus less than or equal to one. Thus, if each diagonal element in Z_F is positive, then necessarily $r(Z_F) \geq q(Z_F)$. On the other hand, the squaring involved in the objective $G(U, S, V)$ cannot guarantee that diagonal elements in Z_G are all of one sign. It is the cancellation that causes the behavior of $r(Z_G(t))$ shown in Figure 8.16. A special notice should be made at this point. It appears in Figure 8.16 that $q(Z_F(t)) \geq q(Z_G(t))$, at least in the transit stage. This appearance is somewhat disturbing because the flow $Z_G(t)$ is made to increase the values of $G(U, S, V)$ more rapidly whereas, in contrast, $Z_F(t)$ is to increase the values of $F(U, S, V)$. A closeup shown in the right drawing of Figure 8.16 indicates that it is indeed the case that

$$\left.\frac{dq(Z_G(t))}{dt}\right|_{t=0} \geq \left.\frac{dq(Z_F(t))}{dt}\right|_{t=0},$$

because $Z_G(0) = Z_F(0)$. However, as integration continues, $Z_G(t)$ departs from $Z_F(t)$ as they follow different dynamical systems. It happens that the superiority of q over r is lost at about $t = 0.5$. Nevertheless, we stress that at convergence it happens that $\lim_{t\to\infty} q(Z_G(t)) \approx 0.923893 \geq \lim_{t\to\infty} q(Z_F(t)) \approx 0.923833$.

For the case $k = 3$, we plot the evolution of the three singular values of the numerical solution during the integration in Figure 8.17. It is important to note that the curves never cross the time line, indicating that the rank is safely maintained throughout the integration.

Finally, we conclude this example by mentioning that we have experimented with many other randomly generated matrices of different and much larger sizes.

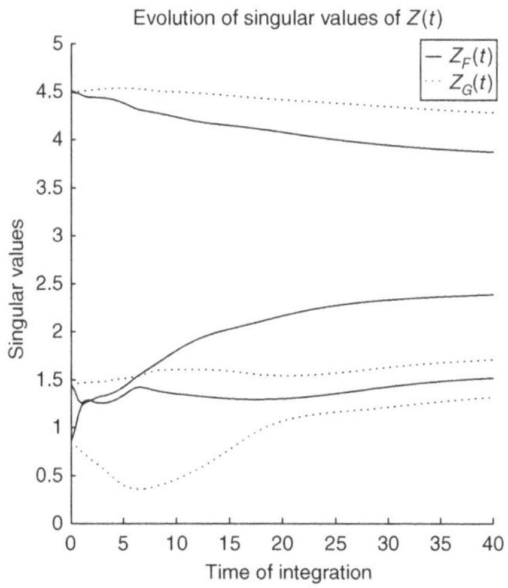

Figure 8.17. *Behavior of the singular values for* $k = 3$

While the computation becomes much more involved, the general behavior of solution flows is observed to be similar.

Example 8.26. A matrix A of size $n \times 3$ can be considered as n points distributed over the unit sphere S^2 in $\mathbb{R}^3$. These points can be conveniently represented in $\mathbb{R}^2$ by their azimuths θ and elevations ϕ. A rank 2 approximation of A with unit row length then can be considered as the intersection of a plane with S^2. That is, rows of the low rank approximation denote points on a big circle. Recall that the fidelity test is equivalent to the nearness test. In the context of nearness test, the approximation problem we are considering therefore can be construed as finding the nearest big circle to the original points on S^2 in the sense of least squares. We find this geometric interpretation interesting.

In the left drawing of Figure 8.18, we randomly generate $n = 120$ uniformly distributed angles $0 \leq \theta \leq 2\pi$ and $0 \leq \phi \leq \pi$. These angles are used as azimuths and elevations to produce the 120×3 matrix A. The dynamical system (8.86) is used to calculate the rank 2 approximation matrix Z_F which is then converted into azimuths and elevations. The resulting more regular pattern (sinuous curve) of the angles for Z_F clearly demonstrate a big circle approximation.

Suppose we limit the random points on S^2 to be in between the cones $\pi/6 \leq \phi \leq \pi/4$. These points are within a circular band in the northern hemisphere. It would be difficult to predict how a big circle would fit these data. In the middle drawing of Figure 8.18, it is interesting to observe how the angles of the best big circle approximation are distributed.

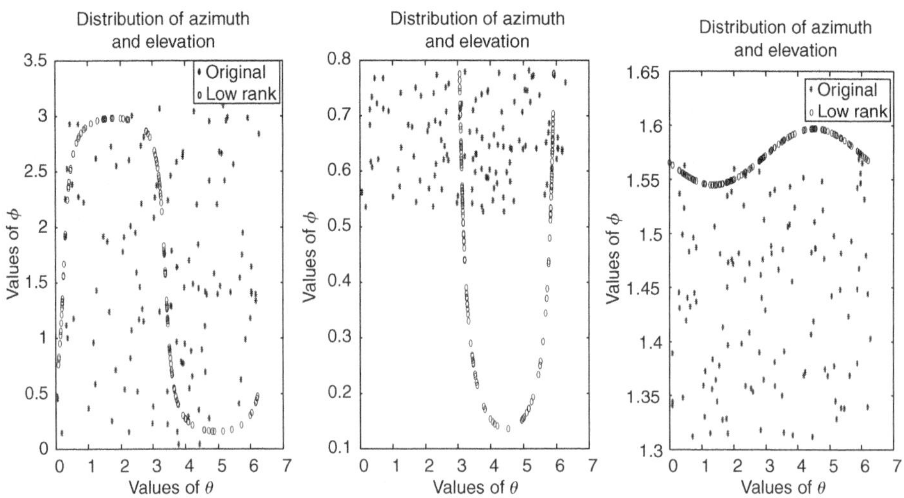

Figure 8.18. *Big circle approximation of points on the unit sphere S^2*

In contrast, suppose we limit the random points to be in between the cones $5\pi/12 \leq \phi \leq \pi/2$. These points are within a circular band close to the unit circle in the xy-plane. We expect that the best fitting big circle should also be nearby. The right drawing of Figure 8.18 clearly indicates that elevations ϕ are in a neighborhood of $\pi/2$ while azimuths θ rotates around.

Example 8.27. Projected gradient flows offer a globally convergent method for tracking down the locally best low rank approximations. To follow any of these flows, an initial value must be provided. This example explores the results by using normalized TSVDs as the starting values.

Given a matrix A and a desirable low rank k, it is known that the (globally) nearest approximation of rank k to A is given by the truncated SVD $A^{(k)}$ of A. It is clear that $A^{(k)}$ generally does not satisfy the normalization condition. Let $Z_0^{(k)}$ denote the row-wise normalized $A^{(k)}$ and $Z_*^{(k)}$ denote the (nearly) limit point of the fidelity flow using $Z_0^{(k)}$ as the starting value. Table 8.7 summarizes the distances $\|A^{(k)} - A\|_F$, $\|Z_0^{(k)} - A\|_F$, $\|Z_0^{(k)} - A\|_F$, and $\|Z_0^{(k)} - Z_*^{(k)}\|_F$ of a 10×10 random matrix A with controlled singular values. It is known that the $(k+1)$-th singular value of A should be precisely equal to $\|A^{(k)} - A\|_2$.

Because all rows of A are of unit length, it is always true that $\sum_{i=1}^{10} \sigma_i^2 = 10$. We control the singular values of A so that five singular values are relatively larger than the other five. From the table it is clear that the difference $\|Z_0^{(k)} - Z_*^{(k)}\|_F$ is relatively significant up to $k = 4$. Since the local tolerance in the ODE integrator is set only at 10^{-6}, the difference between $Z_0^{(k)}$ and $Z_*^{(k)}$ is negligible when $k > 5$. Most importantly, we note from this experiment that the normalized TSVD generally does not maintain the highest fidelity.

Table 8.7. *Comparison of nearness of TSVD and fidelity flow*

Rank k	$\|A^{(k)} - A\|_2$	$\|A^{(k)} - A\|_F$	$\|Z_0^{(k)} - A\|_F$	$\|Z_*^{(k)} - A\|_F$	$\|Z_0^{(k)} - Z_*^{(k)}\|_F$
1	1.5521	2.7426	3.4472	3.3945	0.9227
2	1.4603	2.2612	2.6551	2.5330	2.1866
3	1.3056	1.7264	1.9086	1.8588	0.8085
4	1.1291	1.1296	1.1888	1.1744	0.3341
5	0.0237	0.0343	0.0343	0.0343	5.5969×10^{-7}
6	0.0183	0.0248	0.0248	0.0248	2.1831×10^{-7}
7	0.0141	0.0167	0.0167	0.0167	1.1083×10^{-7}
8	0.0083	0.0089	0.0089	0.0089	6.4180×10^{-9}
9	0.0034	0.0034	0.0034	0.0034	2.7252×10^{-9}

8.6.5 *Summary*

Data mining involves many aspects of work. This section touches upon the issue that under certain circumstances the raw data intrinsically acquiesce to some normalization conditions. Through a linear model we want to demonstrate that the loading matrix in factor analysis and the indexing matrix in latent semantic indexing should have rows (or columns, depending upon how the matrix is positioned) of unit length. Any representation or approximations of these normalized raw data should inherit similar properties. As such, the problem of low rank approximation subject to the normalization condition is considered. We have described, after properly parameterizing the feasible set, explicit formulas for the project gradients and whence a numerical means to compute the best approximation can be proposed.

The capacity of the proposed differential equation approach is limited only by the current development of numerical ODEs. It can handle matrices of sizes up to hundreds within a reasonable frame of time, say, a few seconds. However, it may not be efficient enough to handle very large-scale problem which often is the case in real applications. Even so, it is our hope that this study will have shed some light on the interesting low rank approximation problem. The availability of a projected gradient in explicit form makes it possible to apply other optimization techniques including iterative methods. It might be worth further investigation in this direction.

8.7 Low rank nonnegative factorization

The linear model (8.56) described in Section 8.6.1 has many other interpretations. Within the air pollution research community, for example, the *receptor model* is an observational technique which makes use of the ambient data and source profile data to apportion sources or source categories. The fundamental principle in this model is that mass conservation can be assumed and a mass balance analysis can be used to identify and apportion sources of airborne particulate matter in the atmosphere. One approach to obtaining a data set for receptor modelling is to determine a large number of chemical constituents such as elemental concentrations in a number of samples. The relationships between k sources which contribute m chemical species to n samples, therefore, lead to a mass balance equation,

$$y_{ij} = \sum_{\ell=1}^{k} a_{i\ell} f_{\ell j}, \tag{8.106}$$

where y_{ij} is the elemental concentration of the i-th chemical measured in the j-th sample, $a_{i\ell}$ is the gravimetric concentration of the i-th chemical in the k-th source, and $f_{\ell j}$ is the airborne mass concentration that the k-th source has contributed to the j-th sample. In a typical scenario, only values of y_{ij} are observable whereas neither the sources are known nor the compositions of the local particulate emissions are measured. Thus, a critical question is to estimate the number p, the compositions $a_{i\ell}$, and the contributions $f_{\ell j}$ of the sources. Tools that have been employed to analyze the linear model include principal component analysis, factor analysis, cluster analysis, and other multivariate statistical techniques. In this modelling, however, there is a physical constraint imposed upon the data. That is, the source compositions $a_{i\ell}$ and the source contributions $f_{\ell j}$ must all be nonnegative. The identification and apportionment, therefore, becomes a nonnegative matrix factorization problem of Y. Classical tools cannot guarantee to maintain the nonnegativity.

As another example, in a recent article (Lee and Seung, 1999) the notion of linear model is proposed as a way to find a set of basis functions for representing nonnegative data. It is argued that the notion is particularly applicable to image articulation libraries made up of images showing a composite object in many articulations and poses. It is suggested that the factorization would enable the identification and classification of intrinsic "parts" that make up the object being imaged by multiple observations. More specifically, each column $\mathbf{y}_j$ of a nonnegative matrix Y now represents m pixel values of one image. The columns $\mathbf{a}_k$ of A are basis elements in $\mathbb{R}^m$. The columns of F, belonging to $\mathbb{R}^k$, can be thought of as coefficient sequences representing the n images in the basis elements. In other words, the relationship,

$$\mathbf{y}_j = \sum_{\ell=1}^{k} \mathbf{a}_\ell f_{\ell j}, \tag{8.107}$$

can be thought of as that there are *standard parts* $\mathbf{a}_\ell$ in a variety of positions and that each image $\mathbf{y}_j$ is made by superposing these parts together in certain ways. Those parts, being images themselves, are necessarily nonnegative. The superposition coefficients, each part being present or absent, are also necessarily nonnegative.

In either case above, we see that the p factors, interpreted as either the sources or the basis elements, play a vital role. In practice, there is a need to determine as fewer factors as possible and, hence, a low rank nonnegative matrix factorization (**NNMF**) of the data matrix Y. The mathematical problem can be stated as follows.

Problem 8.7 *(NNMF)*
Given a nonnegative matrix $Y \in \mathbb{R}^{m\times n}$ and a positive integer $p < \min\{m, n\}$, find nonnegative matrices $U \in \mathbb{R}^{m\times k}$ and $V \in \mathbb{R}^{k\times n}$ so as to minimize the functional

$$f(U, V) := \tfrac{1}{2}\|Y - UV\|_F^2. \tag{8.108}$$

The product UV is called a nonnegative matrix factorization of Y. Clearly the product UV is of rank at most p.

It is mentioned in the survey (Tropp, 2003) that, despite the many algorithms proposed in the literature for the NNMF, they generally seem to lack a firm theoretical foundation. The difficulty lies in the fact that nonnegative matrices form a cone with many facets which make it hard to characterize which and when a facet is active or not in the optimization. Indeed, Donoho and Stodden (2003) interpret the NNMF geometrically as the problem of finding a simplicial cone which contains a cloud of data points and which is contained in the positive orthant.

In this section we briefly describe a first-order optimality condition which can be regarded as the Kuhn–Tucker condition in closed form. We then discuss some general ideas for solving the NNMF numerically. Some of the approaches can automatically detect which facet is active along the integration while others have apparent simplicity for computation. We also want to demonstrate by some real-world examples the limitation and difficulty in interpreting the NNMF results.

8.7.1 *First-order optimality condition*

The cone $\mathbb{R}_+^{m\times p}$ of nonnegative matrices in $\mathbb{R}^{m\times k}$ can be written as

$$\mathbb{R}_+^{m\times p} = \{E \circ E | E \in \mathbb{R}^{m\times k}\}, \tag{8.109}$$

where $M \circ N = [m_{ij}n_{ij}]$ denotes the Hadamard product of two matrices. In a sense, the expression $E \circ E$ is one way to parameterize nonnegative matrices over the open set $\mathbb{R}^{m\times k}$. Indeed, it is a good one because the parametrization is differentiable and the problem of nonnegative matrix factorization can now be expressed as the minimization of

$$g(E, F) := \tfrac{1}{2}\|Y - (E \circ E)(F \circ F)\|_F^2 \tag{8.110}$$

where variables (E, F) are from the open set $\mathbb{R}^{m\times k} \times \mathbb{R}^{k\times n}$. This parametrization effectively transformed the constrained optimization over the cones into a problem with no constraint at all.

Consider g as a differentiable functional over the space $\mathbb{R}^{m\times k} \times \mathbb{R}^{k\times n}$ with Frobenius inner product,

$$\langle (X_1, Y_1), (X_2, Y_2)\rangle = \langle X_1, X_2\rangle + \langle Y_1, Y_2\rangle, \tag{8.111}$$

whenever $X_1, X_2 \in \mathbb{R}^{m\times k}$ and $Y_1, Y_2 \in \mathbb{R}^{k\times n}$. The Fréchet derivative of g therefore can be calculated component by component. In particular, the partial derivative of g with respect to E acting on an arbitrary $H \in \mathbb{R}^{m\times k}$ is given by

$$\begin{aligned}\frac{\partial g}{\partial E} \cdot H &= \langle -(H \circ E + E \circ H)(F \circ F), \alpha(E, F)\rangle \\ &= \langle -2H, E \circ (\alpha(E, F)(F \circ F)^\top)\rangle,\end{aligned} \tag{8.112}$$

where, for convenience, we have adopted the notation

$$\alpha(E, F) := Y - (E \circ E)(F \circ F). \tag{8.113}$$

Similarly, the partial derivative of g with respect to F acting on an arbitrary $K \in \mathbb{R}^{k \times n}$ is given by

$$\frac{\partial g}{\partial F}.K = \langle -2K, F \circ ((E \circ E)^\top \alpha(E, F)) \rangle. \tag{8.114}$$

By the Riesz representation theorem, the gradient of g at (E, F) can be expressed as

$$\nabla g(E, F) = (-2E \circ (\alpha(E, F)(F \circ F)^\top), -2F \circ ((E \circ E)^\top \alpha(E, F))). \tag{8.115}$$

We are now ready to characterize the first-order optimality condition for the nonnegative matrix factorization problem as follows.

Theorem 8.15. If (E, F) is a local minimizer of the objective functional g in (8.110), then necessarily the equations

$$E \circ (\alpha(E, F)(F \circ F)^\top) = 0 \in \mathbb{R}^{m \times k}, \tag{8.116}$$

$$F \circ ((E \circ E)^\top \alpha(E, F)) = 0 \in \mathbb{R}^{k \times n}. \tag{8.117}$$

are satisfied. The optimal solution to the nonnegative matrix factorization problem is given by $U = E \circ E$ and $V = F \circ F$.

Corollary 8.16. The necessary condition for $(U, V) \in \mathbb{R}_+^{m \times p} \times \mathbb{R}_+^{p \times n}$ to solve the nonnegative matrix factorization problem is

$$U \cdot *((Y - UV)V^\top) = 0 \in \mathbb{R}^{m \times p}, \tag{8.118}$$

$$V \cdot *(U^\top (Y - UV)) = 0 \in \mathbb{R}^{p \times n}. \tag{8.119}$$

It is not surprising to see that the complementarity condition of zeros in (8.118) and (8.119) effectively characterize the Kuhn–Tucker conditions for the minimization of (8.108) subject to the nonnegative constraint (Lawson and Hanson, 1995). Indeed, the following inequalities are also necessary.

Corollary 8.17. The two matrices $-(Y - UV)V^\top$ and $-U^\top(Y - UV)$ are precisely the Lagrangian multipliers specified in the Kuhn–Tucker condition. At a solution (U, V) of the nonnegative matrix factorization problem, it is necessary that

$$(Y - UV)V^\top \leq 0, \tag{8.120}$$

$$U^\top (Y - UV) \leq 0. \tag{8.121}$$

8.7.2 *Numerical methods*

Several existing numerical methods for solving the NNMF have already been reviewed in (Liu and Yi, 2003; Tropp, 2003). Although schemes and approaches are different, any numerical method is essentially centered around satisfying the first-order optimality condition. That is, at the local solution either the nonlinear systems (8.116) and (8.117) for general E and F or the systems (8.118) and (8.119) for nonnegative U and V must be satisfied. It should be cautioned, however, that merely satisfying the first-order optimality condition is not enough to guarantee that the critical point be a minimizer. Various additional mechanisms, such as the Hessian information or some descent properties, are built into the different schemes to ensure that a critical point is a solution to (8.108). We outline some general ideas, old and new, for solving the NNMF in this section.

Constrained quasi-Newton methods The Kuhn–Tucker conditions form the basis for many nonlinear programming algorithms. The article (Hock and Schittkowski, 1983), for example, compares the performance of 27 computer codes which are all designed to solve the general constrained nonlinear optimization problem. Among these, one of the most efficient techniques is the Sequential Quadratic Programming (SQP) method. The SQP methods solve successively a sequence of quadratic programming subproblems obtained by linearizing the original nonlinear problems at various approximate solutions. One unique feature in the SQP methods is that they accumulate second-order information via a quasi-Newton updating procedure. For that reason, the SQP methods are also known as constrained quasi-Newton methods (Lawson and Hanson, 1995). There are many established results concerning the SQP technique, including its superlinear convergence. An overview of the SQP methods can be found in Fletcher (1987) and Gill et al. (1981).

We shall not discuss this often very elaborate SQP implementation here. We only point out that when applied to the NNMF problem, the Kuhn–Tucker conditions are explicitly given by (8.118), (8.119), (8.120), and (8.121). We believe that the SQP methods can take further advantage of the underlying structure in a similar way as the ADI Newton method which we outline below.

ADI Newton iteration The matrix size involved in the NNMF problem is usually very large. Trying to tackle the system directly would be quite computationally extensive, if possible at all. One possible approach is to alternate between U and V by fixing the other. This idea of alternating direction iteration (ADI) has been used in many applications.

In our situation, we may start by fixing V in (8.118) and solve the system,

$$U \cdot *[B - UC] = 0, \tag{8.122}$$

for a nonnegative matrix $U \in \mathbb{R}_+^{m\times p}$, where both $B = YV^\top \in \mathbb{R}^{m\times p}$ and $C = VV^\top \in \mathbb{R}^{p\times p}$ are fixed and nonnegative matrices. We then fix U and next solve the system

$$V \cdot *[R - SV] = 0, \tag{8.123}$$

for a nonnegative matrix $V \in \mathbb{R}_+^{p\times n}$ with fixed $R = U^\top Y \in \mathbb{R}^{p\times n}$ and $S = U^\top U \in \mathbf{R}^{p\times p}$. We call this one sweep of the *outer loop iteration.* Note that because p is low, the sizes of the square matrices C and S are relatively small.

It is obvious that merely taking $U = BC^{-1}$ is not good enough for (8.122) because U could have negative entries. Extra effort is needed to satisfy the complementary condition in Corollary 8.16 and the inequalities in Corollary 8.17. Also, it is not clear under what conditions the outer loop iteration will converge. We suggest two schemes to tackle some of the concerns.

In the first scheme, consider the fact that the objective function (8.108) is separable in columns. For a fixed $U \in \mathbb{R}^{m\times p}$, each single column of V amounts to a least squares minimization for an objective function of the form

$$\phi(\mathbf{v}) = \tfrac{1}{2}\|\mathbf{y} - U\mathbf{v}\|_2^2, \tag{8.124}$$

subject to the constraint that $\mathbf{v} \in \mathbb{R}^p$ is nonnegative. Such a nonnegative least squares problem has been studied extensively in the literature. For example, the MATLAB routine LSQNONNEG using a scheme that essentially is a projected Newton method (Lawson and Hanson, 1995, Chapter 23) is readily available for exactly this type of least squares problem. Alternating between U and V and employing the projected Newton method or the existing LSQNONNEG for each column of V and each row of U, we now have another numerical method for the NNMF problem.

In the second scheme, consider that the solution U in the matrix equation (8.122) could be solved row by row. Note that each row of U is not related to any other rows. Each *row* gives rise to a nonlinear system of equations of the form

$$\mathbf{u}^\top \cdot *[\mathbf{b}^\top - \mathbf{u}^\top C] = \mathbf{0}. \tag{8.125}$$

Likewise, if U is fixed, then V in (8.123) can be solved column by column in a similar manner. Though there are m rows for U and n columns for V to be solved, respectively, note that the coefficient matrices involved is either $C = VV^\top$ or $S = U^\top U$. These coefficient matrices need to be updated once per sweep of the outer loop iteration and, more importantly, are of the much smaller size $p \times p$.

To guarantee the nonnegativity of $\mathbf{u}^\top$, we rewrite (8.125) as the equation

$$\psi(\mathbf{e}) = (C(\mathbf{e} \cdot * \mathbf{e}) - \mathbf{b}) \cdot * \mathbf{e} = \mathbf{0}, \tag{8.126}$$

by the fact that C is symmetric and by taking $\mathbf{e} \cdot * \mathbf{e} = \mathbf{u}$ with $\mathbf{e} \in \mathbb{R}^p$. It is easy to see that the Frèchet derivative of ψ acting on an arbitrary vector $\mathbf{h} \in \mathbb{R}^p$ can be calculated as

$$\psi'(\mathbf{e}) \cdot \mathbf{h} = 2(C(\mathbf{e} \cdot * \mathbf{h})) \cdot * \mathbf{e} + (C(\mathbf{e} \cdot * \mathbf{e}) - \mathbf{b}) \cdot * \mathbf{h}. \tag{8.127}$$

This expression is equivalent to the matrix–vector multiplication

$$\psi'(\mathbf{e}) \cdot \mathbf{h} = \{2\text{diag}(\mathbf{e})C\text{diag}(\mathbf{e}) + \text{diag}(C(\mathbf{e} \cdot *\mathbf{e}) - \mathbf{b})\}\,\mathbf{h}. \tag{8.128}$$

In other words, we have calculated the Jacobian matrix of ψ. A standard Newton iteration scheme can be applied now to solve $\psi(\mathbf{e}) = 0$ as follows.

Algorithm 8.4 (ADI Newton iteration for NNMF)
Given $\mathbf{e}^{(0)}$ such that $C(\mathbf{e}^{(0)} \cdot * \, \mathbf{e}^{(0)}) - \mathbf{b} \geq 0$, do the following for $k = 0, 1, \ldots$ until convergence:

1. Compute $\mathbf{r}^{(k)} = C(\mathbf{e}^{(k)}. * \mathbf{e}^{(k)}) - \mathbf{b}$.
2. Solve for $\mathbf{h}$ from the linear system

$$\left\{2\text{diag}(\mathbf{e}^{(k)})C\text{diag}(\mathbf{e}^{(k)}) + \text{diag}(\mathbf{r}^{(k)})\right\}\mathbf{h} = -\mathbf{r}^{(k)}. * \mathbf{e}^{(k)}. \tag{8.129}$$

3. Update $\mathbf{e}^{(k+1)} = \mathbf{e}^{(k)} + \alpha^{(k)}\mathbf{h}$.

With $|\mathbf{e}^0|$ large enough, the step size $\alpha^{(k)}$ is adapted so as to maintain $\mathbf{r}^{(k)} \geq 0$ for all k. Obviously, at convergence, the row vector $\mathbf{u} = \mathbf{e} \cdot *\mathbf{e}$ is a nonnegative solution to (8.125). Repeating the process for each row of U (and indeed these rows can be processed in parallel), we obtain a nonnegative solution U to (8.118) which also satisfies the inequality requirement (8.120), for each fixed V. Exchanging the roles of U and V, we can obtain a nonnegative solution V to (8.119), for each fixed U. This completes one sweep of the outer loop iteration.

We stress again that it is not clear under what conditions the outer loop iteration converges. Even if the outer loop converges, we have to point out that the Newton iteration only finds critical points satisfying the first-order optimality condition. The iteration does not distinguish a minimizer from a maximizer unless additional information is brought in. It is possible that the iteration converges to, for example, a saddle point.

Reduced quadratic model In contrast to the Newton-type approach outlined above, the notion of reduced quadratic model approach is very simple to use. Similar to the SQP methods where the original nonlinear programming problem is approximated by a sequence of quadratic programming subproblems, the idea is to replace the quadratic function $\phi(\mathbf{v})$ defined in (8.124) by a sequence of simpler quadratic functions. More specifically, near any given $\mathbf{v}^c$, the quadratic function $\phi(\mathbf{v})$ which can be rewritten as

$$\phi(\mathbf{v}) = \phi(\mathbf{v}^c) + (\mathbf{v} - \mathbf{v}^c)^\top \nabla\phi(\mathbf{v}^c) + \tfrac{1}{2}(\mathbf{v} - \mathbf{v}^c)^\top U^\top U(\mathbf{v} - \mathbf{v}^c) \tag{8.130}$$

is approximated by a *simpler* quadratic model of the form

$$\varphi(\mathbf{v}; \mathbf{v}^c) = \phi(\mathbf{v}^c) + (\mathbf{v} - \mathbf{v}^c)^\top \nabla\phi(\mathbf{v}^c) + \tfrac{1}{2}(\mathbf{v} - \mathbf{v}^c)^\top D(\mathbf{v}^c)(\mathbf{v} - \mathbf{v}^c), \tag{8.131}$$

where $D(\mathbf{v}^c)$ is a diagonal matrix depending on $\mathbf{v}^c$. The minimizer of $\phi(\mathbf{v})$ is approximated by the minimizer $\mathbf{v}^+$ of $\varphi(\mathbf{v};\mathbf{v}^c)$, near which a new quadratic model is created. The definition of $D(\mathbf{v}^{(c)})$ is quite intriguing. We describe below two possible models.

Let the entries be denoted by $\mathbf{v}^c = [v_i^c] \in \mathbb{R}^p$, $D(\mathbf{v}^c) = \operatorname{diag}\{d_1(\mathbf{v}^c), \ldots, d_p(\mathbf{v}^c)\}$, and so on. The first model introduced by Lee and Seung (2001) defines the diagonal entries by

$$d_i(\mathbf{v}^c) := \frac{(U^\top U\mathbf{v}^c)_i}{v_i^c}, \quad i = 1, \ldots, p. \tag{8.132}$$

It is important to observe that four consequences follow from this choice of $D(\mathbf{v}^{(c)})$. First, it can be shown that (Lee and Seung, 2001)

$$(\mathbf{v} - \mathbf{v}^c)^\top \left(D(\mathbf{v}^c) - U^\top U\right)(\mathbf{v} - \mathbf{v}^c) \geq 0 \tag{8.133}$$

for all $\mathbf{v}$. In other words, the matrix $D(\mathbf{v}^c) - U^\top U$ is positive semidefinite, implying that φ dominates ϕ in the sense that $\phi(\mathbf{v}) \leq \varphi(\mathbf{v};\mathbf{v}^c)$ for all $\mathbf{v}$. Secondly, the minimum of any quadratic function always has a closed form solution, but with $D(\mathbf{v}^c)$ being diagonal the close form solution is easy. In fact, the minimum $\mathbf{v}^+$ of $\varphi(\mathbf{v};\mathbf{v}^c)$ is given by

$$\mathbf{v}^+ := \mathbf{v}^c - D^{-1}(\mathbf{v}^c)(U^\top U\mathbf{v}^c - U^\top\mathbf{y}). \tag{8.134}$$

Thirdly, note from the definition of $D(\mathbf{v}^c)$ that the entries of $\mathbf{v}^+$ are precisely

$$v_i^+ = v_i^c \frac{(U^\top\mathbf{y})_i}{(U^\top U\mathbf{v}^c)_i}, \quad i = 1, \ldots, p, \tag{8.135}$$

and, hence, remain nonnegative if $\mathbf{v}^c$ is nonnegative. Finally, it is important to note that

$$\phi(\mathbf{v}^+) \leq \varphi(\mathbf{v}^+;\mathbf{v}^c) \leq \varphi(\mathbf{v}^c;\mathbf{v}^c) = \phi(\mathbf{v}^c), \tag{8.136}$$

showing that $\mathbf{v}^+$ is an improved update from $\mathbf{v}^c$.

Repeating the above process for each individual column and assembling all columns together, the updated matrix $V^+ = [v_{ij}^+]$ for (8.108) from a given nonnegative matrix $V^c = [v_{ij}^c]$ and a fixed nonnegative matrix U can be defined by the multiplicative rule:

$$v_{ij}^+ := v_{ij}^c \frac{(U^\top Y)_{ij}}{(U^\top UV^c)_{ij}}, \quad i = 1, \ldots, p, \; j = 1, \ldots, n. \tag{8.137}$$

In terms of the element-by-element multiplication $\cdot *$ and division $\cdot /$, the relationship (8.137) can simply be written as

$$V^+ := V^c \cdot * \, (U^\top Y) \cdot /(U^\top U V^c). \tag{8.138}$$

In a similar way, the update $U^+ = [u_{ij}^+]$ for (8.108) from a given nonnegative matrix $U^c = [u_{ij}^c]$ and a fixed nonnegative matrix V can be defined by the rule:

$$U^+ := U^c \cdot *(Y V^\top) \cdot /(U^c V V^\top). \tag{8.139}$$

Alternating these multiplicative update rules between U and V has been proposed in Lee and Seung (2001) as a means of solving (8.108).

Distinguishing itself from the Newton-type approach, note that the descent property (8.136) of the Lee and Seung method ensures that the objective function $f(U, V)$ is nonincreasing under the update rules.

The choice of $D(\mathbf{v}^c)$ as defined in (8.132) deserves further comment. It is clear that there are many other ways to set forth the simpler model (8.131). For example, if all diagonal entries of D are sufficiently large, say, larger than the spectral radius of $U^\top U$, then $D - U^\top U$ is positive definite. Nonetheless, the larger the D, the smaller the D^{-1} and, hence, the less difference between $\mathbf{v}^+$ and $\mathbf{v}^c$ according to (8.134). The challenge thus lies in finding a diagonal matrix D that is large enough to make $D - U^\top U$ positive definite, yet is also small enough to signify the difference between $\mathbf{v}^+$ and $\mathbf{v}^c$. It is possible to propose a different quadratic model in the form of (8.131) where the diagonal matrix $D = \operatorname{diag}\{d_1, \ldots, d_k\}$ carries the additional property that its trace is minimized. The notion is based on the semidefinite programming (SDP) technique (Vandenberghe and Boyd, 1996).

We outline the idea by working on the least squares Problem (8.124) where $U \in \mathbb{R}^{m \times p}$ is fixed. Denote $S = U^\top U \in \mathbb{R}^{p \times p}$ and $\mathbf{g}^c = [g_i^c] := U^\top U \mathbf{v}^c - U^\top \mathbf{y} \in \mathbb{R}^p$. Let $\lambda_i(D)$ denote the ith eigenvalue of $D - S$. Consider the function

$$\omega(D) := \sum_{i=1}^{p} \ln \frac{1}{\lambda_i(D)} + \sum_{i=1}^{p} \ln \frac{1}{d_i} + \sum_{i=1}^{p} \ln \frac{1}{d_i v_i^c - g_i^c}. \tag{8.140}$$

Because the (real-valued) logarithm is defined only for positive arguments, the function ω can be defined only for diagonal matrices D such that $D - S$ is positive definite, D has positive diagonal entries, and $D\mathbf{v}^c - \mathbf{g}^c$ is a positive vector. The barrier function $\omega(D)$ is introduced because its level curves serve as reasonable approximations to the boundary of the desirable feasible domain.

The following two results give rise to the gradient $\nabla \omega(D)$ and the Hessian $\nabla^2 \omega(D)$. The proofs can be found in Chu and Wright (1995). More general results can be found in Boyd and El Ghaoui (1993) and Nesterov and Nemirovskii (1994).

Lemma 8.18. The gradient vector of $\omega(D)$ with $D = \text{diag}\{d_1, \ldots, d_p\}$ is given by

$$\nabla\omega(D) = \text{diag}\big((D - S)^{-1} - D^{-1}\big) - \begin{bmatrix} \dfrac{v_1^c}{d_1 v_1^c - g_1^c} \\ \vdots \\ \dfrac{v_k^c}{d_k v_k^c - g_k^c} \end{bmatrix}. \tag{8.141}$$

Lemma 8.19. The Hessian matrix $H(D)$ of $\omega(D)$ is given by

$$\begin{aligned} H(D) &= (D - S)^{-1} \cdot *(D - S)^{-1} + D^{-1} \cdot *D^{-1} \\ &\quad + \text{diag}\left\{ \left(\frac{v_1^c}{d_1 v_1^c - g_1^c} \right)^2, \ldots, \left(\frac{v_k^c}{d_k v_k^c - g_k^c} \right)^2 \right\}. \end{aligned} \tag{8.142}$$

We note from the well-known Schur product theorem (Horn and Johnson, 1991, Theorem 7.5.3) that $H(D)$ is positive definite if D is feasible, which also shows that the function ω is strictly convex over its feasible domain.

Recall that an ellipsoid $\mathcal{E} \subset \mathbb{R}^p$ can best be characterized by its center $\boldsymbol{\gamma} \in \mathbb{R}^p$ and a symmetric and positive definite matrix $\Gamma \in \mathbb{R}^{p\times p}$ in such a way that

$$\mathcal{E} = \mathcal{E}(\Gamma, \boldsymbol{\gamma}) := \left\{ \mathbf{x} \in \mathbb{R}^p | (\mathbf{x} - \boldsymbol{\gamma})^\top \Gamma^{-1} (\mathbf{x} - \boldsymbol{\gamma}) \leq 1 \right\}. \tag{8.143}$$

Within the feasible domain of ω, we can approximate its level curves by a sequence of inscribed ellipsoids determined by the Hessians of ω in the following sense due to Dikin (1967).

Theorem 8.20. Suppose $D^c = \text{diag}(\mathbf{d}^c)$ is a strictly feasible point with respect to (8.140). Then every diagonal matrix $D^+ = \text{diag}(\mathbf{d}^+)$ with $\mathbf{d}^+$ from the ellipsoid $\mathcal{E}(H(D^c)^{-1}, \mathbf{d}^c)$ is also strictly feasible.

Proof It has been argued in Chu and Wright (1995) that D^+ is positive and that $D^+ - S$ is positive definite. It only remains to show that $D^+\mathbf{v}^c - \mathbf{g}^c$ is positive. Denote $\Delta := D^+ - D^c = \text{diag}\{\delta_1, \ldots, \delta_p\}$. It follows that

$$D^+\mathbf{v}^c - \mathbf{g}^c = (D^c\mathbf{v}^c - \mathbf{g}^c) + \Delta\mathbf{v}^c.$$

Since

$$\sum_{i=1}^{p} \frac{(\delta_i v_i^c)^2}{(d_i^c v_i^c - g_i^c)^2} < 1,$$

it is clear that $|\delta_i v_i^c| < d_i^c v_i^c - g_i^c$ for all $i = 1, \ldots, p$. □

Given a feasible D^c, any point from the ellipsoid $\mathcal{E}(H(D^c)^{-1}, d^c)$ will carry the four properties that Lee and Seung's choice (8.132) possesses. It is a matter of which point $\mathbf{d}^+$ on $\mathcal{E}(H(D^c)^{-1}, d^c)$ will serve the "goal" better. For instance, in attempting to make D^+ small, one possible objective is to minimize the trace of D^+, that is,

$$\text{Minimize} \quad \mathbf{1}^\top \mathbf{d}, \tag{8.144}$$

$$\text{Subject to} \quad \mathbf{d} \in \mathbf{E}(H(D^c)^{-1}, \mathbf{d}^c) \tag{8.145}$$

where $\mathbf{1} := [1, \ldots, 1]^\top$. Clearly, one can choose to weight the diagonal entries of D differently and end up with a different linear objective functional. Such an optimization of linear objective functional over ellipsoids has a closed form solution (Grötschel et al., 1993, p. 68).

Lemma 8.21. For $\mathbf{p} \neq 0$, the minimal value of the linear functional $\mathbf{p}^\top \mathbf{x}$ subject to the condition $\mathbf{x} \in \mathcal{E}(\Gamma, \gamma)$ occurs at

$$\mathbf{x}^* := \gamma - \frac{1}{\sqrt{\mathbf{p}^\top \Gamma \mathbf{p}}} \Gamma \mathbf{p}. \tag{8.146}$$

Using (8.142) and the fact that p is low, it is extremely easy to implement a basic Dikin algorithm as follows.

Algorithm 8.5 (Basic Dikin method)
Given $\mathbf{d}^{(0)} \in \mathbb{R}^p$ strictly feasible, do for $k = 0, 1, \ldots$ the following:

1. If $D^{(k)} - S$ is singular, then stop,
2. Otherwise,
   (a) Solve $H(D^{(k)})\mathbf{d} = \mathbf{1}$ for $\mathbf{d}$;
   (b) Update $\mathbf{d}^{(k+1)} := \mathbf{d}^{(k)} - \frac{1}{\sqrt{\mathbf{1}^\top \mathbf{d}}} \mathbf{d}$.

Theorem 8.20 guarantees that $\mathbf{d}^{(k)}$ is strictly feasible and hence $D^{(k)} - S$ is never singular in exact arithmetic. However, in floating point arithmetic, one has to settle the singularity (or rank deficiency) of a matrix for an eigenvalue (or a singular value) less than a prescribed tolerance. A usual choice of tolerance for zero is $\epsilon \|S\|$ where ϵ is the machine dependent floating point relative accuracy. For this reason it is possible that the algorithm stops at a point where $D - S$ is *numerically* semidefinite yet trace(D) may have not reached its minimal value. To reduce the risk of hitting the boundary of the feasible domain too soon, we find it is a good idea to start the Dikin method from a sufficiently large scalar matrix.

The main difference between the Lee and Seung diagonal matrix $D_{\text{Lee\&Seung}}$ defined by (8.132) and the Dikin diagonal matrix D_{Dikin} defined by Algorithm 8.5 is that $D_{\text{Lee\&Seung}}$ is *always* on the boundary of the feasible domain because $D_{\text{Lee\&Seung}} - S$ has a zero eigenvalue with eigenvector $\mathbf{v}^c$. While the Dikin algorithm produces a diagonal matrix that has minimal trace, the Lee and Seung algorithm is remarkably cheap for computation.

Gradient flow The dynamical system

$$\frac{dE}{dt} = E \circ (\alpha(E,F)(F \circ F)^\top) \in \mathbb{R}^{m\times k}, \tag{8.147}$$

$$\frac{dF}{dt} = F \circ ((E \circ E)^\top \alpha(E,F)) \in \mathbb{R}^{k\times n}, \tag{8.148}$$

moves in the space $\mathbb{R}^{m\times k} \times \mathbb{R}^{k\times n}$ along the steepest descent direction of the objective functional g. In the event that an iterative method becomes too expensive an endeavor because of the matrix sizes involved in the system, integrating (8.147) and (8.148) by existing ODE solvers can serve as an alternative numerical means for solving the NNMF. It is easy to check that along the solution flow $(E(t), F(t))$,

$$\begin{aligned}\frac{d\phi(E(t),F(t))}{dt} &= -\langle(\alpha(E,F)(F\circ F)^\top)\circ E, (\alpha(E,F)(F\circ F)^\top)\circ E\rangle \\ &\quad - \langle F\circ((E\circ E)^\top\alpha(E,F)), F\circ((E\circ E)^\top\alpha(E,F))\rangle \leq 0.\end{aligned}$$

The objective functional $\phi(E, F)$ therefore can be used as the Lyapunov function for the dynamical system and global convergence is guaranteed. At the limit point, it is necessary that the first-order optimality conditions (8.116) and (8.117) are satisfied.

Steepest descent method Instead of integrating (8.147) and (8.148) by high precision ODE integrators, the Euler method with appropriate step size selection is another way of making use of the gradient information. One way to implement the steepest descent scheme is to update E and F in the following iterations:

$$E^{(k+1)} := E^{(k)} + \mu_k E^{(k)} \cdot *(\delta(E^{(k)}, F^{(k)})(F^{(k)} \cdot * F^{(k)})^\top), \tag{8.149}$$

$$F^{(k+1)} := F^{(k)} + \mu_k F^{(k)} \cdot *((F^{(k)} \cdot * F^{(k)})^\top \delta(E^{(k)}, F^{(k)})). \tag{8.150}$$

Recently, Shepherd (2004) has proposed an update scheme as follows:

$$U^{(k+1)} = U^{(k+1)}(\mu_k) := \max\{0, U^{(k)} + \mu_k(Y - U^{(k)}V^{(k)})(V^{(k)})^\top\}, \tag{8.151}$$

$$V^{(k+1)} = V^{(k+1)}(\mu_k) := V^{(k)} + \mu_k(U^{(k)})^\top(Y - U^{(k)}V^{(k)}), \tag{8.152}$$

where max is taken component by component. In either case, the selection of μ_k is critical. In general practice, a backtracking line search using, say, a cubic interpolation and a merit function, is performed to determine the step length μ_k (Fletcher, 1987; Gill et al., 1981). For the NNMF problem, the selection of step length is easier. For example, the function,

$$\Theta(\mu) := F(U^{(k+1)}(\mu), V^{(k+1)}(\mu)), \tag{8.153}$$

with $U^{(k+1)}(\mu)$ and $V^{(k+1)}(\mu)$ defined by (8.152) is a quartic polynomial in μ. It was suggested that we use the Tartaglia formula (Shepherd, 2004) to compute directly the roots of $\Theta^{'}(\mu)$ and hence locate the optimal μ.

8.7.3 *An air pollution and emission example*

Despite the fact that there are many numerical ways to solve the NNMF problem, we apply some of the techniques to one real-world problem to demonstrate the limits and difficulties in interpreting the factorizations. We stress that proper interpretations or additional constraints on the factors are needed for NNMF applications.

The 8×15 matrix Y in Table 8.8 represents the annual total masses (in thousand short tons) of eight pollutants estimated by the EPA over 15 years (EPA, 2001). The blanks at the lower left corner indicate that no data were collected during those years and are assumed zero (and hence bias the analysis). The 4×15 matrix F in Table 8.9 represents the annual total emissions by four principal sectors across the national economy, each of which contains a spectrum of many more pertinent subsectors. Details can be found in the report (EPA, 2001).

In our first scenario, suppose that both Y and F are available. The problem is to determine a nonnegative matrix A of size 8×4 that solves the following optimization problem:

$$\begin{aligned}
&\text{Minimize} \quad \tfrac{1}{2}\|Y - AF\|_F^2, \\
&\text{Subject to } A \geq 0, \text{ and } \sum_{i=1}^{8} a_{ij} = 1, \quad j = 1, \ldots, 4.
\end{aligned}$$

Each column of A represents the best fitting percentage distribution of pollutants from the emission of the corresponding sector. This is a convex programming problem and the global minimizer is unique.

Using existing software, such as FMINCON in MATLAB, we find that the optimal distribution A_{opt} to Problem (8.154) is given in Table 8.14. This best fitting distribution is in contrast to the average distribution A_{avg} in Table 8.10 that would have to be obtained, otherwise, by extensive efforts in gathering *itemized* pollutant emissions of each sector per year (EPA, 2001). There are several discrepancies that warrant attention. For example, it is estimated in A_{opt} that 32.70% emissions from fuel burning contribute to the volatile organic compounds whereas A_{avg} counts only 2.65%. It is estimated in A_{opt} that only 6.31% of emissions from the fuel goes to nitrogen oxides whereas A_{avg} count 27.54%. It is clear that the estimates from A_{opt}, though best fitting the data, are inconsistent with the scientific truth.

In our second scenario, suppose that only Y is available. The problem is to determine four sectors, *not necessarily in any order or any definition*, and their

Table 8.8. *Annual pollutants estimates (in thousand short tons)*

	1970	1975	1980	1985	1989	1990	1991	1992	1993	1994	1995	1996	1997	1998	1999
Carbon monoxide	129444	116756	117434	117013	106438	99119	101797	99307	99790	103713	94057	101294	101459	96872	97441
Lead	221	160	74	23	5	5	4	4	4	4	4	4	4	4	4
Nitrogen oxides	20928	22632	24384	23197	23892	24170	24338	24732	25115	25474	25052	26053	26353	26020	25393
Volatile organic	30982	26080	26336	24428	22513	21052	21249	11862	21100	21682	20919	19464	19732	18614	18145
PM_{10}	13165	7677	7109	41397	40963	27881	27486	27249	27502	28756	25931	25690	25900	26040	23679
Sulfur dioxide	31161	28011	25906	23658	23294	23678	23045	22814	22475	21875	19188	18859	19366	19491	18867
$PM_{2.5}$						7429	7317	7254	7654	7012	6909	7267	7065	6773	6773
Ammonia						4355	4412	4483	4553	4628	4662	4754	4851	4929	4963

Table 8.9. *Annual emissions estimates (in thousand short tons)*

	1970	1975	1980	1985	1989	1990	1991	1992	1993	1994	1995	1996	1997	1998	1999
Fuel	41754	40544	43512	41661	40659	39815	39605	40051	38926	38447	36138	36018	35507	34885	34187
Industrial	48222	32364	29615	22389	21909	21120	20900	21102	21438	21467	21190	17469	17988	17868	20460
Transportation	125637	121674	117527	119116	107978	100877	106571	105114	106328	108125	99642	106069	104748	103523	100783
Miscellaneous	10289	6733	10589	46550	46560	45877	42572	40438	41501	45105	39752	43829	46487	42467	39836

Table 8.10. *Average distribution of pollutants from sectors*

	Fuel	Industrial	Transportation	Miscellaneous
Carbon monoxide	0.1535	0.3116	0.7667	0.3223
Lead	0.0001	0.0002	0.0002	0
Nitrogen oxides	0.2754	0.0417	0.1177	0.0113
Volatile organic	0.0265	0.4314	0.0908	0.0347
PM_{10}	0.0368	0.0768	0.0074	0.4911
Sulfur dioxide	0.4923	0.0996	0.0112	0.0012
$PM_{2.5}$	0.0148	0.0272	0.0043	0.0761
Ammonia	0.0007	0.0115	0.0016	0.0634

Table 8.11. *NNMF distribution estimates of pollutants from sectors (Lee and Seung algorithm)*

	Sector 1	Sector 2	Sector 3	Sector 4
Carbon monoxide	0.2468	0.0002	0.7969	0.0001
Lead	0	0.0008	0	0.0000
Nitrogen oxides	0.0000	0	0.1641	0.1690
Volatile organic	0.3281	0.2129	0.0391	0
PM_{10}	0.0000	0.5104	0.0000	0.5532
Sulfur dioxide	0.4251	0.2757	0.0000	0
$PM_{2.5}$	0.0000	0.0000	0	0.1680
Ammonia	0.0000	0	0	0.1097

corresponding percentage distributions U and total emissions per year V so as to best fit the observed data Y. This is precisely an NNMF problem.

By using the Lee and Seung algorithm, we obtain local solutions U and V indicated in Tables 8.11 and 8.12, respectively. We stress that we do not know what each column of U really stands for. It requires a careful interpretation to identify what *factor* is being represented. It is likely that a single column could represent a mixture of two or more known economy sectors. We have noted the improvement in the objective functions, that is,

$$\begin{aligned}\tfrac{1}{2}\|Y - UV\|_F^2 = 1.5873 \times 10^7 &< \|Y - A_{\text{opt}}F\|_F^2 \\ = 2.7017 \times 10^8 &< \tfrac{1}{2}\|Y - A_{\text{avg}}F\|^2 = 7.1548 \times 10^8.\end{aligned}$$

However, the somewhat unevenness in the NNMF emission estimates per sector given in Table 8.12 seems to make it more difficult to predict the estimate.

Similarly, by using the constrained quasi-Newton method, we obtain another percentage distribution of pollutants from sectors in Table 8.13. (To save space,

Table 8.12. *NNMF emission estimates (in thousand short tons)*

	1970	1975	1980	1985	1989	1990	1991	1992	1993	1994	1995	1996	1997	1998	1999
Sector 1	58705	57455	57162	3718	4974	47464	46314	47175	47864	43630	44643	42657	43578	42926	43585
Sector 2	25487	11755	7431	81042	75327	10313	10784	8313	6848	12613	4069	3403	3541	3159	1
Sector 3	143614	128945	130225	145512	132349	109442	113118	109881	110295	116521	104440	113926	113910	108437	108828
Sector 4	0	3139	6254	2	4702	40785	39618	41539	43358	40302	43236	43319	43599	44239	42832

Table 8.13. *NNMF distribution estimates of pollutants from sectors (constrained quasi-Newton method) (Lee and Seung algorithm)*

	Sector 1	Sector 2	Sector 3	Sector 4
Carbon monoxide	0.3124	0.4468	0.5426	0.6113
Lead	0	0	0.0000	0.0007
Nitrogen oxides	0.1971	0.1299	0.0366	0.1412
Volatile organic	0.0239	0.0654	0.1720	0.1191
PM_{10}	0.1936	0.3101	0.0401	0.0220
Sulfur dioxide	0.0287	0.0477	0.2087	0.1058
$PM_{2.5}$	0.1480	0.0000	0	0
Ammonia	0.0963	0	0.0000	0.0000

Table 8.14. *Optimal distribution of pollutants from sectors with fixed emission estimates*

	Fuel	Industrial	Transportation	Miscellaneous
Carbon monoxide	0.1925	0.3400	0.8226	0.0090
Lead	0	0.0000	0	0.0000
Nitrogen oxides	0.0631	0	0.1503	0.1524
Volatile organic	0.3270	0.2759	0.0272	0
PM_{10}	0.0000	0.1070	0.0000	0.6198
Sulfur dioxide	0.4174	0.2771	0.0000	0
$PM_{2.5}$	0.0000	0.0000	0	0.1326
Ammonia	0.0000	0	0	0.0862

the corresponding emission estimates are not listed.) This much more sophisticated method is computationally more expensive but is able to find local solutions that give smaller objective values (1.0645×10^7). Again, it is not clear how to identify the sectors and to interpret the distributions of pollutants.

8.7.4 *Summary*

The nonnegative matrix factorization is another kind of low rank approximation and has been desired by many important applications. We have specified in closed form the first-order optimality condition and suggested a number of numerical procedures that can be employed to obtain a factorization that is at least locally optimal.

Nonetheless, proper interpretations or additional constraints on the factors are needed for NNMF applications. To demonstrate our points, we have used two real-world problems to illustrate that the factorization itself does not necessarily provide immediate interpretation of the real data – the basic parts of the irises are themselves complicated images (and sometimes with overlapped irises); and the percentage distributions of pollutants from economical sectors are not always consistent with data obtained by other means (and could represent mixtures across several sectors.)

9

GROUP ORBITALLY CONSTRAINED APPROXIMATION

9.1 Overview

Thus far, the notion of isospectrality for either eigenvalues or singular values seems to predominate in all the discussion for inverse eigenvalue problems. A larger picture, however, is that inverse eigenvalue problems are but a special case of the more general canonical form identification problem. To conclude this book, it is perhaps fitting to deviate from the traditional context of inverse eigenvalue problems and navigate into a more abstract setting by actions of matrix groups. We hope that this abstraction and generalization will offer a clue to open up many more new ideas in the near future.

It is fair to say that linear transformation is one of the simplest yet most profound ways to describe the relationship between two vector spaces. To further deduce the essence of a linear transformation, it is often desired to identify a matrix by its *canonical form*. Eigenvalue computation, for example, is to identify a matrix by its Jordan canonical form or, for the sake of computational stability, its Schur pseudo-triangular form. The inverse eigenvalue problem, on the other hand, intends to identify a matrix with known spectrum, say, a diagonal matrix or a triangular matrix, isospectrally by a matrix with a desired structure. In this sense, both forward and inverse eigenvalue problems are two sides of the coin of the identification problem. For years researchers have made great effort to describe, analyze, and modify algorithms to achieve this goal. Thus far, this identification process is mostly done via iterative procedures of which success is evidenced by the many available discrete methods.

Recall the interesting question V. I. Arnold raised in Arnold (1988, p. 239), "What is *the simplest form* to which a family of matrices depending smoothly on the parameters can be reduced by *a change of coordinates* depending smoothly on the parameters?" In this quotation, we italicize the two phrases to further accentuate the important inquiries:

- What is qualified as the simplest form?
- What kind of continuous change can be employed?

Recently it has been observed that the use of differential equations for issues in computation can afford fundamental insights into the structure and behavior of existing discrete methods and, sometimes, can suggest new and improved numerical methods. In certain cases, there are remarkable connections between smooth flows and discrete numerical algorithms. In other cases, the flow approach seems advantageous in tackling very difficult problems.

What is being talked about thus far can be categorically characterized as a realization process. A realization process, in a sense, means any deducible procedure that we use to rationalize and solve problems. The simplest form therefore refers to the ability to think and draw conclusions out of the form. In mathematics, the steps taken in a realization process often appear in the form of an iterative procedure or a differential equation.

Altogether, it appears that both the discrete approach and flow approach actually follow the *orbit* of a certain matrix groups action on the underlying matrix. The question to ask, therefore, is to what canonical form a matrix or a family of matrices can be linked by the orbit of a group action. The choice of the group, the definition of the action, and the canons intended to reach it will affect the various transitions that have been studied in the literature. This chapter will attempt to expound the various aspects of the recent development and application in this direction. Some possible new research topics will also be proposed. The earliest discussion along these lines seems to be in Della-Dora (1975).

A realization process usually consists of three components. First of all, we will have two abstract problems of which one is made-up and is easy to solve, while the other is the real problem and is difficult. The basic idea is to realize the solution of the difficult problem through a bridge or a path that starts from the solution of the easier problem. Once the plan is set, we then need a numerical method to move along the path to obtain the desired solution.

To build the bridge, sometimes we are given specific construction guidelines. For example, the bridge could be constructed by monitoring the values of certain specified function. In this case, the path is guaranteed to work. The projected gradient method is a typical representation of this approach. Sometimes the guidelines are not so specific and hence we are kind of constructing the bridge instinctively. We can only hope that the bridge will connect to the other end. It takes effort to prove that the bridge actually makes the desired connection. A typical continuation method in this class is the homotopy method. Another situation is that the bridge seems to exist unnaturally. But in fact, usually much deeper mathematics or physics is involved. When we begin to understand the theory, we are often amazed to see that these seemingly unnatural bridges exist by themselves naturally.

We shall be more specific about how a bridge could be built. For now, it is sufficient to say that a bridge, if it exists, usually is characterized by an ordinary differential equation and that the discretization of a bridge, or a numerical method in travelling along a bridge, usually produces an iterative scheme. Known as geometric integration, research into numerical integrators that respect the underlying geometric structure has been attracting considerable attention recently, but is beyond the scope of this survey.

9.2 A case study

Before we move into more details, we use two classical examples and their connections to demonstrate the main points made above. We also briefly describe their generalization with the hope that this discussion would serve as a stepping stone and shed light on the thrust throughout this section. More details can be found in Chu (1988).

9.2.1 *Discreteness versus continuousness*

Consider first the symmetric eigenvalue problem that has been of critical importance in many applications. Given a symmetric matrix A_0, the problem is to find all scalars λ so that the equation

$$A_0\mathbf{x} = \lambda\mathbf{x}, \tag{9.1}$$

has a nontrivial solution $\mathbf{x}$. Currently, one of the most crucial techniques for eigenvalue computation is by iteration. Recall the fact that any matrix A enjoys the QR decomposition:

$$A = QR,$$

where Q is orthogonal and R is upper triangular. The basic QR algorithm defines a sequence of matrices $\{A_k\}$ via the recursion relationship (Golub and Van Loan, 1996):

$$\begin{cases} A_k = Q_k R_k, \\ A_{k+1} = R_k Q_k. \end{cases} \tag{9.2}$$

Because $A_{k+1} = Q_k^\top A_k Q_k$, every matrix A_k in the sequence has the same eigenvalues as A_0. More importantly, it can be proved that the sequence $\{A_k\}$ converges to a diagonal matrix and, hence, eigenvalues are found.

In contrast, let a symmetric matrix X be decomposed as

$$X = X^o + X^- + X^+,$$

where X^o, X^-, and X^+ denote the diagonal, the strictly lower triangular, and the strictly upper triangular parts of X, respectively. Define

$$\Pi_0(X) := X^- - X^{-\top}. \tag{9.3}$$

The Toda lattice is an initial value problem defined by Deift and Nanda (1984) and Symes (1981/82):

$$\begin{cases} \dfrac{dX(t)}{dt} = [X(t), \Pi_0(X(t))], \\ X(0) = X_0, \end{cases} \tag{9.4}$$

where $[M, N] := MN - NM$ denotes the commutator bracket. It is known that when the solution of the Toda lattice (9.4) is sampled at integer times, the sequence $\{X(k)\}$ gives rise to the same iterates as does the QR algorithm (9.2) applied to the matrix $A_0 = \exp(X_0)$. Furthermore, it can be proved that the evolution of $X(t)$ starts from X_0, converges to the limit point of Toda flow, which is a diagonal matrix, and that $X(t)$ maintains the same spectrum as that of X_0 for all t. The Toda lattice is a Hamiltonian system constructed from some physics settings. Certain physical quantities are kept constant, that is, the Toda lattice is a *completely integrable* system.

In both approaches, the eigenvalue computation is cast as a realization process that starts from the "easy" matrix whose eigenvalues are to be found and ends at the "difficult" matrix which is the diagonal matrix carrying the eigenvalue information. The bridges that connect the original matrix to the final diagonal matrix are the QR algorithm in the discrete process and the Toda lattice in the continuous process, respectively. Maintaining isospectrality everywhere along the bridges is the most important property inhered in both bridges. It is remarkable that the QR algorithm and the Toda lattice are so closely related.

Consider next the least squares matrix approximation problem which also plays a significant role in disciplines of various areas. Given a symmetric matrix N, the problem is to find a least squares approximation of N while maintaining a prescribed set of eigenvalues $\{\lambda_1, \ldots, \lambda_n\}$. The problem can be formulated as an equality constrained minimization problem as follows:

$$\begin{array}{ll} \text{Minimize} & F(Q) := \frac{1}{2}||Q^\top \Lambda Q - N||^2 \\ \text{subject to} & Q^T Q = I. \end{array}$$

While iterative optimization techniques such as the augmented Lagrangian methods or the sequential quadratic programming methods could readily be applied to solve this problem, none of these techniques respects the matrix structure. For example, the constraint carries lots of algebraic redundancies and actually defines only a $n(n-1)/2$-dimensional manifold.

In contrast, the projected gradient of F can easily be calculated and a projected gradient flow can be defined via the double bracket differential equation (Brockett, 1991; Chu, 1990):

$$\begin{cases} \dfrac{dX(t)}{dt} = [X(t), [X(t), N]], \\ X(0) = \Lambda. \end{cases} \tag{9.5}$$

The solution $X(t) := Q(t)^\top \Lambda Q(t)$ moves in a descent direction to reduce $\|X(t) - N\|^2$ as t goes to infinity. It can be proved that the optimal solution X can be fully characterized in terms of the spectral decomposition of N and is unique.

In this setting, the evolution starts from the "easy" matrix which is the diagonal matrix of the prescribed eigenvalues and converges to the limit point which solves the least squares problem. The flow is built on the basis of systematically reducing the difference between the current position and the target position. This differential system is a gradient flow.

At first glance, the eigenvalue computation and the least squares matrix approximation seem to be two unrelated problems. However, when X is tridiagonal and

$$N = \mathrm{diag}\{n, \ldots, 2, 1\},$$

it can easily be verified that (Bloch et al., 1990)

$$[X, N] = \Pi_0(X).$$

In other words, the gradient flow (9.5) in fact is also a Hamiltonian flow (9.4).

9.2.2 *Generalization*

Both differential systems (9.4) and (9.5) are special cases of the more general Lax dynamical system:

$$\begin{cases} \dfrac{dX(t)}{dt} := [X(t), k_1(X(t))], \\ X(0) \quad := X_0, \end{cases} \tag{9.6}$$

where $k_1 : \mathbb{R}^{n\times n} \to \mathbb{R}^{n\times n}$ is some selected matrix-valued function to be specified later. Associated with (9.6), we define two parameter dynamical systems:

$$\begin{cases} \dfrac{dg_1(t)}{dt} := g_1(t)k_1(X(t)), \\ g_1(0) \quad := I, \end{cases} \tag{9.7}$$

and

$$\begin{cases} \dfrac{dg_2(t)}{dt} := k_2(X(t))g_2(t), \\ g_2(0) \quad := I, \end{cases} \tag{9.8}$$

with the property that

$$k_1(X) + k_2(X) = X. \tag{9.9}$$

The following theorem has been established earlier (Chu and Norris, 1988).

Theorem 9.1. For any t within the interval of existence, the solutions $X(t)$, $g_1(t)$, and $g_2(t)$ of the systems (9.6), (9.7), and (9.8), respectively, are related to each other by the following three properties:

(i) (Similarity property)

$$X(t) = g_1(t)^{-1}X_0g_1(t) = g_2(t)X_0g_2(t)^{-1}. \tag{9.10}$$

(ii) (Decomposition property)

$$\exp(tX_0) = g_1(t)g_2(t). \tag{9.11}$$

(iii) (Reversal property)

$$\exp(tX(t)) = g_2(t)g_1(t). \tag{9.12}$$

The consequences of Theorem 9.1 are quite broad and significant. The similarity property, for example, immediately implies that the solution $X(t)$ to (9.6) is isospectral. The theorem also suggests an abstraction of the QR algorithm in the following sense. It is known that the (Lie) group $Gl(n)$ of real-valued $n \times n$ nonsingular matrices can be decomposed as the product of two (Lie) subgroups in the neighborhood of I if and only if the corresponding tangent space (Lie algebra) $gl(n)$ of real-valued $n \times n$ matrices can be decomposed as the sum of two (Lie) subalgebras. By the decomposition property and the reversal property, the Lie structure apparently is not needed in our setting. It suffices to consider a factorization of a *one-parameter semigroup* in the neighborhood of I as the product of two nonsingular matrices, that is, the decomposition as indicated in (9.11). Then, correspondingly there is a subspace decomposition of $gl(n)$ as is indicated in (9.9). Some of the jargon will be clarified in the next section. Our point at present is to bring forth the intriguing structure that is underneath the transformations.

Analogous to the QR decomposition, the product $g_1(t)g_2(t)$ will be called the *abstract g_1g_2 decomposition* of $\exp(X_0t)$. By setting $t = 1$, we see that

$$\begin{cases} \exp(X(0)) = g_1(1)g_2(1), \\ \exp(X(1)) = g_2(1)g_1(1). \end{cases} \tag{9.13}$$

Since the dynamical system for $X(t)$ is autonomous, it follows that the phenomenon characterized by (9.13) will occur at every feasible integer time. Corresponding to the abstract g_1g_2 decomposition, the above iterative process for all feasible integers will be called the *abstract g_1g_2 algorithm*. It is thus seen that the curious iteration in the QR algorithm is completely generalized and abstracted via the mere subspace splitting (9.9).

9.3 General Framework

What is being manifested in the preceding section is but a special case of a more general framework for the realization process. We have already seen a resemblance between the Toda flow $X(t)$ defined by (9.4) and the isospectral flow $X(t)$ defined by (9.6). The former can be considered as a curve on the "orbit" $Q(t)^\top X_0 Q(t)$ under the orthogonal similarity transformation by all $n \times n$ orthogonal matrices Q which form a group while the latter can be considered as a curve on the orbit $g_1(t)^{-1}X_0g_1(t)$ under the similarity transformation by whatever matrices $g_1(t)$ that can be defined but not necessarily form a group. It is natural to exploit what other transformations or sets from which a transformation is taking place can be characterized in general.

9.3.1 *Matrix group and actions*

We begin the generalization with the notion of matrix groups. We shall eventually drop the requirement of group structure in the later part of this discussion. Many manifolds arising from applications are not groups themselves, but can still be "parameterized" by groups.

Recall that a subset of nonsingular matrices (over any field) which are closed under matrix multiplication and inversion is called a *matrix group* (Baker, 2002; Belinfante and Kolman, 1989; Curtis, 1984). A smooth manifold which is also a group where the multiplication and the inversion are smooth maps is called a *Lie group*. It has been shown that every matrix group is in fact a Lie group (Howe, 1983, 1984). The reason that Lie groups are interesting is because this particular entity inherits both algebraic and geometric structures. The most remarkable feature of a Lie group is that the structure is the same in the neighborhood of each of its elements. We shall elaborate more on this point later.

Matrix groups are central in many parts of mathematics and applications. Lots of realization processes used in numerical linear algebra are the results of actions of matrix groups. For the convenience of later references, we tabulate in Table 9.1 some classical matrix groups and their subgroups over $\mathbb{R}$.

Any of these matrix groups could be used to transform matrices. To be more precise, we define the action of a group according to the following rule.

Definition 9.2. Given a group G and a set $\mathbb{V}$, a function $\mu : G \times \mathbb{V} \longrightarrow \mathbb{V}$ is said to be a group action of G on $\mathbb{V}$ if and only if

1. $\mu(gh, \mathbf{x}) = \mu(g, \mu(h, \mathbf{x}))$ for all $g, h \in G$ and $\mathbf{x} \in \mathbb{V}$.
2. $\mu(e, \mathbf{x}) = \mathbf{x}$, if e is the identity element in G.

In our applications, the set $\mathbb{V}$ is used to specify the principal characteristics of the matrices that we are interested. The action, in turn, is used to specify the transformations that are allowed. Given an arbitrary $\mathbf{x} \in \mathbb{V}$, two important sets are associated with a group action. These are the *stabilizer* of $\mathbf{x}$,

$$Stab_G(\mathbf{x}) := \{g \in G | \mu(g, \mathbf{x}) = \mathbf{x}\}, \tag{9.14}$$

which is a subgroup of G, and the *orbit* of $\mathbf{x}$,

$$Orb_G(\mathbf{x}) := \{\mu(g, \mathbf{x}) | g \in G\}. \tag{9.15}$$

The orbit of a matrix under a certain group action is particularly relevant to the central theme of this chapter. To demonstrate how some of the conventional transformations used in practice can be cast as group actions, we list in Table 9.2 some of the useful actions by matrix groups.

The groups listed in Table 9.1 and actions listed in Table 9.2 represent only a small collection of wild possibilities. They already indicate a wide open area for further research because, except for the orthogonal group $\mathcal{O}(n)$, not many other groups nor actions have been employed in practice. In numerical analysis, it is customary to use actions of the orthogonal group to perform the change of coordinates for the sake of cost efficiency and numerical stability. It becomes interesting to ask, for example, what conclusion could be drawn if actions of the isometry group $\mathcal{I}som(n)$ are used instead. The isometry group is appealing for

Table 9.1. *Examples of classical matrix groups over* $\mathbb{R}$

Group	Subgroup	Notation	Characteristics
General linear		$\mathcal{G}l(n)$	$\{A \in \mathbb{R}^{n\times n} \mid \det(A) \neq 0\}$
	Special linear	$\mathcal{S}l(n)$	$\{A \in \mathcal{G}l(n) \mid \det(A) = 1\}$
Upper triangular		$\mathcal{U}(n)$	$\{A \in \mathcal{G}l(n) \mid A$ is upper triangular$\}$
	Unipotent	$\mathcal{U}nip(n)$	$\{A \in \mathcal{U}(n) \mid a_{ii} = 1$ for all $i\}$
Orthogonal		$\mathcal{O}(n)$	$\{Q \in \mathcal{G}l(n) \mid Q^\top Q = I\}$
Generalized orthogonal		$\mathcal{O}_S(n)$	$\{Q \in \mathcal{G}l(n) \mid Q^\top S Q = S\}$, S is a fixed symmetric matrix
	Symplectic	$\mathcal{S}p(2n)$	$\mathcal{O}_J(2n)$, $J := \begin{bmatrix} 0 & I \\ -I & 0 \end{bmatrix}$
	Lorentz	$\mathcal{L}or(n,k)$	$\mathcal{O}_L(n+k)$, $L := \mathrm{diag}\{\underbrace{1,\ldots,1}_{n}, \underbrace{-1,\ldots,-1}_{k}\}$
Affine		$\mathcal{A}ff(n)$	$\left\{\begin{bmatrix} A & \mathbf{t} \\ \mathbf{0} & 1 \end{bmatrix} \mid A \in \mathcal{G}l(n), \mathbf{t} \in \mathbb{R}^n\right\}$
	Translation	$\mathcal{T}rans(n)$	$\left\{\begin{bmatrix} I & \mathbf{t} \\ \mathbf{0} & 1 \end{bmatrix} \mid \mathbf{t} \in \mathbb{R}^n\right\}$
	Isometry	$\mathcal{I}som(n)$	$\left\{\begin{bmatrix} Q & \mathbf{t} \\ \mathbf{0} & 1 \end{bmatrix} \mid Q \in \mathcal{O}(n), \mathbf{t} \in \mathbb{R}^n\right\}$
Center of G		$Z(G)$	$\{z \in G \mid zg = gz$, for every $g \in G\}$, G is a given group
Product of G_1 and G_2		$G_1 \times G_2$	$\{(g_1, g_2) \mid g_1 \in G_1, g_2 \in G_2\}$, $(g_1, g_2) * (h_1, h_2) := (g_1 h_1, g_2 h_2)$, G_1 and G_2 are given groups
Quotient		G/N	$\{Ng \mid g \in G\}$, N is a fixed normal subgroup of G
	Hessenberg	$\mathcal{H}ess(n)$	$\mathcal{U}nip(n)/\mathcal{Z}_n$

at least three reasons: the inverse of an isometry matrix is easy to compute since

$$\begin{bmatrix} Q & \mathbf{t} \\ \mathbf{0} & 1 \end{bmatrix}^{-1} = \begin{bmatrix} Q^\top & -Q^\top \mathbf{t} \\ \mathbf{0} & 1 \end{bmatrix};$$

the numerical stability by isometric transformation is guaranteed; and the isometry group contains the orthogonal group as its subgroup and hence offers more flexibility. To our knowledge, however, we have seen the usage of such an

Table 9.2. *Examples of group actions and their applications*

Set $\mathbb{V}$	Group G	Action $\mu(g, A)$	Application
$\mathbb{R}^{n\times n}$	Any subgroup	$g^{-1}Ag$	Conjugation
$\mathbb{R}^{n\times n}$	$\mathcal{O}(n)$	$g^{\top}Ag$	Orthogonal similarity
$\underbrace{\mathbb{R}^{n\times n}\times\cdots\times\mathbb{R}^{n\times n}}_{k}$	Any subgroup	$(g^{-1}A_1g,\ldots,g^{-1}A_kg)$	Simultaneous reduction
$\mathbb{S}(n)\times\mathbb{S}_{PD}(n)$	Any subgroup	$(g^{\top}Ag, g^{\top}Bg)$	Symm. positive definite pencil reduction
$\mathbb{R}^{n\times n}\times\mathbb{R}^{n\times n}$	$\mathcal{O}(n)\times\mathcal{O}(n)$	$(g_1^{\top}Ag_2, g_1^{\top}Bg_2)$	QZ decomposition
$\mathbb{R}^{m\times n}$	$\mathcal{O}(m)\times\mathcal{O}(n)$	$g_1^{\top}Ag_2$	Singular value decomposition
$\mathbb{R}^{m\times n}\times\mathbb{R}^{p\times n}$	$\mathcal{O}(m)\times\mathcal{O}(p)\times\mathcal{G}l(n)$	$(g_1^{\top}Ag_3, g_2^{\top}Bg_3)$	Generalized singular value decomposition

isometric transformation only in the discipline of physics, but rarely in numerical computation.

Some more exotic actions include, for example, the orthogonal conjugation plus shift by the product group of the orthogonal group $\mathcal{O}(n)$ and the additive group $\mathbb{R}_+$ of real numbers,

$$\mu((Q,s),A) := Q^{\top}AQ + sI, \quad Q\in\mathcal{O}(n), s\in\mathbb{R}_+;$$

the orthogonal group with scaling by the product group of $\mathcal{O}(n)$ and the multiplicative group $\mathbb{R}_\times$ of nonzero real numbers,

$$\mu((Q,s),A) := sQ^{\top}AQ, \quad Q\in\mathcal{O}(n), s\in\mathbb{R}_\times;$$

or even the orthogonal conjugation with multiple scalings,

$$\mu((Q,\mathbf{s},\mathbf{t}),A) := \operatorname{diag}\{\mathbf{s}\}Q^{\top}AQ\operatorname{diag}\{\mathbf{t}\}, \quad Q\in\mathcal{O}(n), \mathbf{s},\mathbf{t}\in\mathbb{R}^n_\times.$$

It is important to note that, by using the group inverse, the information about A can be retrieved at any point along any of the orbits defined above. The issue is how to define curves on these orbits so that the limit points would be useful for us to retrieve information about A. We do not think preposterous group actions by arbitrary groups are all useful. We do want to point out, however, that there are many unanswered questions that deserve further study.

9.3.2 *Tangent space and projection*

Given a group G and its action μ on a set $\mathbb{V}$, the associated orbit $\mathrm{Orb}_G(\mathbf{x})$ characterizes the rule by which $\mathbf{x}$ is to be changed in $\mathbb{V}$. Depending on the group G, an orbit is often a high dimensional manifold that is too "wild" to be readily traced for finding the "simplest form" of $\mathbf{x}$. Therefore, depending on the applications, it is desired to build a path, bridge, curve, or differential equation on the orbit so as to connect $\mathbf{x}$ to its simplest form. We have yet to define what is meant by the canonical form, but we first understand the notion of a vector field on a manifold.

A differential equation on the orbit $\mathrm{Orb}_G(\mathbf{x})$ is equivalent to a differential equation on the group G. We have seen this relationship in Section 9.2. The Lax dynamics (9.6) describes a bridge for the isospectral curve $X(t)$ on the orbit under orthogonal conjugation. Correspondingly, the parameter dynamics (9.7) and (9.8) characterize the flows for $g_1(t)$ in $\mathcal{O}(n)$ and $g_2(t)$ in $\mathcal{U}(n)$, respectively.

To stay in either the orbit or the group, the vector field of the dynamical system must be distributed over the tangent space of the corresponding manifold. Most of the tangent spaces for matrix groups can be calculated explicitly. If some kind of objective function has been used to control the connecting bridge, the gradient of the objective function should be projected to the tangent space.

Again, for completeness, we briefly review the notion of tangent space for a matrix group below. Given a matrix group G which is a subgroup in $\mathcal{G}l(n)$, the *tangent space* to G at an element $A \in G$ can be defined as the set

$$\mathcal{T}_A G := \{\gamma'(0) | \gamma \text{ is a differentiable curve in } G \text{ with } \gamma(0) = A\}. \tag{9.16}$$

The tangent space of a matrix group G at the identity I is so critical that it deserves special notation. We shall state the following facts without proof.

Theorem 9.3. Let $\mathfrak{g}$ denote the tangent space $\mathcal{T}_I G$ of a matrix group G at the identity I. Then

(i) The set $\mathfrak{g}$ is a Lie subalgebra in $\mathbb{R}^{n\times n}$. That is, if $\alpha'(0)$ and $\beta'(0)$ are two elements in $\mathfrak{g}$, then the Lie bracket $[\alpha'(0), \beta'(0)]$ is also an element $\mathfrak{g}$.

(ii) The tangent space of a matrix group has the same structure everywhere in the sense that the tangent space at any elements A in G is a translation of $\mathfrak{g}$ via

$$\mathcal{T}_A G = A\mathfrak{g}. \tag{9.17}$$

(iii) The Lie subalgebra $\mathfrak{g}$ can be characterized as the logarithm of G in the sense that

$$\mathfrak{g} = \{M \in \mathbb{R}^{n\times n} | \exp(tM) \in G, \text{ for all } t \in \mathbb{R}\}. \tag{9.18}$$

Tangent spaces for some of the groups mentioned earlier are listed in Table 9.3.

The exponential map $\exp : \mathfrak{g} \to G$, as we have seen in (9.11), is a central step from a Lie algebra $\mathfrak{g}$ to the associated Lie group G (Celledoni and Iserles, 2000). Since exp is a local diffeomorphism, mapping a neighborhood of the zero matrix O in $\mathfrak{g}$ onto a neighborhood of the identity matrix I in G, any dynamical system

Table 9.3. *Example of tangent spaces*

Group G	Algebra $\mathfrak{g}$	Characteristics
$\mathcal{G}l(n)$	$gl(n)$	$\mathbb{R}^{n\times n}$
$\mathcal{S}l(n)$	$sl(n)$	$\{M \in gl(n) \mid \text{trace}(M) = 0\}$
$\mathcal{A}ff(n)$	$aff(n)$	$\left\{\begin{bmatrix} M & \mathbf{t} \\ \mathbf{0} & 0 \end{bmatrix} \mid M \in gl(n), \mathbf{t} \in \mathbb{R}^n\right\}$
$\mathcal{O}(n)$	$o(n)$	$\{K \in gl(n) \mid \text{K is skew-symmetric}\}$
$\mathcal{I}som(n)$	$isom(n)$	$\left\{\begin{bmatrix} K & \mathbf{t} \\ \mathbf{0} & 0 \end{bmatrix} \mid K \in o(n), \mathbf{t} \in \mathbb{R}^n\right\}$
$G_1 \times G_2$	$\mathcal{T}_{(e_1,e_2)} G_1 \times G_2$	$\mathfrak{g}_1 \times \mathfrak{g}_2$

in G in the neighborhood of I therefore would have a corresponding dynamical system in $\mathfrak{g}$ in the neighborhood of O. It is easy to see that for any fixed $M \in \mathfrak{g}$, the map

$$\gamma(t) := \exp(tM) \tag{9.19}$$

defines a one-parameter subgroup in G.

Thus far we have drawn only an outline suggesting that different transformations can be unified under the same framework of tracing orbits associated with corresponding group actions. Under this framework it is plausible that more sophisticated actions could be composed that, in turn, might offer the design of new numerical algorithms. We point out specifically that, because of the Lie structure, the tangent space structure of a matrix group is the same at each of its element. It is yet to be determined how a dynamical system should be defined over an orbit or, equivalently, over a group so as to locate a certain canonical matrix on the orbit. Toward that end, the notion of "simplicity" of a matrix must be addressed. We shall demonstrate that such a conception varies according to the applications in the next section. Once the goal is set, we shall also discuss how various objective functions could be used to control the dynamical systems that drive curves on the orbits to a desired limit point. The vector field of such a dynamical system, if it can be defined, must lie in the tangent space. This requisite usually can be assured by a proper projection onto the tangent space. If the tangent space of a matrix group is known, then the projection usually can be calculated explicitly.

We have already employed the idea of projection earlier in Section 7.2. So that this chapter is self-contained, we illustrate the idea of projection again as follows.

Example 9.1. By Theorem 9.3, the tangent space of $\mathcal{O}(n)$ at any orthogonal matrix Q is

$$\mathcal{T}_Q\mathcal{O}(n) = Qo(n).$$

The normal space of $\mathcal{O}(n)$ at any orthogonal matrix Q is

$$\mathcal{N}_Q\mathcal{O}(n) = Qo(n)^{\perp},$$

where the orthogonal complement $o(n)^{\perp}$ is precisely the subspace $\mathcal{S}(n)$ of all symmetric matrices. The space $\mathbb{R}^{n\times n}$ can be split as the direct sum of

$$\mathbb{R}^{n\times n} = Qo(n) \oplus Qo(n)^{\perp}.$$

Any $X \in \mathbb{R}^{n\times n}$ therefore has a unique orthogonal splitting as

$$X = Q(Q^{\top}X) = Q\left\{\tfrac{1}{2}(Q^{\top}X - X^{\top}Q)\right\} + Q\left\{\tfrac{1}{2}(Q^{\top}X + X^{\top}Q)\right\}.$$

The projection of X onto the tangent space $\mathcal{T}_Q\mathcal{O}(n)$ is given by the formula

$$\mathrm{Proj}_{\mathcal{T}_Q\mathcal{O}(n)}X = Q\left\{\tfrac{1}{2}(Q^{\top}X - X^{\top}Q)\right\}. \tag{9.20}$$

9.4 Canonical form

We pointed out earlier that one of the main purposes in a realization process is to identify the simplest representation of a given linear transformation. The superlative adjective "simplest" is a relative term which should be interpreted broadly. Roughly speaking, canonical form refers to a "specific structure" by which a certain conclusion can be drawn or a certain goal can be achieved. Thus, depending on the applications, the specific structure could refer to a matrix with a specified pattern of zeros, such as a diagonal, tridiagonal, or triangular matrix. It could also refer to a matrix with a specified construct, such as Toeplitz, Hamiltonian, stochastic, or other linear varieties. It could even refer to a matrix with a specified algebraic constraint, such as low rank or nonnegativity. To reach each of the different canonical forms, different group actions and different curves on the orbits must be taken. We find that continuous group actions often enable us to tackle existence problems that are seemingly impossible to solve by conventional discrete methods.

Listed in Table 9.4 are some commonly used canonical forms together with desired actions to reach these forms.

It should be noted that in the last example above we have slightly extended the meaning of an orbit. The map

$$(U, S, V) \hookrightarrow \left(\mathrm{diag}\left(USS^{\top}U^{\top}\right)\right)^{-1/2} USV^{\top}$$

is not a group action. However, it is a natural way to define the set of rank k matrices whose rows are normalized to unity. The manifold of images under such a nonlinear map certainly is nontrivial but, analogous to a group orbit, is still computational traceable if we know how to manage the changes of U, S, and V in their respective groups. Generalizations such as this offer considerably more

Table 9.4. *Examples of canonical forms used in practice*

Canonical form	Also known as	Action
Bidiagonal J	Quasi-Jordan decomp., $A \in \mathbb{R}^{n\times n}$	$P^{-1}AP = J$, $P \in \mathcal{G}l(n)$
Diagonal Σ	Sing. value decomp., $A \in \mathbb{R}^{m\times n}$	$U^\top AV = \Sigma$, $(U,V) \in \mathcal{O}(m) \times \mathcal{O}(n)$
Diagonal pair (Σ_1, Σ_2)	Gen. sing. value decomp., $(A,B) \in \mathbb{R}^{m\times n} \times \mathbb{R}^{p\times n}$	$(U^\top AX, V^\top BX) = (\Sigma_1, \Sigma_2)$, $(U,V,X) \in \mathcal{O}(m) \times \mathcal{O}(p) \times \mathcal{G}l(n)$
Upper quasi-triangular H	Real Schur decomp., $A \in \mathbb{R}^{n\times n}$	$Q^\top AQ = H$, $Q \in \mathcal{O}(n)$
Upper quasi-triangular H Upper triangular U	Gen. real Schur decomp., $A, B \in \mathbf{R}^{n\times n}$	$(Q^\top AZ, Q^\top BZ) = (H, U)$, $Q, Z \in \mathcal{O}(n)$
Symmetric Toeplitz T	Toeplitz inv. eigenv. prob., $\{\lambda_1, \ldots, \lambda_n\} \subset \mathbb{R}$ is given	$Q^\top \mathrm{diag}\{\lambda_1, \ldots, \lambda_n\} Q = T$, $Q \in \mathcal{O}(n)$
Nonnegative $N \geq 0$	Nonneg. inv. eigenv. prob., $\{\lambda_1, \ldots, \lambda_n\} \subset \mathbb{C}$ is given	$P^{-1}\mathrm{diag}\{\lambda_1, \ldots, \lambda_n\} P = N$, $P \in \mathcal{G}l(n)$
Linear variety X with fixed entries at fixed locations	Matrix completion prob., $\{\lambda_1, \ldots, \lambda_n\} \subset \mathbb{C}$ is given $X_{i_\nu, j_\nu} = a_\nu, \nu = 1, \ldots, \ell$	$P^{-1}\{\lambda_1, \ldots, \lambda_n\} P = X$, $P \in \mathcal{G}l(n)$
Nonlinear variety with fixed singular values and eigenvalues	Test matrix construction, $\Lambda = \mathrm{diag}\{\lambda_1, \ldots, \lambda_n\}$ and $\Sigma = \mathrm{diag}\{\sigma_1, \ldots, \sigma_n\}$ are given	$P^{-1}\Lambda P = U^\top \Sigma V$ $P \in \mathcal{G}l(n), \quad U, V \in \mathcal{O}(n)$
Maximal fidelity	Structured low rank approx. $A \in \mathbb{R}^{m\times n}$	$\left(\mathrm{diag}\left(USS^\top U^\top\right)\right)^{-1/2} USV^\top$, $(U,S,V) \in \mathcal{O}(m) \times \mathbb{R}^k_\times \times \mathcal{O}(n)$

flexibility in applications. We shall see a few more similar generalizations in the sequel.

9.5 Objective functions

The orbit of a selected group action only defines the rule by which a transformation is to take place. To connect the current point and the desired canonical form on a given orbit, a bridge in the form of either an iterative scheme or a differential equation must be properly formulated. With the help of some objective functions, it is easy to control the construction of such a bridge. In this section, we outline some ideas on how objective functions could be designed.

9.5.1 *Least squares and projected gradient*

This section extends the notion of Section 7.2. We begin with the most general setting of projected gradient flows on a given continuous matrix group $G \subset \mathcal{G}l(n)$. Let X be a fixed matrix in a subset $\mathbb{V} \subset \mathbb{R}^{n\times n}$. Let $f : \mathbb{V} \longrightarrow \mathbb{R}^{n\times n}$ be a differentiable map with certain "inherent" properties such as symmetry, isospectrum, low rank, or other algebraic constraints. Suppose that a group action $\mu : G \times \mathbb{V} \longrightarrow \mathbb{V}$ has been specified. Suppose also that a projection map P from $\mathbb{R}^{n\times n}$ onto a singleton, a linear subspace, or an affine subspace $\mathbb{P}$ in $\mathbb{R}^{n\times n}$ is available. Elements in the set $\mathbb{P}$ carry a certain desired structure, for example, the canonical form. Consider the functional $F : G \longrightarrow \mathbb{R}$

$$F(g) := \frac{1}{2} \left\| f(\mu(g, X)) - P(\mu(g, X)) \right\|_F^2 . \tag{9.21}$$

The goal is to minimize F over G. The meaning of this constrained minimization is that, while staying in the orbit of X under the action of μ and maintaining the inherent property guaranteed by the function f, we look for the element $g \in G$ so that the matrix $f(\mu(g, X))$ best realizes the desired canonical structure in the sense of least squares.

Though iterative methods based on conventional optimization techniques for (9.21) are possible, we find that the projected gradient flow approach can conveniently be formulated as follows.

Algorithm 9.1 (Projected gradient method)
A generic projected gradient flow can be constructed via three steps:

1. Compute the gradient $\nabla F(g)$.
2. Project $\nabla F(g)$ onto the tangent space $\mathcal{T}_g G$ of G at g.
3. Follow the projected gradient numerically until convergence.

We note that many new techniques have been developed recently for dynamical systems on Lie groups, including the RK–MK methods (Munthe-Kaas, 1998), Magnus and Fer expansions (Blanes et al., 1998) and so on. A partial list of references for Lie structure preserving algorithms can be found in the seminal review paper by Iserles et al. (2000) and the book by Hairer et al. (2002). These new ODE techniques can certainly help the computations described in this paper, but we shall avoid discussing computational specifics in the current presentation.

Applications embracing the framework of the projected gradient approach are plenty. The following differential equations rehash what has been done in Chapter 7 and exemplify various dynamical systems on some appropriate orbits after computing gradients of some appropriate objective functions and performing projections onto some appropriate tangent spaces. We shall not repeat all the details for each of the cases. Readers are referred to the references for a more careful layout of the settings.

Example 9.2. Given a symmetric matrix Λ and a desirable structure $\mathbb{V}$, find a symmetric matrix X in $\mathbb{V}$ or closest to $\mathbb{V}$ that has the same spectrum as Λ (Chu, 1990).

$$\begin{cases} \dot{X} = [X, [X, P(X)]], \\ X(0) = \Lambda. \end{cases} \tag{9.22}$$

There are many important applications of (9.2). We remind readers specifically that, by taking $\Lambda = \operatorname{diag}\{\lambda_k\}_{k=1}^n$ and by choosing the structure retained in $\mathbb{V}$, the flow (9.2) can be used to tackle various kinds of very difficult structured inverse eigenvalue problems.

Example 9.3. Given a sequence of matrices $A_1, \ldots, A_k \in \mathbb{R}^{n\times n}$ and desirable structures in each individual $\mathbb{V}_1, \ldots, \mathbb{V}_k$, reduce each A_i to X_i that is closest to the structure $\mathbb{V}_i$ simultaneously by orthogonal similarity transformations (Chu and Driessel, 1991)

$$\begin{cases} \dot{X}_i = \left[X_i, \sum_{j=1}^{p} \frac{[X_j, P_j^{\top}(X_j)] - [X_j, P_j^{\top}(X_j)]^{\top}}{2}\right], \\ X_i(0) = A_i, \quad i = 1, \ldots, k. \end{cases} \tag{9.23}$$

A dynamical system similar to (9.3) can be derived if orthogonal equivalence transformations are used. Simultaneous reduction problems arise from areas such as system identification where it is desirable to model complicated phenomena by as fewer variables as possible.

Example 9.4. Given a matrix $A \in \mathbb{C}^{n\times n}$, find its nearest normal matrix approximation $W \in \mathbb{C}^{n\times n}$ (Chu and Driessel, 1991; Ruhe, 1987)

$$\begin{cases} \dot{W} = \left[W, \frac{[W, \operatorname{diag}(W^*)] - [W, \operatorname{diag}(W^*)]^*}{2}\right], \\ W(0) = A. \end{cases} \tag{9.24}$$

We note that this interesting approximation problem can be solved by using the unitary group over the complex domain.

Example 9.5. Construct a symmetric matrix with prescribed diagonal entries **a** and spectrum $\{\lambda_k\}_{k=1}^n$ (Chu, 1995a)

$$\begin{cases} \dot{X} = [X, [\mathrm{diag}(X) - \mathrm{diag}(\mathbf{a}), X]], \\ X(0) = \mathrm{diag}\{\lambda_k\}_{k=1}^n. \end{cases} \tag{9.25}$$

Recall that this inverse problem SHIEP is the harder part of the Schur–Horn theorem (Horn and Johnson, 1991).

Example 9.6. Given a matrix pencil $A - \lambda B$ where A is symmetric and B is symmetric and positive definite, and two desirable structures $\mathbb{V}_1$ and $\mathbb{V}_2$, find a matrix pencil $X - \lambda Y$ in or closest to the structure under the congruence transformation (Chu, 1998)

$$\begin{cases} \dot{X} = -\left((XW)^\top + XW\right), \\ \dot{Y} = -\left((YW)^\top + YW\right), \end{cases} \quad W := X(X - P_1(X)) + Y(Y - P_2(Y)). \tag{9.26}$$

This problem is a generalization of Example 9.2 to matrix pencils. The congruence transformation guarantees that the spectrum of the pencil $X - \lambda Y$ is the same as that of $A - \lambda B$.

9.5.2 *Systems for other objectives*

The least squares formulation (9.21) is but one way to define a dynamical system on an orbit. In this section, we briefly review some systems that are impelled by other objectives in mind. Some are defined even under no palpable objectives.

Example 9.7. The solution $X(t)$ to the Toda lattice (9.4) starting with a symmetric and tridiagonal initial value $X(0)$ will remain tridiagonal and symmetric for all $t \in \mathbb{R}$. The lattice actually arises from the dynamics of a special mass–spring system (Deift and Nanda, 1984; Symes, 1981/82).

It is the setup of the physical configuration that induces the tridiagonal structure. It is the law of Hamiltonian equations that governs the definition of the vector field. Other than these, there was no specific objective function used.

The extension of (9.4) to general matrices,

$$\dot{X} = [X, \Pi_0(G(X))], \tag{9.27}$$

where $G(z)$ is an analytic function over the spectrum of $X(0)$, appears to be done totally by brute force and blind reasoning (Chu, 1988; Watkins, 1984). Nevertheless, the differential equation approach nicely explains the pseudo-convergence and convergence behavior of the classical QR algorithm for general and normal matrices, respectively. The sorting of eigenvalues at the limit point had been observed, but did not seem to be clearly understood until Brockett's double bracket flow (Brockett, 1991),

$$\dot{X} = [X, [X, N]], \tag{9.28}$$

where N is a fixed and symmetric matrix, was developed.

The differential system (9.28) is a special case of the projected gradient flow (9.21) where $G = \mathcal{O}(n)$, $\mu(Q, X) = Q^\top \Lambda Q$, $P(\mu(Q, X)) \equiv N$, and f is the identity map. It follows that sorting is necessary in the first-order optimality condition (Chu, 1988). Furthermore, by taking the special diagonal matrix $N = \operatorname{diag}\{n, n-1, \ldots, 2, 1\}$ in the symmetric and tridiagonal case, it can be checked that the double bracket flow is exactly the same as the Toda lattice (Bloch et al., 1990). Thus, here comes the sudden understanding that the classical Toda lattice does have an objective function in mind.

The component-wise scaled Toda lattice (Chu, 1995b),

$$\dot{X} = [X, K \circ X], \tag{9.29}$$

where K is fixed and skew-symmetric and $\circ$ denotes the Hadamand product, is yet another venture into abstract mathematics where no explicit objective function is in sight. By varying K, it offers considerable flexibilities in component-wise scaling and enjoys very general convergence behavior.

Consider further some flows on the orbit $Orb_{\mathcal{O}(m)\times\mathcal{O}(n)}(X)$ under equivalence actions. Any flow on the orbit $Orb_{\mathcal{O}(m)\times\mathcal{O}(n)}(X)$ under equivalence must be of the form

$$\dot{X} = X(t)h(t) - k(t)X(t),$$

where $h(t) \in o(n)$ and $k(t) \in o(m)$ are functions of skew-symmetric matrices.

Example 9.8. Mimicking the way that the Toda lattice is related to the QR algorithm, the differential system

$$\begin{cases} \dot{X}_1 = X_1 \Pi_0(X_2^{-1}X_1) - \Pi_0(X_1 X_2^{-1})X_1, \\ \dot{X}_2 = X_2 \Pi_0(X_2^{-1}X_1) - \Pi_0(X_1 X_2^{-1})X_2, \end{cases} \tag{9.30}$$

is a continuous realization of the discrete QZ algorithm that has been used for solving the generalized eigenvalue problem (Chu, 1986a).

The original objective in deriving (9.30) was simply to fill in the natural generalization of the Toda lattice in the same way as the QZ algorithm generalizes the QR algorithm.

Example 9.9. In contrast to Example 9.8, the SVD flow (Chu, 1986b),

$$\begin{cases} \dot{Y} = Y \Pi_0\left(Y(t)^\top Y(t)\right) - \Pi_0\left(Y(t)Y(t)^\top\right) Y, \\ Y(0) = Y_0, \end{cases} \tag{9.31}$$

where Y_0 is a bidiagonal matrix was designed with two specific objectives in mind: one is that both skew-symmetric matrices $h(t)$ and $k(t)$ are kept tridiagonal and the other is that $Y(t)$ is kept bidiagonal throughout the continuous transition.

Component-wise, the dynamical system (9.31) can be represented simply by

$$\dot{x}_{i,i} = x_{i,i}\left(x_{i,i+1}^2 - x_{i-1,i}^2\right), \quad 1 \le i \le n,$$
$$\dot{x}_{j,j+1} = x_{j,j+1}\left(x_{j+1,j+1}^2 - x_{j,j}^2\right), \quad 1 \le j \le n-1,$$

but we think that the matrix equation (9.31) clearly manifests the group actions. It turns out that the flow (9.31) gives rise to the Toda flows for the tridiagonal matrices $Y^\top Y$ and $YY^\top$.

9.6 Generalization to non-group structures

Thus far, it appears that the orthogonal group $\mathcal{O}(n)$ is the most frequently employed group for actions. Though orthogonal group actions seem to be significant enough, it is at least of theoretical interest to ask the following questions:

- Can any advantages or applications be taken out of actions by other matrix groups?
- What can be said of the limiting behavior of dynamical systems defined on orbits by using different groups and actions?
- Does the isometry group, as a larger group, offer any serious alternatives to the orthogonal group?

As many as there are unexplored groups or actions, so are there many of these questions yet to be answered. Not all these groups or actions will be useful, but we think that there is a wide open area in this direction that deserves further study.

Furthermore, we think that the idea of group actions, least squares, and the corresponding gradient flows can be extended to other structures that have the structure of groups but are not groups. Some of the structures that have found applications in practice include, for example, the Stiefel manifold,

$$\mathcal{O}(p,q) := \{Q \in \mathbb{R}^{p\times q} | Q^\top Q = I_q\}; \tag{9.32}$$

the manifold of oblique matrices,

$$\mathcal{OB}(m,n) := \{Q \in \mathbb{R}^{m\times n} | \mathrm{diag}(QQ^\top) = I_m\}; \tag{9.33}$$

the cone of nonnegative matrices,

$$\mathcal{C}(n) := \{R \circ R | R \in \mathcal{R}^{n\times n}\}; \tag{9.34}$$

the manifold of low rank matrices,

$$\mathcal{L}(m,n,k) := \mathcal{O}(m,k) \times \mathbb{R}_\times^k \times \mathcal{O}(n,k); \tag{9.35}$$

and semigroups such as the standard symplectic matrices,

$$\mathcal{Z}(n) := \left\{ \begin{bmatrix} \alpha & \alpha t \\ s\alpha & \alpha^{-\top} + s\alpha t \end{bmatrix} \in \mathbb{R}^{2n\times 2n} \mid s,\, t \text{ symmetric and positive definite} \right\}. \tag{9.36}$$

These sets do not meet the criteria of being a Lie group, but can be characterized by using the product topology of some separate groups. In other words, groups can be used as some superficial coordinate systems to describe the motions on these manifolds. The framework discussed earlier can thus still be applied.

We illustrate how to use the above ideas to tackle some very challenging problems. So far as we know, no effective iterative methods are available yet that can address these difficulties.

Example 9.10. Recall that the StIEP discussed in Section 4.5 is solved by recasting the inverse problem through a minimization problem:

$$\begin{aligned} &\text{Minimize} \quad F(g, R) := \tfrac{1}{2}\|gJg^{-1} - R \circ R\|^2, \\ &\text{Subject to} \quad g \in \mathcal{G}l(n),\ R \in gl(n), \end{aligned}$$

where J is a real-valued matrix carrying the prescribed spectral information. Note that the constraints literally are immaterial because both $\mathcal{G}l(n)$ and $gl(n)$ are open sets. No projection onto the constraints is needed. Thus

$$\begin{cases} \dot{g} = [(gJg^{-1})^\top, \alpha(g, R)]g^{-\top} \\ \dot{R} = 2\alpha(g, R) \circ R, \end{cases} \tag{9.37}$$

with $\alpha(g, R) := gJg^{-1} - R \circ R$, represents the steepest descent flow in $\mathcal{G}l(n) \times gl(n)$.

To further stabilize the computation and avoid g^{-1}, recall that we have introduced a "parametrization" of g by its analytic singular value decomposition in the product group $\mathcal{O}(n) \times \mathbb{R}^n_\times \times \mathcal{O}(n)$. Suppose $g(t) = X(t)S(t)Y(t)^\top$ is the singular value decomposition of $g(t)$ where $S(t)$ is a diagonal matrix with elements from $\mathbb{R}^n_\times$ and $X(t)$ and $Y(t)$ are elements from $\mathcal{O}(n)$. Then the relationship of derivatives

$$X^\top \dot{g} Y = \underbrace{X^\top \dot{X}}_{Z} S + \dot{S} + S \underbrace{\dot{Y}^\top Y}_{W}, \tag{9.38}$$

is clearly true. Define $\Upsilon := X^\top \dot{g} Y$ with $\dot{g}$ given by (9.37). Then we now have a differential system,

$$\begin{cases} \dot{S} = \operatorname{diag}(\Upsilon), \\ \dot{X} = XZ, \\ \dot{Y} = YW, \end{cases} \tag{9.39}$$

that governs how the triplet $(X(t), S(t), Y(t))$ should be varied in the group $\mathcal{O}(n) \times \mathbb{R}^n_\times \times \mathcal{O}(n)$. Note that in the above Z and W are skew-symmetric matrices obtainable from off-diagonal elements of Υ and S. Together, the objective function F is now a function of the four variables (X, S, Y, R) in $\mathcal{O}(n) \times \mathbb{R}^n_\times \times \mathcal{O}(n) \times gl(n)$.

Example 9.11. Given a matrix $A \in \mathbb{R}^{n\times m}$ whose rows are of unit length, recall that its low rank approximation with rows of unit length is a least squares problem in the oblique space $\mathcal{OB}(n, m)$ (Chu et al., 2003). We recast the problem as minimizing the functional,

$$E(U, S, V) := \tfrac{1}{2} \left\| \left(\operatorname{diag}(USSU^\top)\right)^{-1/2} USV^\top - A \right\|_F^2, \tag{9.40}$$

with $U \in \mathcal{O}(n, k)$, $S \in \mathbb{R}^k_\times$, and $V \in \mathcal{O}(m, k)$, where $\mathcal{O}(p, q)$ denotes the Stiefel manifold. By construction, the product $Z = \left(\operatorname{diag}(USSU^\top)\right)^{-1/2} USV^\top$ is guaranteed to be of rank k and in $\mathcal{OB}(n, m)$. We can thus control the feasible matrix Z by controlling the variables (U, S, V).

The Stiefel manifold is not a group, but its tangent space can be explicitly calculated. It can be shown that the projection $P_{\mathcal{O}(p,q)}(M)$ of any matrix $M \in \mathbb{R}^{p\times q}$ onto the tangent space $\mathcal{T}_Q\mathcal{O}(p, q)$ is given by

$$\mathcal{P}_{\mathcal{O}(p,q)}(M) = Q\frac{Q^\top M - M^\top Q}{2} + (I - QQ^\top)M. \tag{9.41}$$

Replacing M in (9.41) by the appropriate partial derivatives of E with respect to U and V, respectively, we have established the projected gradient flow. Detailed calculations and numerical evidence are given in Chu et al. (2003).

9.7 Summary

There is a close relationship between matrix groups and linear transformations. Group actions together with properly formulated objective functions can offer a channel to tackle various classical or new and challenging problems arising from applied linear algebra. This chapter outlines some basic ideas and examples with the hope of bringing together the notions of group theory, linear transformations, and dynamical systems as a tool to undertake the task of system identification by canonical forms. More sophisticated actions can be composed that might offer the design of new numerical algorithms. The list of applications continues to grow. New computational techniques for structured dynamical systems on matrix group will further extend and benefit the scope of this interesting topic.

REFERENCES

The numbers in brackets following a reference indicate page numbers where the reference is cited.

Adamjan, V. M., D. Z. Arov, and M. G. Kreĭn. 1971. Analytic properties of the Schmidt pairs of a Hankel operator and the generalized Schur–Takagi problem, *Mat. Sb. (N.S.)* **86(128)**, 34–75. [251]

Adler, M., L. Haine, and P. van Moerbeke. 1993. Limit matrices for the Toda flow and periodic flags for loop groups, *Math. Ann.* **296**, no. 1, 1–33. [75]

Alexander, J. C. 1978. The additive inverse eigenvalue problem and topological degree, *Proc. Amer. Math. Soc.* **70**, no. 1, 5–7. [56, 63]

Alfakih, A. Y., A. Khandani, and H. Wolkowicz. 1999. Solving Euclidean distance matrix completion problems via semidefinite programming, *Comput. Optim. Appl.* **12**, no. 1–3, 13–30. Computational optimization — a tribute to Olvi Mangasarian, Part I. [288, 290]

Ames, W. F. 1992. *Numerical methods for partial differential equations*, Third edn, Academic Press, Boston, MA. [2]

Ammar, G., W. Gragg, and L. Reichel. 1991. Constructing a unitary Hessenberg matrix from spectral data, in *Numerical linear algebra, digital signal processing and parallel algorithms* (Leuven, 1988), pp. 385–395. [110, 112]

Ammar, G. S. and C. Y. He. 1995. On an inverse eigenvalue problem for unitary Hessenberg matrices, *Linear Alg. Appl.* **218**, 263–271. [110, 111, 112]

Anderson, E., Z. Bai, C. Bischof, S. Blackford, J. Demmel, J. Dongarra, J. Du Croz, A. Greenbaum, S. Hammarling, A. McKenney, and S. Sorensen. 1999. *Lapack users' guide*, Third edn, SIAM, Philadelphia. See also http://www.netlib.org/lapack. [3]

Andrea, S. A. and T. G. Berry. 1992. Continued fractions and periodic Jacobi matrices, *Linear Alg. Appl.* **161**, 117–134. [75]

Andrew, A. L. 1994. Some recent developments in inverse eigenvalue problems, in *Computational techniques and applications: CTAC93*, ed. D. Stewart, H. Gardner, and D. Singleton, pp. 94–102. [18]

Arnold, B. C. 1987. *Majorization and the Lorenz order: a brief introduction*, Lecture Notes in Statistics, vol. 43, Springer-Verlag, Berlin. [113]

Arnold, V. I. 1988. *Geometrical methods in the theory of ordinary differential equations*, Springer-Verlag, New York. Translated from the Russian by Joseph Szücs, Translation edited by Mark Levi. [339]

Aulbach, B. 1984. *Continuous and discrete dynamics near manifolds of equilibria*, Lecture Notes in Mathematics, vol. 1058, Springer-Verlag, Berlin. [224]

Baker, A. 2002. *Matrix groups*, Springer Undergraduate Mathematics Series, Springer-Verlag London Ltd., London. An introduction to Lie group theory. [345]

Bakonyi, M. and C. R. Johnson. 1995. The Euclidean distance matrix completion problem, *SIAM J. Matrix Anal. Appl.* **16**, 645–654. [287, 288]

Barcilon, V. 1974a. On the solution of inverse eigenvalue problems with high orders, *Geophys. J. Royal Astronomical Society* **39**, 153–154. [19]

Barcilon, V. 1974b. On the uniqueness of inverse eigenvalue problems, *Geophys. J. Royal Astronomical Society* **38**, 287–298. [19]

Barcilon, V. 1979. On the multiplicity of solutions of the inverse problems for a vibrating beam, *SIAM J. Appl. Math.* **37**, no. 3, 119–127. [17]

Barcilon, V. 1986. Inverse eigenvalue problems, in *Inverse problems* (Montecatini Terme, 1986), pp. 1–51, Lecture Notes in Mathematics, vol. 1225. Springer, Berlin. [2]

Barcilon, V. 1990. A two-dimensional inverse eigenvalue problem, *Inverse Problems* **6**, no. 1, 11–20. [17]

Barrett, W. W. and C. R. Johnson. 1984. Possible spectra of totally positive matrices, *Linear Alg. Appl.* **62**, 231–233. [94]

Bauer, F. L. 1963. Optimally scaled matrices, *Numer. Math.* **5**, 73–87. [66]

Bayer, D. A. and J. C. Lagarias. 1989. The nonlinear geometry of linear programming, I and II, *Trans. Amer. Math. Soc.* **314**, 499–581. [222]

de Beer, R. 1995. Quantitative in vivo nmr (nuclear magnetic resonance on living objects), Ph.D. Thesis. Available at http://dutnsic.tn.tudelft.nl:8080/c59_to_html/c59.html. [25, 250, 251, 260]

Belinfante, J. G. F. and B. Kolman. 1989. *A survey of Lie groups and Lie algebras with applications and computational methods*, Classics in Applied Mathematics, vol. 2, Society for Industrial and Applied Mathematics (SIAM), Philadelphia, PA. Reprint of the 1972 original. [345]

Berman, A. and A. Ben-Israel. 1971. A note on pencils of Hermitian or symmetric matrices, *SIAM J. Appl. Math.* **21**, 51–54. [232]

Berman, A. and R. J. Plemmons. 1979. *Nonnegative matrices in the mathematical sciences*, Academic Press, New York. Also Classics in Applied Mathematics 9, SIAM, Philadelphia, 1994. [93, 94, 95]

Berry, M. W., Z. Drmač, and E. R. Jessup. 1999. Matrices, vector spaces, and information retrieval, *SIAM Rev.* **41**, no. 2, 335–362 (electronic). [28, 301, 305]

Biegler-König, F. W. 1981a. Construction of band matrices from spectral data, *Linear Alg. Appl.* **40**, 79–87. [73]

Biegler-König, F. W. 1981b. Sufficient conditions for the solubility of inverse eigenvalue problems, *Linear Alg. Appl.* **40**, 89–100. [37, 58]

Bini, D. A., G. Heinig, and E. Tyrtyshnikov. 2003. Special issue on structured matrices: analysis, algorithms and applications, Elsevier Science B.V., Amsterdam. Papers from the workshop held in Cortona, September 21–28, 2000, *Linear Alg. Appl.* **366** (2003). [3]

Bischof, C., A. Carle, P. Hovland, P. Khademi, and A. Mauer. 1995. *Adifor 2.0 user's guide*, ANL/MCS-TM-192, Argonne National Laboratory. Available at http://www.mcs.anl.gov/adifor/. [259]

Blanes, S., F. Casas, J. A. Oteo, and J. Ros. 1998. Magnus and Fer expansions for matrix differential equations: the convergence problem, *J. Phys. A* **31**, no. 1, 259–268. [352]

Bloch, A. M. 1990. Steepest descent, linear programming, and Hamiltonian flows, *Contemporary Math.* **114**, 77–88. [220, 222]

Bloch, A. M. 1994. *Hamiltonian and gradient flows, algorithms and control*, Fields Institute Commun., vol. III, AMS. [223]

Bloch, A. M., R. W. Brockett, and R. S. Ratiu. 1990. A new formulation of the generalized toda lattice equations and their fixed point analysis via the momentum map, *Bulletin AMS* **23**, 477–485. [113, 222, 343, 355]

Boley, D. L. and G. H. Golub. 1978. The matrix inverse eigenvalue problem for periodic Jacobi matrices, in *Proceedings of the fourth symposium on basic problems of numerical mathematics* (Plzeň, 1978), pp. 63–76. [75]

Boley, D. L. and G. H. Golub. 1984. A modified method for reconstructing periodic Jacobi matrices, *Math. Comput.* **24**, 143–150. [75]

Boley, D. L. and G. H. Golub. 1987. A survey of matrix inverse eigenvalue problems, *Inverse Problems* **3**, no. 4, 595–622. [6, 45, 73, 74, 75, 82, 84, 85]

Boley, D. L., F. T. Luk, and D. Vandevoorde. 1997. Vandermonde factorization of a Hankel matrix, in *Scientific computing* (Hong Kong, 1997), pp. 27–39. [252]

Bolt, B. A. 1980. What can inverse theory do for applied mathematics and the sciences?, *Austral. Math. Soc. Gaz.* **7**, no. 3, 69–78. [1]

de Boor, C. and G. H. Golub. 1978. The numerically stable reconstruction of a Jacobi matrix from spectral data, *Linear Alg. Appl.* **21**, no. 3, 245–260. [74, 75, 80, 84]

Boyd, S. and L. El Ghaoui. 1993. Method of centers for minimizing generalized eigenvalues, *Linear Alg. Appl.* **188/189**, 63–111. [328]

Boyle, M. and D. Handelman. 1991. The spectra of nonnegative matrices via symbolic dynamics, *Ann. Math.* **133**, 249–316. [94, 95]

Brockett, R. W. 1989. Least squares matching problems, *Linear Alg. Appl.* **122/123/124**, 761–777. [222, 269]

Brockett, R. W. 1991. Dynamical systems that sort lists, diagonalize matrices and solve linear programming problems, *Linear Alg. Appl.* **146**, 79–91. [200, 220, 342, 355]

Brünger, A. T. and M Nilges. 1993. Computational challenges for macromolecular structure determination by x-ray crystallography and solution nmr-spectroscopy, *Q. Rev. Biophysics* **26**, 49–125. [289]

Bunse-Gerstner, A., R. Byers, V. Mehrmann, and N. K. Nichols. 1991. Numerical computation of an analytic singular value decomposition of a matrix valued function, *Numer. Math.* **60**, 1–40. [39, 108]

Bunse-Gerstner, A., R. Byers, and V. Mehrmann. 1993. Numerical methods for simultaneous diagonalization, *SIAM J. Matrix Anal. Appl.* **14**, no. 4, 927–949. [238, 239]

Burer, S. and R. D. C. Monteiro. 2001. A projected gradient algorithm for solving the maxcut sdp relaxation, *Optim. Methods Softw.* **15**, 175–200. [290]

Burer, S., R. D. C. Monteiro, and Y. Zhang. 1999. *Solving semidefinite programs via nonlinear programming, part 1: Transformations and derivative*, Department of Computational and Applied Mathematics, Rice University. [290]

Burg, J. P., D. G. Luenberger, and D. L. Wenger. 1982. Estimation of structured covariance matrices, *Proceedings of the IEEE* **70**, 963–974. [25]

Burkard, R. E. and U. Derigs. 1980. *Assignment and matching problems: solution methods with FORTRAN programs*, Springer-Verlag, Berlin. With the assistance of T. Bonniger and G. Katzakidis. [203]

Byers, R. 1988. A bisection method for measuring the distance of a stable matrix to the unstable matrices, *SIAM J. Sci. Stat. Comput.* **9**, no. **5**, 875–881. [33]

Byers, R. and S. G. Nash. 1989. Approaches to robust pole assignment, *Internat. J. Control* **49**, no. **1**, 97–117. [9, 145]

Byrnes, C. I. 1989. *Pole assignment by output feedback*, Three decades of mathematical system theory, pp. 31–78. [9, 12, 33]

Cadzow, J. A. 1988. Signal enhancement: A composite property mapping algorithm, *IEEE Trans. on Acoust., Speech, Signal Process.* **36**, 39–62. [25, 246, 248, 254, 260]

Cadzow, J. A. and D. M. Wilkes. 1990. Signal enhancement and the SVD, *Proceedings of the 2nd international workshop on SVD and signal processing*, pp. 144–151. [246, 250, 251, 260]

Calvetti, D., G. H. Golub, W. B. Gragg, and L. Reichel. 2000. Computation of Gauss–Kronrod quadrature rules, *Math. Comput.* **69**, no. 231, 1035–1052. [25, 55]

Calvo, M. P., A. Iserles, and A. Zanna. 1997. Numerical solution of isospectral flows, *Math. Comput.* **66**, 1461–1486. [93]

Cantoni, A. and F. Bulter. 1976. Eigenvalues and eigenvectors of symmetric centrosymmetric matrices, *Linear Alg. Appl.* **13**, 275–288. [3, 86, 87, 148, 258]

Carvalho, J., B. N. Datta, W. W. Lin, and C. S. Wang. 2001. Eigenvalue embedding in a quadratical pencil using symmetric low rank updates, Technical Report 2001-8, National Center for Theoretical Sciences, Mathematics Div. [161]

Causey, R. L. 1964. On closest normal matrices, Ph.D. Thesis, Department of Computer Science, Stanford University. [242]

Celledoni, E. and A. Iserles. 2000. Approximating the exponential from a Lie algebra to a Lie group, *Math. Comput.* **69**, no. 232, 1457–1480. [340]

Chadan, K., D. Colton, L. Päivärinta, and W. Rundell. 1997. *An introduction to inverse scattering and inverse spectral problems*, Society for Industrial and Applied Mathematics (SIAM), Philadelphia, PA. [2, 18]

Chan, R. H., J. G. Nagy, and R. J. Plemmons. 1994. Circulant preconditioned Toeplitz least squares iterations, *SIAM J. Matrix Anal. Appl.* **15**, no. 1, 80–97. [265]

Chan, R. H., S.-F. Xu, and H.-M. Zhou. 1999. On the convergence of a quasi-Newton method for inverse eigenvalue problems, *SIAM J. Numer. Anal.* **36**, no. 2, 436–441 (electronic). [51]

Chan, T. F. 1988. An optimal circulant preconditioner for Toeplitz systems, *SIAM J. Sci. Statist. Comput.* **9**, no. 4, 766–771. [265, 267]

Cheney, W. and A. A. Goldstein. 1959. Proximity maps for convex sets, *Proc. Amer. Math. Soc.* **10**, 448–450. [199, 201, 254]

Chu, E. K.-W. and B. N. Datta. 1996. Numerical robust pole assignment for second-order systems, *Internat. J. Control* **64**, 1113–1127. [161]

Chu, M. T. 1986a. A continuous approximation to the generalized Schur decomposition, *Linear Alg. Appl.* **78**, 119–132. [355]

Chu, M. T. 1986b. A differential equation approach to the singular value decomposition of bidiagonal matrices, *Linear Alg. Appl.* **80**, 71–79. [355]

Chu, M. T. 1988. On the continuous realization of iterative processes, *SIAM Rev.* **30**, no. 3, 375–387. [47, 234, 343, 354, 355]

Chu, M. T. 1990. Solving additive inverse eigenvalue problems for symmetric matrices by the homotopy method, *IMA J. Numer. Anal.* **10**, no. 3, 331–342. [56, 63, 269, 342, 353]

Chu, M. T. 1992a. Matrix differential equations: A continuous realization process for linear algebra problems, *Nonlinear Anal.*, TMA **18**, no. 12, 1125–1146. [134, 200, 224]

Chu, M. T. 1992b. Numerical methods for inverse singular value problems, *SIAM J. Numer. Anal.* **29**, no. 3, 885–903. [129, 131]

Chu, M. T. 1993. *On a differential equation* $\frac{dX}{dt} = [x, k(x)]$ *where* k *is a Toeplitz annihilator*. Available at http://www4.ncsu.edu/~mtchu. [93]

Chu, M. T. 1994. *On the refinement of a Newton method for the inverse Toeplitz eigenvalue problem*. Available at http://www4.ncsu.edu/~mtchu. [90]

Chu, M. T. 1995a. Constructing a Hermitian matrix from its diagonal entries and eigenvalues, *SIAM J. Matrix Anal. Appl.* **16**, no. 1, 207–217. [354]

Chu, M. T. 1995b. Scaled Toda-like flows, *Linear Alg. Appl.* **215**, 261–273. [355]

Chu, M. T. 1998. Inverse eigenvalue problems, *SIAM Rev.* **40**, no. 1, 1–39 (electronic). [2, 6, 8, 73, 341, 354]

Chu, M. T. 1999. On constructing matrices with prescribed singular values and diagonal elements, *Linear Alg. Appl.* **288**, no. 1–3, 11–22. [116]

Chu, M. T. 2000. A fast recursive algorithm for constructing matrices with prescribed eigenvalues and singular values, *SIAM J. Numer. Anal.* **37**, no. 3, 1004–1020 (electronic). [135, 136, 138, 139, 289, 301, 305]

Chu, M. T. and K. R. Driessel. 1990. The projected gradient method for least squares matrix approximations with spectral constraints, *SIAM J. Numer. Anal.* **27**, 1050–1060. [47, 50]

Chu, M. T. and K. R. Driessel. 1991. Constructing symmetric nonnegative matrices with prescribed eigenvalues by differential equations, *SIAM J. Math. Anal.* **22**, 1372–1387. [96]

Chu, M. T. and G. H. Golub. 2002. Structured inverse eigenvalue problems, *Acta Numerica*, 1–71. [2, 6]

Chu, M. T. and Q. Guo. 1998. A numerical method for the inverse stochastic spectrum problem, *SIAM J. Matrix Anal. Appl.* **19**, no. **4**, 1027–1039. [106]

Chu, M. T. and L. K. Norris. 1988. Isospectral flows and abstract matrix factorizations, *SIAM J. Numer. Anal.* **25**, no. 6, 1383–1391. [214, 225, 236, 343]

Chu, M. T. and J. L. Watterson. 1993. Multivariate eigenvalue problem: I. algebraic theory and power method, *SIAM J. Sci. Comput.* **14**, no. 5, 1089–1106. [199]

Chu, M. T. and J. W. Wright. 1995. The educational testing problem revisited, *IMA J. Numer. Anal.* **15**, no. 1, 141–160. [328, 329]

Chu, M. T. and S. F. Xu. 2005. On computing minimal realizable spectral radii of nonnegative matrices, *Numer. Linear Alg. Appl.* 12, 77–86. [99]

Chu, M. T., N. Del Buono, L. Lopez, and T. Politi. 2003. *On the low rank approximation of data on the unit sphere*, North Carolina State University. [357, 358]

Chu, M. T., F. Diele, and I. Sgura. 2003. *On the robust matrix completion with prescribed eigenvalues,* Future Generation Computing Systems **19**, 1139–1153. [124]

Chu, M. T., F. Diele, and I. Sgura. 2004. Gradient flow methods for matrix completion with prescribed eigenvalues, *Linear Alg. Appl.* **379**, 85–112. [119, 126]

Chu, M. T., R. E. Funderlic, and G. H. Golub. 1995. A rank-one reduction formula and its applications to matrix factorizations, *SIAM Rev.* **37**, no. 4, 512–530. [253]

Chu, M. T., R. E. Funderlic, and R. J. Plemmons. 2003. Structured low rank approximation, *Linear Alg. Appl.* **366**, 157–172. Special issue on structured matrices: analysis, algorithms and applications (Cortona, 2000). [311]

Chu, M. T., Y. C. Kuo, and W. W. Lin. 2004. On inverse quadratic eigenvalue problems with partially prescribed eigenstructure, *SIAM J. Mat. Anal. Appl.* **25**, 995–1020. [162]

Chugunov, V. N. 2001. *Maple procedure for the construction of a matrix with prescribed eigenvalues and n prescribed entries.* Personal communication. [117]

Commandeur, J. J. F., P. J. F. Groenen, and J. J. Meulman. 1999. A distance-based variety of nonlinear multivariate data analysis, including weights for objects and variables, *Psychometrika* **64**, no. 2, 169–186. [287]

Conn, A. R., N. I. M. Gould, and Ph. L. Toint. 1992. *LANCELOT*, Springer-Verlag, Berlin. A Fortran package for large-scale nonlinear optimization (release A). [259]

Corless, R. M., P. M. Gianni, B. M. Trager, and S. M. Watt. 1995. The singular value decomposition for polynomial system, *Proc. int. symp. symbolic and algebraic computation,* pp. 195–207. [26]

Crippen, G. M. and T. F. Havel. 1988. *Distance geometry and molecular conformation,* Chemo-metrics Series, vol. 15, Research Studies Press Ltd., Chichester. [286, 287, 290]

Curtis, M. L. 1984. *Matrix groups*, Second edn, Universitext, Springer-Verlag, New York. [345]

Cybenko, G. 1984. On the eigenstructure of Toeplitz matrices, *IEEE Trans. Acoust., Speech, Signal Process.* **ASSP-32**, 918–920. [148, 149, 150, 190]

Cybenko, G. 1982. Moment problems and low rank Toeplitz approximations, *Circuits Systems Signal Process.* **1**, no. 3–4, 345–366. [25, 248]

Czyzyk, J., M. Mesnier, and J. Moré. 1996. *The network-enabled optimization system (neos) server.* MCS-P615-1096, Argonne National Laboratory. [14, 260]

Dai, H. 1988. Inverse eigenvalue problems for real symmetric banded matrices, *Math. Numer. Sinica* **10**, no. 1, 107–111. [73]

Dai, H. 1995. About an inverse eigenvalue problem arising in vibration analysis, *RAIRO Modél. Math. Anal. Numér.* **29**, no. 4, 421–434. [17]

Datta, B. N. 2002. *Finite element model updating, eigenstructure assignment and eigenvalue embedding techniques for vibrating systems*, Vibration Control, pp. 83–96. [17, 146, 161]

Datta, B. N. and D. R. Sarkissian. 2001. Theory and computations of some inverse eigenvalue problems for the quadratic pencil, in *Structured matrices in mathematics, computer science, and engineering, i* (Boulder, CO, 1999), pp. 221–240. [16, 161]

Datta, B. N., S. Elhay, and Y. M. Ram. 1997. Orthogonality and partial pole assignment for the symmetric definite quadratic pencil, *Linear Alg. Appl.* **257**, 29–48. [75]

Datta, B. N., S. Elhay, Y. M. Ram, and D. R. Sarkissian. 2000. Partial eigenstructure assignment for the quadratic pencil, *J. Sound Vibration* **230**, no. 1, 101–110. [16, 161]

Dattorro, J. 2004. *Euclidean distance matrix*, Stanford University. http://www.stanford.edu/~dattorro/EDM.pdf. [287]

Davis, P. J. 1979. *Circulant matrices*, John Wiley & Sons, New York-Chichester-Brisbane. A Wiley-Interscience Publication, Pure and Applied Mathematics. [265, 266]

De Moor, B. 1994. Total least squares for affinely structured matrices and the noisy realization problem, *IEEE Trans.* Signal Processing **42**, 3104–3113. [25, 261]

Deakin, A. S. and T. M. Luke. 1992. On the inverse eigenvalue problem for matrices, *J. Phys. A* **25**, no. 3, 635–648. [19]

Deift, P. and T. Nanda. 1984. On the determination of a tridiagonal matrix from its spectrum and a submatrix, *Linear Alg. Appl.* **60**, 43–55. [341, 354]

Della-Dora, J. 1975. Numerical linear algorithms and group theory, *Linear Alg. Appl.* **10**, 267–283. [340]

Delsarte, P. and Y. Genin. 1984. *Spectral properties of finite Toeplitz matrices*, Mathematical theory of networks and systems, pp. 194–213. Lecture Notes in Control and Information Sciences vol. 58. [89, 91, 148]

Dendrinos, M., S. Bakamidis, and G. Carayannis. 1991. Speech enhancement from noise: A regenerative approach, *Speech Communication* **10**, 45–57. [25, 260]

Deutsch, F. 2001. Accelerating the convergence of the method of alternating projections via a line search: a brief survey, in *Inherently parallel algorithms in feasibility and optimization and their applications* (Haifa, 2000), pp. 203–217. [199, 201, 254]

Dhillon, I. S. 2001. Concept decompositions for large sparse text data using clustering, *Machine Learning J.* **42**, 143–175. [306, 311]

Dieci, L. and T. Eirola. 1999. On smooth decompositions of matrices, *SIAM J. Matrix Anal. Appl.* **20**, no. 3, 800–819 (electronic). [39]

Dieci, L., R. D. Russel, and E. S. V. Vleck. 1994. Unitary integrators and applications to continuous orthonormalization techniques, *SIAM J. Numer. Anal.* **31**, 261–281. [54, 93, 214]

Diele, F. and I. Sgura. 1999. Isospectral flows and the inverse eigenvalue problem for Toeplitz matrices, *J. Comput. Appl. Math.* **110**, no. 1, 25–43. [93, 225]

Dikin, I. I. 1967. Iterative solution of problems of linear and quadratic programming, *Dokl. Akad. Nauk SSSR* **174**, 747–748. [329]

Donoho, D. and V. Stodden. 2003. *When does nonnegative matrix factorization give a correct decomposition into parts*, Department of Statistics, Stanford University. Available at http://www-stat.stanford.edu/~donoho. [322]

Downing, A. C. and A. S. Householder. 1956. Some inverse characteristic value problems, *J. Assoc. Comput. Mach.* **3**, 203–207. [65, 73]

Driessel, K. R. 1987a. On finding the eigenvalues and eigenvectors of a matrix by means of an isospectral gradient flow, Technical Report 541, Clemson University. [225]

Driessel, K. R. 1987b. On finding the singular values and singular vectors of a matrix by means of an isospectral gradient flow, Technical Report 87-01, Idaho State University. [229]

Duarte, A. L. 1989. Construction of acyclic matrices from spectral data, *Linear Alg. Appl.* **113**, 173–182. [18]

Eisenstat, S. C., J. W. Lewis, and M. H. Schultz. 1982. Optimal block diagonal scaling of block 2-cyclic matrices, *Linear Alg. Appl.* **44**, 181–186. [67]

Elishakoff, I. 2000. A selective review of direct, semi-inverse and inverse eigenvalue problems for structures described by differential equations with variable coefficients, *Arch. Comput. Methods Engrg.* **7**, no. 4, 451–526. [2, 9, 17]

Elsner, L. and Kh. D. Ikramov. 1998. Normal matrices: an update, *Linear Alg. Appl.* **285**, no. 1–3, 291–303. [241]

EPA. 2001. *National air quality and emissions trends report*, EPA 454/R-01-004, Office of Air Quality Planning and Standards, EPA, Research Triangle Park. [332]

Erra, R. and B. Philippe. 1997. On some structured inverse eigenvalue problems, *Numer. Algorithms* **15**, no. 1, 15–35. [74]

Farebrother, R. W. 1987. Three theorems with applications to Euclidean distance matrices, *Linear Alg. Appl.* **95**, 11–16. [287]

Faßbender, H. 1997. Inverse unitary eigenproblems and related orthogonal functions, *Numer. Math.* **77**, no. 3, 323–345. [110]

Faßbender, H. 2000. *Symplectic methods for the symplectic eigenproblem*, Kluwer Academic/Plenum Publishers, New York. [3]

Faybusovich, L. 1991a. Dynamical systems which solve optimization problems with linear constraints, *IMA J. Math. Control Inform.* **8**, 135–149. [223]

Faybusovich, L. 1991b. Hamiltonian structure of dynamical systems which solve linear programming problems, *Physics D* **53**, 217–232. [223]

Ferguson, W. 1980. The construction of Jacobi and periodic Jacobi matrices with prescribed spectra, *Math. Comput.* **35**, 79–84. [24, 75, 80]

Ferng, W. R., W.-W. Lin, D. J. Pierce, and C.-S. Wang. 2001. Nonequivalence transformation of λ-matrix eigenproblems and model embedding approach to model tuning, *Numer. Linear Alg. Appl.* **8**, no. 1, 53–70. [161]

Fiedler, M. 1974. Eigenvalues of nonnegative symmetric matrices, *Linear Alg. Appl.* **9**, 119–142. [93]

Fletcher, R. 1987. *Practical methods of optimization*, Second edn, A Wiley-Interscience Publication, John Wiley & Sons Ltd., Chichester. [324, 331]

Foltete, E., G. M. L. Gladwell, and G. Lallement. 2001. On the reconstruction of a damped vibrating system from two complex spectra, part ii — experiment, *J. Sound Vib.* **240**, 219–240. [79]

Forsythe, G. E. and E. G. Straus. 1955. On best conditioned matrices, *Proc. Amer. Math. Soc.* **6**, 340–345. [21, 66, 194]

Friedland, S. 1972. Matrices with prescribed off-diagonal elements, *Israel J. Math.* **11**, 184–189. [117]

Friedland, S. 1975. On inverse multiplicative eigenvalue problems for matrices, *Linear Alg. Appl.* **12**, 127–137. [65]

Friedland, S. 1977. Inverse eigenvalue problems, *Linear Alg. Appl.* **17**, no. 1, 15–51. [56]

Friedland, S. 1978. On an inverse problem for nonnegative and eventually nonnegative matrices, *Israel J. Math.* **29**, 43–60. [94]

Friedland, S. 1983. Simultaneous similarity of matrices, *Adv. in Math.* **50**, no. 3, 189–265. [230]

Friedland, S. and A. A. Melkman. 1979. On the eigenvalues of nonnegative Jacobi matrices, *Linear Alg. Appl.* **25**, 239–254. [94, 95, 96]

Friedland, S., J. Nocedal, and M. L. Overton. 1986. Four quadratically convergent methods for solving inverse eigenvalue problems, in *Numerical analysis* (Dundee, 1985), pp. 47–65. [32, 39, 46, 133]

Friedland, S., J. Nocedal, and M. L. Overton. 1987. The formulation and analysis of numerical methods for inverse eigenvalue problems, *SIAM J. Numer. Anal.* **24**, no. 3, 634–667. [51, 213]

Frieze, A. M., R. Kannan, and S. Vempala. 1998. Fast Monte-Carlo algorithms for finding low-rank approximations, *in IEEE symposium on foundations of computer science*, pp. 370–378. Available at http://www.cs.yale.edu/~kannan. [246]

Gabriel, R. 1979. Matrizen mit maximaler Diagonale bei unitärer Similarität, *J. Reine Angew. Math.* **307/308**, 31–52. [242, 243]

Gabriel, R. 1987. The normal δH-matrices with connection to some Jacobi-like methods, *Linear Alg. Appl.* **91**, 181–194. [243]

Gantmacher, F. P. and M. G. Krein. 2002. *Oscillation matrices and kernels and small vibrations of mechanical systems*, Revised, AMS Chelsea Publishing, Providence, RI. Translation based on the 1941 Russian original, Edited and with a preface by Alex Eremenko. [14, 17]

Gantmacher, F. R. 1959. *The theory of matrices. Vols. 1, 2*, Chelsea Publishing Co., New York. [224, 238]

Gel′fand, I. M. and B. M. Levitan. 1955. On the determination of a differential equation from its spectral function, *Amer. Math. Soc. Transl.* (2) **1**, 253–304. [19]

Gel′fand, I. M. and V. A. Ponomarev. 1969. Remarks on the classification of a pair of commuting linear transformations in a finite-dimensional space, *Funkcional. Anal. i Priložen.* **3**, no. 4, 81–82. [230]

Gill, P. E., W. Murray, and M. H. Wright. 1981. *Practical optimization*, Academic Press Inc. [Harcourt Brace Jovanovich Publishers], London. [213, 214, 216, 235, 325, 331]

Gladwell, G. M. L. 1984. The inverse problem for the vibrating beam, *Proc. R. Soc. a* **393**, 277–295. [2, 17]

Gladwell, G. M. L. 1986a. The inverse mode problem for lumped-mass systems, *Quart. J. Mech. Appl. Math.* **39**, no. 2, 297–307. [17, 79, 146, 147]

Gladwell, G. M. L. 1986b. *Inverse problems in vibration*, Martinus Nijhoff, Dordrecht, Netherlands. [6, 9, 17, 28, 77, 78]

Gladwell, G. M. L. 1986c. Inverse problems in vibration, *Appl. Mech. Rev* **39**, 1013–1018. [8]

Gladwell, G. M. L. 1996. Inverse problems in vibration, II, *Appl. Mech. Rev* **49**, no. 10, 2–27. [2, 6, 8, 17, 79, 212]

Gladwell, G. M. L. 1997. Inverse vibration problems for finite-element models, *Inverse Problems* **13**, no. 2, 311–322. [17]

Gladwell, G. M. L. 1999. Inverse finite element vibration problems, *J. Sound Vib.* **211**, 309–324. [17]

Gladwell, G. M. L. 2001. On the reconstruction of a damped vibrating system from two complex spectra, part i — theory, *J. Sound Vib.* **240**, 203–217. [79, 80]

Gladwell, G. M. L. 2004. *Minimal mass solution to inverse eigenvalue problem*, University of Waterloo. [14]

Gladwell, G. M. L. and J. A. Gbadeyan. 1985. On the inverse problem of the vibrating string or rod, *Q. J. Mech. Appl. Math.* **38**, 169–174. [17]

Gladwell, G. M. L. and A. Morassi. 1999. Estimating damage in a rod from changes in node position, *Inverse Problems in Engineering* **7**, 215–233. [19]

Gladwell, G. M. L. and B. R. Zhu. 1992. Inverse problems for multidimensional vibrating systems, *Proc. Roy. Soc. London Ser.* A **439**, no. 1907, 511–530. [17]

Glunt, W., T. L. Hayden, S. Hong, and J. Wells. 1990. An alternating projection algorithm for computing the nearest Euclidean distance matrix, *SIAM J. Matrix Anal. Appl.* **11**, no. 4, 589–600. [27, 254, 287, 290]

Glunt, W., T. L. Hayden, and M. Raydan. 1993. Molecular conformations from distance matrices, *J. Comput. Chemistry* **14**, 114–120. [290]

Gohberg, I., P. Lancaster, and L. Rodman. 1982. *Matrix polynomials*, Academic Press Inc. [Harcourt Brace Jovanovich Publishers], New York. Computer Science and Applied Mathematics. [160, 161]

Gohberg, I., M. A. Kaashoek, and F. van Schagen. 1995. *Partially specified matrices and operators: classification, completion, applications*, Birkhäuser Verlag, Basel. [54, 63, 70]

Goldstine, H. H. and L. P. Horwitz. 1959. A procedure for the diagonalization of normal matrices, *J. Assoc. Comput. Mach.* **6**, 176–195. [234, 244]

Golub, G. H. and J. M. Varah. 1974. On a characterization of the best l_2-scaling of a matrix, *SIAM J. Numer. Anal.* **11**, 472–479. [66]

Golub, G. H. and C. F. Van Loan. 1996. *Matrix computations*, Third edn, Johns Hopkins University Press, Baltimore, MD. [3, 213, 228, 233, 234, 246, 341]

Gower, J. C. 1982. Euclidean distance geometry, *Math. Sci.* **7**, no. 1, 1–14. [27, 287]

Gower, J. C. 1985. Properties of Euclidean and non-Euclidean distance matrices, *Linear Alg. Appl.* **67**, 81–97. [287]

Gragg, W. B. and W. J. Harrod. 1984. The numerically stable reconstruction of Jacobi matrices from spectral data, *Numer. Math.* **44**, no. 3, 317–335. [74, 85]

Gray, L. J. and D. G. Wilson. 1976. Construction of a Jacobi matrix from spectral data, *Linear Alg. Appl.* **14**, 131–134. [73]

Gray, R. M. 2002. *Toeplitz and circulant matrices: A review*, Stanford University. Available at http://ee.stanford.edu/~gray/toeplitz.pdf. [85]

Greenbaum, A. 1992. Diagonally scalings of the Laplacian as preconditioners for other elliptic differential operators, *SIAM J. Matrix Anal. Appl.* **13**, 826–846. [66, 193, 194]

Greenbaum, A. and G. H. Rodrigue. 1989. Optimal preconditioners of a given sparsity pattern, *BIT* **29**, 610–634. [21, 193, 194]

Grone, R., C. R. Johnson, E. M. Sa, and H. Wolkowicz. 1987. Normal matrices, *Linear Alg. Appl.* **87**, 213–225. [241]

Gropp, W. and J. J. Moré. 1997. Optimization environments and the NEOS server, in *Approximation theory and optimization* (Cambridge, 1996), pp. 167–182. [259, 260]

Grötschel, M., L. Lovász, and A. Schrijver. 1993. *Geometric algorithms and combinatorial optimization*, Second edn, Algorithms and Combinatorics, vol. 2, Springer-Verlag, Berlin. [330]

Guillemin, V. and A. Pollack. 1974. *Differential topology*, Prentice-Hall Inc., Englewood Cliffs, NJ. [216]

Guo, W. 1996. An inverse eigenvalue problem for nonnegative matrices, *Linear Alg. Appl.* **249**, 67–78. [94]

Guo, W. 1997. Eigenvalues of nonnegative matrices, *Linear Alg. Appl.* **266**, 261–270. [97, 98, 99]

Hadeler, K. P. 1968. Ein inverse Eigenwertproblem, *Linear Alg. Appl.* **1**, 83–101. [58]

Hairer, E., C. Lubich, and G. Wanner. 2002. *Geometric numerical integration*, Springer Series in Computational Mathematics, vol. 31, Springer-Verlag, Berlin. Structure-preserving algorithms for ordinary differential equations. [54, 109, 214, 353]

Hald, O. H. 1972. On discrete and numerical inverse Sturm–Liouville problems, Ph.D. Thesis, New York University. [19, 73]

Hald, O. H. 1976. Inverse eigenvalue problems for Jacobi matrices, *Linear Alg. Appl.* **14**, no. 1, 63–85. [73, 74, 77, 80, 81, 82]

Hald, O. H. and J. R. McLaughlin. 1989. Solutions of inverse nodal problems, *Inverse Problems* **5**, no. 3, 307–347. [146]

Hald, O. H. and J. R. McLaughlin. 1996. Inverse nodal problems: finding the potential from nodal lines, *Mem. Amer. Math. Soc.* **119**, no. 572, viii+148. [2, 146]

Halmos, P. R. 1974. *Finite dimensional vector spaces*, Second edn, Springer-Verlag, New York. Reprinting of the 1958 second edition, Undergraduate Texts in Mathematics. [vi]

Han, S.-P. 1988. A successive projection method, *Math. Programming* **40**, no. 1, (Ser. A), 1–14. [199, 201]

Hanke, M., J. Nagy, and R. Plemmons. 1993. Preconditioned iterative regularization for ill-posed problems, in *Numerical linear algebra* (Kent, OH, 1992), pp. 141–163. [265, 278]

Hansen, P. C. 1987. The truncated SVD as a method for regularization, *BIT* **27**, no. 4, 534–553. [28, 246]

Hansen, P. C. 1990. Truncated singular value decomposition solutions to discrete ill-posed problems with ill-determined numerical rank, *SIAM J. Sci. Statist. Comput.* **11**, no. 3, 503–518. [28]

Havel, T. F. and K. Wüthrich. 1984. A distance geometry program for determining the structures of small proteins and other macromolecules from nuclear magnetic resonance mesaurements of intermolecular ^1H-^1H proximities in solution, *Bull. Math. Bio.* **46**, 673–698. [287]

Haykin, S. 1996. *Adaptive filter theory*, Third ed., Prentice Hall Information and System Sciences Series, Prentice Hall, Upper Saddle River, NJ. [85, 86]

Helton, W., J. Rosenthal, and X. Wang. 1997. Matrix extensions and eigenvalue completions, the generic case, *Trans. Amer. Math. Soc.* **349**, no. 8, 3401–3408. [35]

Hershkowitz, D. 1978. Existence of matrices satisfying prescribed conditions, Master's Thesis, Haifa. [45, 119]

Hershkowitz, D. 1983. Existence of matrices with prescribed eigenvalues and entries, *Linear and Multi-linear Algebra* **14**, 315–342. [76, 119]

Higham, N. J. 1989. Matrix nearness problems and applications, in *Applications of matrix theory* (Bradford, 1988), pp. 1–27. [241, 243]

Higham, N. J. 2002. Computing the nearest correlation matrix — a problem from finance, *IMA J. Numer. Anal.* **22**, no. 3, 329–343. [254]

Hill, R. D., R. G. Bates, and S. R. Waters. 1990. On centrohermitian matrices, *SIAM J. Matrix Anal. Appl.* **11**, 128–133. [86]

Hochstadt, H. 1967. On some inverse problems in matrix theory, *Arch. Math.* **18**, 201–207. [73, 74, 77]

Hochstadt, H. 1974. On construction of a Jacobi matrix from spectral data, *Linear Alg. Appl.* **8**, 435–446. [73, 81]

Hochstadt, H. 1979. On the construction of a Jacobi matrix from mixed given data, *Linear Alg. Appl.* **28**, 113–115. [74, 75]

Hock, W. and K. Schittkowski. 1983. A comparative performance evaluation of 27 nonlinear programming codes, *Computing* **30**, no. 4, 335–358. [324]

Hoffman, A. J. and H. W. Wielandt. 1953. The variation of the spectrum of a normal matrix, *Duke Math. J.* **20**, 37–39. [222]

Horn, A. 1954a. Doubly stochastic matrices and the diagonal of a rotation matrix, *Amer. J. Math.* **76**, 620–630. [113]

Horn, A. 1954b. On the eigenvalues of a matrix with prescribed singular values, *Proc. Amer. Math. Soc.* **5**, 4–7. [135]

Horn, R. A. and C. R. Johnson. 1991. *Matrix analysis*, Cambridge University Press, New York. [3, 50, 198, 222, 329, 354]

Howe, R. 1983 (and 1984). Very basic Lie theory, *Amer. Math. Monthly* **90 (and 91)**, no. 9 (and 4), 600–623 (and 247). [345]

Hu, Y. H. and S. Y. Kung. 1985. Toeplitz eigensystem solver, *IEEE Trans. Acoustics, Speech, Signal Process.* **33**, 1264–1271. [86]

Huang, H. X., Z. A. Lian, and P. M. Pardalos. 2003. Some properties for the Euclidean distance matrix and positive semidefinite matrix completion problems, *J. Global Optim.* **25**, 3–21. [288]

Ikramov, K. D. and V. N. Chugunov. 2000. Inverse matrix eigenvalue problems, *J. Math. Sci.* (New York) **98**, no. 1, 51–136. [6, 45, 76, 112, 114, 116, 117, 118, 119]

Iserles, A., H. Munthe-Kaas, S. P. Nørsett, and A. Zanna. 2000. Lie group methods, *Acta Numerica* **9**, 1–151. [54, 93, 109, 214, 353]

Johnson, C. R. and P. Tarazaga. 1995. Connections between the real positive semidefinite and distance matrix completion problems, *Linear Alg. Appl.* **223/224**, 375–391. [287]

Johnson, C. R., T. J. Laffey, and R. Loewy. 1996. The real and the symmetric nonnegative inverse eigenvalue problems are different, *Proc. Amer. Math. Soc.* **124**, no. 12, 3647–3651. [94, 95, 102]

Joseph, K. T. 1992. Inverse eigenvalue problem in structural design, *AIAA J.* **30**, 2890–2896. [17, 69]

Kac, M. 1966. Can one hear the shape of a drum?, *Amer. Math. Monthly* **73**, no. 4, part II, 1–23. [146]

Kailath, T., A. H. Sayed, and B. Hassibi. 2000. *Linear estimation*, Prentice Hall, New York. [265]

Karmarkar, N. K. and Y. N. Lakshman. 1998. On approximate GCDs of unvariate polynomials, *J. Symbolic Comput.* **26**, 653–666. [26]

Karpelevič, F. I. 1951. On the characteristic roots of matrices with nonnegative elements, *Izv. Akad. Nauk SSSR Ser. Mat.* **15**, 361–383 (in Russian). [3, 104, 105]

Kato, T. 1966. *Perturbation theory for linear operators*, Springer-Verlag New York, Inc., New York. [39]

Kautsky, J. and S. Elhay. 1984. Gauss quadratures and Jacobi matrices for weight functions not of one sign, *Math. Comp.* **43**, no. 168, 543–550. [24]

Kautsky, J., N. K. Nichols, and P. Van Dooren. 1985. Robust pole assignment in linear state feedback, *Internat. J. Control* **41**, no. 5, 1129–1155. [9, 12, 13, 33, 122, 144, 145]

Kleinberg, J., C. Papadimitriou, and P. Raghavan. 1998. A microeconomic view of data mining, *Data Mining and Knowledge Discovery* **2**, 311–324. [301]

Knobel, R. and J. R. McLaughlin. 1994. A reconstruction method for a two-dimensional inverse eigenvalue problem, *Z. Angew. Math. Phys.* **45**, no. 5, 794–826. [17]

Kolda, T. G. and D. P. O'Leary. 1998. A semi-discrete matrix decomposition for latent semantic indexing in information retrieval, *ACM Transact. Inform. Systems* **16**, 322–346. [304]

Kosowski, P. and A. Smoktunowicz. 2000. On constructing unit triangular matrices with prescribed singular values, *Computing* **64**, no. 3, 279–285. [135]

Kreĭn, M. G. 1933. On the spectrum of a Jacobian matrix, in connection with the torsional oscillations of shafts, *Mat. Sbornik* **40**, 455–478. [17]

Lai, E. and K. Ananthasuresh. 2003. On the design of bars and beams for desired mode shapes, *J. Sound and Vibration* **254**, no. 3, 393–406. [146]

Lambert, J. D. 1991. *Numerical methods for ordinary differential systems*, John Wiley & Sons Ltd., Chichester. The initial value problem. [22]

Lancaster, P. 1964. On eigenvalues of matrices dependent on a parameter, *Numer. Math.* **6**, 377–387. [39, 202]

Lancaster, P. and U. Prells. 2003. *Inverse problems for damped vibration systems*, University of Calgary. [161, 164, 170]

Lancaster, P. and M. Tismenetsky. 1985. *The theory of matrices*, Second edn, Computer Science and Applied Mathematics, Academic Press Inc., Orlando, FL. [180, 182, 221]

Landau, H. J. 1994. The inverse eigenvalue problem for real symmetric Toeplitz matrices, *J. Amer. Math. Soc.* **7**, no. 3, 749–767. [89]

Laurent, M. 1998. A connection between positive semidefinite and Euclidean distance matrix completion problem, *Linear Alg. Appl.* **273**, 9–22. [287, 290]

Laurie, D. P. 1988. A numerical approach to the inverse Toeplitz eigenproblem, *SIAM J. Sci. Stat. Comput.* **8**, 405–410. [90, 148, 213]

Laurie, D. P. 1991. Solving the inverse eigenvalue problem via the eigenvector matrix, *J. Comput. Appl. Math.* **35**, no. 1–3, 277–289. [90]

Laurie, D. P. 1997. Calculation of Gauss–Kronrod quadrature rules, *Math. Comp.* **66**, no. 219, 1133–1145. [24, 54]

Lawson, C. L. and R. J. Hanson. 1995. *Solving least squares problems, Classics in Applied Mathematics*, vol. 15, Society for Industrial and Applied Mathematics (SIAM), Philadelphia, PA. Revised reprint of the 1974 original. [213, 323, 324, 325]

Lee, D. D. and H. S. Seung. 1999. Learning the parts of objects by nonnegative matrix factorization, *Nature* **401**, 788–791. [321]

Lee, D. D. and H. S. Seung. 2001. *Algorithms for non-negative matrix factorization*, Advances in Neural Information Processing 13, pp. 556–562. [327, 328]

de Leeuw, J. and W. Heiser. 1982. Theory of multidimensional scaling, Classification, pattern recognition and reduction of dimensionality, North Holland, Amsterdam. pp. 285–316. [287]

Levitan, B. M. 1987. *Inverse Sturm–Liouville problems*, VSP, Zeist. Translated from the Russian by O. Efimov. [18]

Li, C. K. and R. Mathias. 2001. Construction of matrices with prescribed singular values and eigenvalues, *BIT* **41**, no. 1, 115–126. [135]

Li, H. B., P. Stoica, and J. Li. 1999a. Computationally efficient maximum likelihood estimation of structured covariance matrices, *IEEE Trans. Signal Process.* **47**, 1314–1323. [25]

Li, H., P. Stoica, and J. Li. 1999b. Computationally efficient maximum likelihood estimation of structured covariance matrices, *IEEE Trans. Signal Process.* **47**, no. 5, 1314–1323. [25]

Li, L. 1997a. On solvability of inverse eigenvalue problems with Hermitian matrices, *Linear Alg. Appl.* **253**, 89–101. [25, 260]

Li, N. 1997b. A matrix inverse eigenvalue problem and its application, *Linear Alg. Appl.* **266**, 143–152. [20, 145]

Liu, W. and J. Yi. 2003. *Existing and new algorithms for nonnegative matrix factorization*, University of Texas at Austin. Available at http://www.cs.utexas.edu/users/liuwg/383CProject/final_report.pdf. [324]

Loewy, R. and D. London. 1978. A note on an inverse problems for nonnegative matrices, *Linear and Multilinear Alg.* **6**, 83–90. [94]

London, D. and H. Minc. 1972. Eigenvalues of matrices with prescribed entries, *Proc. Amer. Math. Soc.* **34**, 8–14. [76, 116]

Luenberger, D. G. 1969. *Optimization by vector space methods*, John Wiley & Sons Inc., New York. [210, 281]

Majindar, K. N. 1979. Linear combinations of Hermitian and real symmetric matrices, *Linear Alg. Appl.* **25**, 95–105. [232]

Makhoul, J. 1981. On the eigenvectors of symmetric Toeplitz matrices, *IEEE Trans. Acoust., Speech Signal Process.* **29**, no. 4, 868–872. [148]

Marshall, A. W. and I. Olkin. 1979. *Inequalities: Theory of majorization and its applications*, Academic Press, New York. [113]

Mathar, R. 1985. The best Euclidian fit to a given distance matrix in prescribed dimensions, *Linear Alg. Appl.* **67**, 1–6. [287, 290]

McLaughlin, J. R. and O. H. Hald. 1995. A formula for finding a potential from nodal lines, *Bull. AMS*, New Series **32**, 241–247. [18]

McLaughlin, J. R. and G. H. Handelman. 1980. Sturm–Liouville inverse eigenvalue problems, *Mechanics Today* **5**, 281–295. [18]

McLaughlin, J. R., P. L. Polyakov, and P. E. Sacks. 1994. Reconstruction of a spherically symmetric speed of sound, *SIAM J. Appl. Math.* **54**, 1203–1223. [18]

Melsa, J. L. and D. L. Cohn. 1978. *Decision and estimation theory*, McGraw-Hill Book Co., New York. [281]

Millhauser, G. L., A. A. Carter, D. J. Schneider, J. H. Freed, and R. E. Oswald. 1989. Rapid singular value decomposition for time-domain analysis of magnetic resonance signals by use of the Lanczos algorithm, *J. Magn. Reson.* **82**, 150–155. [246]

Minc, H. 1988. *Nonnegative matrices*, Wiley, New York. [94, 104, 105]

Mirsky, L. 1958. Matrices with prescribed characteristic roots and diagonal elements, *J. London Math. Soc.* **33**, 14–21. [113]

Mittelmann, H. D. and J. A. Cadzow. 1987. Continuity of closest rank-p approximations to matrices, *IEEE Trans. Acoust., Speech, Signal Process.* **35**, 1211–1212. [250]

Moler, C. B. and G. W. Stewart. 1973. An algorithm for generalized matrix eigenvalue problems, *SIAM J. Numer. Anal.* **10**, 241–256. Collection of articles dedicated to the memory of George E. Forsythe. [233]

Moré, J. J. and S. J. Wright. 1993. *Optimization software guide*, Society for Industrial and Applied Mathematics (SIAM), Philadelphia, PA. [259]

Moré, J. J. and Z. Wu. 1999. Distance geometry optimization for protein structures, *J. Global Optim.* **15**, no. 3, 219–234. [290]

Morgan, A. P., A. J. Sommese, and L. T. Watson. 1989. Finding all isolated solutions to polynomial systems using HOMPACK, *ACM Trans. Math. Software* **15**, no. 2, 93–122. [47]

Mottershead, J. E. and M. I. Friswell. 2001 Model updating-euromech, special issue of Mech. Sys. Signal Process., on *Inverse Methods in Structural Dynamics* (Liverpool, 1999), 15, no. 1. [17]

Müller, M. 1992. An inverse eigenvalue problem: computing B-stable Runge-Kutta methods having real poles, *BIT* **32**, no. 4, 676–688. [22]

Munthe-Kaas, H. 1998. Runge-Kutta methods on Lie groups, *BIT* **38**, no. 1, 92–111. [352]

Nagy, J. G., V. P. Pauca, R. J. Plemmons, and T. C. Torgersen. 1997. Space-varying restoration of optical images, *J. Opt. Soc. Amer. A.* **14**, 3162–3174. [265, 278]

Nesterov, Y. and A. Nemirovskii. 1994. *Interior-point polynomial algorithms in convex programming*, SIAM Studies in Applied Mathematics, vol. 13, Society for Industrial and Applied Mathematics (SIAM), Philadelphia, PA. [328]

Neumaier, A. 1997. Molecular modeling of proteins and mathematical prediction of protein structure, *SIAM Rev.* **39**, no. 3, 407–460. [289]

Nichols, N. K. and J. Kautsky. 2001a. Robust eigenstructure assignment in quadratic matrix polynomials: nonsingular case, *SIAM J. Matrix Anal. Appl.* **23**, no. 1, 77–102 (electronic). [122]

Nichols, N. K. and J. Kautsky. 2001b. Robust eigenstructure assignment in quadratic matrix polynomials: nonsingular case, *SIAM J. Matrix Anal. Appl.* **23**, no. 1, 77–102 (electronic). [161]

Nievergelt, Y. 1997. Schmidt–Mirsky matrix approximation with linearly constrained singular values, *Linear Alg. Appl.* **261**, 207–219. [21]

Nylen, P. 1999. Inverse eigenvalue problem: existence of special mass-damper-spring systems, *Linear Alg. Appl.* **297**, no. 1–3, 107–132. [16, 18, 79]

Nylen, P. and F. Uhlig. 1997a. Inverse eigenvalue problem: existence of special spring-mass systems, *Inverse Problems* **13**, no. 4, 1071–1081. [18, 79]

Nylen, P. and F. Uhlig. 1997b. Inverse eigenvalue problems associated with spring-mass systems, in *Proceedings of the fifth conference of the International Linear Alg. Soc.* (Atlanta, GA, 1995), pp. 409–425. [18, 79]

de Oliveira, G. N. 1970. Note on an inverse characteristic value problem, *Numer. Math.* **15**, 345–347. [58, 59]

de Oliveira, G. N. 1973a. Matrices with prescribed entries and eigenvalues, I, *Proc. Amer. Math. Soc.* **37**, no. 2, 380–386. [76, 114]

de Oliveira, G. N. 1973b. Matrices with prescribed entries and eigenvalues, II, *SIAM J. Appl. Math.* **24**, 414–417. [76, 114, 115, 116]

de Oliveira, G. N. 1975. Matrices with prescribed entries and eigenvalues, III, *Arch. Math. (Basel)* **26**, 57–59. [76, 114]

de Oliveira, G. N. 1983. Nonnegative matrices with prescribed spectrum, *Linear Alg. Appl.* **54**, 117–121. [94]

Osborne, M. R. 1971. On the inverse eigenvalue problem for matrices and related problems for difference and differential equations, in *Conference on applications of numerical analysis* (University of Dundee, 1971), pp. 155–168. Lecture Notes in Math., vol. 228. [2, 19]

Paige, C. C. and M. A. Saunders. 1981. Towards a generalized singular value decomposition, *SIAM J. Numer. Anal.* **18**, no. 3, 398–405. [233]

Paine, J. 1984. A numerical method for the inverse Sturm–Liouville problem, *SIAM J. Sci. Stat. Comput.* **5**, 149–156. [18]

Park, H., L. Zhang, and J. B. Rosen. 1999. Low rank approximation of a Hankel matrix by structured total least norm, *BIT* **39**, no. 4, 757–779. [25, 250, 251, 257, 260]

Park, H., M. Jeon, and J. B. Rosen. 2001. *Lower dimensional representation of text data in vector space based information retrieval,* Computational information retrieval (Raleigh, NC, 2000), pp. 3–23. [306, 311]

Parlett, B. N. 1998. *The symmetric eigenvalue problem,* Society for Industrial and Applied Mathematics (SIAM), Philadelphia, PA. Corrected reprint of the 1980 original. [vi, 32, 74, 82]

Perfect, H. 1953. Methods of constructing certain stochastic matrices, *Duke Math. J.* **20**, 395–404. [93, 106]

Perfect, H. 1955. Methods of constructing certain stochastic matrices. II, *Duke Math. J.* **22**, 305–311. [93, 106]

Pryce, J. D. 1993. *Numerical solution of Sturm–Liouville problems,* Oxford University Press, New York. [2, 18]

Ram, Y. M. 1994a. An inverse mode problem for the continuous model of an axially vibrating rod, *Trans. ASME J. Appl. Mech.* **61**, no. 3, 624–628. [146, 147]

Ram, Y. M. 1994b. Inverse mode problems for the discrete model of a vibrating beam, *J. Sound and Vibration* **169**, 239–252. [147]

Ram, Y. M. 1995. *Pole-zero assignment of vibratory system by state feedback control,* University of Adelaide. [75, 161]

Ram, Y. M. 2003. http://me.lsu.edu/~ram/PAPERS/publications.html, Louisiana State University. [16, 17, 76, 161]

Ram, Y. M. and J. Caldwell. 1992. Physical parameters reconstruction of a free-free mass-spring system from its spectra, *SIAM J. Appl. Math.* **52**, no. 1, 140–152. [17]

Ram, Y. M. and S. Elhay. 1996. An inverse eigenvalue problem for the symmetric tridiagonal quadratic pencil with application to damped oscillatory systems, *SIAM J. Appl. Math.* **56**, no. 1, 232–244. [76, 80]

Ram, Y. M. and G. M. L. Gladwell. 1994. Constructing a finite element model of a vibratory rod from eigendata, *J. Sound and Vibration* **169**, 229–237. [17, 147]

Ravi, M. S., J. Rosenthal, and X. A. Wang. 1995. On decentralized dynamic pole placement and feedback stabilization, *IEEE Trans. Automat. Control* **40**, no. 9, 1603–1614. [13]

Reams, R. 1996. An inequality for nonnegative matrices and the inverse eigenvalue problem, *Linear and Multilinear Alg.* **41**, no. 4, 367–375. [94]

Rosenthal, J. and J. C. Willems. 1999. Open problems in the area of pole placement, in *Open problems in mathematical systems and control theory,* Springer, London, pp. 181–191. [12]

Ruhe, A. 1987. Closest normal matrix finally found!, *BIT* **27**, no. 4, 585–598. [234, 241, 243, 244, 245, 353]

Scascighini, A. and Troxler, A. 2001. Design of diffusers by an inverse method (Swansea, Wales, UK, September 2001) [146]

Scharf, L. L. and D. W. Tufts. 1987. Rank reduction for modeling stationary signals, *IEEE Trans. Acoustics, Speech, and Signal Process.* **35**, 350–355. [25]

Schulz, T. J. 1997. Penalized maximum-likelihood estimation of covariance matrices with linear structure, *IEEE Trans. Signal Process.* **45**, 3027–3038. [25]

Shampine, L. F. and M. K. Gordon. 1975. *Computer solution of ordinary differential equations*, W. H. Freeman and Co., San Francisco, CA. The initial value problem. [223]

Shampine, L. F. and M. W. Reichelt. 1997. The MATLAB ODE suite, *SIAM J. Sci. Comput.* **18**, no. 1, 1–22. Dedicated to C. William Gear on the occasion of his 60th birthday. [315]

Shapiro, A. 1983. On the unsolvability of inverse eigenvalues problems almost everywhere, *Linear Alg. Appl.* **49**, 27–31. [38]

Shaw, A. K. and R. Kumaresan. 1988. Some structured matrix approximation problems, *IEEE International Conference on Acoustics, Speech and Signal Process.* 2324–2327. [25]

Shepherd, S. J. 2004. *Private communication.* [331, 332]

Silva, F. C. 1987a. Matrices with prescribed characteristic polynomial and submatrices, *Portugal. Math.* **44**, 261–264. [76]

Silva, F. C. 1987b. Matrices with prescribed eigenvalues and principal submatrices, *Linear Alg. Appl.* **92**, 241–250. [76]

Dias da Silva, J. A. 1974. Matrices with prescribed entries and characteristic polynomial, *Proc. Amer. Math. Soc.* **45**, 31–37. [76, 116]

Dias da Silva, J. A. and T. J. Laffey. 1999. On simultaneous similarity of matrices and related questions, *Linear Alg. Appl.* **291**, no. 1–3, 167–184. [230]

Simon, H. D. and H. Zha. 2000. Low-rank matrix approximation using the Lanczos bidiagonalization process with applications, *SIAM J. Sci. Comput.* **21**, no. 6, 2257–2274 (electronic). [246]

Sing, F. Y. 1976. Some results on matrices with prescribed diagonal elements and singular values, *Canad. Math. Bull.* **19**, no. 1, 89–92. [115]

Sivan, D. D. and Y. M. Ram. 1999. Physical modifications to vibratory systems with assigned eigendata, *ASME J. Appl. Mech.* **66**, 427–432. [161]

van der Sluis, A. 1969/1970a. Condition numbers and equilibration of matrices, *Numer. Math.* **14**, 14–23. [66]

van der Sluis, A. 1969/1970b. Condition numbers and equilibration of matrices, *Numer. Math.* **14**, 14–23. [194]

Soules, G. W. 1983. Constructing symmetric nonnegative matrices, *Linear and Multilinear Alg.* **13**, 241–251. [94, 106]

Starek, L. and D. J. Inman. 1997. A symmetric inverse vibration problem for nonproportional underdamped systems, *Trans. ASME J. Appl. Mech.* **64**, no. 3, 601–605. [76]

Starek, L. and D. J. Inman. 2001. Symmetric inverse eigenvalue vibration problem and its applications, *Mechanical Systems and Signal Process.* **15**, no. 1, 11–29. [17, 76, 79, 161]

Starek, L., D. J. Inman, and A. Kress. 1992. A symmetric inverse vibration problem, *Trans. ASME, J. Vibration and Acoustics* **114**, no. 3, 564–568. [76, 79]

Stewart, G. W. 1985. A Jacobi-like algorithm for computing the Schur decomposition of a non-Hermitian matrix, *SIAM J. Sci. Statist. Comput.* **6**, no. 4, 853–864. [233]

Stiefel, E. 1935. Richtungsfelder und fernparallelismus in n-dimensionalel manning faltigkeiten, *Commentarii Mathematici Helvetici* **8**, 305–353. [313]

Suleĭmanova, H. R. 1949. Stochastic matrices with real characteristic numbers, *Doklady Akad. Nauk SSSR (N.S.)* **66**, 343–345. [101, 102]

Sun, J. G. 1985. Eigenvalues and eigenvectors of a matrix dependent on several parameters, *J. Comput. Math.* **3**, 351–364. [39]

Sun, J. G. 1996. Perturbation analysis of the pole assignment problem, *SIAM J. Matrix Anal. Appl.* **17**, no. 2, 313–331. [9]

Sun, J. G. 1999. Backward errors for the inverse eigenvalue problem, *Numer. Math.* **82**, no. 2, 339–349. [44, 60]

Sun, J. G. and Q. Ye. 1986. The unsolvability of inverse algebraic eigenvalue problems almost everywhere, *J. Comput. Math.* **4**, no. 3, 212–226. [44]

Symes, W. W. 1981/82. The QR algorithm and scattering for the finite nonperiodic Toda lattice, *Phys. D* **4**, no. 2, 275–280. [225, 234, 341, 354]

Syrmos, V. L., C. T. Abdallah, P. Dorato, and K. Grigoriadis. 1997. Static output feedback — a survey, *Automatica J. IFAC* **33**, no. 2, 125–137. [12]

Thompson, R. C. 1977. Singular values, diagonal elements, and convexity, *SIAM J. Appl. Math.* **32**, no. 1, 39–63. [115]

Thompson, R. C. and P. McEnteggert. 1968. Principal submatrices. II. The upper and lower quadratic inequalities, *Linear Alg. Appl.* **1**, 211–243. [83]

Tisseur, F. and K. Meerbergen. 2001. The quadratic eigenvalue problem, *SIAM Review* **43**, 235–286. Available at http://www.ma.man.ac.uk/~ftisseur. [17, 160]

Trench, W. F. 1989. Numerical solution of the eigenvalue problem for Hermitian Toeplitz matrices, *SIAM J. Matrix Anal. Appl.* **10**, no. 2, 135–146. [148]

Trench, W. F. 1997. Numerical solution of the inverse eigenvalue problem for real symmetric Toeplitz matrices, *SIAM J. Sci. Comput.* **18**, no. 6, 1722–1736. [90]

Tropp, J. A. 2003. *Literature survey: Nonnegative matrix factorization*, University of Texas at Austin. [322, 324]

Trosset, M. W. 1997. Distance matrix completion by numerical optimizaion, Technical Report TR95-31, Rice University. [288, 290, 295, 299]

Tufts, D. W. and A. A. Shah. 1993. Estimation of a signal wave-form from noisy data using low rank approximation to a data matrix, *IEEE Trans. Signal Process.* **41**, 1716–1721. [25]

Uhlig, F. 1973. Simultaneous block diagonalization of two real symmetric matrices, *Linear Alg. Appl.* **7**, 281–289. [232]

Uhlig, F. 1976. A canonical form for a pair of real symmetric matrices that generate a nonsingular pencil, *Linear Alg. Appl.* **14**, no. 3, 189–209. [232]

Uhlig, F. 1979. A recurring theorem about pairs of quadratic forms and extensions: a survey, *Linear Alg. Appl.* **25**, 219–237. [232]

Van Loan, C. F. 1976. Generalizing the singular value decomposition, *SIAM J. Numer. Anal.* **13**, no. 1, 76–83. [232]

Van Loan, C. F. 1992. *Computational frameworks for the fast Fourier transform,* Society for Industrial and Applied Mathematics (SIAM), Philadelphia, PA. [4, 265, 267]

Vandenberghe, L. and S. Boyd. 1996. Semidefinite programming, *SIAM Rev.* **38**, no. 1, 49–95. [328]

Vanderbei, R. J. and H. Y. Benson. 1999. *On formulating semidefinite programming problems as smooth convex nonlinear optimization problems*, Department of Statistics and Operation Research, Princeton University. [290]

van der Veen, A.-J. 1996. A Schur method for low-rank approximation, *SIAM J. Matrix Anal. Appl.* **17**, no. 1, 139–160. [247]

Watkins, D. S. 1984. Isospectral flows, *SIAM Rev.* **26**, no. 3, 379–391. [354]

Weyl, H. 1949. Inequalities between the two kinds of eigenvalues of a linear transformation, *Proc. Nat. Acad. Sci. USA* **35**, 408–411. [135]

Wilhelmi, W. 1974. Ein Algorithmus zur losung eines inversen Eigenwertproblems, *Z. Angew. Math. Mech.* **54**, 53–55. [32]

Wilkinson, J. H. 1965. *The algebraic eigenvalue problem*, Clarendon Press, Oxford. [2, 41, 123, 222]

Williams, D. E. and D. H. Johnson. 1993. Robust estimation on structured covariance matrices, *IEEE Trans. Signal Process.* **41**, 2891–2906. [25]

Wright, K. 1992. Differential equations for the analytic singular value decomposition of a matrix, *Numer. Math.* **3**, no. 2, 283–295. [108, 109]

Wu, Q. 1990. Determination of the size of an object and its location in a cavity by eigenfrequency shifts, Ph.D. Thesis. [17]

Xu, S. F. 1993. A stability analysis of the Jacobi matrix inverse eigenvalue problem, *BIT* **33**, no. 4, 695–702. [84]

Xu, S. F. 1998. *An introduction to inverse algebraic eigenvalue problems,* Peking University Press, and Friedr. Vieweg & Sohn Verlagsgesellschaft mbH, Braunschweig/Wiesbaden, Beijing. [3, 6, 35, 38, 42, 80, 82, 84, 97, 98, 106]

Xue, G. and Y. Ye. 1997. An efficient algorithm for minimizing a sum of Euclidean norm with applications, *SIAM J. Optim.* **7**, 1017–1036. [290]

Yang, X.-F. 1997. A solution of the inverse nodal problem, *Inverse Problems* **13**, no. 1, 203–213. [146]

Zayed, E. M. E. 1993. An inverse eigenvalue problem for an arbitrary multiply connected bounded region: an extension to higher dimensions, *Internat. J. Math. Sci.* **16**, no. 3, 485–492. [18]

Zha, H. and H. D. Simon. 1999. On updating problems in latent semantic indexing, *SIAM J. Sci. Comput.* **21**, no. 2, 782–791 (electronic). [28]

Zha, H. and Z. Zhang. 1995. A note on constructing a symmetric matrix with specified diagonal entries and eigenvalues, *BIT* **35**, 448–452. [113]

Zhornitskaya, L. A. and V. S. Serov. 1994. Inverse eigenvalue problems for a singular Sturm–Liouville operator on $[0,1]$, *Inverse Problems* **10**, no. 4, 975–987. [18]

Zhou, S. and H. Dai. 1991. *Daishu tezhengzhi fanwenti (in Chinese), the algebraic inverse eigenvalue problem*, Henan Science and Technology Press, Zhengzhou, China. [2, 6, 38, 39, 41, 79]

Zimmerman, D. C. and M. Widengren. 1990. Correcting finite element models using a symmetric eigenstructure assignment technique, *AIAA* **28**, no. 9, 1670–1676. [161]

INDEX

The manufacturer's authorised representative in the EU for product safety is Oxford University Press España S.A. of El Parque Empresarial San Fernando de Henares, Avenida de Castilla, 2 - 28830 Madrid (www.oup.es/en or product.safety@oup.com). OUP España S.A. also acts as importer into Spain of products made by the manufacturer.

Printed and bound by CPI Group (UK) Ltd, Croydon, CR0 4YY

14/09/2026

02216306-0012